Circuit, Device and Process Simulation

Circuit, Device and Process Simulation

Mathematical and Numerical Aspects

G. F. Carey
University of Texas at Austin, TX, USA

W. B. Richardson
Visiting Researcher, University of Texas at Austin, and Motorola

C. S. Reed
Supercomputing Research Center, Bowie, TX, USA

B. Mulvaney
Motorola, Austin, TX, USA

JOHN WILEY & SONS
Chichester • New York • Brisbane • Toronto • Singapore

National 01243 779777
International (+44) 1243 779777

Other Wiley Editorial Offices

John Wiley & Sons, Inc., 605 Third Avenue,
New York, NY 10158-0012, USA

Jacaranda Wiley Ltd, 33 Park Road, Milton,
Queensland 4064, Australia

John Wiley & Sons (Canada) Ltd, 22 Worcester Road,
Rexdale, Ontario M9W 1L1, Canada

John Wiley & Sons (Asia) Pte Ltd, 2 Clementi Loop #02-01,
Jin Xing Distripark, Singapore 0512

Library of Congress Cataloguing in Publication Data

Circuit, device, and process simulation : mathematical and numerical
aspects / G.F. Carey ... [et al.].
p. cm.
Includes bibliographical references and index.
ISBN 0 471 96019 5 (hb : alk. paper)
1. Electronic circuit design – Data processing. 2. Computer-aided
design. 3. Semiconductors – Mathematical models – Data processing.
4. Numerical analysis – Data processing. I. Carey, Graham F.
TK7867. C4973 1996
621.3815' 01' 1 – dc20. 95–49408
CIP

British Library Cataloguing in Publication Data

A catalogue record for this book is available from the British Library

ISBN 0 471 96019 5

Produced from camera-ready copy supplied by the authors.
Printed and bound in Great Britain by Bookcraft (Bath) Ltd.
This book is printed on acid-free paper responsibly manufactured from sustainable forestation,
for which at least two trees are planted for each one used for paper production.

Contents

Preface

This book is intended to present an integrated approach to the physical processes, mathematical models, and numerical techniques used in semiconductor circuit, device, and process simulation. Although scarcely twenty years old, the subject area is already vast, and the present treatment reflects the authors' own interests and experience. Many of the underlying physical processes and certainly the mathematical and numerical models apply equally well to a broad class of problems in the semiconductor industry. Therefore, whenever possible, similarities to other basic transport problems in chemical engineering, fluid dynamics, and dynamical systems are emphasized. A common thread of the text is the time evolution and steady-state solution of a physical system as described by differential equations. This includes the reaction-diffusion partial differential equations of process modeling, the drift-diffusion models of device simulation, and the coupled nonlinear ordinary differential equations of circuit simulation. The mathematical background assumed of the reader makes the book particularly suitable for a graduate level electrical engineering course in simulation and as a basic reference for researchers in the semiconductor industry. However, it will also serve to introduce analysts working in related areas of engineering and computational mathematics to semiconductor applications.

For several reasons, we have elected to organize the material to begin with circuit simulation, followed by device and process modeling. First, the circuit is the end-product of interest to the engineer and therefore the raison d'être that motivates the underlying device and process technologies. Second, circuits are composed of discrete components characterized by constitutive relations involving node currents and voltages. These relations, together with Kirchoff's laws, lead logically to a system of differential algebraic equations. Hence, the reader is "gently" introduced to mathematical models involving discrete systems as well as the theory of ordinary differential equations and numerical methods for their solution, before encountering partial differential equations. It also permits us to illustrate in the ODE setting certain numerical themes, which recur in the discretized PDE systems governing device transport and impurity diffusion. For pedagogical reasons we have elected to keep the circuit examples in the early chapters relatively simple so that ideas of stiffness, stability, roundoff error and accuracy can be easily demonstrated. We then build on these concepts throughout the remaining chapters in more mathematically complex treatments of partial differential equations. The material can be restructured in several ways to meet different goals. For example, we have lectured to industrial participants covering selected topics in essentially the order presented here. Drs. Carey, Richardson and Mulvaney have also taught a graduate course on mathematical and numerical aspects of device and process simulation from the relevant chapters. The introductory material on the circuit problem can be used in conjunction with selected papers or a text such as *Computer Methods for Circuit Analysis and Design* by Vlach and Singhal [541]. Similarly, the mathematical approach to device and process simulation offered here can be complemented by papers and more detailed treatments of physical models from the cited references or books such as *IC Processes and Devices* by Dutton and Yu [148]. Exercises are provided to help consolidate the material or as more extensive simulation and software assignments.

The main editorial responsibilities for integrating the book into a cohesive whole were assumed by Graham Carey and Rich Richardson. The responsibilities for the technical material are as follows: Chapter 1 on modeling was written primarily by Graham Carey and Rich Richardson; Chapters 2–5 on circuit simulation are due to Coke Reed; Chapter 6 on parameter extraction and Chapter 10 dealing with ion implantation were written by Brian Mulvaney; Chapters 7–9,15 and 16 on device simulation, grids, oxidation and crystal growth are by Graham Carey; Chapters 11–14 on reaction-diffusion processes were written by Rich Richardson. Finally, Rich Richardson wrote the bulk of Chapter 17, with contributions from Brian Mulvaney and Graham Carey.

The authors would like to express their appreciation to a number of colleagues and students. Rich Richardson and Brian Mulvaney thank Tim Crandle and Greg Siebers for helping to make the process simulator PEPPER a reality. Several graduate students – Paul Murray, Mahesh Sharma, Rodney Greene and Sudentra Tatti – spent summers working with the simulation group at MCC. Rich also extends special thanks to Alfonso Castro, Shair Ahmad, Bill Taylor and Karen Baker for their assistance in his work on extending advanced diffusion models to 3D. Coke Reed acknowledges his colleagues Bill Read, Joe Eyles and Steve Hamm on the MUSIC project. Brian Mulvaney expresses his appreciation to Bob Garbs for discussions regarding parameter extraction and Monte Carlo analysis. Graham Carey would like to thank several former and current graduate students: some of the material on oxidation and device modeling given here is based on research with Paul Murray, Mahesh Sharma, Steve Bova, Bruce Davis, and Anand Pardhanani. The interaction with faculty colleagues Al Tasch and Chris Maziar and their students has been of great value. He would also like to thank Varis Carey for assisting with technical proofing and for useful suggestions related to the mathematical treatment. Several of the research projects that contributed to the ideas in this monograph were supported by the Semiconductor Research Consortium, the National Science Foundation and the Texas Advanced Technology Program. Varis Carey, Gurcan Bicken, Yun Shen, Cathy Check, and Tija Carey assisted in the type setting of this manuscript.

Graham F. Carey
Walter B. Richardson, Jr.
Coke Reed
Brian J. Mulvaney

Austin, Texas 1995

Chapter 1

Modeling and Simulation

1.1 Introduction

Since the release of the programs SUPREM I for process modeling, PISCES for device modeling and SPICE for circuit simulation there has been an exponential increase in the use of Computer Aided Design (CAD) tools at all levels of integrated circuit design and manufacture. Prior to 1975 an engineer with a new idea for an individual device or an entire integrated circuit would have been justified in making several runs through the fabrication line until a successful device was fabricated. Since then, the cost of a wafer lot has increased to the point that it is now essential that a design be thoroughly verified by simulation before it is committed to silicon. As device dimensions shrink below 0.5 μm and packing densities reach millions of devices per chip, there will be an even greater need for computer simulation at every stage of the design process. Over a billion dollars are invested in building one of today's fabrication lines. It is therefore essential that new technologies be implemented quickly to recoup these huge investments. Perhaps more than in other industries, in chip design and manufacture, substantial profits go to those who are first to market their product. The relatively short lifetime of a given technology also necessitates a much reduced design cycle. Higher level CAD tools such as timing verification, routing, layout, synthesis and logic design, all rest upon data from three basic areas: process, device, and circuit simulation. This chapter will discuss the overall philosophy and methodology of modeling, before the specific areas of simulation are discussed in detail.

The manufacturing of silicon integrated circuits consists of a complex series of processes that build and pattern layers of insulators, conductors, and semiconductor materials. The starting point is a silicon wafer, a thin disk of crystalline silicon. The basic processes for building a circuit on the wafer include deposition, photolithography, etching, ion implantation, diffusion, and oxidation [559]. The deposition process places a layer of new material on top of the existing structure, generally by chemical vapor deposition or sputtering the material in a low pressure chamber. In the photolithography process, a photo-sensitive material is deposited on the wafer, and this material is exposed to high intensity light through an opaque mask. The exposed areas of photosensitive material are then washed away, leaving on the wafer an image of the mask. Photolithography is typically followed by an etch process, which chemically removes the layer of material not covered by the photosensitive material. The etch can be either a wet chemical etch or a plasma etch. In this way, patterns are transfered onto the wafer. The photolithography step is also often followed by an ion implantation, which selectively dopes regions of the wafer to produce the desired conduction regions within the semiconductor. Ion implantation is followed by a high temperature (of the order of 900°K or more) diffusion, which is necessary to anneal the damage caused by implantation and to electrically activate the ions. Oxidation is a high temperature diffusion that is performed in the presence of an oxidizing atmosphere to form a layer of silicon dioxide on the exposed silicon regions. Silicon dioxide is an insulator, and the electrical properties of the silicon/silicon dioxide interface are excellent for making insulated gate devices.

These basic processing steps are repeated a number of times in various combinations to create the final integrated circuit. A typical complementary metal-oxide-semiconductor (CMOS) process might require over twenty photolithography steps, a dozen implantations and diffusions, several oxidations, and dozens of deposisitions and etches. The performance, sensitivity, and reliability of the resulting device and, in turn, of the integrated circuit containing the device must then be assessed under a range of operating conditions. Ideally, the problem is one of design optimization where, for instance, a device design may be upgraded by changes in the geometry, layout, etc.

1.2 Engineering Design

The development of large-scale integrated circuits on microchips at low cost has been the keystone of the communications and computer revolution. This integrated circuit problem is particularly unusual with respect to scale. That is, a typical circuit may consist of many thousands of components and the overall performance will depend upon the nature and interconnection of individual components operating at a much smaller scale. The performance of these individual devices will, in turn, depend upon their specific design characteristics and the applied voltages. Finally, these devices themselves result from local processes such as growing oxide layers and diffusing impurities into subregions. Hence, there is a natural decomposition of the problem to finer scales at the level of (1) the circuit, (2) the individual device and (3) the process. This text treats the mathematical modeling and numerical simulation of these three problem classes. There are other areas such as packaging and interconnection of circuits that are important and involve related modeling concepts. We shall confine our treatment, however, to representative circuit, device and process models since these are inherently coupled and are of fundamental interest.

Conceptually, the development of a numerical simulation capability involves an appreciation of the engineering problems and design issues. Figure 1.1 shows a schematic of both the "system" and "technology" design processes. On the left is the system design which begins with the initial specification of the system and ends with the final integrated circuit layout for custom and semi-custom blocks. Computer Aided Design has had a profound impact on all areas of system design: partitioning, logic design and testing, circuit design, layout, routing, and timing verification. At the system level, today's integrated circuits may contain hybrid digital/analog subcircuits, bipolar compatible MOS and perhaps combined Si/GaAs technology on the same chip. This increasing complexity means that the actual system design process is far less linear than Figure 1.1 suggests and underscores the need for increased use of CAD.

Technology design deals with the design of the individual devices used to implement the logic on the right in Figure 1.1 on an integrated circuit. The device design portion deals with how the device operates electrically under various applied voltages after it is made, while process design is concerned with the physical and chemical steps used to manufacture a device. Three major interfaces exist between the two paths shown in Figure 1.1. At the device level, designers and technologists communicate critical performance information through the mechanism of device models for circuit simulation. At the layout level, communication occurs via the specification of design rules. Of increasing importance is the projection phase that occurs at the end of the last generation of system/technology design. At this interface the needs and expectations of both groups for the next generation of integrated circuits must be enunciated and reconciled.

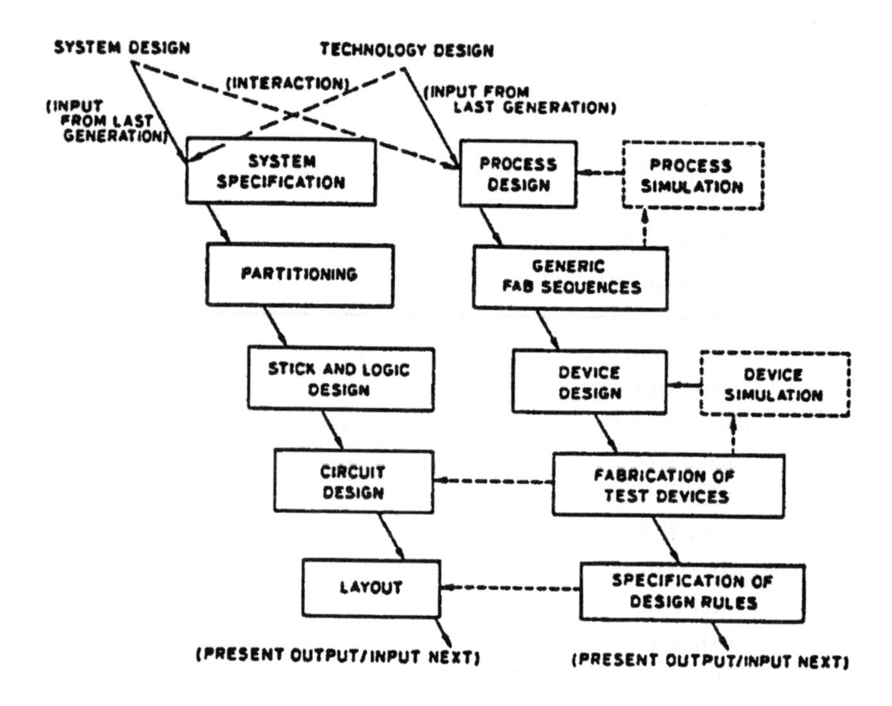

Figure 1.1: Schematic of "system" and "technology" design processes and their relationships (from [149]).

1.3 Circuit, Device and Process Models

The problem of scale is a dominant feature of integrated circuit technology. At the microscopic level, the performance of a circuit reflects the movement of electrons through a complex heterogeneous domain. It is neither computationally feasible nor in fact meaningful to model the entire circuit at this most primitive level. Instead, macroscopic physical models (resistors, capacitors, etc.) are introduced that describe the performance of individual components in the circuit and the overall circuit response at a level sufficient to characterize the circuit behavior. To be useful for computer simulation, these physical models for circuit components must be described mathematically and this yields a mathematical model for circuit analysis and design studies. In particular, at the circuit level the result is a differential–algebraic system with tens of thousands of unknowns that must be integrated numerically to determine circuit response to a given input.

The circuit components at the fine scale are composed of individual devices and at this level it is possible to consider physical models that attempt to treat electron transport. Here the objective is to describe, for instance, the current–voltage characteristics of a device so that these can be incorporated into the circuit description at the coarser scale. A continuum mechanics approach is justified, provided the representative volume is still sufficiently large. This is the case for devices down to moderate submicron dimensions. Applying the transport theorem and conservation laws together with appropriate constitutive assumptions leads to a mathematical model for the device. The precise nature of the resulting model will depend on detailed assumptions related to constitutive behavior, such as relaxation times, etc.,that may in turn depend on operating voltages, device size, geometry, composition, and other factors. Typically the resulting mathematical description consists of a system of nonlinear partial differential equations. The complexity of this system makes analytic solution impossible, so discrete finite difference or finite element schemes on numerical grids are introduced to yield a family of numerical models that are designed to approximate the continuum model. In the limit of decreasing grid size, the solution to the approximate problem should converge to the mathematical solution of the continuum device problem. Of course, this requires that the discrete approximate systems must be solved on very fine grids, which leads to the consideration of efficient numerical solution techniques.

Finally, the composition of the device consists of different layers of materials such as silicon dioxide or silicon nitride deposited on silicon, and regions that have been "doped" by diffusion of various impurities to create *pn* junctions which are the building blocks of diodes and transistors. These features are introduced by different types of processes involving, for instance, diffusion, chemical reaction and heat transfer. Since the very nature of the device and its integrity depend upon the success and reliability of these processes, it is important that they be represented faithfully in process design. To model the growth of oxides and dopant diffusion, reaction-diffusion equations can be derived, again using conservation of mass and the transport theorem. The resulting partial differential equations can be discretized in a manner similar to the device problem to generate a large algebraic system or ordinary differential equation system for numerical solution.

1.4 Simulation

The simulator is an important complement to analysis and experiment since it allows a spectrum of conditions to be tested inexpensively before actual fabrication and sensitivity studies are conducted. Simulation also lends itself to the optimal design problem, including maximizing yield while minimizing defects in a statistical sense. We also emphasize that simulation can be used to explore the fundamental underlying physical models to enhance our understanding and demonstrate the shortcomings of existing physical models. For example, the effect of drastically reduced device dimensions may be to violate some of the assumptions, involving relaxation times and constitutive relations essential to present models. Comparison of simulation and experiment can help pinpoint weaknesses in the old models and suggest improvements. In this way, simulation can aid our understanding of the underlying physical processes and lead to new advances in process, device and circuit technology.

Simulation relies on the development of mathematical and numerical models that incorporate the dominant physics. The conceptual framework is indicated in Figure 1.2 for a general system and can be applied to the process, device or circuit design of interest here. The first step is the conception of the engineering system or process. Step 2 then involves identification of the underlying physical laws (conservation of mass in dopant diffusion, charge transport in devices, conservation of energy, etc.) followed by the association of appropriate constitutive theory and models (e.g., Fick's law for diffusion, the parabolic band assumption or Fourier assumptions in charge transport, etc.). These physical laws and constitutive assumptions lead to a mathematical model for the process, device or circuit, usually in the form of a system of differential or integro-differential equations. The complexity of the resulting mathematical system in step 3 is such that the problems are not analytically tractable.

Approximate mathematical models based on, for instance, discretization using finite difference or finite element approaches may then be introduced in step 4. The resulting discretized or semi-discretized systems lead to large, sparse, nonlinear algebraic systems or to ordinary differential equation systems. Numerical analysis techniques such as successive approximation, Newton's method, linear system solution, stiff and nonstiff integration, etc. and associated algorithms may then be implemented in step 5 to provide an approximate solution. Finally, post-processing with graphics permits convenient characterization of the solution and interpretation of the results.

As the problem conditions change the various components in Figure 1.2 may need to be revised to include new physical models (e.g., to incorporate hot carrier effects as device size shrinks). These changes will then impact the mathematical model and the final numerical solution schemes. For example, the stiffness of an ordinary differential equation (ODE) system may

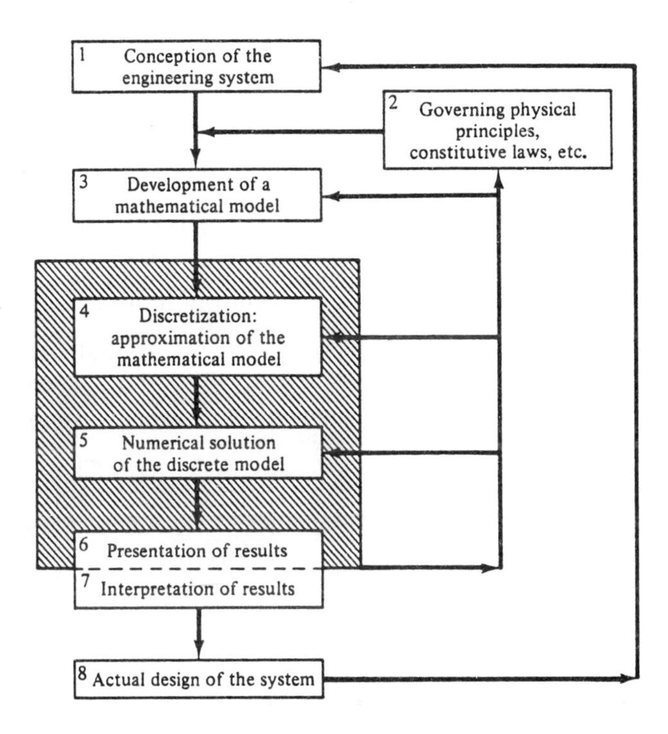

Figure 1.2: Conceptual Framework for Analysis, Design, and Simulation (from [89]).

dramatically increase and necessitate an implicit integrator. Hence, there are "feedback" loops in Figure 1.2 that relate the various components from the physical model to the final numerical algorithm and also to modify the system or process design. The main object of this book is to present an integrated treatment of the mathematical and numerical aspects for the established physical models that are relevant to circuit, device and process modeling.

Process design encompasses many steps including oxidation, diffusion, ion implantation, continuous vapor deposition, epitaxy, lithography and rapid thermal processing. Although the sophistication of the physical models used to simulate the various steps varies, there are computer codes that allow the designer to accurately model each of the steps in process design and thereby "inspect" quantities such as junction depth and oxide thickness.

Not only is the need for many expensive trial fabrications and measurements obviated, but the simulators provide a wealth of data that cannot be measured directly with experimental methods: e.g. vacancy and interstitial profiles from a 3D Monte Carlo ion implant simulation can be obtained and stress from a Navier–Stokes oxidation calculation. Simulation can provide feedback and cross-checking during process design and manufacturing, as well as careful design and process control over entire sequences of steps, not just a simple diffusion or oxidation.

Just above process simulation in the technology hierarchy is device simulation, which deals with how a device operates under various applied voltages after it is manufactured. Device simulation is a more mature discipline than its process counterpart for two reasons. First, the physics of device operation are better understood and provide a hierarchy of increasingly accurate, but more CPU intensive models including drift-diffusion, hydrodynamic, Boltzmann transport, and finally quantum-mechanical models. Second, the impossibility of making direct measurements of the scalar fields for potential, electron and hole concentrations at selected points within the transistor during operation, force the designer to make extensive use of computer simulation. These scalar fields for potential and carrier concentrations within the device boundaries uniquely determine the voltages and currents measured on the contacts, but only the latter can be determined experimentally. That is, the boundary values of the solution can be verified, not its values throughout the interior of the device. Coupling between process and device modeling occurs chiefly via the impurity profile that is output from the former and supplies input data to the latter. Also of importance is the geometrical description of oxide regions, polysilicon gates, and metal interconnects.

Many of the key objectives in our treatment are related to methodology and numerical techniques for discretizing the mathematical systems and for computer simulation, respectively. For example, we formulate the main finite difference and finite element schemes for discretizing partial differential equation systems for device and process problems. Questions of accuracy, stability, and the utility of numerical grids are addressed. Similarly, we develop the fundamental numerical integration techniques for circuit and process simulation as well as associated linear and nonlinear sparse solution algorithms.

An integrated CAD system for semiconductor analysis and design would involve, as a basic modeling capability process simulators, device carrier simulators and circuit simulators. The focus in the present treatment is the associated mathematical and numerical aspects. However, other technology and software would interface with these simulation components or be integrated into them. This might include geometric modeling or computing, grid generation with refinement, visualization and optimization software.

For example, geometric modeling deals with the specification of the geometry of an object and facilitates manipulation of surfaces and projections. This general capability removes the need for the user to develop specific geometric computing software for a given application. Instead, the problem reduces to manipulation of data and objects within the given geometric computing environment. As an example, the evolving surfaces in an oxidation process can be monitored for surface intersection, loops, etc. using the geometric modeler. Similarly, a semiconductor device consisting of several layers of different materials with complex surface shapes and topology can be described conveniently in this framework. A solid object is thereby defined as interior to boundary faces; the intersections of boundary faces define edges and edge intersections define vertices. Orientation is also attached as a property to distinguish inside and outside surface views, etc. Within this setting surfaces and curves can be further described using standard analytic shapes, by splines or by certain simple operations such as translation.

1.5 Preview and Organization

As indicated in the preface, we have organized the presentation in a "top-down" mode beginning with the circuit problem, then device simulation and finally process modeling. There are two main reasons for adopting this order. First, by beginning with the circuit we are able to focus on the end product and then examine its component devices and the processes by which they are fabricated. Secondly, the circuit model as connected individual components leads easily to the idea of discrete systems of equations to be solved numerically. Simple circuit examples can be easily introduced to illustrate the formulation and different concepts. Accordingly, we have kept the treatment in these early chapters elementary and emphasize that the purpose here is not to provide a state-of-the-art treatment or survey for circuit simulation. Similarly, the presentation of theoretical results and theorems has been stated at a level requiring only a knowledge of advanced calculus. More advanced modeling and numerical analysis schemes, such as the Krylov subspace methods, advanced integrators, etc., are introduced later.

The chapter on parameter extraction provides the "bridge" between the circuit and device problem and motivates the subsequent treatment for device simulation. The governing equations for analysis of carrier transport and devices are typically formulated as a coupled system of partial differential equations so this adds a significant level of complexity to the mathematical and numerical treatment. For example, we develop both finite difference schemes for the PDE systems and also Galerkin finite element schemes for the associated weak integral formulation.

These themes and the numerical analysis techniques for solving the resulting discretized PDE problem are interwoven throughout the chapters on device and process modeling. Other issues such as the grid generation problem, adaptive grid refinement strategies, and Monte Carlo simulations are covered, together with a more advanced treatment of sparse solver techniques and integrators.

The basic organization of the technical material in the following chapters is as follows: Chapters 2 through 5 deal primarily with the circuit simulation problem. We begin with the circuit laws and construct simple algebraic and differential-algebraic systems for elementary representative problems and then extend this to include the transistor model and nonlinear diode. From this foundation the basic approaches to numerically integrating circuit ordinary differential equations are taken up. Finally we briefly consider Newton iteration for nonlinear systems solution. This treatment also provides an introduction to several algorithms that are applied later for the device and process problem. In Chapter 6, the important connection between the device problem and the circuit simulation problem is identified. Device models for circuit simulation are discussed along with parameter extraction and statistical variation of model parameters.

The semiconductor device transport problem is then developed in Chapters 7 and 8 beginning with the drift-diffusion model that is extensively used by device designers with such codes as PISCES. We investigate different techniques for treating the dominant drift term and the associated problem of numerical oscillations in standard finite difference and finite element methods explaining the source of these numerical difficulties. These problems are exacerbated when the device shrinks to submicron level and hot carrier effects become important, as seen in Chapter 8 where a Taylor–Galerkin finite element scheme is introduced to regularize the numerical problem. The importance of grading the grid into the extreme solution layers is also discussed briefly here and developed further in Chapter 9.

Grid generation is a fundamental component of the process and device simulation problems since these involve spatial discretization of partial differential equations. The output from the geometric modeling software can be used to drive the grid generator and also to update the definition of the problem domain for process problems with moving surfaces such as the oxidation problem. The accuracy and reliability of the simulation depends closely on the mesh which should contain well-shaped cells and be graded into regions where the solution has large gradients or layer structures. The grid also impacts on the efficiency of the computation since it is directly related to the size of the resulting discrete systems that must be solved or integrated. Finally, other numerical issues such as matrix conditioning and stiffness of ODE systems are influenced by the grid.

Monte Carlo techniques for modeling scattering phenomena in ion implantation are described in Chapter 10 together with an algorithm for computing an ion trajectory and supporting numerical studies. Chapters 11 to 14 deal with the physical and numerical aspects of diffusion during semiconductor processing. Used for redistributing impurities and reforming the lattice after implantation, diffusion also occurs during any high temperature step, such as oxidation or nitridation. The Fokker–Planck equation is derived, first using a random walk and then conservation of mass and Fick's law. The linear diffusion equation accurately models intrinsic or low concentration impurity diffusion in silicon. Closed-form solutions are presented for the special cases of predeposition and "drive-in" in 1-D. At higher concentrations, it becomes necessary to use a concentration dependent diffusivity, whose exact form can be inferred on physical grounds or by using the Boltzmann–Matano technique. Several modifications to the basic theory are discussed including field-aided diffusion, segregation and transport boundary conditions at material interfaces, and clustering of impurities when concentrations approach the solid solubility level. Numerical methods for solving the diffusion equation are discussed in the last section of Chapter 11. They follow immediately from the integration techniques discussed for circuit simulation: parabolic partial differential equations (PDEs) can be considered as ODEs in a function space.

In the quest for shallower junctions, very high concentration diffusions are required. This leads to profiles which depart radically from simple complementary error function or Gaussian profiles. To properly account for these "anomalies", a more sophisticated physical model of diffusion is needed. This leads in Chapter 12 to a discussion of multiple species models, which consider interactions between impurity atoms and the rich array of point defects which exist in the silicon lattice, to be the chief mechanism for dopant diffusion. The hierarchy of models developed over the last twenty years is reviewed; followed by a detailed discussion of an eight-species model for phosphorus diffusion. Mathematically, these physical models are all characterized by reaction-diffusion PDEs. Their implementation requires careful determination of the diffusivities of the mobile species and rate constants for all reactions.

Although it is possible to use the same numerical methods for the multiple as for the single–species models, the size of the discretized PDE systems suggests the use of multistep integration techniques. The method of lines and a backward differentiation formula algorithm have proven to be very effective on stiff reaction-diffusion PDEs. System integrators such as LSODE offer convenient implementations of these variable-stepsize, variable-order methods. The quasi-Newton method used to solve the nonlinear system works well with a difference-quotient approximation to the Jacobian. The associated linear system is usually so large that both iterative and projection methods are competitive with LU-decomposition. Derivations of the conjugate gradient algorithm and more general Krylov subspace methods are given.

Chapter 14 treats specialized diffusion topics such as rapid thermal annealing, diffusion during epitaxy and oxidation, and movement of impurities through grain boundaries of polysilicon. The multiple species models of Chapter 12 are shown by straightforward extensions to handle diffusion in III-IV compounds as well. In Chapters 15 and 16 we consider two other representative process problems associated with silicon oxidation at elevated temperatures and the horizontal Bridgman process for crystal growth, respectively. These problems are typical of coupled fluid mechanics and transport phenomena that must be modeled in the semiconductor industry. As such they may be viewed as problems of computational mechanics and, again, similar questions concerning discretization techniques, sparse system properties, and efficient reliable algorithms must be confronted. Here we construct some appropriate methods and give sample results to illustrate the approach. Finally, in Chapter 17 we close with an overview of representative simulators in all three areas and demonstrate the actual simulation of a CMOS oscillator.

1.6 Summary

The three inter-related areas of circuit, device and process modeling can be described using appropriate physical models based on the underlying continuum processes at the scale in question. A valid mathematical model and numerical simulation capability permits the analysis of different process treatments for devices, the current-voltage characteristics for new or improved devices and the performance of more complex circuits. These three areas are clearly inter-related and each is a formidable numerical modeling challenge in its own right. Given a mathematical model which correctly characterizes the underlying physics, an accurate approximation scheme and reliable efficient solution algorithms are needed to permit a variety of processes and designs to be considered in a timely manner.

Chapter 2

The Circuit Equations

2.1 Introduction

Circuit simulation involves numerically solving a coupled system of nonlinear ordinary differential equations. This subject has remained an active research topic in the design community over the past thirty years because very large scale integrated (VLSI) circuits have become more and more complex. As device geometries shrink and new technologies are developed, it has become necessary to use more physically sophisticated models for the transistors in order to accurately predict circuit behavior. The increased complexity of the transistor models requires that they be implemented mathematically with great care. Numerical difficulties can also arise when simulating circuits which contain only a few transistors, when these devices are built using certain submicron CMOS technologies or III-V compounds such as gallium arsenide and indium phosphide. Not only is it more difficult to simulate a given circuit in a new technology, it has become necessary to simulate larger and larger circuits. The early simulators were designed to analyze circuits containing hundreds of transistors, while today circuits containing tens of thousands of transistors are modeled. Because of the importance of obtaining first-pass working silicon and because circuit simulation is one of the major users of computer cycles in the semiconductor industry, there has been a large expenditure of time and money on the problem of improving today's simulators.

During the 1960s Research at the University of California at Berkeley led to the development of the circuit simulation program CANCER [399], which was followed in 1972 by SPICE [400]. There are a number of computer aided design companies that now market commercial circuit simulation programs.

Many of these are SPICE derivatives which have been modified to execute on a number of architectures including personal computers, workstations, mainframes, and supercomputers [190]. Each of the major integrated circuit companies have simulation groups that maintain an in-house circuit simulator tailored to their specific circuit needs and which is often quite different from SPICE in several important aspects. A commercial stand-alone, multiprocessor computer dedicated to the circuit simulation problem [139, 140] has been constructed, and one integrated circuit company has built a special purpose, parallel machine solely to model its own IC devices [401].

Historically, a close relationship has existed between circuit analysis and network theory: each circuit can be represented conceptually by a network consisting of nodes and branches. Each branch connects two nodes in a directed sense and contains a specific circuit element such as a resistor or capacitor. However, it is erroneous to think that transistor models used in today's simulators necessarily fit into this simple structure. Instead the model might be considered a "black box" having the property that the currents flowing into the four terminals (source, gate, drain, and bulk) are functions of the four terminal voltages and their derivatives with respect to time. In contrast to the classical theory, current flowing into the drain, for example, is not decomposed into currents coming from source, gate, and bulk. These more general models do not even classify current sources as those depending only on nodal voltages and those depending only upon their time derivatives.

In the following chapters, the concepts of circuit theory will be introduced via a set of simple examples, which also exhibit various pathological properties normally found only in larger circuits. These canonical examples will also illustrate the subtleties of the various hypotheses of associated theorems. For clarity of exposition, the main theorems needed to lay the foundation of circuit simulation will be proved in two-dimensional Euclidean space. The proofs, moreover, will have the property that they contain all of the main ideas necessary to construct an argument in the more abstract, n-dimensional setting.

2.2 Ohm's and Kirchoff's Laws

In what follows, it is convenient to suppose the existence of a reference or ground node denoted by n_0. For a node n_i, the expressions "voltage at node n_i relative to the reference node n_0" and "voltage at node n_i" will be used interchangeably. If n_i and n_j are two nodes of a given circuit, and v_i is the voltage at node n_i with v_j the voltage at node n_j, then $v_{i,j}$ denotes the difference $v_i - v_j$, called the voltage drop from node n_i to n_j.

Suppose that R is a resistor on the branch connecting nodes n_i and n_j; let r denote the resistance of R. The voltage drop across the resistor will cause a current I_R to flow through the resistor R. The amount of current is a function of v_i and v_j, given by Ohm's law: the current I_R through R flowing into node n_j is $[v_i - v_j]/r$ or

$$I_R = v_{i,j}/r \tag{2.1}$$

It is very important to note that the current is a signed quantity, i.e. the current through R flowing into node n_i is $[v_j - v_i]/r$. Because the voltage drop across R determines the current flow, R is referred to as a voltage-controlled current source. Of course the voltage drop $v_{i,j} = rI_R$ is a function of the current, so it is also possible to consider the ideal resistor as a current-controlled voltage source. The nomenclature actually depends upon one's choice for the unknowns in the circuit equations: currents, nodal voltages, or branch voltages. Other resistive devices have a more complicated functional relationship between I_R and $v_{i,j}$.

The equations that describe the circuit's behavior are based on the Kirchoff voltage and current laws:

1. Kirchoff's voltage law: The sum of the signed voltage drops around a loop is zero.

2. Kirchoff's current law: The sum of the currents into a node is zero.

For example, to understand Kirchoff's voltage law, consider a loop connecting three nodes n_i, n_j, and n_k; i.e., a connected path which travels from n_i to n_j, then from n_j to n_k, and finally from n_k back to n_i. The voltage drops around the loop are $v_{i,j}$, $v_{j,k}$, and $v_{k,i}$, so that Kirchoff's voltage law implies

$$v_{i,j} + v_{j,k} + v_{k,i} = (v_i - v_j) + (v_j - v_k) + (v_k - v_i) = 0 \tag{2.2}$$

Now it is an elementary mathematical observation that, given a finite sequence $s_1, s_2, \ldots, s_n$, if the differences $d_k = s_{k+1} - s_k$ are formed, then the sum $\sum_{k=1}^{n-1} d_k$ collapses to $s_n - s_1$. As seen in (2.2), the voltage drops are in fact differences, and in the case of a loop we return to our starting point so $s_{last} - s_{first} = v_i - v_i = 0$. Recall that a conservative vector field is one which arises as the gradient of a scalar potential ϕ and $\oint \nabla\phi \, ds = 0$. Clearly, there are close connections between Kirchoff's voltage law and this property of conservative vector fields.

The Kirchoff current law holds because a node does not have the ability to store electrons: the current is defined as a constant times the net directional flow of electrons per unit time, so the sum of the currents into a node must be zero, otherwise electrons would accumulate or be depleted at the node. Thus, Kirchoff's current law is a discrete statement of the conservation laws that will be discussed later in the chapters on process and device modeling. It is interesting to note that conservation laws can be derived by applying the divergence theorem and suitable limiting processes to a control volume.

As an elementary example of the Kirchoff current law, consider the circuit composed of resistors $\{R_m\}$ as illustrated in Figure 2.1. For each m,

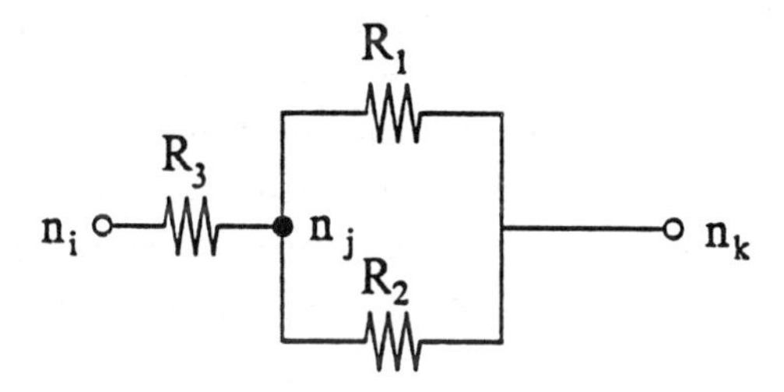

Figure 2.1: A simple resistive circuit.

let the resistance of R_m be r_m. Now Ohm's law may be applied to calculate the sum of the currents into node n_j (recall that the currents are signed quantities): the representative current into node n_j through R_3 is $v_{i,j}/r_3$; the current into n_j through R_1 is $v_{k,j}/r_1$; and the current into n_j through R_2 is $v_{k,j}/r_2$. Therefore, Kirchoff's current law applied to node n_j yields

$$\frac{v_{i,j}}{r_3} + \frac{v_{k,j}}{r_1} + \frac{v_{k,j}}{r_2} = 0 \tag{2.3}$$

which can be rewritten as

$$\frac{(v_i - v_j)}{r_3} + \frac{(v_k - v_j)}{r_1} + \frac{(v_k - v_j)}{r_2} = 0 \tag{2.4}$$

If the resistances r_1, r_2, and r_3 and the voltages v_i and v_k are known, then it is possible to solve (2.4) for v_j to obtain

$$v_j = \left(\frac{v_i}{r_3} + \frac{v_k}{r_1} + \frac{v_k}{r_2}\right)\left(\frac{r_1 r_2 r_3}{r_2 r_3 + r_1 r_3 + r_1 r_2}\right) \tag{2.5}$$

The process used to solve the circuit of Figure 2.1 gives an outline of the methodology of circuit simulation in general. First Kirchoff's current law is used to derive a system of equations that describe the flow of current through the circuit. (Actually, a slightly generalized version of Kirchoff's current law to be discussed later will be necessary.) Next, the system, which here is algebraic but later will contain time derivatives as well, is solved for the unknown voltages. It must be remembered that Kirchoff's laws represent algebraic constraints that must be satisfied by the currents and voltages of the circuit — mathematically they describe the surface or manifold on which the solution vector must lie.

2.3 A Simple Resistive Circuit

Consider the circuit of Figure 2.2 containing 98 resistors, 10 nodes of known voltage, and 25 nodes of unknown voltage. The nodes are located at the points (i,j) of the xy-plane where $0 \leq i \leq 4$ and $-1 \leq j \leq 5$, while subscripts $j = -1$ or $j = 5$ identify the independent voltage sources. Nodes with unknown voltages are connected by resistors to their horizontal, vertical, and diagonal neighbors. The nodes are numbered such that the node at position (i,j) has subscript $k = i + 5j$ giving the node list n_{-5}, n_{-4}, ..., n_{29}. As before, let v_i denote the voltage at node n_i. Then $(v_{-5}, v_{-4}, \ldots, v_{-1})$ and $(v_{25}, v_{26}, \ldots, v_{29})$ are known voltages while the components of the vector $\boldsymbol{x} = (v_0, v_1, \ldots, v_{24})$ are still to be determined. If there is a resistor between node n_i and node n_j, then its resistance is denoted by the positive quantity $r_{i,j}$ and $r_{i,j} = r_{j,i}$.

Kirchoff's current law can be used to derive an equation for each node n_k of unknown voltage as follows: the sum of the signed currents into node n_5 of Figure 2.2 is zero, and using Ohm's law the nodal voltages satisfy

$$\frac{v_0 - v_5}{r_{0,5}} + \frac{v_1 - v_5}{r_{1,5}} + \frac{v_6 - v_5}{r_{6,5}} + \frac{v_{10} - v_5}{r_{10,5}} + \frac{v_{11} - v_5}{r_{11,5}} = 0 \qquad (2.6)$$

This procedure can be repeated for the other nodes of the circuit to define an algebraic system. The terms involving known voltages are transposed to the right-hand-side to yield the resulting system of 25 equations in 25 unknowns. In matrix form we have

$$\boldsymbol{A}\boldsymbol{x} = \boldsymbol{b} \qquad (2.7)$$

where $\boldsymbol{A}$ is a known 25×25 matrix, $\boldsymbol{b}$ is a vector of 25 known values and $\boldsymbol{x}$ is the vector of 25 unknown voltages. Notice that the non-zero components in $\boldsymbol{b}$ correspond to nodes connected to independent voltage sources.

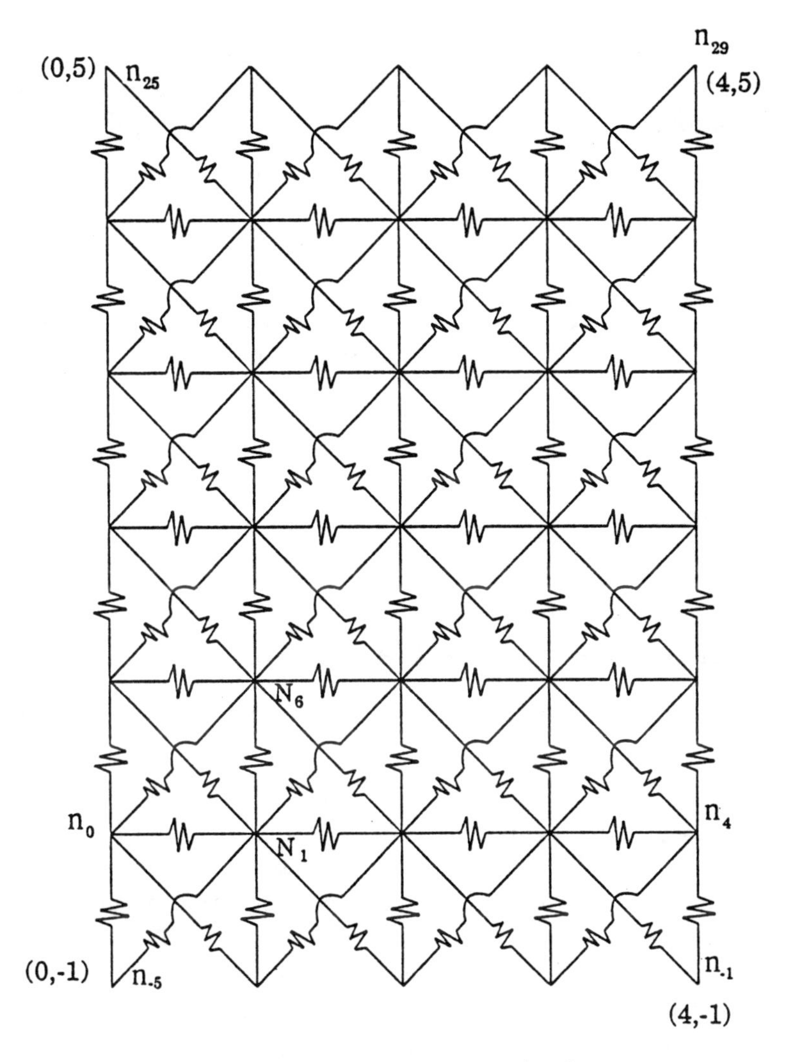

Figure 2.2: A resistive circuit.

Since the network connections are local, i.e. each node is connected to the adjacent nodes in the network, there are few non-zero entries in $\boldsymbol{A}$. Such a matrix is said to be sparse. Moreover, with the natural ordering of the network as defined here, the non-zero entries occur along diagonals of $\boldsymbol{A}$, as indicated in Figure 2.3. The sparsity pattern of $\boldsymbol{A}$, can be obtained simply from the connectivity of the circuit and is represented by the connectivity matrix $\boldsymbol{M}$, which is 1 where $\boldsymbol{A}$ is non-zero and 0 otherwise. This connectivity matrix can be used to form a dependency tree that is useful in parallel matrix factorization routines [17, 145, 337]. The banded structure of $\boldsymbol{A}$ is familiar to students of numerical methods for partial differential equations and will be encountered again later in the discretized PDE models for process and device simulation.

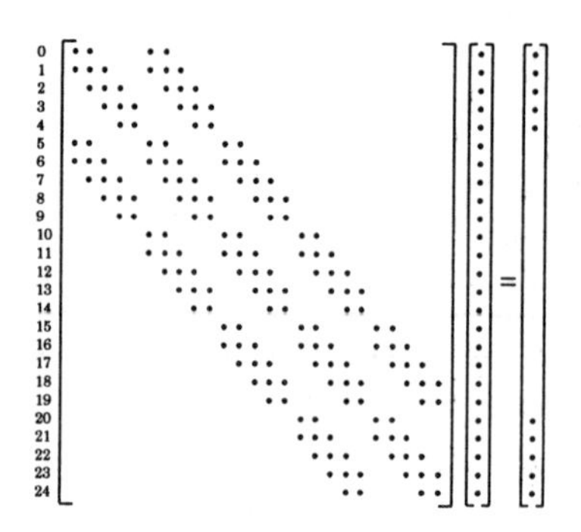

Figure 2.3: The sparsity structure of the matrix $\boldsymbol{A}$.

2.4 Charge in a Linear Capacitor

In order to solve for the voltages at the interior nodes of the resistive circuit in Figure 2.2, it is necessary to characterize the flow of current through a resistor as a function of the terminal voltages. The linear resistor is a voltage-controlled current source that obeys Ohm's law. In order to solve for the voltages at the interior nodes of circuits that contain more complicated voltage-controlled devices such as diodes, capacitors, and transistors, it will be necessary to derive the equations that give the flow of current through these devices in terms of both nodal voltages and their time derivatives.

To accomplish this task, we will solve a simple example problem, namely that of determining the electrostatic potential for a two-plate capacitor. It will be shown that, under certain simplifying assumptions, the exact solution to the circuit problem of section (2.3) can be viewed as an approximate solution to this capacitor problem. There will be no proofs here, only a preview of coming attractions in the chapters on device modeling.

Choose a reference point $\boldsymbol{p}_0$ (also called the ground point). The electrostatic potential field ϕ is the scalar field such that, for each point $\boldsymbol{p}$ of $I\!R^3$, $\phi(\boldsymbol{p})$ is the work required to move a unit charge from $\boldsymbol{p}$ to $\boldsymbol{p}_0$. The electric field $\boldsymbol{E}$ is the vector field such that for each point $\boldsymbol{p}$ of $I\!R^3$, $\boldsymbol{E}(\boldsymbol{p})$ is the force that would be exerted on a point charge placed at $\boldsymbol{p}$, by all the charges in the space. Units for ϕ and $\boldsymbol{E}$ are joule/coulomb = volt and newton/coulomb, respectively. The electric field $\boldsymbol{E}$ is related to the electrostatic potential field

ϕ by

$$\boldsymbol{E} = -\nabla\phi \tag{2.8}$$

If two scalar fields, F and G differ by a constant, then they have the same gradient (another consequence of the fundamental theorem of calculus). Therefore, the electric field does not depend on the choice of the reference point $\boldsymbol{p}_0$. To find the electric field $\boldsymbol{E}$, we pick an arbitrary point $\boldsymbol{p}_0$, determine the potential field ϕ relative to $\boldsymbol{p}_0$, and then use (2.8).

Let ρ denote the charge density function such that for each point $\boldsymbol{p}$ of the space, $\rho(\boldsymbol{p})$ is the amount of charge per unit volume at $\boldsymbol{p}$. Charge conservation implies $\oint \boldsymbol{E} \cdot \boldsymbol{n}\, ds = \rho/\varepsilon$ where ε is the material permittivity and $\boldsymbol{n}$ is the unit outward normal to the boundary of domain Ω in $\mathbf{R}^3$ [179]. Applying the divergence theorem and using (2.8), the electrostatic potential field ϕ satisfies Poisson's equation

$$\frac{\partial^2\phi}{\partial x^2} + \frac{\partial^2\phi}{\partial y^2} + \frac{\partial^2\phi}{\partial z^2} = -\frac{\rho}{\varepsilon} \tag{2.9}$$

subject to certain conditions on ϕ at the boundary $\partial\Omega$ of Ω. The operator $\partial^2/\partial x^2 + \partial^2/\partial y^2 + \partial^2/\partial z^2$ (as well as its two-dimensional counterpart $\partial^2/\partial x^2 + \partial^2/\partial y^2$) is called the Laplacian and is denoted by Δ. If the right hand side of (2.9) is zero, Poisson's equation is referred to as Laplace's equation and is the canonical elliptic partial differential equation. Poisson's equation will now be used to find the electrostatic potential field of a capacitor.

The circuit of the last section can in fact be viewed as an analog computer designed to find approximate solutions to (2.9). Consider a capacitor consisting of two thin rectangular plates parallel to the xz-plane, having width W and length L, and separated by a distance d. One plate labeled α is at a distance five units above the xz-plane while the other plate labeled β is at distance one unit below the xz-plane, so that $d = 6$. The plates are positioned such that the bottom surface of α consists of those points (x, y, z) with $0 \le x \le 4$, $y = 5$, and $-\frac{L}{2} \le z \le \frac{L}{2}$, and the top surface of β consists of those points (x, y, z) with $0 \le x \le 4$, $y = -1$, and $-\frac{L}{2} \le z \le \frac{L}{2}$, as illustrated in Figure 2.4. Suppose that α is positively charged, β is negatively charged, and that the capacitor is in a vacuum far removed from any other charges. Since opposite charges attract, all of the charge on each plate will be on its interior surface. Three simplifying assumptions will be made:

1. The charge is distributed uniformly on the parallel plates in the x and z directions (Dirichlet boundary conditions).

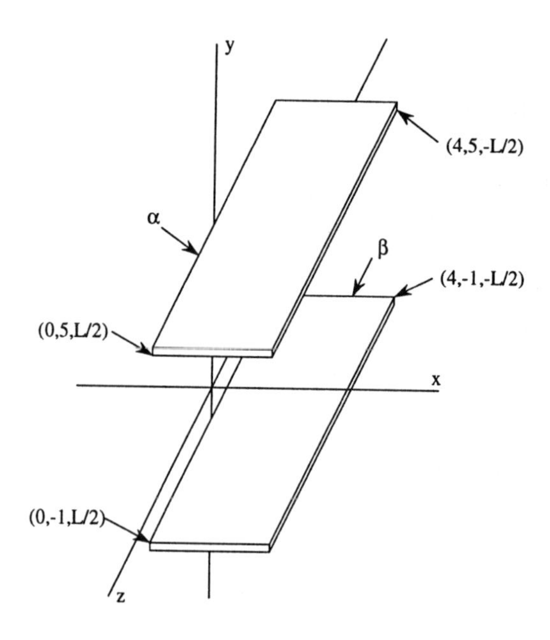

Figure 2.4: An idealized parallel-plate capacitor.

2. For every point $\boldsymbol{p} = (x, y, z)$ with $x = 0$ or $x = 4$, $\frac{\partial \phi}{\partial x}(\boldsymbol{p})$ is 0 (Neumann boundary conditions).

3. For every point $\boldsymbol{p}$, $\frac{\partial \phi}{\partial z}(\boldsymbol{p}) = 0$; in other words ϕ is effectively only a function of x and y, allowing us to reduce the dimension of the problem.

Let σ denote the charge per unit area on the surface of α so that the charge per unit area on β is $-\sigma$. Figure 2.5 shows two cross sections with $z = 0$. On the left is an illustration of the actual electrical field, while the figure on the right represents the electrical field obtained using the three assumptions.

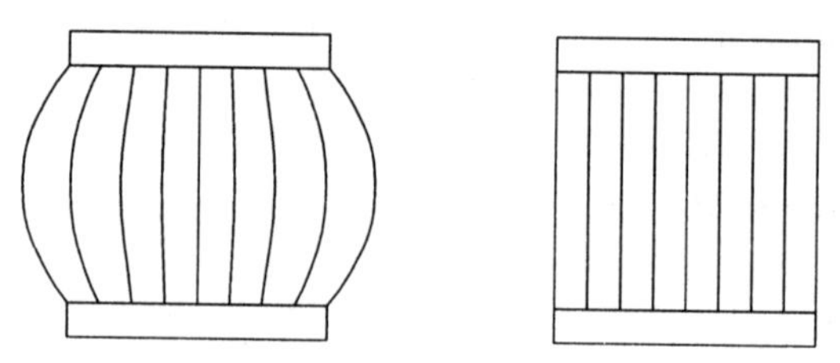

Figure 2.5: The actual and idealized electric field of a capacitor.

Notice that the resistive circuit of the last section just fits between the plates in this cross section. Although no current flows between the plates of the capacitor, there is a flow of current through the circuit. Nevertheless, the voltage at each node n_i of the circuit approximates the electrostatic potential between the plates at the location of n_i. Let w denote a twice differentiable function whose domain consists of all points (a, b) such that $0 \le a \le 4$ and $-1 \le b \le 5$, and such that if (x, y) is the position of node n_k then $w(x, y)$ is the voltage v_k. The function w can be viewed as a numerical approximation to ϕ restricted to the xy-plane. Choose an interior node, n_7, and let s be the sum of the currents flowing into this node through the four horizontal and vertical resistors,

$$s = v_{12,7} + v_{2,7} + v_{6,7} + v_{8,7} \tag{2.10}$$

Set the network node spacing $h = 1$ so that $v_{12,7} = w(2,2) - w(2,1) = [w(2, 1+h) - w(2,1)]/h$, and therefore $v_{12,7}$ can be viewed as the standard discretization of $w_y(2, (1+\frac{h}{2}))$. Similarly, $v_{2,7} = w(2,0) - w(2,1) = -[w(2, 0+h) - w(2,0)]/h$ and $v_{2,7}$ is the standard approximation of $-w_y(2, h/2)$. Therefore $v_{12,7} + v_{12,7} = [w_y(2, 1.5) - w_y(2, .5)]/h$, which approximates $w_{yy}(2,1)$. A similar argument shows that $v_{6,7} + v_{8,7}$ approximates $w_{xx}(2,1)$. If the circuit contained only the horizontal and vertical branches, Kirchoff's current law would demand that s in (2.10) is zero, and since for $0 \le m \le 4$ and $1 \le n \le 4$, $w_{xx}(m,n) + w_{yy}(m,n) \approx 0$, the solution to the circuit problem provides an approximate solution to Laplace's equation. Laplace's equation is invariant under rotation and, by a similar argument, a circuit containing only the diagonal branches would provide another approximate solution to Laplace's equation. In either case the sum s depends upon the voltages at five nodes: in the language of numerical methods for solving PDEs, these discretizations are based on five-point stencils. Combining the cartesian and rotated diagonal discretizations gives an approximate solution based on a nine-point stencil.

There is a rich theory that guarantees convergence of the approximate solutions using either a five- or nine-point stencil to the desired exact potential function under suitable hypotheses. In the present case, since the electrostatic potential function ϕ is planar, the solution to the discrete problem defined by the stencil is exact. Since the gradient of a planar function is a constant function, the electrostatic field function $\boldsymbol{E}$ is constant between the plates and zero elsewhere, as illustrated in Figure 2.5.

Let A be the area of α, d be the distance from α to β, and let V be the electrostatic potential difference between α and β. Since σ is the charge per unit area on the surface of α, the total charge Q on α is σA, and

$$V = |\boldsymbol{E}|\, d = \frac{\sigma}{\varepsilon} d = \frac{d}{\varepsilon A} Q \tag{2.11}$$

which implies

$$Q = cV \tag{2.12}$$

where the constant $c = \frac{\varepsilon A}{d}$ is the capacitance. The coulomb/volt is the unit of capacitance and is referred to as the farad.

Equation (2.12) only holds under the three simplifying assumptions made previously, none of which are satisfied exactly in practice. Notice that charges near the end of a plate have more force exerted upon them from one side than the other and would migrate toward the edge of the plate, which contradicts the first assumption. To show that the Neumann boundary assumption and the simplifying assumption of dimension reduction are incorrect, first solve for the correct steady state charge distribution on the plates, then solve Laplace's equation with the plates serving as the only boundary conditions, and notice that these assumptions do not hold for the actual electrical field depicted in Figure 2.5.

The above development naturally gives rise to several questions. How is Laplace's equation with the more complicated boundary conditions solved? If there is an equivalent circuit that can be used to obtain a numerical approximation to the electrostatic potential field, what is the best location of the nodes and the branches? Should different values of resistance be used in the various branches? Or, in the terminology of numerical methods for solving PDEs, what is the optimal grid placement for solving this system? For the idealized boundary conditions, it was demonstrated that one could use the entire circuit to produce a nine-point stencil or two different subcircuits each involving five-point stencils. Is there a weighted average of the two five-point stencil approximations that is more accurate than that of the standard nine-point stencil? When is the extra work of computing the nine-point approximation justified by the increased accuracy? These types of questions will be considered later in the chapters on device and process modeling.

The previous equivalent circuit illustrates an interesting interplay that occurs between the discrete and continuous formulations of a problem in both physics and mathematics. Many partial differential equations of solid-state physics are based upon conservation laws and either Boltzmann or Fermi–Dirac statistics [46, 529, 547]. The derivation of these equations often begins with a discrete process, which after a suitable limiting procedure is applied yields a differential equation in a continuum. Finding an approximate numerical solution, on the other hand, starts with the continuous equation, introduces an appropriate discretization, and finally solves the resulting discrete system.

What are the consequences of substituting the actual electrostatic field for the idealized one? Equation (2.12) will still hold in the sense that there exists a $c > 0$ such that $Q = cV$, but the error introduced at the edges causes the actual value of c to be larger than the one calculated for (2.12). Equations that more accurately characterize the device can be obtained by removing the simplifying assumptions. The improved device equations better describe the circuit's operation, but it often then becomes necessary to develop more advanced numerical techniques to provide accurate approximate solutions of the resulting circuit equations.

Note that $Q = cV$ was derived after simplification from Maxwell's laws of electrostatics; however, in circuit simulation this formula is only of interest when V is changing with time. The electrostatic law (2.8) is a special case of the more general relation

$$\boldsymbol{E} = -\nabla\phi - \frac{\partial \boldsymbol{A}}{\partial t} \tag{2.13}$$

where $\boldsymbol{A}$ is the vector potential field, while for the transient case Poisson's equation (2.9) must be replaced by

$$\Delta\phi = \frac{1}{C^2}\frac{\partial^2\phi}{\partial t^2} - \frac{\rho}{\varepsilon} \tag{2.14}$$

where C is the speed of light. In the range of operation of most integrated circuits, $\frac{\partial \boldsymbol{A}}{\partial t}$ and $\frac{1}{\gamma^2}\frac{\partial^2\phi}{\partial t^2}$ are small enough to be ignored.

Nonlinear capacitors actually found in integrated circuits are far more complicated than the simple two-plate linear capacitor discussed here. In our example, the reduction to two dimensions becomes a less valid assumption as L is decreased; i.e. as the minimum feature size decreases, three-dimensional effects become more important. Moreover, as switching times are decreased and electromagnetic fields change more rapidly, the electrostatic assumption also loses validity. As technology advances it is important to introduce more accurate, physically-based (as opposed to empirical) device models when formulating the circuit equations, at the cost of more sophisticated mathematical and numerical methods. For instance, hot carrier effects are important in submicron devices and advanced transport models must be used. This topic is discussed in Chapters 7 and 8.

2.5 Current in the Linear Capacitor

The resistor is a two-terminal voltage-controlled current source such that the current passing through it is a linear function of its terminal voltages.

The idealized linear capacitor is a two-terminal voltage-controlled current source in which the current is a linear function of the time derivatives of its terminal voltages. This follows from the discussion of the continuum model in Section 2.4. Let C be a parallel plate capacitor on the branch connecting nodes n_i and n_j, with capacitance c. Let p_i be the plate of C nearest node n_i and p_j the plate of C nearest node n_j. If $v_\alpha(t)$ is the voltage at node n_α, $\alpha = i, j$, at time t, then the charge $q_i(t)$ on p_i is

$$q_i(t) = c(v_i(t) - v_j(t)) = cv_{i,j}(t) \tag{2.15}$$

The charge function $q_i(t)$ is a signed quantity; to verify that the sign in (2.15) is correct, suppose that $v_i(t) > v_j(t)$. From the definition of potential, it follows that at time t, the work required to move a positive charge from p_j to p_i is positive. Since like charges repel, it follows that $q_i(t)$ must be a positive quantity of charge. Since the charges on the plates have equal magnitude but opposite signs, the charge $q_j(t)$ on the other plate p_j is simply $c(v_j(t)-v_i(t))$.

The current flowing through C is the derivative of the charge with respect to time. Increasing the charge $q_i(t)$ causes electrons to flow toward node n_i and, by convention, current to flow away from n_i. Therefore, the current $I(t)$ flowing into n_i through C is

$$I(t) = -q_i{}'(t) = -cv'_{i,j}(t) \tag{2.16}$$

so that the current in the capacitor is a linear function of the time derivative of the voltage drop across it. Note that the current is a signed quantity. The current into node n_j through C is $cv'_{i,j}(t)$.

A Linear RC Circuit

In a circuit containing only linear resistors, the node voltages can be found by solving a linear system of algebraic equations. When a circuit

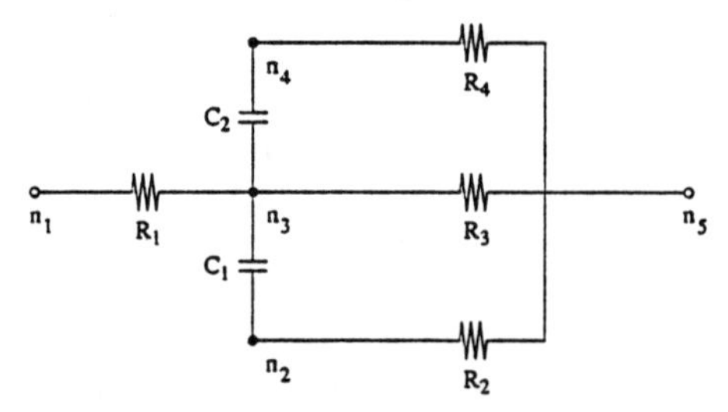

Figure 2.6: A circuit composed of resistors and capacitors.

contains both linear resistors and capacitors, a system of linear ordinary differential equations must be solved to determine nodal voltages. As an elementary example, consider the circuit illustrated in Figure 2.6. For simplicity, suppose that each capacitor has capacitance c and each resistor has resistance r, while $v_1(t)$ and $v_5(t)$ are two known functions of time. The problem is to solve for the unknown voltages $v_2(t)$, $v_3(t)$, and $v_4(t)$. The linear ODE system corresponding to this circuit can be derived from Kirchoff's current law: for each node n_i of unknown voltage, the sum of the currents flowing into n_i at time t is zero, which implies

$$\begin{aligned} c(v_3'(t) - v_2'(t)) + \frac{v_5(t) - v_2(t)}{r} &= 0 \\ c(v_2'(t) - v_3'(t)) + c[v_4'(t) - v_3'(t)] + \frac{v_1(t) - v_3(t)}{r} + \frac{v_5(t) - v_3(t)}{r} &= 0 \\ c(v_3'(t) - v_4'(t)) + \frac{v_5(t) - v_4(t)}{r} &= 0 \end{aligned} \tag{2.17}$$

Numerical techniques for integrating such a first-order system are discussed in the next chapter.

2.6 The Linear Inductor

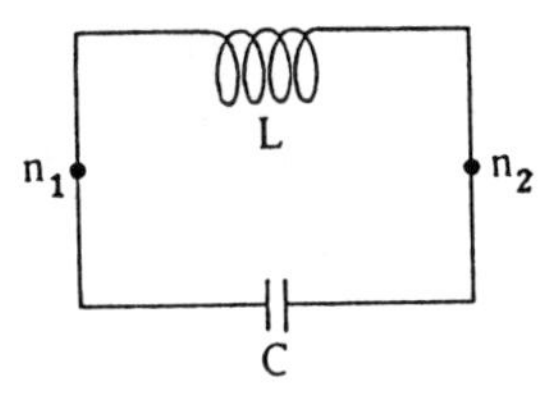

Figure 2.7: Simple circuit containing a capacitor/inductor pair.

The ideal linear inductor is a two-terminal device such that the voltage drop across its terminals is a linear function of the time derivative of the current passing through it. Let L be an inductor with inductance l henries connecting nodes n_i and n_j, which at time t are at voltages $v_i(t)$ and $v_j(t)$ respectively. The current $I(t)$ through L into node n_j satisfies

$$I'(t) = l[v_i(t) - v_j(t)] \tag{2.18}$$

Next, consider the simple two-node circuit in Figure 2.7 containing a linear capacitor C (with $c = 1$ farad) and a linear inductor L (with $l = 1$

henry). Let $v_k(t)$ denote the voltage at node n_k and let $I_C(t)$ and $I_L(t)$ denote the current flowing into node n_2 through the capacitor and inductor, respectively. Suppose that the following initial conditions are imposed

$$v_1(0) = \alpha, \; v_2(0) = \beta, \text{ and } I_L(0) = \gamma \tag{2.19}$$

Since the system is closed, for all time t, $v_1(t) + v_2(t) = \alpha + \beta$ and if we define the function $V(t) = v_1(t) - v_2(t)$ then $I_L'(t) = V(t)$ in (2.18) can be integrated to yield

$$I_L(t) = V(0) + \int_0^t V(s)\,ds \tag{2.20}$$

Kirchoff's current law requires that

$$I_C(t) + I_L(t) = 0 \tag{2.21}$$

and therefore, the function V can be found by solving the integro-differential equation

$$V'(t) + V(0) + \int_0^t V(s)\,ds = 0 \tag{2.22}$$

Differentiating (2.22) gives the following second-order differential equation

$$V''(t) + V(t) = 0 \tag{2.23}$$

which can be converted to a system of two first-order equations by the introduction of an additional variable, $w(t) = V'(t)$,

$$\begin{aligned} W'(t) + V(t) &= 0 \\ W(t) - V'(t) &= 0 \end{aligned} \tag{2.24}$$

First-order systems of this type will be considered in Chapter 3.

2.7 Summary

We begin the development of circuit equations with the simplest case of circuits consisting of linear resistors. Here the classical discrete model based on Kirchoff's laws leads to a sparse linear system of algebraic equations with sparsity determined by network connectivity. Related system properties are briefly described and the connection to discretized stencils for PDE problems in device or process modeling is noted. The next step of including linear capacitors and inductors is then considered and shown to lead to linear systems of ODEs. Basic techniques for integrating these systems are introduced later in Chapter 4. The extension to circuits containing nonlinear components is considered next in Chapter 3 and the nonlinear algorithmic system solution in Chapter 5.

2.8 Exercises

1. Remove the diagonal branches from the circuit in Figure 2.2. Draw the sparsity structure for the matrix associated with the new circuit.

2. Remove the horizontal and vertical branches from the circuit in Figure 2.2. Draw the sparsity structure for the matrix associated with the new circuit.

3. Once again refer to the circuit of Figure 2.2. The Kirchoff voltage law applied to the loop from n_{26} to n_{22} to n_{16} to n_{20} and then back to n_{26} yields an equation involving the voltages at three nodes of unknown voltage and one node of known voltage. Find 25 such "tours", such that each node is in at least one tour and only diagonal branches are used. Can these 25 equations be solved to obtain the values for all of the node voltages?

4. Assign the value of one farad to each capacitor and one ohm to each resistor of the circuit in Figure 2.6. Find numbers a, b, c, d, e, and f so that the following three equations hold: $v_2 = a \sin t + b \cos t$, $v_3 = c \sin t + d \cos t$, and $v_4 = e \sin t + f \cos t$. Does this require solving three equations in six unknowns?

5. In the circuit of Figure 2.2 suppose that for each of the ten nodes n_k of known voltage, $v_k = k \sin kt$. Write a program to find the unknown voltages as functions of time. When you have completed this problem, you will have produced a simulator to obtain the transient characteristics of the circuit of Figure 2.2.

6. Assign the ten known voltages for Figure 2.2 as in the last problem. Use Kirchoff's voltage law to derive a system of 25 independent equations. Solve the system using the program of the previous problem. Discuss your results.

7. There are 98 resistors in the circuit of Figure 2.2. Each resistor has a resistance and is connected to two nodes. Write a list that is of length 98 so that the nth item of the list corresponds to the nth resistor in the circuit. The nth resistor is on a branch connecting two nodes a and b. The nth item on the list is of the form [resistance, a, b]. Write a program that takes this list as input and produces matrix $\boldsymbol{A}$ of equation (2.7). When you have completed this program, you will have produced the admittance matrix from a SPICE type data set for a resistive circuit.

Chapter 3

Transistors and Semiconductor Circuits

3.1 Introduction

In the preceding discussion of current in resistors and capacitors, there was no consideration as to how electrons actually move through a solid. This section will give some insight into that part of solid-state physics which explains electron movement in semiconductors and metals. Soon after the discovery of the electron, physicists sought to interpret Ohm's law at the atomic level. Attempts were made to explain the observed relationships between thermal and electrical conductivity as well as between electrical resistance and temperature. It was conjectured that in a metallic atom the electrons in the inner shell of an atom are tightly bound to the nucleus, but some of the outermost electrons are able to move freely throughout the solid. This view suggests that the number of conducting electrons would be an integral multiple of the number of atoms and electrical resistance would occur as a result of scattering of electrons by collisions with nuclei and other electrons.

In an early attempt to explain these experimental results, Drude [295] proposed a model in which all of the electrons possessed a classical momentum of $\sqrt{3m_0kT}$, where m_0 is the mass of an electron, k is Boltzmann's constant, and T is the absolute temperature. Lorentz [514] attempted to improve this model by assuming that electron velocities were distributed in accordance with Maxwell–Boltzmann statistics. Neither model proved entirely satisfactory. In fact, no model based on classical physics completely describes the movement of electrons in a semiconductor because the system is quantum mechanical in nature.

Fermi–Dirac statistics provide the first effective model for the energy distribution of particles in a crystal such as silicon. These statistics generalize the Pauli exclusion principle from a single atom to a collection of atoms arranged in a periodic crystalline structure. Consider a crystal in equilibrium at a constant temperature T. Let N be the number of electrons in the crystal and W be the sum of their energies, which is constant. It can be shown that, at each instant, there exists a finite collection S of M states (each state consists of three quantum numbers and a spin number), where M is much larger than N, such that each electron in the crystal must occupy one of the states of S, and no two electrons can simultaneously occupy a single state. Moreover, there exists a finite, increasing sequence of energies $E_1, E_2, \ldots, E_s$ with s much smaller than N such that each member of S has one of the energies in the sequence E_j. For $1 \leq i \leq s$, let A_i denote the number of states in the collection S which are at energy E_i.

Now fix a particular instant of time t, and let R denote the state distribution that the electrons occupy at time t. Notice that R is an N-member subcollection of S. For $1 \leq i \leq s$, let B_i denote the number of states of R at energy E_i, so that $\sum_i B_i E_i = W$ and $\sum_i B_i = N$. Two configurations of electrons that correspond to the same subcollection R are considered indistinguishable; that is, the swapping of two electrons does not change the state of the system. There are a number of physically possible distributions Q, distinct from R, so that the sequence $B(Q)$ associated with Q is identical to $B(R)$. In fact, for a given energy E_i, the number of ways $D_i(B_i)$ to choose the B_i states of R at energy E_i from the number of states A_i of S at energy E_i is given by

$$D_i(B_i) = \frac{A_i!}{B_i!(A_i - B_i)!} \tag{3.1}$$

Therefore, the total number of distributions $G(B)$ that produce the sequence $B_1, B_2, \ldots, B_s$ is given by $G(B) = D_i$. Consider the following problem: find a sequence $C = C_1, C_2, \ldots, C_s$ having the properties that (1) $\sum_i C_i E_i = W$, (2) $\sum_i C_i = N$, and (3) the maximum value attained by G is $G(C)$. Fermi and Dirac were able to solve for this "most likely" sequence C as follows: first embed the discrete function G into a continuous function F with the aid of Stirling's formula then find the global maximum of F by setting its partial derivatives to zero. This system of equations is solved using Lagrange multipliers subject to the condition that the resulting Fermi–Dirac distribution must approximate the Maxwell–Boltzmann distribution for large values of temperature [46].

The collection S has the property that there exists an integer $j > 1$ such that, $A_1 + A_2 + \ldots + A_j < N$. Therefore, even when the temperature is at absolute zero, some electrons in the system must have positive energies. The Fermi–Dirac distribution gives the probability that a given state of energy E is occupied by an electron as

$$f(E) = \frac{1}{1 + \exp\left(\frac{E - E_F}{kT}\right)} \tag{3.2}$$

where k is Boltzmann's constant, T is the temperature in Kelvin, and E_F is a temperature-dependent constant called the Fermi energy or Fermi potential, such that $f(E_F(T)) = \frac{1}{2}$. Each substance has a unique Fermi potential. The reason why some energies are more predominant than others follows from elementary probability theory in the same sense that in a throw of a pair of dice, sevens are more likely than elevens, and in a very large number of rolls, the ratio of the number of sevens to the number of elevens can be accurately calculated. The distribution (3.2) is a radical departure from those of classical physics in that there exists a distribution of electron energies even at $T = 0$. The importance of the Fermi–Dirac statistic lies in its far-reaching consequences: using (3.2) Summerfield [2] showed that almost all of the electrical current is carried by electrons with energies quite close to $E_F(T)$. Because there are very few electrons with energies in this region, the classical assumption that the number of carriers is an integral multiple of the number of atoms is false. Since the number of carriers is small, it follows that in order to produce the observed current, individual carriers have to be able to move quite freely through the lattice. The most important consequence of the Fermi–Dirac statistic is that it can be used to derive the band theory of semiconductors, which played a key role in the development of the transistor and is used extensively in the design and simulation of today's microelectronic devices.

3.2 Band Theory for Semiconductors

This section is written as an introductory overview of the physics and chemistry of semiconducting materials and provides some fundamentals for the treatment in later chapters on device models. The term "semiconductor" is applied to a group of materials which are insulators at very low temperatures, but conduct a small amount of current at room temperature. There is a large number of known semiconductors, including the elements silicon

III	IV	V
5 B Boron	6 C Carbon	7 N Nitrogen
13 Al Aluminum	14 Si Silicon	15 P Phosphorus
31 Ga Gallium	32 Ge Germanium	33 As Arsenic
49 In Indium	50 Sn Tin	51 Sb Antimony

Table 3.1: A portion of the periodic table

and germanium, as well as the III-V compounds gallium arsenide and indium phosphide. In Table 3.1 a portion of the periodic table is reproduced, showing the location of the elements of interest in semiconductor physics: silicon (Si), germanium (Ge), phosphorus (P), arsenic (As), antimony (Sb), boron (B), gallium (Ga), and indium (In). Because of its widespread use in current technology, silicon will be used in the following discussion.

Arguments based on quantum mechanics, and in particular Fermi–Dirac statistics, have been given to show that the energies of electrons in the silicon lattice are restricted to lie in definite bands. We refer to [46, 295, 514] for complete derivations. For a less detailed, but insightful account, consult [179]. For a summary of the debate over the validity of certain of the simplifying assumptions behind the band model see [46] and [179]. There are energies E_v and E_c, $E_c > E_v$, such that electrons can have energies above E_c and below E_v, but not in the "forbidden" zone between E_v and E_c. The range of energies below E_v is referred to as the valence band. Electrons with energies in this band are tightly bound to atoms in the lattice and cannot move about the crystal freely. The range of energies above E_c comprises the conduction band: electrons with such energies are free to move about the crystal carrying current. In pure silicon at room temperature the thermal vibration of the lattice imparts enough energy to a very few electrons so that they can enter the conduction band.

The actual mechanisms of current flow in a semiconductor lattice are quite complicated; however, a simplified yet useful way of viewing the current is in terms of two types of carriers, electrons and holes. When an electron attains enough energy to enter the conduction band and move freely about the crystal, it leaves behind a positively charged site that is referred to as a "hole". The atom that is short of an electron can acquire one from a nearest neighbor, thereby changing the position of the positively charged site in the crystal. This neighboring atom that is now short of a valence electron can take one from yet another nearby atom, causing the positive site to move again. By this mechanism, holes can effectively move through the lattice. The total current in the crystal is therefore due to the movement of both negatively charged electrons and positively charged holes. In a later chapter on device modeling it will be seen that for some transistor devices there is only one "majority" carrier that dominates the current flow, while for other devices it is necessary to solve for both types of carriers. Although a hole is not a true particle, as is the electron, the concept of a positively charged entity is very convenient. Without it, we would be forced to solve for concentrations of both conduction and valence electrons, constantly recalling that there is an effective movement of valence electrons as they acquire energy to enter the conduction band.

When a voltage is applied across a semiconductor crystal, the holes move toward the negative charge while the electrons move toward the positive charge. When a hole and conducting electron "meet", they annihilate one another, a process referred to as recombination. New electron–hole pairs are constantly being produced by crystal vibrations, which impart enough energy to a valence electron to move it into the conduction band, a mechanism termed generation. At a given time, only one electron in 10^{13} is to be found in the conduction region. This implies that there are roughly a billion conducting electron–hole pairs per cubic centimeter of silicon and it explains why pure silicon is such a poor conductor. In a pure semiconductor the number of free electrons per unit volume is called the intrinsic carrier concentration n_I and is equal to the number p_I of holes per unit volume.

The situation changes when impurities are introduced into the pure crystal, a process referred to as "doping." The mathematical formulation and numerical treatment of doping by diffusion are considered in Chapters 11 and 12. In a doped semiconductor the number of holes is not necessarily equal to the number of free electrons. Suppose that an atom with one more bonding electron than silicon replaces a silicon atom in the lattice. Since this electron is not needed for bonding, the thermal vibration of the crystal can easily set it free to travel throughout the lattice. When it leaves its parent atom, the electron leaves behind a positively charged atom, or ion, which unlike a hole, is not free to move. An atom with an extra bonding electron is referred to as a donor because it donates an electron to the conduction band. Silicon doped with donor atoms is referred to as n-type, since there are more negative carriers than positive carriers (the word carrier refers to either a hole or electron). Elements from column V of the periodic table — phosphorus, arsenic, and antimony — are the most common donors for silicon. In practice, the donor concentration N_D in doped silicon ranges from 10^{16} to 10^{21} atoms cm^{-3}, which is small compared with the number of lattice sites (6×10^{22} atoms cm^{-3}), but still several orders of magnitude higher than n_I. Since a majority of the donor atoms release one electron into the conduction band, whenever $N_D > n_I$, the number n of free electrons per unit volume is approximately equal to the donor concentration, and the semiconductor is said to be extrinsic. As the electron concentration is increased above n_I the probability that a hole will meet an electron and be annihilated is increased. Therefore, in n-type extrinsic material the concentration p of holes is less than p_I, the intrinsic hole concentration,while for p-type material $n < n_I$. It can be shown [2] that at equilibrium the product of electron and hole concentrations in intrinsic material equals the corresponding product for extrinsic silicon

$$np = n_I p_I \tag{3.3}$$

Silicon can also be doped with an element from column III such as boron, which has one fewer bonding electron than silicon. In this case, there is a tendency for the impurity atom to capture or accept a conduction electron from a neighboring lattice atom in order to have a complete set of bonding electrons. This capture results in a shortage of an electron at another location in the lattice and a hole is thereby formed. Such an ionized impurity atom is termed an acceptor and the doped silicon is referred to as p-type material because the majority of free carriers are holes. The fact that, in a substance where the only free particles are electrons, a current could be conducted by entities that behaved like positive particles, was at first viewed as strange. This phenomenon was discovered by an experiment that measured the charge on the surface of conducting beryllium. The charge is due to the Hall effect and would be of opposite sign for positive and negative carriers [179]. The interplay between theoretical, experimental, and computational physics has been particularly productive in the field of solid-state physics. The relationship (3.3) holds for p-type material just as it does for n-type. Common acceptor impurities for silicon include boron, gallium, and indium. The presence of dopants in the lattice influences the energy diagram as shown in Figure 3.1. E_d is the energy of the extra electron in a donor atom, while E_a is the energy acquired by an electron when captured by an acceptor atom. Since E_d is close to E_c, the vibration of the lattice at room temperature is sufficient to lift nearly all of the donor electrons into

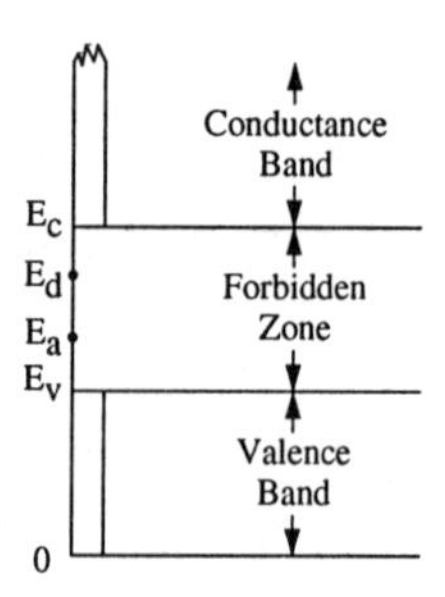

Figure 3.1: Energy bands for extrinsic semiconductors.

the conduction band. Similarly, most acceptor atoms are able to capture a valence band electron, because E_a is only slightly above E_v.

3.3 Drift and Diffusion

This section gives a brief introduction to how drift and diffusion contribute to the total current flow in a semiconductor. Later in Chapter 7 we develop the drift-diffusion mathematical model and construct approximation strategies. Consider a block of semiconductor material of dimensions L, W, and D. Suppose that a voltage V is applied across the material so that free electrons will tend to move toward the positive terminal. The average velocity of the electrons, v_d, is referred to as the drift velocity and the average time τ for an electron to travel the distance L, $\tau = \frac{L}{v_d}$, is called the transit time. If Q is the sum of the charges in the block, then the current is the time rate of change of Q, so that $I = -\frac{Q}{\tau}$. For silicon, v_d depends upon the magnitude of the electrical field $\boldsymbol{E}$. Hence v_d depends on V, as shown in Figure 3.2 which graphs $v_d = f(|\boldsymbol{E}|)$ as a function of $|\boldsymbol{E}|$. For low electrical fields, v_d is proportional to $|\boldsymbol{E}|$ and the proportionality constant $f'(x)$ is referred to as the mobility μ_n (similarly there is a mobility μ_p for holes).

This linear relationship at low fields is quite plausible: as long as the mean time between collisions for an electron is large, the greater the applied field the more energy the electron will acquire during its flight. Beyond a certain point, however, collisions between the electron and the lattice atoms, as well as with other electrons, become so frequent and the mean free path so short, that increasing $|\boldsymbol{E}|$ has little effect on v_d — the velocity "saturates." The function f can be determined both by experiment and by solving a coupled system of partial differential equations (the drift-diffusion equations) as discussed in Chapter 7. It must be remembered that current flow due to drift can only occur in the presence of a non-zero electric field.

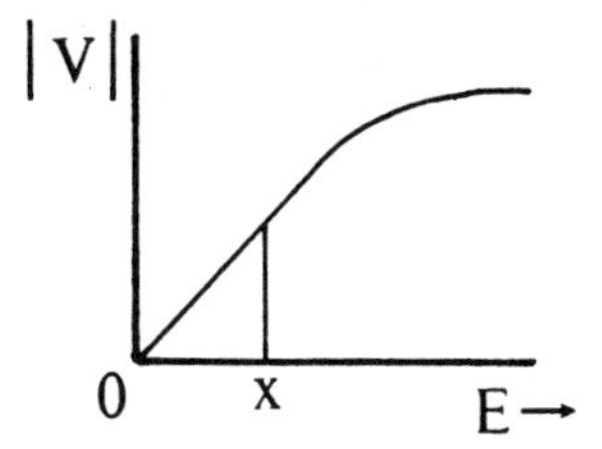

Figure 3.2: The magnitude of drift velocity as a function of $|\boldsymbol{E}|$.

Another mechanism, diffusion, can contribute to current flow even in the absence of an electric or magnetic field. Diffusion occurs when the carrier concentration is not constant throughout the semiconductor, in other words when the gradient is nonzero. This could result from uneven photon bombardment at the surface, variations in temperature, or uneven doping. Figure 3.3 illustrates the situation in which the concentration of conduction band electrons decreases from left to right in the semiconductor. Suppose that no electric or magnetic field is applied. Since the drift velocity is zero, the electrons are equally likely to move in either direction. Because there are more electrons on the left, a "random walk" by many electrons will result in more of the carriers crossing to the right. This net migration of the negatively charged electrons from left to right results in a net current flow from right to left, called a diffusion current. The interplay between drift and diffusion processes will be studied in greater detail later. The concepts are used now to explain how a *pn*-junction operates as a diode.

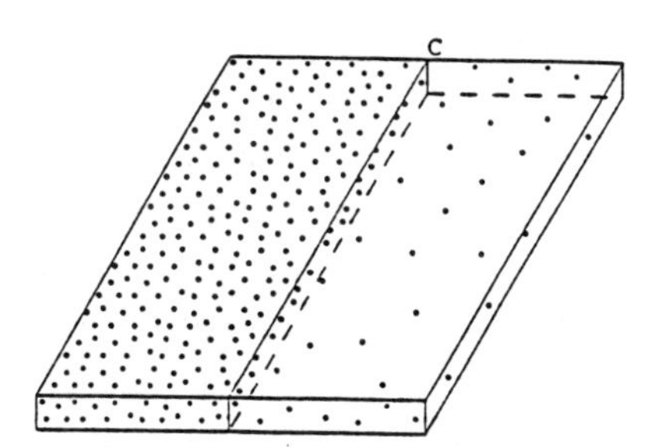

Figure 3.3: Uneven concentration of conducting electrons.

3.4 The *pn*-junction

Doping a semiconductor as *p*-type in one area and *n*-type in an adjacent area forms a *pn*-junction at the interface. Suppose that donor impurities are introduced uniformly into a slab of silicon to the left of a point β, while the silicon to the right of β is uniformly doped with acceptor impurities. Let N_D be the dopant concentration in the n-type material, N_A that of the p-type material, and suppose that $N_A > N_D$. To the left of β there is an excess of electrons which will therefore tend to diffuse to the right of the pn-junction as illustrated in Figure 3.4. Similarly, holes will diffuse from right to left across the junction. This movement of carriers will cause the region to the right of the junction to become negatively charged, while a net positive charge will exist to the left of β. The difference in charge creates an internal electric field, which in turn produces a drift current in the direction opposite to the diffusion current. As diffusion continues the electric field increases in strength until finally no net current flow occurs and a steady-state is achieved.

In Chapter 7 the partial differential equations associated with drift and diffusion will be studied. These equations show that when a steady state is reached, there exist numbers α, β and γ, $\alpha < \beta < \gamma$, such that nearly all of the free electrons in the region $[\alpha, \beta]$ diffuse to the right of β, but there is almost no net diffusion of electrons that lie to the left of α. Similarly, the region $[\beta, \gamma]$ is depleted of holes, while there is almost no net diffusion of holes that lie to the right of γ. Moreover, it can be shown that the number of holes that diffuse across the junction from the left must equal the number of electrons diffusing across the junction from the right. The interval $[\alpha, \gamma]$ defines the "depletion region", as illustrated in Figure 3.4, where the + symbols represent holes that have diffused into the n-type material and the − signs denote electrons that have diffused into the p-type material. There is a net positive charge on the n-type material, while the p-type material has a net negative charge, so that the potential is higher to the left of the junction. A plot of potential as a function of position is also included in Figure 3.4. It can be shown that the potential difference ΔE is equal to the difference in Fermi potentials of the n-type and p-type materials [529].

The system is in equilibrium because there is no net flow of current: drift current due to a non-constant field is canceled by diffusion current due to the non-uniform distribution of carriers. There are a very few holes to the left of α and occasionally one of these will randomly cross the junction. When this occurs, the hole will be swept across the depletion region and down the potential hill, thus generating a current denoted by I_0. There is a higher density of holes to the right of γ and many of them approach the region $[\alpha, \gamma]$, but only a tiny fraction possess sufficient energy to climb the potential hill and cross the depletion region. The current caused by these holes is exactly $-I_0$. Similarly, electrons crossing the depletion region from

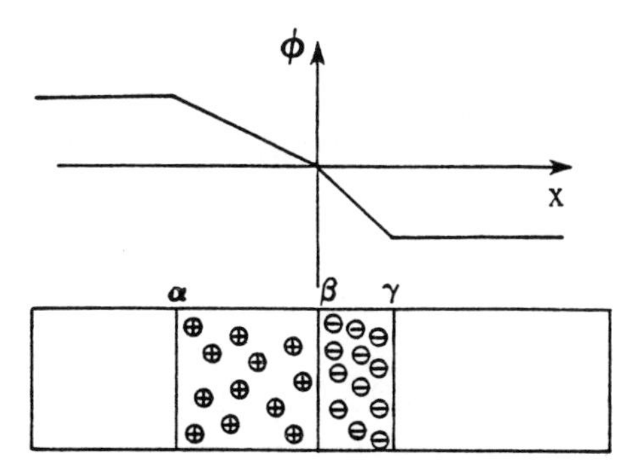

Figure 3.4: Migration of carriers across a *pn*-junction and associated plot of potential.

γ to α create a current I_0 that is canceled by an equal flow of electrons traveling in the opposite direction.

Suppose now that a forward bias is applied to the *pn*-junction as shown in the lower graph of Figure 3.5. Electrons will be attracted toward the positive charge and move out of the depletion region through the *p*-type material; similarly, holes will move from the depletion region and across the *n*-type material toward the negative charge. The potential across the region is reduced and equilibrium destroyed, since there is no longer sufficient charge to keep electrons and holes from entering the depletion region. The drift-diffusion equations now show that the size of the depletion region will be reduced in proportion to the size of the applied bias. In the case of forward bias, current that now flows is due to the majority carriers. If a negative bias is applied, as shown in the upperr part of Figure 3.5, the opposite effect occurs and the size of the depletion region increases because minority carriers are swept across it. Since there are so few carriers of this type, the resulting current will be extremely small. It can be shown that there is a positive number a such that, if a voltage v is applied across the junction, then the current I through the diode is [179]

$$I(v) = I_0(e^{av} - 1) \tag{3.4}$$

When v is negative but very large in magnitude, the assumptions used to derive (3.4) no longer hold and a large current begins to flow. This situation is referred to as breakdown and it causes the destruction of the diode due to overheating. For silicon diodes the breakdown voltages depend upon the specific method of fabrication, the dimensions of the device in the plane perpendicular to current flow and the ambient temperature. Actually it is not the voltage that determines when breakdown occurs, but the charge per

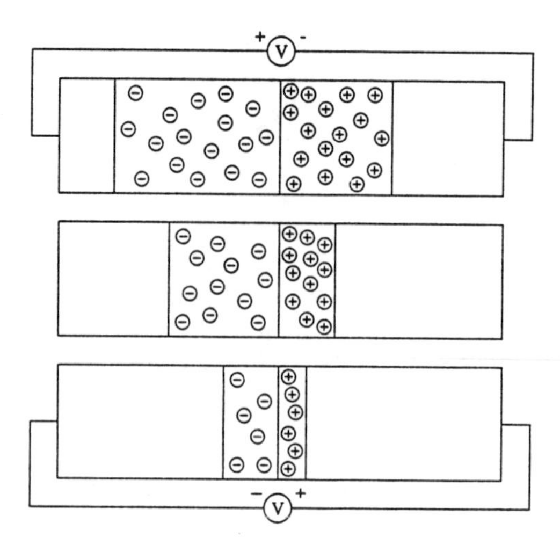

Figure 3.5: Depletion region size as a function of voltage .

unit area in the plane of the junction. Most integrated circuits have supply voltages of 3 to 5 volts, but as device dimensions shrink to less than 0.5 μm, the charge per cm^2 will exceed the critical value. A major goal of processing simulation as described in later chapters is to assist in the design of devices that are as small as possible, yet avoid such breakdown. In the normal ranges of operation, a *pn*-junction acts as a diode or rectifier, essentially allowing current to flow in only one direction.

3.5 The Ideal Diode

The resistor is a device in which current is a linear function of the voltage drop across its two terminals. Circuits containing only resistors can be simulated by solving a system of linear algebraic equations. Recall that to model an RC circuit consisting of only resistors and capacitors, we must solve a system of linear ordinary differential equations because the current in each capacitor is a linear function of the time derivative of the voltage drop across it. It will be shown in this section that circuits containing only linear resistors and ideal diodes can be modeled by a system of nonlinear algebraic equations. To simulate circuits containing all three types of devices requires solving a system of ordinary differential equations which are nonlinear in the nodal voltages and linear in the derivatives of the nodal voltages.

Assume that two circuit nodes, n_i and n_j, are connected by an ideal diode D; let $\Delta v = v_{i,j} = v_i - v_j$, and suppose that the diode is oriented so that when Δv is positive, the diode is forward-biased. The current I

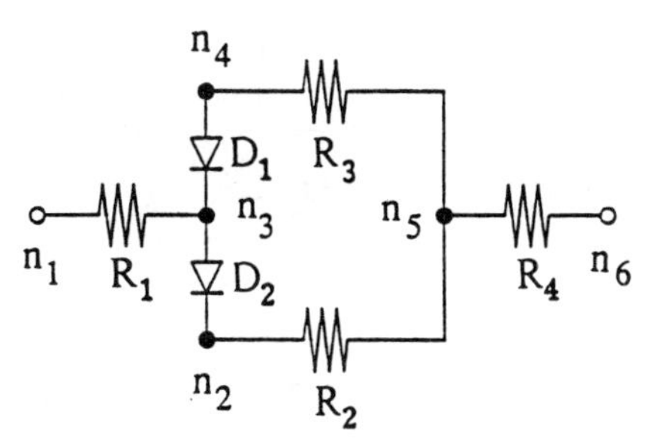

Figure 3.6: A circuit containing diodes and resistors.

through D from n_i to n_j is $I = I_0(e^{a\Delta v} - 1)$ where the parameters I_0 and a are determined by experiment or device simulation and are functions of temperature. Figure 3.6 illustrates a circuit consisting of six nodes $(n_1, n_2, \ldots, n_6)$, four linear resistors (R_1, R_2, R_3, R_4) and two ideal diodes (D_1, D_2). Let $\boldsymbol{v} = (v_1, v_2, \ldots, v_6)$ be the vector of node voltages with v_k the voltage at node n_k; assume that v_1 and v_6 are known and the remaining voltages unknown. Let r_k denote the resistance of R_k; for each diode D_k, let I_k and a_k denote the corresponding parameters in equation (3.4). For $2 \leq k \leq 5$, let $f_k(\boldsymbol{v})$ denote the current flowing into n_k and let $f(\boldsymbol{v})$ be the vector $(f_2(\boldsymbol{v}), f_3(\boldsymbol{v}), f_4(\boldsymbol{v}), f_5(\boldsymbol{v}))$. By Ohm's law and equations (3.4) we have

$$
\begin{aligned}
f_2(\boldsymbol{v}) &= I_2(e^{a_2(v_3-v_2)} - 1) + \frac{v_5 - v_2}{r_2} \\
f_3(\boldsymbol{v}) &= I_1(e^{a_1(v_4-v_3)} - 1) - I_2(e^{a_2(v_3-v_2)} - 1) + \frac{v_1 - v_3}{r_1} \\
f_4(\boldsymbol{v}) &= \frac{v_5 - v_4}{r_3} - I_1(e^{a_1(v_4-v_3)} - 1) \\
f_5(\boldsymbol{v}) &= \frac{v_6 - v_5}{r_4} + \frac{v_4 - v_5}{r_3} + \frac{v_2 - v_5}{r_2} \qquad (3.5)
\end{aligned}
$$

Methods for solving such a system of nonlinear algebraic equations, and therefore simulating the circuit in Figure 3.6, will be studied in Chapter 4.

The circuit of Figure 3.7 is identical to that of Figure 3.6 except that an ideal capacitor C with capacitance c has been inserted between nodes n_3 and n_5. This slight change will have a profound effect upon the way the circuit is modeled mathematically. Using the notation developed previously, the equations for $f_2(v)$ and $f_4(v)$ are unchanged, while those for $f_3(\boldsymbol{v})$ and $f_5(\boldsymbol{v})$ are modified to

$$
f_3(\boldsymbol{v}) = I_1(e^{a_1(v_4-v_3)} - 1) - I_2(e^{a_2(v_3-v_2)} - 1) + \frac{v_1 - v_3}{r_1} + \frac{d}{dt}(c(v_5 - v_3))
$$

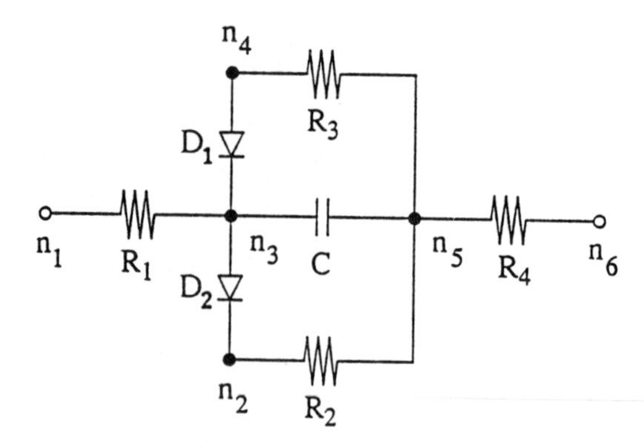

Figure 3.7: A circuit with diodes, resistors, and a capacitor.

$$f_5(v) \quad = \quad \frac{v_6 - v_5}{r_4} + \frac{v_4 - v_5}{r_3} + \frac{v_2 - v_5}{r_2} + \frac{d}{dt}(c(v_3 - v_5)) \tag{3.6}$$

Note that if the voltages at nodes n_3 and n_5 are held constant their time derivatives are zero and system (3.5) is recovered. To summarize, for a circuit consisting only of diodes and resistors, simulation involves solving a system of nonlinear algebraic equations. If capacitors are added it becomes necessary to solve a system of nonlinear ordinary differential equations, which are linear in the derivative term.

3.6 Nonlinear Capacitance in a Diode

In the ideal diode current is expressed as a function of the terminal voltages. However, in the true diode the current is a function of terminal voltages and their derivatives. To illustrate this fact, a true diode may be drawn as two devices – in one of the devices current is a function of terminal voltages and in the other it is a function of derivatives of voltage. This "two"-device model is indicated in Figure 3.8.

More specifically, it was noted previously that a diode stores charge in an amount which is a function of the terminal voltages, as illustrated in Figure 3.5. A change in terminal voltage causes the amount of stored charge to vary, which in turn causes a current to flow. This current passing through diode D from node n_0 to n_1 may be modeled as

$$I_D(t) = \alpha \left(e^{b(v_0(t) - v_1(t))} - 1 \right) \tag{3.7}$$

A more physically accurate model that includes the nonlinear capacitance may be constructed by combining the previous diode with a parallel capacitor to obtain an improved model of a diode as indicated in Figure 3.8. The charge $q(t)$ at time t is of the form [561]

$$q(t) = \rho e^{b(v_0(t)-v_1(t))} - \rho + \beta \int_0^{v_0(t)-v_1(t)} (1-\gamma x)^\delta dx \tag{3.8}$$

where the parameters ρ, β, γ, and δ are functions of the device parameters and temperature. The current flowing through the improved diode is the time derivative of its charge, $I_C(t) = q'(t)$ so that

$$I_C(t) = \left(\rho b e^{b(v_0(t)-v_1(t))} + \beta(1-\gamma(v_0(t)-v_1(t)))^\delta\right)(v_0'(t) - v_1'(t)) \tag{3.9}$$

Note that for the linear capacitor, the capacitance can be defined in terms of charge using either $q = cv$ or $\frac{dq}{dv} = c$. For a nonlinear capacitor in which c is a function of the voltage, $q = c(v)v$ and the product rule gives $\frac{dq}{dv} = c(v) + \frac{dc}{dv}v$. Observe that the current through such a capacitor, however, can be obtained without discussing the device's capacitance.

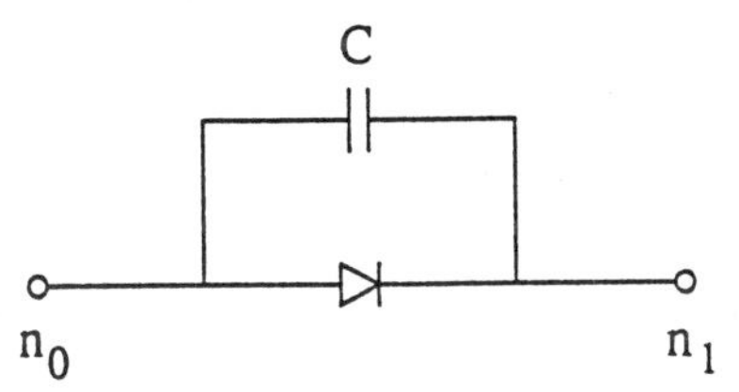

Figure 3.8: Improved model of a diode which includes internal capacitance.

If the ideal diodes in Figure 3.7 are replaced by the improved model in Figure 3.8, the circuit in Figure 3.9 is obtained. As before, let $\boldsymbol{v}(t) = (v_1(t), v_2(t), \ldots, v_6(t))$ be the vector of nodal voltages with $v_1(t)$ and $v_6(t)$ known functions. For $1 \leq k \leq 6$, let $f_k(\boldsymbol{v}_t)$ denote the current flowing into node n_k at time t, and $\boldsymbol{f}(\boldsymbol{v}(t)) = [f_2(\boldsymbol{v}(t)), f_3(\boldsymbol{v}(t)), f_4(\boldsymbol{v}(t)), f_5(\boldsymbol{v}(t))]^T$. The equations for $\boldsymbol{f}(\boldsymbol{v}(t))$ can be obtained by including the additional current terms for the nonlinear capacitor in the equations that were derived earlier using Kirchoff's current law. The first equation of (3.5) now becomes

$$\begin{aligned} f_2(\boldsymbol{v}(t)) = I_2(e^{a_2(v_3(t)-v_2(t))} - 1) + \frac{v_5(t) - v_2(t)}{r_2} + \\ (\alpha_2 b_2 e^{b_2(v_3(t)-v_2(t))} + \beta_2(1-\gamma_2(v_3(t)-v_2(t)))^\delta)(v_3(t) - v_2(t))' \end{aligned} \tag{3.10}$$

If the change of voltage with respect to time is small, then the circuit of Figure 3.7 is a good approximation to that of Figure 3.9. If the voltages are changing rapidly, however, then the more advanced model of the diode is necessary for accurate simulation. The next chapter will show how to obtain a numerical solution to these more complicated systems of equations.

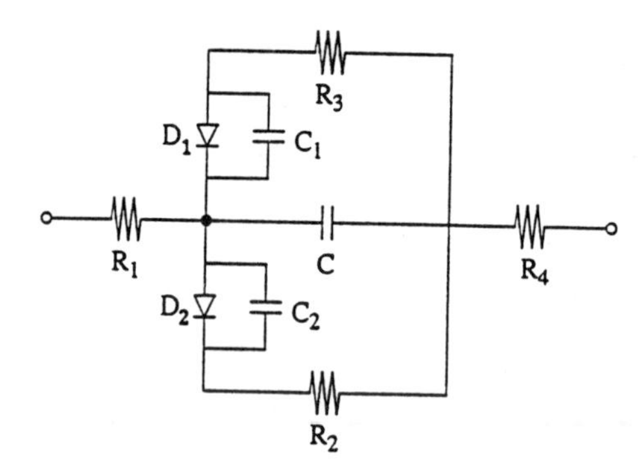

Figure 3.9: A circuit with improved-model diodes, resistors, and a capacitor.

3.7 Modeling the MOS Transistor

It is instructive to consider the "true" diode as a circuit containing two elements and two nodes. Current through one element is a function of the voltage drop across the terminals, while current through the other element is a function of the time derivative of the voltage drop across the nodes. Similarly, it is possible to model a metal-oxide semiconductor (MOS) transistor at the circuit level as a subcircuit composed of various elements such as resistors and capacitors. In the simplest model, the transistor is just a three-terminal device (source, gate, and drain) with only one element, as shown in Figure 3.10. Let the voltage at the source S be lower than the voltage at the drain D, so that current will flow from drain to source only if the voltage drop from gate to source is above a certain threshold. The gate will therefore serve as a switch in the current I_{SD} between source and drain, which can be viewed as a function of the three terminal voltages.

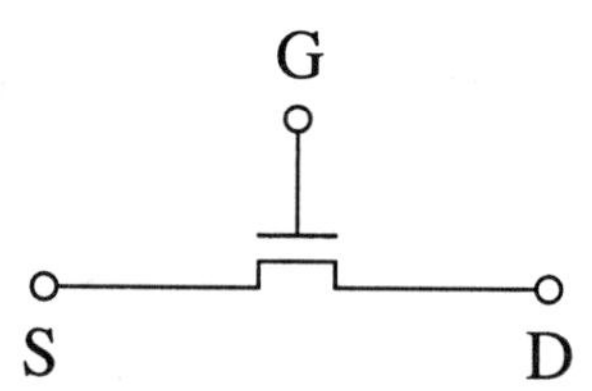

Figure 3.10: Simple model of a transistor.

This model of the transistor can be refined in various ways. For example, the transistor can be interpreted as a four-terminal device (source, drain, gate and bulk), or more generally as a circuit with seven elements, as shown in Figure 3.11, where the "black box" X in the center represents a device that carries the DC component of the current between source and drain. The current through X is assumed to be a function of all four terminal voltages v_S, v_B, v_G, and v_D. The current through each diode is modeled by (3.4), while for each capacitor the current is the time derivative of its charge, which is a nonlinear function of the four terminal voltages. The mathematical models of the devices for the circuit in Figure 3.11 are derived from various physical considerations. Even the simple models summarized here lead to considerable circuit complexity, providing more accurate and reliable simulations, but at the cost of additional computation.

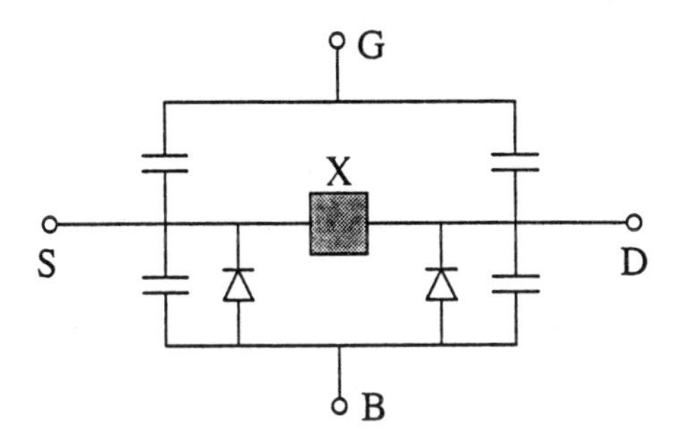

Figure 3.11: Model of a MOS transistor using several circuit elements.

Let $\mathcal{U}$ be the subset of $\mathbf{R}^4$ consisting of all vectors (V_S, V_D, V_G, V_B) of possible voltages at source, drain, gate, and bulk. It is common practice to partition $\mathcal{U}$ into "regions of operation" of the transistor, enforcing various simplifying assumptions in each region. An early model [476] for the flow of current through the "black box" of Figure 3.11 is used in the circuit simulator SPICE I [400] and is given in Table 3.2. The DC current is a function of five voltage drops $v_{D,S}$,$v_{B,S}$, $v_{G,S}$, $v_{G,D}$, and $v_{B,D}$ where $v_{D,S} = v_D - v_S$, etc. The model postulates two threshold voltages, v_{tho} and v_{th}. The threshold voltage v_{tho} is a constant such that if $v_{B,S} = 0$, then the device carries current only when $v_{G,S} < v_{tho}$. The threshold voltage v_{th} is referred to as the effective threshold voltage for the forward region ($v_{D,S} \geq 0$),

$$v_{th} = v_{tho} + \gamma(\sqrt{\phi - v_{B,S}} - \phi) \tag{3.11}$$

where γ and ϕ are positive constants. The current I is then a function of v_S, v_D, v_G, and v_B. The forward region of operation is divided into three subregions, depending upon the magnitude of $v_{G,S}$:

$$\textit{Subthreshold Region}: \quad 0 \leq v_{G,S} \leq v_{th}$$

$$
\begin{aligned}
I_1 &= 0 \\
\textit{Saturation Region}: \quad & v_{th} \leq v_{G,S} \leq v_{D,S} + v_{th} \\
I_2 &= \beta(v_{G,S} - v_{th})^2(1 + \lambda v_{D,S}) \\
\textit{Linear Region}: \quad & v_{D,S} + v_{th} \leq v_{G,S} < \infty \\
I_3 &= \beta v_{D,S}(2(v_{G,S} - v_{th}) - v_{D,S})(1 + \lambda v_{D,S})
\end{aligned}
\tag{3.12}
$$

Note that the first two formulas agree at $v_{G,S} = v_{th}$ and the second two agree at $v_{G,S} = v_{th} + v_{D,S}$, so that $I(v_{G,S})$ is in fact a well-defined function, which is continuous on its domain. Taking the partial derivative of current with respect to $v_{G,S}$,

$$
\begin{aligned}
\frac{\partial I_1}{\partial v_{G,S}} &= 0 \\
\frac{\partial I_2}{\partial v_{G,S}} &= 2\beta(v_{G,S} - v_{th})(1 + \lambda v_{D,S}) \\
\frac{\partial I_3}{\partial v_{G,S}} &= 2\beta v_{D,S}(1 + \lambda v_{D,S})
\end{aligned}
\tag{3.13}
$$

Notice that $\frac{\partial I}{\partial v_{G,S}}$ is continuous. Similar calculations show that $\frac{\partial I}{\partial v_S}$, $\frac{\partial I}{\partial v_D}$, and $\frac{\partial I}{\partial v_B}$ are continuous. The importance of such smoothness in the model equations will become apparent in the next chapter.

Modeling the four capacitors shown in Figure 3.11 has evolved considerably over the years. Early formulations resulted in an inconsistency referred to as the "charge conservation problem". It appeared possible to construct continuously differentiable functions f and g such that:

1. f defines a simple closed curve; i.e., $f : [0,1] \rightarrow \mathbf{R}^2$, $f(0) = f(1)$, and f does not intersect itself

2. g is a real-valued function defined on $\mathbf{R}^2$

3. $\int_0^1 (g(f(x)))'\,dx$ is a positive number.

Recall that the fundamental theorem of calculus implies

$$
\int_0^1 (g(f(x)))'\,dx = g(f(1)) - g(f(0)) \tag{3.14}
$$

and since $f(1) = f(0)$, this integral must be zero. By repeatedly integrating around the curve defined by f, it appeared that one could increase the value of g without bound, in effect building up charge on the capacitor without limit by an influx of electrons, which seemingly materialized from nowhere. This inconsistency was circumvented by assuming that the charge on each terminal is a function of the four terminal voltages [561]. Current into a terminal is then the derivative of that charge as a function of time.

A representative model for charge in a MOSFET is indicated in Table 3.2 and assumes the existence of six constants α, β, γ, δ, ϕ, and ω, together with three auxiliary functions

$$\begin{aligned} v_{th} &= \beta + \phi + \delta\sqrt{\phi - v_{B,S}} \\ f &= \alpha + \gamma(v_{G,S} - v_{th}) \\ v_{gsat} &= v_{th} + f v_{D,S} \end{aligned} \tag{3.15}$$

The charges q_G, q_S, q_D, and q_B on the gate, source, drain and bulk nodes, respectively, are defined piecewise in four regions (linear, , subthreshold and accumulation). Here v_{gsat} represents a saturation voltage. Other models as well as techniques for parameter extraction are discussed in Chapter 6.

The functions q_G, q_S, q_D, and q_B are continuous at the boundaries of their regions of definition. However, in the region $\frac{\partial q_S}{\partial v_G} = \frac{2\omega}{3}$ and in the subthreshold region, $\frac{\partial q_S}{\partial v_G} = 0$, so that $\frac{\partial q_S}{\partial v_G}$ is not continuous. In developing a viable model, a compromise must be made between the amount of physics that is incorporated into the model and the mathematical and computational complexity that result from fewer simplifying assumptions. The following questions arise for the modeler:

1. Is there a realistic model which is smooth, i.e. one for which all partial derivatives of charge with respect to terminal voltages are continuous?

2. Is $\frac{\partial q_S}{\partial v_G}$ actually a discontinuous function in a physical sense or is this merely a result of the mathematical formulation?

3. Given these discontinuities, is the mathematical problem of solving the corresponding differential equation well-posed in the sense of existence, uniqueness, and regularity?

4. If the answer to the above question is yes, what is the effect of the discontinuities on the numerical algorithms employed to obtain an approximate solution?

5. Is the assumption that charge is a function of terminal voltages only, and not their time derivatives, correct? If not, for what geometries and

1. *Linear Region* : $v_{G,S} < v_{gsat}$

$$q_G = \omega\left(v_{G,S} - \beta - \phi - \frac{v_{D,S}}{2} + \frac{f v_{D,S}^2}{12(v_{G,S} - v_{th} - \frac{1}{2} f v_{D,S})}\right)$$

$$q_S = -\omega\left(\frac{v_{G,S} - v_{th}}{2} + \frac{f v_{D,S}}{4} + \frac{f^2 v_{D,S}^2}{24(v_{G,S} - v_{th} - \frac{1}{2} f v_{D,S})}\right)$$

$$q_D = -\omega\left(\frac{v_{G,S} - v_{th}}{2} - \frac{3 v_{D,S} f}{4} + \frac{f^2 v_{D,S}^2}{8(v_{G,S} - v_{th} - \frac{1}{2} f v_{D,S})}\right)$$

$$q_B = \omega\left(\beta + \phi - v_{th} - \frac{1}{2}(1-f) v_{D,S} - \frac{(1-f) f v_{D,S}^2}{12(v_{G,S} - v_{th} - \frac{1}{2} f v_{D,S})}\right)$$

2. *Saturation Region* : $v_{th} < v_{G,S} < v_{gsat}$

$$q_G = \omega\left(v_{G,S} - \beta - \phi - \frac{v_{G,S} - v_{th}}{3f}\right)$$

$$q_B = \omega\left(\beta + \phi - v_{th} + \frac{(1-f)(v_{G,S} - v_{th})}{3f}\right)$$

$$q_S = \omega\left(\frac{2(v_{G,S} - v_{th})}{3}\right)$$

$$q_D = 0$$

3. *Subthreshold Region* : $\beta + v_{B,S} < v_{G,S} < v_{th}$

$$q_G = \omega\frac{\delta^2}{2}\left(-1 + \left(1 + \frac{4(v_{G,S} - \beta - v_{B,S})}{\delta^2}\right)^{1/2}\right)$$

$$q_B = -q_G$$

$$q_S = q_D = 0$$

4. *Accumulation Region* : *Otherwise*

$$q_G = \omega(v_{G,S} - \beta - v_{B,S})$$

$$q_B = -q_G$$

$$q_S = q_D = 0$$

Table 3.2: Regions of operation for a MOS transistor.

at what switching speeds is the error introduced by this assumption significant?

Methods for extracting the model parameters are discussed in Chapter 6 and issues of computational efficiency and stability of solutions are also examined.

3.8 Overview of a Circuit Simulator

Although circuit simulators vary widely in details of implementation, because they solve a nonlinear system of ordinary differential equations, there are a number of features common to all. The following steps summarize the hierarchy of actions taken by a "generic" circuit simulator to solve the circuit ODE system:

1. Form the circuit equations using a library of device models and some form of modified nodal analysis, which is based on the Kirchoff current laws [400, 554]. These equations are a coupled system of nonlinear ordinary differential equations of the form $\boldsymbol{f}(t, \boldsymbol{v}(t), \boldsymbol{v}'(t)) = \mathbf{0}$, where $\boldsymbol{v}(t)$ is the vector of nodal voltages at time t. The free parameters in the models [7, 46, 476, 529, 547, 561] are chosen to closely approximate either experimental data or results of device simulations. It is also possible to avoid the use of models altogether by constructing tables of data and then finding values for current flow by performing linear B-spline interpolation/extrapolation of the table values [31, 545].

2. Employ a stiffly-stable integration method such as the trapezoidal scheme [400], backward differentiation formula [67, 204] (used in the Livermore ODE Solver LSODE, for example), implicit Runge-Kutta [74], or TR-BDF [545] methods, to convert the system of nonlinear ODEs into a system of nonlinear algebraic equations. This step is often called the "outer loop".

3. Employ a Newton or secant method to construct a sequence of systems of linear equations, the solutions to which converge to the desired solution of the nonlinear algebraic system [545, 400]. This step is termed the "inner loop".

4. Solve the large, sparse linear system [37, 231, 252, 350, 399, 402] that results from the previous step. For large scale applications solving these systems involves parallel and vector computers [17, 145, 239, 337, 574].

The first step has been covered in Chapters 2 and 3; numerical integration is discussed in Chapter 4; and steps 3 and 4, solving nonlinear and linear systems, are covered in Chapter 5.

In order for a circuit simulator to accurately model a new technology, several tasks must first be accomplished. More accurate physical models for diffusion, implantation, oxidation, and transport must be incorporated at the process and device level as described in later chapters. This in turn allows for the development of new transistor models which are more realistic physically, but often harder to implement numerically. In order for the circuit simulator to work efficiently with these new models, or with lookup tables having finer grids and using more accurate curve fitting techniques, it may be necessary to improve the simulator in fundamental ways. To accomplish this task, the subtle relationships that exist between the device models and the numerics of the simulator, as well as between the various levels of the simulation program, must be thoroughly understood.

Increased accuracy can be achieved by careful time step control and improved integration techniques. With stiff systems (see Chapter 4), the issue of robustness is of prime importance. Even though a great deal of work has been done over the last thirty years by many researchers, existing codes still have problems with the newest, most complex devices. It is often difficult to isolate the cause of convergence failure, particularly when the simulator performs well on a medium sized circuit, but fails on a large problem.

The inner loop of a circuit simulator involves solving a system of N linear equations in N unknowns, where N is at least as great as the number of nodes. For dense linear systems the amount of work required to solve such a system by direct methods such as Gaussian elimination is proportional to N^3. For a sparse system, the work is reduced, depending upon the amount of sparsity, but is still usually super-linear. There has been a great deal of research [239] on the development of parallel algorithms for matrix computations using both direct and iterative methods (see Chapter 10).

One important idea that applies specifically to circuit simulation, is that of "tearing" or partitioning the circuit into parts, simulating each of the parts separately, and then "glueing" the pieces back together. In addition to attacking the super-linearity problem of the inner loop, these methods allow for different time steps for the various subcircuits. This can result in a substantial payoff when some subcircuits are latent, i.e. undergo no change, or nearly latent for a period of time. The price for these gains is an additional outer loop that must update the inputs and outputs to the various subcircuits, based on the waveforms obtained from the last step of the simulation [554, 562].

3.9 Summary

Some fundamental concepts from solid state physics are briefly introduced here to set the stage for an appropriate model of the transistor. This also serves to provide background material for device simulation discussed in Chapters 7 and 8. For example, concepts underlying the energy band assumptions, drift and diffusion for advanced transport models are introduced. Some extensions of the circuit models in Chapter 2 to include effects such as nonlinear capacitance in a diode and transistor modeling conclude the treatment. In the next two chapters we consider elementary numerical techniques for integrating the resulting ODE systems and for solving associated sparse nonlinear algebraic systems of equations.

3.10 Exercises

1. Show that $\frac{\partial I}{\partial v_s}$ in (3.12) is continuous. Show that q_s and $\frac{\partial q_s}{\partial v_s}$ in Table 3.1 are continuous.

2. Suppose that in Figure 3.8, $v_0(t) = 0$ and $v_1(t) = \sin(\omega t)$. Explain why there exists a value of ω such that the maximum current through the capacitor exceeds the maximum current through the ideal diode.

3. Again in Figure 3.8, suppose that $v_0(t) = 0$ and $v_1(t) = \sin \omega t$. Explain why the current through the ideal diode is at a maximum when $\omega t = 3\frac{\pi}{2}$ and why the current through the capacitor is zero there. Is the current through the capacitor at a maximum when $\omega t = 0$?

Chapter 4

Numerical Integration of Circuit ODEs

4.1 Introduction

In the preceding two chapters it has been shown that Kirchoff's laws together with constitutive relations for the various circuit elements, give a concise mathematical model of an integrated circuit. Mathematically, this model consists of a very large differential-algebraic equation (DAE) system. There are several features of this system which are peculiar to the circuit problem. Its size typically ranges from tens of thousands to several million equations and requires advanced numerical methods for efficient solution. It is usually impossible to put this first-order system into normal form in which the time derivatives appear explicitly on the left hand side of each equation. Parasitic capacitances which are required for greater accuracy in the constitutive relations for transistors also cause extreme stiffness with attendant numerical difficulties.

This chapter begins by considering a canonical example for which there is a closed-form solution. This will introduce the concept of an abstract dynamical system, as well as the fundamental existence and uniqueness theorem for ODEs. The forward Euler method provides a starting point for a discussion of numerical techniques for integrating a dynamical system. An explicit method, it is conceptually simple, but can suffer from numerical instability unless extremely small timesteps are used. For this reason, modern circuit simulators employ implicit methods such as the backward Euler or trapezoidal method, as discussed in the final sections of this chapter.

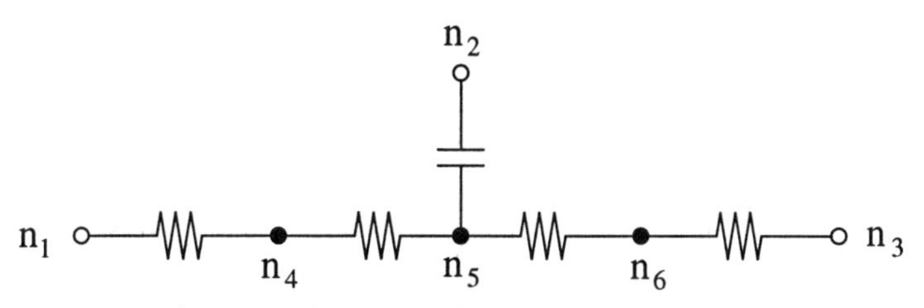

Figure 4.1: An analytic circuit model.

4.2 An Analytic Circuit Model

Before analyzing circuits modeled by differential equations that must be solved numerically, consider a simple circuit with normalized voltage for which the corresponding ODE system is linear and can be solved in closed form. The circuit in Figure 4.1 contains three nodes of known voltages (v_1, v_2, v_3) and three nodes of unknown voltage (v_4, v_5, v_6). The capacitor is linear with capacitance c and each resistor has a resistance of one ohm. Assume that for $t \leq 0$, the voltages at the three input pins are constant with $(v_1(t), v_2(t), v_3(t)) = (0, 1, 1)$. Since all voltages are constant for negative time, no current flows prior to $t = 0$ and the circuit behaves as if the capacitor had been removed. Applying Kirchoff's current law to this steady-state problem yields an algebraic system whose solution is $(0.25, 0.50, 0.25)$. This DC solution gives the initial conditions to the system of ordinary differential equations that model the transient behavior of the system.

Next, consider time-varying voltages applied to each of the three pins, so that for $t \geq 0$, $(v_1(t), v_2(t), v_3(t)) = (\sin t, e^{-t}, \cos t)$. Since the voltages change with time, a current flows through the capacitor. The corresponding differential-algebraic system is seen from Kirchoff's laws to be

$$\begin{aligned} \sin t - v_4(t) + v_5(t) - v_4(t) &= 0 \\ v_5(t) - v_6(t) + (\cos t - v_6(t)) &= 0 \\ v_4(t) - v_5(t) + c(e^{-t} - v_5(t))' + (v_6(t) - v_5(t)) &= 0 \end{aligned} \tag{4.1}$$

Note that the time derivative enters only in the last equation. From the first two equations of the system (4.1),

$$v_4(t) = \frac{1}{2}(\sin t + v_5(t)) \ , \quad v_6(t) = \frac{1}{2}(\cos t + v_5(t)) \tag{4.2}$$

and substituting these into the third equation yields

$$cv_5'(t) + v_5(t) = \frac{1}{2}(\sin t + \cos t) - ce^{-t}, \ v_5(0) = 1/2 \tag{4.3}$$

where the initial condition is obtained from the DC solution. Therefore, the voltage at node n_5 is obtained by solving an initial value problem of the general form

$$a_1 f'(t) + a_0 f(t) = g(t), \quad f(0) = r \tag{4.4}$$

where f is the unknown voltage function, g is a given forcing function and a_0, a_1, and p are specified constants. In mathematical modeling of the circuit, it is important that there exists a solution, that it be unique, and finally that it should depend continuously on the initial data. Mathematically, these questions of existence, uniqueness and regularity must be addressed in order to insure that the problem is well-posed. The linear ODE system (4.4) has a unique solution which can be written in closed form, showing continuous dependence on the initial data. Let L be the linear differential operator defined by $L(f) = cf' + f$ and let g be the forcing function on the right hand side of (4.3). Recall that (4.4) can be solved by a four-step process: (1) find the complementary solution to the associated homogenous problem $L(q) = 0$; (2) find a particular solution p to $L(p) = g$; (3) notice that the general solution to $L(f) = g$ must be of the form $p + q$; and (4) find the function $f = p + q$ that satisfies the initial condition.

The complementary solution to (4.3) has the form $q(t) = ke^{-t/c}$ and a particular solution p may be determined by assuming that it has the same general form as g, namely $p(t) = \alpha e^{-t} + \beta \sin t + \gamma \cos t$ for constants α, β, γ. Substituting in (4.3) yields

$$\alpha = \frac{c}{c-1}, \quad \beta = \frac{1+c}{2(c^2+1)}, \quad \gamma = \frac{1-c}{2(c^2+1)} \tag{4.5}$$

The general solution to (4.3) is therefore $f(t) = ke^{-t/c} + p(t)$; setting $f(0) = \frac{1}{2}$ and solving gives $k = \frac{1}{2} - \beta - \alpha$. The closed-form solution for the voltage $v_5(t) = f(t)$ is then

$$v_5(t) = ke^{-t/c} + \alpha e^{-t} + \beta \sin t + \gamma \cos t \tag{4.6}$$

The exponential terms in (4.6) decay to zero as t increases while the remaining two terms oscillate. For sufficiently large t, the first two terms will be negligible, and the solution will asymptotically approach the oscillatory behavior of the function $h(t) = \beta \sin t + \gamma \cos t$. Since β and γ depend on c, the amplitude $\sqrt{\beta^2 + \gamma^2}$ of the limit response function h depends on c, as well. Note also that because α and k depend on c, the rate at which f approaches h is a function of c. In particular, increasing c causes the amplitude of h to decrease, while f converges more slowly to h.

4.3 Dynamical Systems

The circuit of Figure 2.7 is now used to introduce some of the fundamental concepts of dynamical systems. In this circuit, determining the nodal voltages requires solving the system

$$x'(t) = -y(t), \quad y'(t) = x(t), \quad x(0) = a, \quad y(0) = b \tag{4.7}$$

Define the function $\boldsymbol{G}$ from $\mathbf{R}^2$ into $\mathbf{R}^2$ by

$$\boldsymbol{G}(x, y) = (-y, x) \tag{4.8}$$

If $\boldsymbol{p}(t)$ denotes the point $(x(t),\ y(t))$ then (4.8) can be written in the vector form

$$\boldsymbol{p}'(t) = \boldsymbol{G}(\boldsymbol{p}(t)) \tag{4.9}$$

The function $\boldsymbol{G}$ is said to be the infinitesimal generator of the differential equation. Some circuits are modeled by differential equations whose infinitesimal generator $\boldsymbol{G}$ cannot be written in closed form, while in other cases $\boldsymbol{G}$ is only defined on a proper subset of $\mathbf{R}^2$. Conceptually, it is useful to think of the vector function $\boldsymbol{p}(t)$ as describing the position at time t of a particle moving in the plane $\mathbf{R}^2$, in which case $\boldsymbol{p}'(t) = \boldsymbol{G}(\boldsymbol{p}(t))$ is the velocity of the particle. Thus, the function $\boldsymbol{G}$ defines a velocity field on $\mathbf{R}^2$ (mathematically a vector field) and the differential equation is called a dynamical system because it can be used to describe the dynamics of a body in motion.

This setting can be generalized to study a wide class of circuit problems by letting $\boldsymbol{G}$ be a transformation from $\mathbf{R}^n$ into $\mathbf{R}^n$. Then $\boldsymbol{G}$ defines a dynamical system in $\mathbf{R}^n$ by means of the initial-value problem

$$\boldsymbol{f}'(t) = \boldsymbol{G}(\boldsymbol{f}(t)), \quad \boldsymbol{f}(t_0) = \boldsymbol{p}_0 \tag{4.10}$$

Again one may consider that $\boldsymbol{f}$ represents the trajectory of a particle which at time t_0 is located at $\boldsymbol{p}_0$. Under suitable conditions on $\boldsymbol{G}$, (4.10) has a unique solution for every $\boldsymbol{p}_0$ in $\mathbf{R}^n$, and hence it defines a whole family of trajectories indexed by the vector parameter $\boldsymbol{p}_0$.

The ODE system modeling the circuit in Figure 4.1 is said to be autonomous because the vector field $\boldsymbol{G}$ does not depend explicitly upon time. When t appears explicitly in the coefficients or in the forcing function, the corresponding system of differential equations is nonautonomous. The general form for a nonautonomous dynamical system is

$$\boldsymbol{f}'(t) = \boldsymbol{G}(t, \boldsymbol{f}(t)), \quad \boldsymbol{f}(t_0) = \boldsymbol{p}_0 \tag{4.11}$$

where $G : \mathbf{R}^{n+1} \to \mathbf{R}^n$ is a given vector field which is sufficiently smooth. Although not in the form (4.10), it is possible to imbed (4.11) into a higher dimensional autonomous problem. Define $h : \mathbf{R} \to \mathbf{R}^{n+1}$ by $h(t) = (t, f(t))$ where f solves (4.11); the function h is then a solution to the autonomous system $h'(t) = (1, G(t, f(t)))$. In essence, t is now considered as both an independent and a dependent variable. A nonautonomous system can always be written in the form (4.10) by this technique, and therefore both the theory and numerical methods developed for autonomous systems extend to the nonautonomous case. A property that will often be used is the following: suppose that S is a subset of $\mathbf{R}^n$; the vector field G is said to have a Lipschitz constant K on S if for each p and q in S,

$$||G(p) - G(q)|| \leq K||p - q|| \tag{4.12}$$

In one dimension, if $|f'(u) - f'(v)| < K|u - v|$, then the mean value theorem for derivative implies K is a bound on the second derivative of f. If G is defined on all of $\mathbf{R}^n$ and (4.12) holds for all p and q, then K is said to be a global Lipschitz constant.

4.4 Existence and Uniqueness

Regularity of solutions is important from a practical standpoint because for integrated circuits, similar inputs to the pins should cause similar performance of the device. Additionally, regularity with respect to changes in G is also important, in the sense that the behavior of the chip should be predictable for variations in the process that produces it. These considerations of existence, uniqueness and regularity apply not just to ODEs, but also to the partial differential equations which arise in process and device modeling.

Suppose now that G is a vector field on $\mathbf{R}^n$, p_0 is a point in $\mathbf{R}^n$, and $t_0 \in \mathbf{R}$.

Theorem 4.1 (Fundamental Existence and Uniqueness):

(i) Existence: If G is continuous on a neighborhood of p_0, then there is an open interval I containing t_0 and a function f with domain I such that $f(t_0) = p_0$ and for each number t in I, $f'(t) = G(f(t))$.

(ii) Uniqueness: Suppose that G is Lipschitz in a neighborhood U of p_0; that is, there is a $K > 0$ such that for p and q in U, (4.12) holds. Then there is an open interval I containing t_0 such that there is exactly one function f with domain I such that $f(t_0) = p_0$ and for each number t in I, $f'(t) = G(f(t))$.

(iii) Regularity of Solutions: If $\boldsymbol{G}$ *has a global Lipschitz constant, then the set of all solutions forms a "flow": there exists a continuous transformation* $\boldsymbol{T}$ *from* $\mathbf{R} \times \mathbf{R}^n$ *into* $\mathbf{R}^n$ *such that* $\boldsymbol{T}(0, \boldsymbol{p}) = \boldsymbol{p}$ *and* $\boldsymbol{T}(t, \boldsymbol{T}(s, \boldsymbol{p})) = \boldsymbol{T}(t+s, \boldsymbol{p})$ *for all* t *and* s *in* $\mathbf{R}$, $\boldsymbol{p}$ *in* $\mathbf{R}^n$. □

The mathematical properties of the vector field $\boldsymbol{G}$ will determine which numerical methods are most suitable for its simulation. Of course, physically existence is immediate since the circuit works in some sense. If, however, the various constitutive models discussed in Chapters 2 and 3 give rise to a $\boldsymbol{G}$ which fails to be continuous, then caution is in order becuase the hypotheses of the theorem are not satisfied. Note that the theorem gives sufficient but not necessary conditions, so that a solution may in fact exist mathematically even for a discontinuous $\boldsymbol{G}$.

4.5 Numerical Solution of the Circuit Equations

As in Section 4.3, let $\boldsymbol{G}$ denote the vector field defined by $\boldsymbol{G}(x, y) = (-y, x)$. The function $\boldsymbol{f}$ to be approximated satisfies the initial value problem

$$\boldsymbol{f}'(t) = \boldsymbol{G}(\boldsymbol{f}(t)), \quad \boldsymbol{f}(0) = (1, 0) \tag{4.13}$$

Note that $\boldsymbol{G}$ is the composition of two reflections, the first through the line $y = x$ and the second through the line $x = 0$. Since reflections are norm-preserving, it follows that $\boldsymbol{G}$ has a global Lipschitz constant $K = 1$. Equation (4.13) can be solved in closed form by considering the characteristic equation $\lambda^2 + 1 = 0$ of the associated second order equation (2.23). Note that $\forall t$ the velocity vector $\boldsymbol{f}'(t)$ is orthogonal to the position vector $\boldsymbol{f}(t)$; hence, a trajectory starting at a point $\boldsymbol{p} = (r\cos\theta, r\sin\theta)$ must remain on the circle of radius r centered at the origin. The vector field $\boldsymbol{G}$ generates the flow $\boldsymbol{T}$ defined by

$$\boldsymbol{T}(t, (r\cos\theta, r\sin\theta)) = (r\cos(\theta + t), r\sin(\theta + t)) \tag{4.14}$$

The solution has period 2π, which is reasonable since the circuit represents a perfect oscillator.

There are many techniques for constructing numerical approximations to the solution $\boldsymbol{f}$ of (4.13). Perhaps the simplest is the forward Euler method with a fixed timestep Δt, which produces an approximate, piecewise-linear solution. For n a nonnegative integer, let $\boldsymbol{y}_n$ denote the approximation to $\boldsymbol{f}(n\Delta t)$. Forward Euler recursively builds the sequence $\{\boldsymbol{y}_n\}$ using the formula

$$\boldsymbol{y}_{n+1} = \boldsymbol{y}_n + \Delta t \boldsymbol{G}(\boldsymbol{y}_n), \quad \boldsymbol{y}_0 = (1, 0) \tag{4.15}$$

The approximate solution is then the piecewise-linear function that passes through the set of points $\{(n\Delta t, \boldsymbol{y}_n)\}$. A straightforward derivation of the recursion amounts to approximating the derivative in (4.13) with a forward difference quotient and evaluating the right-hand side at the current time. This simple algorithm can, in fact, provide a proof of existence and uniqueness as described in the theorem. Moreover, the procedure serves as an excellent starting point for the analysis of error propagation in numerical integration of ODEs. Computations with $\Delta t = 2\pi/m$ and $m = 10^2$, 10^3, and 10^4 give rise to RMS errors of 0.218, 0.0199, and 0.00197, respectively, at the final time 2π. This demonstrates that the error is approximately linear in Δt. A more detailed discussion of accuracy and stability follows in the next two sections.

4.6 The Forward Euler Method

To examine the accuracy of the recursion (4.15) suppose that $\boldsymbol{G} : \mathbf{R}^n \to \mathbf{R}^n$ satisfies $\|\boldsymbol{G}(\boldsymbol{p})\| \leq A$, $\|\boldsymbol{G}(\boldsymbol{p}) - \boldsymbol{G}(\boldsymbol{q})\| \leq B\|\boldsymbol{p} - \boldsymbol{q}\|$ for all $\boldsymbol{p}$, $\boldsymbol{q} \in \mathbf{R}^n$. Let $\boldsymbol{f}$ be the exact solution to the initial value problem (4.10) on the interval $[t_0, T]$. For n a positive integer, let $\Delta t = (T - t_0)/n$. Set $\boldsymbol{p}_k = \boldsymbol{f}(t_0 + k\Delta t)$ and let $\boldsymbol{q}_k$ be the kth point determined by the forward Euler algorithm. The error at the kth step is defined as $\epsilon_k = \|\boldsymbol{p}_k - \boldsymbol{q}_k\|$. Applying the fundamental theorem of calculus twice gives a bound on ϵ_1 in terms of ϵ_0 as follows. First,

$$\begin{aligned}
\epsilon_1 &= \|\boldsymbol{p}_1 - \boldsymbol{q}_1\| = \|\boldsymbol{f}(t_1) - \boldsymbol{q}_1\| \\
&= \| \boldsymbol{f}(t_0) + \int_{t_0}^{t_1} \boldsymbol{f}'(\alpha)\, d\alpha - (\boldsymbol{q}_0 + \Delta t \boldsymbol{G}(\boldsymbol{q}_0)) \| \\
&= \| \boldsymbol{p}_0 - \boldsymbol{q}_0 + \int_{t_0}^{t_1} \boldsymbol{G}(\boldsymbol{f}(\alpha))\, d\alpha - \int_{t_0}^{t_1} \boldsymbol{G}(\boldsymbol{q}_0)\, d\alpha \| \qquad (4.16)
\end{aligned}$$

since the integral of the constant function whose only value is K over the interval $[a, b]$ is just $K(b - a)$.

Next, by the triangle inequality,

$$\begin{aligned}
\epsilon_1 &\leq \|\boldsymbol{p}_0 - \boldsymbol{q}_0\| + \int_{t_0}^{t_1} \|\boldsymbol{G}(\boldsymbol{f}(\alpha)) - \boldsymbol{G}(\boldsymbol{q}_0)\|\, d\alpha \\
&\leq \epsilon_0 + B \int_{t_0}^{t_1} \|\boldsymbol{f}(\alpha) - \boldsymbol{q}_0\|\, d\alpha \qquad (4.17)
\end{aligned}$$

where we have used the fact that B is a global Lipschitz constant for G. Realizing a "hidden zero", i.e. adding and subtracting p_0 yields

$$\epsilon_1 \leq \epsilon_0 + B \int_{t_0}^{t_1} (\|\boldsymbol{f}(\alpha) - \boldsymbol{p}_0\| + \|\boldsymbol{p}_0 - \boldsymbol{q}_0\|)\, d\alpha$$

$$
\begin{aligned}
&\leq \epsilon_0 + B\int_{t_0}^{t_1} \|\boldsymbol{p}_0 - \boldsymbol{q}_0\|\, d\alpha + B\int_{t_0}^{t_1} \|\boldsymbol{f}(\alpha) - \boldsymbol{f}(t_0)\|\, d\alpha \\
&\leq \epsilon_0 + B\Delta t\|\boldsymbol{p}_0 - \boldsymbol{q}_0\| + B\int_{t_0}^{t_1} \|\int_{t_0}^{\alpha} \boldsymbol{f}'(\gamma)\, d\gamma\|\, d\alpha \\
&\leq (1+B\Delta t)\epsilon_0 + B\int_{t_0}^{t_1}\int_{t_0}^{\alpha} \|\boldsymbol{G}(\boldsymbol{f}(\gamma))\|\, d\gamma\, d\alpha \\
&\leq (1+B\Delta t)\epsilon_0 + B\int_{t_0}^{t_1}\int_{t_0}^{\alpha} A\, d\alpha
\end{aligned}
\tag{4.18}
$$

so that

$$
\epsilon_1 \leq (1+B\Delta t)\epsilon_0 + AB(\Delta t)^2 \tag{4.19}
$$

A similar argument shows that the error at the second step is bounded by

$$
\epsilon_2 \leq (1+B\Delta t)\left((1+B\Delta t)\epsilon_1 + AB(\Delta t)^2\right) + AB(\Delta t)^2 \tag{4.20}
$$

and by continuing in this fashion,

$$
\epsilon_n \leq (1+B\Delta t)^n \epsilon_0 + \left(\sum_{j=0}^{n-1}(1+B\Delta t)^j\right) AB(\Delta t)^2 \tag{4.21}
$$

Noting that $(1+B\Delta t) \leq e^{B\Delta t}$ and that the sum of the geometric series $1+s+s^2+.....+s^{n-1}$ is $(s^n-1)/(s-1)$, inequality (4.21) can be rewritten as

$$
\begin{aligned}
\epsilon_n &\leq e^{nB\Delta t}\epsilon_0 + \frac{e^{nB\Delta t}-1}{B\Delta t}AB(\Delta t)^2 \\
&= e^{B(T-t_0)}\epsilon_0 + (e^{B(T-t_0)}-1)A\Delta t
\end{aligned}
\tag{4.22}
$$

This shows that as $\Delta t \to 0$, the approximate solution produced by forward Euler converges to the exact solution. Of greater importance from a practical standpoint is information concerning the rate of convergence. The local truncation error of an integration method is the magnitude of the difference between the exact and approximate solutions after one time step assuming they agree at the beginning of the step. Similarly, the global truncation error is the accumulated error after n steps. Equation (4.19) shows that the local truncation error of forward Euler is proportional to $(\Delta t)^2$, while from (4.22) the global truncation error is seen to decrease linearly with Δt.

4.7 Stability and the Backward Euler Method

The concept of stability is important to the study of differential equations from both a theoretical and a numerical standpoint. The term stability has different meanings, depending upon context. Some call the regularity statement in the previous theorem stability with respect to a perturbation in initial conditions. More commonly, in a numerical analysis setting stability refers to the asymptotic behavior of a dynamical system as t approaches infinity: trajectories that start out close together remain close for all time. A flow $\boldsymbol{T}(t, \boldsymbol{p})$ is stable at $\boldsymbol{p}_0$ if for every $\epsilon > 0$ there is a $\delta > 0$ such that if $\|\boldsymbol{p} - \boldsymbol{p}_0\| < \delta$, then for all time t, $\|\boldsymbol{T}(t, \boldsymbol{p}) - \boldsymbol{T}(t, \boldsymbol{p}_0)\| < \epsilon$. If it is known that a dynamical system has this property at $\boldsymbol{p}_0$, then it is desirable to use an integration method which preserves stability for the approximate solution. Note that in the analysis of truncation error, the focus is upon finding a solution on a fixed interval $[0, T]$ and letting $\Delta t \to 0$. In contrast, stability is concerned with what happens for a fixed Δt as T increases without bound.

If the forward Euler method is used to integrate the circuit of Figure 4.1, it is found to be numerically unstable unless sufficiently small timesteps are used. To analyze this situation, consider the linear initial value problem

$$cf'(t) + f(t) = 0, \quad f(0) = 1 \tag{4.23}$$

where $c > 0$. This equation can be written $f'(t) = G(f(t))$ where the generating function G is defined by $G(x) = -x/c$ and has a global Lipschitz constant of $1/c$. The exact solution to (4.23) is $f(t) = e^{-t/c}$, which for a small value of c decays very rapidly. Now begin the forward Euler method by setting $t_0 = 0$, $y_0 = 1$ and choosing a timestep Δt. At the first step, $y_1 = (1 - \Delta t/c)\, y_0$. For $0 < \Delta t/c < 1$, y_1 has the same sign as y_0 and smaller magnititude. If $1 < \Delta t/c < 2$, then y_1 is opposite in sign to y_0, still with $|y_1| < |y_0|$. However, if $\Delta t/c > 2$, y_1 is opposite in sign to y_0 and $|y_1| > |y_0|$. Since $y_n = y_0(1 - \Delta t/c)^n$, in the last instance the sequence $\{y_n\}$ is oscillating and unbounded. Analogous results are obtained when the forward Euler scheme is used to simulate the circuit of Figure 4.1 on an interval $[0, T]$ for various values of Δt and c. It may be anticipated that there will be numerical difficulties whenever Δt is greater than $2c$. At the end of Section 3.6 it was observed that the physics of a diode is modeled more accurately if a nonlinear capacitor with very small capacitance is added in parallel with the ideal diode described in Section 3.5. However, it is precisely the addition of a very small capacitance that causes numerical instability. These parasitic elements are a necessary part of more realistic circuit element models, but can have the unwanted side-effect of introducing numerical instability. The issue of stability of ODE integrators is discussed in further detail in Chapters 13 and 15.

Although the above analysis is performed on a linear problem, similar considerations apply to the nonlinear case. Note that as $c \to 0$ in (4.23) the differential equation becomes degenerate; in the limiting case $f(t) = 0$ for $t > 0$, $f(0) = 1$, and the solution is discontinuous. For very small c, (4.23) is a singularly perturbed differential equation. Such equations occur in many areas of engineering; for example, in device simulation the model problem $-\epsilon f'' + f' = 0$, $0 < \epsilon << 1$, serves as an important example for understanding device behavior when a strong electric field is applied as discussed in Chapter 7. Note that for the scalar problem the change of variable $z = t/c$ would circumvent the instability. However, in the vector case some components may have large "time constants" while others have short ones, and such a change of variables is impractical on the coupled system. These considerations lead to the notion of stiffness, as discussed later in Section 13.4.

The implicit backward Euler algorithm for integrating (4.10) is defined by the recursion

$$\boldsymbol{y}_{n+1} = \boldsymbol{y}_n + \Delta t \boldsymbol{G}\left(t_{n+1}, \boldsymbol{y}_{n+1}\right) \tag{4.24}$$

It does not give an explicit formula for $\boldsymbol{y}_{n+1}$ in terms of known quantities, but instead yields an algebraic system that must be solved to obtain $\boldsymbol{y}_{n+1}$. Using techniques similar to those of Section 4.6, it can be shown that this method is also locally second order and globally first order (see Exercise 4.1). The advantage of the backward Euler method, and of implicit methods in general, is stability: applying backward Euler to (4.23) yields the recursion $y_{n+1} = (1 + \Delta t/c)^{-1} y_n$, so that $|y_{n+1}| < |y_n|$ and the method is stable regardless of stepsize.

Even if the differential system is linear, backward Euler integration of a typical circuit ODE system will require solving a large linear system at each timestep which can be quite expensive even when using sparse matrix techniques. For circuits with nonlinear elements, solving for $\boldsymbol{y}_{n+1}$ necessitates solving a very large coupled system of nonlinear algebraic equations at each timestep. In either case, repeated solution of these systems accounts for much of the time spent in a circuit simulation.

4.8 A Differential-Algebraic System

A large class of circuits can be modeled by a nonlinear differential system of the form

$$\boldsymbol{F}(t,\boldsymbol{v}(t),\boldsymbol{v}'(t)) = \boldsymbol{0}, \quad \boldsymbol{v}(0) = \boldsymbol{p}_0 \tag{4.25}$$

It is natural to ask if (4.25) can be rewritten in "normal" form, $\boldsymbol{v}'(t) = \boldsymbol{G}(t,\boldsymbol{v}(t))$, because the theory and examples covered thus far correspond to such systems. In fact, very few circuit equations can be put into normal form, implying that the velocity field $\boldsymbol{G}$ is known only implicitly. How is the solution $\boldsymbol{v}(t)$ determined without knowing $\boldsymbol{G}$ explicitly? Consider the example shown in Figure 4.2, and define the voltage functions $v_1(t)$ and $v_4(t)$ applied to the input pins by

$$v_1(t) = \begin{cases} 1, & t \le 0 \\ \cos t, & t > 0 \end{cases}, \quad v_4(t) = \begin{cases} 0, & t \le 0 \\ \sin t, & t > 0 \end{cases} \tag{4.26}$$

Since v_1 and v_4 are constant for negative time, no current will flow in the circuit and $v_1(t) = v_2(t)$, $v_3(t) = v_4(t)$ for $t < 0$. Application of Kirchoff's laws yields the differential system

$$\begin{aligned} v_1(t) - v_2(t) + (v_3(t) - v_2(t))' &= 0 \\ v_4(t) - v_3(t) + (v_2(t) - v_3(t))' &= 0 \end{aligned} \tag{4.27}$$

Next, define $\boldsymbol{u}$ and $\boldsymbol{g}$, $\boldsymbol{A}$ and $\boldsymbol{B}$ by

$$\boldsymbol{u}(t) = \begin{bmatrix} v_2(t) \\ -v_3(t) \end{bmatrix}, \boldsymbol{g}(t) = \begin{bmatrix} v_1(t) \\ -v_4(t) \end{bmatrix}, \boldsymbol{A} = \begin{bmatrix} -1 & 1 \\ 1 & -1 \end{bmatrix}, \boldsymbol{B} = \begin{bmatrix} -1 & 0 \\ 0 & -1 \end{bmatrix}$$

so that (4.27) can be written in the form

$$\boldsymbol{A}\boldsymbol{u}'(t) + \boldsymbol{B}\boldsymbol{u}(t) = \boldsymbol{g}(t) \tag{4.28}$$

If $\boldsymbol{A}$ were nonsingular then (4.28) could be written in normal form as $\boldsymbol{u}'(t) = -\boldsymbol{A}^{-1}\boldsymbol{B}\boldsymbol{u}(t) + \boldsymbol{A}^{-1}\boldsymbol{g}(t)$ and solved in closed form using higher dimensional analogs of the methods discussed in Section 4.2.

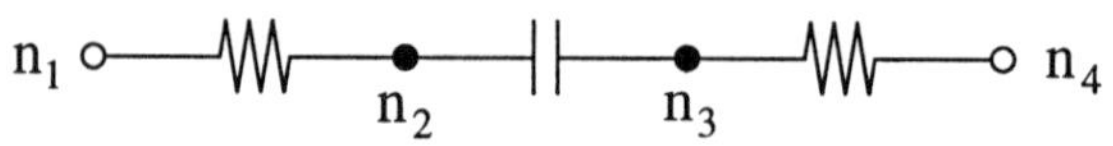

n₁ n₂ n₃ n₄

Figure 4.2: A capacitor between interior circuit nodes.

The matrix $\boldsymbol{A}$ is singular and there does not exist a solution to (4.28) for every initial condition. The only admissible condition is given in (4.26), $\boldsymbol{u}(0) = (1, 0)$, and in fact, there does not exist an explicit representation for the time derivative of $\boldsymbol{u}$. For numerous circuits the ODE system cannot be written in normal form because of the presence of a capacitor between two nodes of unknown voltage. While this presents some difficulties, it is still possible to solve the system both analytically and numerically.

A closed form solution to (4.27) can be obtained by using the change of variables.

$$w(t) = v_2(t) - v_3(t), \quad z(t) = v_2(t) + v_3(t) \tag{4.29}$$

Notice that $z(t) = \sin t + \cos t$ and w satisfies the linear initial-value problem

$$2w'(t) + w(t) = \cos t - \sin t, \qquad w(0) = 1 \tag{4.30}$$

which has the solution $w(t) = 0.4e^{-.5t} + 0.2 \sin t + 0.6 \cos t$. Solving for v_2 and v_3 in terms of z and w gives

$$\begin{aligned} v_2(t) &= \quad 0.2e^{-.5t} + 0.6 \sin t + 0.8 \cos t \\ v_3(t) &= -0.2e^{-.5t} + 0.4 \sin t + 0.2 \cos t \end{aligned} \tag{4.31}$$

Once again, the exponential terms die off with time and v_2 and v_3 converge to periodic functions. The capacitance c affects their magnitude and the rate of convergence to these limit trajectories.

Note that in the above example the forcing function v_4 is continuous but not differentiable at $t = 0$. This discontinuity did not play a role in the analysis, but may adversely affect numerical techniques. Recalling the MOSFET model of Chapter 3, it is possible to have discontinuous derivatives at interior nodes as a result of the discontinuity in $\partial Q_S/\partial V_G$ for the transistor model.

It is not necessary to solve for $\boldsymbol{G}$ explicitly in order to write the equations that step from time t_n to t_{n+1} using either the forward or backward Euler method. Consider the circuit of Figure 4.2 and suppose that (x_n, y_n) is the forward Euler approximation to $(v_2(t_n), v_3(t_n))$. Substituting directly into (4.27)

$$\begin{aligned} \cos t_n - x_n + \frac{1}{\Delta t}(y_{n+1} - y_n) - \frac{1}{\Delta t}(x_{n+1} - x_n) &= 0 \\ \sin t_n - y_n + \frac{1}{\Delta t}(x_{n+1} - x_n) - \frac{1}{\Delta t}(y_{n+1} - y_n) &= 0 \end{aligned} \tag{4.32}$$

which can be rewritten as

$$\begin{bmatrix} -1 & 1 \\ 1 & -1 \end{bmatrix} \begin{bmatrix} y_{n+1}(t) \\ x_{n+1}(t) \end{bmatrix} = \begin{bmatrix} y_n - x_n - \Delta t \ \cos t_n + \Delta t \ x_n \\ x_n - y_n - \Delta t \ \sin t_n + \Delta t \ y_n \end{bmatrix}$$

Since the coefficient matrix is singular, this pair of equations cannot be solved for x_{n+1} and y_{n+1} in terms of known values. Therefore, the forward Euler method cannot be used for this circuit without first making a change of variables. On the other hand, the backward Euler scheme gives

$$\begin{aligned} \cos t_{n+1} - x_{n+1} + \frac{1}{\Delta t}(y_{n+1} - y_n) - \frac{1}{\Delta t}(x_{n+1} - x_n) &= 0 \\ \sin t_{n+1} - y_{n+1} + \frac{1}{\Delta t}(x_{n+1} - x_n) - \frac{1}{\Delta t}(y_{n+1} - y_n) &= 0 \end{aligned} \tag{4.33}$$

which can be rewritten as

$$\begin{bmatrix} -(1+\Delta t) & 1 \\ 1 & -(1+\Delta t) \end{bmatrix} \begin{bmatrix} y_{n+1}(t) \\ x_{n+1}(t) \end{bmatrix} = \begin{bmatrix} y_n - x_n - \Delta t \ \cos t_{n+1} \\ x_n - y_n - \Delta t \ \sin t_{n+1} \end{bmatrix}$$

where the coefficient matrix is now nonsingular. It is interesting that an implicit method such as backward Euler provides a numerical solution to a system which does not have an explicit generating function. Since circuits that contain capacitors connecting interior nodes will always fail to be in normal form, it is not only desirable to employ implicit methods for stability, but is necessary in order to avoid explicit recursive schemes that fail.

4.9 The Trapezoidal Integration Method

Once again consider solving the initial value problem (4.10) using a one-step method. On the interval $[t_0, t_1]$ the exact solution satisfies the integral equation

$$\boldsymbol{f}(t) = \boldsymbol{p}_0 + \int_{t_0}^{t} \boldsymbol{G}(\boldsymbol{f}(\alpha))\, d\alpha \tag{4.34}$$

One step of forward Euler with $\Delta t = t_1 - t_0$ may be viewed as obtaining $\boldsymbol{q}_1 \approx \boldsymbol{f}(t_1)$ by approximating the integral in (4.34) using the left-sided rule $\Delta t \boldsymbol{G}(\boldsymbol{f}(t_0))$, while backward Euler uses the right-sided rule $\Delta t \boldsymbol{G}(\boldsymbol{f}(t_1))$. Greater accuracy could be expected if the trapezoidal rule were employed to approximate the integral. This trapezoidal or Crank–Nicholson method for integrating (4.10) consists of the iteration

$$\boldsymbol{y}_{n+1} = \boldsymbol{y}_n + \frac{\Delta t}{2} \left(\boldsymbol{G}(\boldsymbol{y}_n) + \boldsymbol{G}(\boldsymbol{y}_{n+1})\right) , \quad \boldsymbol{y}_0 = \boldsymbol{p}_0 \tag{4.35}$$

Like backward Euler, this is an implicit method which requires solving a nonlinear algebraic system at each timestep. It too is unconditionally stable, but is also locally third-order and globally second-order accurate, as the following argument shows. As in Section 4.5, write

$$\begin{aligned}
\epsilon_1 &= \|\boldsymbol{p}_1 - \boldsymbol{q}_1\| = \|\boldsymbol{f}(t_1) - \boldsymbol{q}_1\| \\
&= \|\boldsymbol{f}(t_0) + \int_{t_0}^{t_1} \boldsymbol{f}'(\alpha)\, d\alpha - \left(\boldsymbol{q}_0 + \frac{\Delta t}{2}\left(\boldsymbol{G}(\boldsymbol{q}_0) + \boldsymbol{G}(\boldsymbol{q}_1)\right) \right) \| \\
&= \|\boldsymbol{p}_0 - \boldsymbol{q}_0 + \frac{\Delta t}{2}\boldsymbol{G}(\boldsymbol{p}_0) - \frac{\Delta t}{2}\boldsymbol{G}(\boldsymbol{q}_0) + \frac{\Delta t}{2}\boldsymbol{G}(\boldsymbol{p}_1) - \frac{\Delta t}{2}\boldsymbol{G}(\boldsymbol{q}_1) \\
&\qquad + \int_{t_0}^{t_1} \boldsymbol{f}'(\alpha)\, d\alpha - \frac{\Delta t}{2}\left(\boldsymbol{G}(\boldsymbol{p}_0) + \boldsymbol{G}(\boldsymbol{p}_1)\right) \|
\end{aligned}$$

where we have realized two hidden zeros in order to introduce the values of the exact solution, $\boldsymbol{p}_0$ and $\boldsymbol{p}_1$. Using the triangle inequality and the Lipschitz condition on G

$$\epsilon_1 \le \epsilon_0 + \frac{\Delta t}{2}B\epsilon_0 + \frac{\Delta t}{2}B\epsilon_1 + \| \int_{t_0}^{t_1} \boldsymbol{f}'(\alpha)\, d\alpha - \frac{\Delta t}{2}\left(\boldsymbol{f}'(t_0) + \boldsymbol{f}'(t_1)\right) \|$$

which can be rewritten as

$$\left(1 - \frac{\Delta t}{2}B\right)\epsilon_1 \le \left(1 + \frac{\Delta t}{2}B\right)\epsilon_0 + \| \int_{t_0}^{t_1} \boldsymbol{f}'(\alpha)\, d\alpha - \int_{t_0}^{t_1} \boldsymbol{h}(\alpha)\, d\alpha \| \tag{4.36}$$

where $\boldsymbol{h}$ is the piecewise-linear approximation to $\boldsymbol{f}'$ on $[t_0, t_1]$ defined by

$$\boldsymbol{h}(t) = \boldsymbol{f}'(t_0) + \frac{\boldsymbol{f}'(t_1) - \boldsymbol{f}'(t_0)}{t_1 - t_0}(t - t_0) \tag{4.37}$$

Examining the difference function $\boldsymbol{d}(t) = \boldsymbol{f}'(t) - \boldsymbol{h}(t)$ shows that $\boldsymbol{d}(t_0) = \boldsymbol{0}$ and $\boldsymbol{d}'(t) = \boldsymbol{f}''(t) - \left(\boldsymbol{f}'(t_1) - \boldsymbol{f}'(t_0)\right)/\Delta t$, so that we can majorize the normed quantity in (4.36) by

$$\begin{aligned}
\| \int_{t_0}^{t_1} \boldsymbol{d}(\alpha)\, d\alpha \| &\le \int_{t_0}^{t_1} \| \int_{t_0}^{\alpha} \boldsymbol{d}'(\gamma)\, d\gamma \| d\alpha \\
&\le \int_{t_0}^{t_1}\int_{t_0}^{\alpha} \| \frac{1}{\Delta t}\left(\boldsymbol{f}''(\gamma)\Delta t - \int_{t_0}^{t_1} \boldsymbol{f}''(\sigma)\, d\sigma \right) \|\, d\gamma d\alpha \\
&\le \frac{1}{\Delta t}\int_{t_0}^{t_1}\int_{t_0}^{\alpha}\int_{t_0}^{t_1} \|\boldsymbol{f}''(\gamma) - \boldsymbol{f}''(\sigma)\|\, d\sigma d\gamma d\alpha \\
&\le \frac{1}{\Delta t}\int_{t_0}^{t_1}\int_{t_0}^{\alpha}\int_{t_0}^{t_1}\int_{\sigma}^{\gamma} \|\boldsymbol{f}'''(\tau)\|\, d\tau d\sigma d\gamma d\alpha \\
&\le K(\Delta t)^3
\end{aligned} \tag{4.38}$$

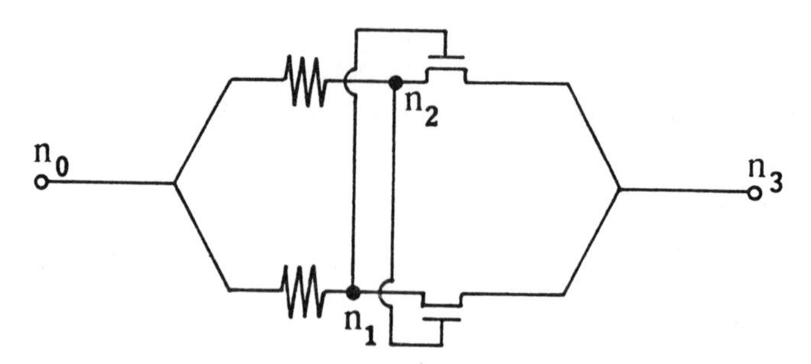

Figure 4.3: A circuit with more than one solution state.

where K is a bound on the third derivative of f. Thus the trapezoidal scheme (4.35) is locally third-order accurate. Continuing the analysis as in Section 4.6, demonstrates that the trapezoidal algorithm produces a global error of order $(\Delta t)^2$. At this point several patterns begin to emerge that will hold for a large class of integration techniques. An algorithm that produces local error of order $(\Delta t)^n$ will produce global error of order $(\Delta t)^{n-1}$. When the previous numerical test case is computed with the trapezoidal rule and $\Delta t = 2\pi/m$ for m=10, 100, 1000 the global error at $t = 2\pi$ is 0.2, 0.002 and 0.00002, respectively. This can be compared with the forward Euler errors computed previously of 4.54, 0.218, and 0.02.

4.10 Remarks on Nonuniqueness

As noted previously, differential equations whose infinitesimal generators satisfy a global Lipschitz property possess a unique solution. In the following examples differential systems which are not Lipschitz are integrated. Consider the circuit of Figure 4.3 that contains two resistors and two transistors. Suppose $v_0(t)$ is zero for $t < 0$, 5 for $t > 5$ and linear on the interval $[0, 5]$ while $v_3(t)$ is held at zero volts. Notice that for $t > 5$, a stable solution occurs when $v_1(t)$ is slightly less than 5 and $v_2(t)$ is greater than 0. However, another stable solution occurs when $v_2(t)$ is a little less than 5 and $v_1(t)$ is slightly greater than 0. The voltage can swing either way and an accurate numerical simulation of this circuit could arrive at either solution while failing to indicate that the other exists. If the two transistors are physically identical and the two resistors nearly so, there would still be a slight tendency for the actual circuit to favor one solution or the other. However, roundoff error could offset this actual physical preference for one solution and simulation would then yield the other.

The following situation occurs very often in transistor circuits. Suppose that a capacitor is originally charged and current extracted from it at one rate for a given time, and then at a different rate. The voltage on a node between the capacitor and the current sink will be continuous, but not differentiable. Notice that this is exactly what occurs in a circuit simulation when using a transistor model of the type described in Section 4.7.

From the above remarks, it is clearly important to study differential equations with discontinuous infinitesimal generators. It is still possible that such systems will have a unique solution and in fact, this is exactly the case with well-designed circuits and accurate models. In order to produce a system with a discontinuous generator, but having a unique solution, the example of Section 4.7 is modified so that now the charge on the capacitor is a discontinuous function of time. The differential equation becomes

$$c(t)f'(t) + f(t) = \frac{1}{2}(\sin t + \cos t) + c(t)ae^{at}, \quad f(0) = \frac{1}{2} \tag{4.39}$$

where the function c is defined by $c(t) = 2$ for $0 < t < \pi/2$ and $c(t) = 5$ for $t > \pi/2$.

4.11 Summary

Mathematically, the ODE system $\boldsymbol{f}'(t) = \boldsymbol{G}(\boldsymbol{f}(t))$ representing a circuit gives rise to an associated dynamical system $\boldsymbol{T}(t, \boldsymbol{p})$. The latter satisfies the causality property, $\boldsymbol{T}(t, \boldsymbol{T}(s, \boldsymbol{p})) = \boldsymbol{T}(t + s, \boldsymbol{p})$, which describes how the system evolves with time. The function $\boldsymbol{G}$ is said to be the generator for $\boldsymbol{T}$; the existence of the flow $\boldsymbol{T}$ given a locally Lipschitz $\boldsymbol{G}$ follows from the fundamental existence and uniqueness theorem of Section 4.3. This theorem is of great practical as well as theoretical value, giving information on when a circuit might have multiple stable states.

The three integration techniques discussed in this chapter can be derived by first replacing $\boldsymbol{f}'(t)$ in the ODE by the difference quotient $(\boldsymbol{f}_{n+1} - \boldsymbol{f}_n)/\Delta t$. Forward Euler then evaluates the right hand side at the old time yielding $\boldsymbol{G}(\boldsymbol{f}_n)$, and is therefore explicit in its representation of $\boldsymbol{f}_{n+1}$. Although simple to implement, it suffers from numerical instability that renders it impractical for circuit simulation. However, it is a useful starting point for analyzing local truncation error, and moreover, provides an elegant, constructive proof of the fundamental theorem, which is quite different from the usual Picard iteration presented in introductory ODE textbooks.

Backward Euler is an implicit method which sets the derivative equal to $\boldsymbol{G}(\boldsymbol{f}_{n+1})$. It is unconditionally stable, but requires the solution of a nonlinear algebraic system at each timestep. Techniques for solving such systems include successive approximation and Newton's method, as discussed in later chapters. Both Euler methods have a global truncation error of $\mathcal{O}(\Delta t)$; for greater accuracy, the trapezoidal method is often used in circuit simulation. It is equivalent to taking one forward Euler step, followed by one backward Euler step, both of length $\Delta t/2$. This scheme is globally second-order accurate and unconditionally stable.

For convenience, in this chapter only fixed stepsizes were considered. A major improvement to any of the methods discussed here is the introduction of variable stepsize: small steps when the solution changes rapidly, more aggressive ones when the circuit is generally quiescent. These ideas, as well as that of variable-order integration techniques, will be explored in Chapter 13. In later chapters dealing with device and process evolution PDEs, discretizing in space by finite difference or finite element methods again leads to sparse systems of ODEs of structure similar to that of the circuit ODE systems. Hence the ideas introduce here regarding dynamical systems, integration techniques, accuracy and stability are again encountered.

4.12 Exercises

1. By modifying the argument for the forward Euler scheme, show that the error after one step of backward Euler satisfies the inequality

$$\epsilon_1 \leq (1 - B\Delta t)^{-1}\{\epsilon_0 + AB(\Delta t)^2\} \tag{4.40}$$

   Using a series expansion for $\frac{1}{z-1}$, show that the local truncation error of the method is $O((\Delta t)^2)$. Prove that backward Euler is globally first order.

2. Use the technique of Section 4.9 to simulate the circuits of Section 4.1 and 4.2. Experiment with various values of K.

3. Consider the circuit of Section 4.2. Define the function $u(t)$ to be -1 for $t < 1$ and 1 for $t \geq 1$. Replace $\sin t$ with $u(t)$ as the known voltage at node 1. Simulate the circuit using forward Euler, backward Euler, Crank–Nicholson, and a variable stepsize method.

4. Solve the singularly perturbed boundary value problem $-\epsilon f'' + f' = 0$, $f(0) = 0$ and $f(1) = 1$, in closed-form. Plot your solution to show that a boundary layer forms for very small values of ϵ.

5. Notice that the argument for convergence of the forward Euler method assumed the existence of the solution f to the differential equation.

By carefully analyzing the proof presented in this chapter see if you can modify it in order to obtain a proof of the existence of a unique solution to the initial value problem. (This is challenging: you will have to show that the sequence of approximate solutions produced by the forward Euler scheme converges to some function, which then must satisfy the ODE.)

Chapter 5

Solving Nonlinear Circuit Equations

5.1 Introduction

The previous chapter showed how electrical circuits and, in particular, integrated circuits, could be simulated by integrating a set of coupled nonlinear ordinary differential equations. The backward Euler and trapezoidal integration methods are implicit, and hence require solving a nonlinear algebraic system at each timestep. Nonlinear equations have several distinguishing characteristics not shared by their linear counterparts. For example, in the absence of roundoff error, Gaussian elimination yields the exact solution to a linear system after finitely many algebraic operations. On the other hand, an algorithm for a nonlinear system is always iterative in nature, giving an ϵ-solution, for which the equations are only satisfied within a user-specified tolerance. Multiple solutions are the rule for nonlinear problems rather than the exception. Generally, in the nonlinear case, the algorithm is only guaranteed to converge if the initial guess is close enough to an actual solution.

The first methods discussed in this chapter are bisection and *regula falsi*. Although restricted to one-dimensional problems, they are useful pedagogically for introducing such concepts as rate of convergence and developing conditions under which convergence is assured. The most general method for solving nonlinear equations in higher dimensions is successive approximation. It provides a unified framework in which to analyze linear iterative methods such as Jacobi as well as certain nonlinear fixed-point iterations used in device simulation. It is easy to use, but generally converges at a much slower rate than Newton's method. A more detailed discussion of successive approximation is given in Chapter 13.

For circuit simulation, some form of Newton–Raphson iteration is the method of choice for solving the associated nonlinear algebraic system. As shown in Section 5.3, it converges quadratically and this property justifies the added expense of forming and factorizing the Jacobian matrix. Several modifications to basic Newton–Raphson are desirable for TCAD applications. Ramping to obtain a good initial guess and damping to reject an update that would cause an increase in the residual, are discussed in the final section of this chapter. Continuation techniques in conjunction with Newton iteration and successive approximation are also considered in the device transport problem of Chapter 7.

As before, the presentation will be via a collection of canonical examples. The notion of rate of convergence here refers to convergence of iterates and therefore has a different meaning than in the previous chapter. Also, round-off error plays a major role here. Both of these topics are introduced via a simple circuit example with no capacitance elements and only one node of unknown voltage.

5.2 A Simple Nonlinear Circuit

The circuit shown in Figure 5.1, consisting of a linear resistor and an ideal diode in series, can be used to introduce the topic of solving nonlinear algebraic equations. Even an example of this simplicity leads to interesting insights because of the exponential nonlinearity of the diode. Suppose that the voltage at node n_0 is held at five volts and the voltage at node n_1 is held at zero volts. Let x denote the unknown voltage at node n_2, and suppose that the current passing through the diode is given by $I_D = e^{100(5-x)} - 1$. If the resistance of R is one ohm, the current passing through it is $I_R = x$.

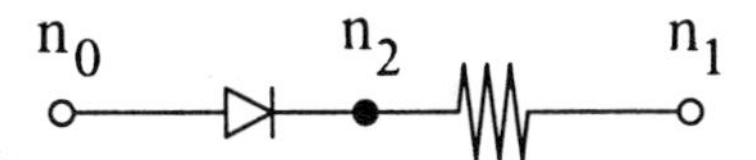

Figure 5.1: A circuit with one diode and one resistor.

Applying Kirchoff's current law, the value of x can be found by solving the equation $I_D = I_R$; that is, by finding a zero of the function

$$f(x) = e^{100(5-x)} - 1 - x \tag{5.1}$$

There is a large family of different algorithms to find a root z of $f(z) = 0$. Some, such as bisection and *regula falsi*, apply to functions of one variable, while others such as fixed-point integration or Newton–Raphson work for functions $f : \mathbf{R}^n \to \mathbf{R}^n$. All methods assume that f is at least continuous, while some presuppose f has a continuous first derivative (i.e. $f \in C^1$). They are iterative in nature and proceed from an initial iterate c_0 to construct a sequence $c_0, c_1, c_2, \ldots$ converging to the solution z.

Bisection and *Regula Falsi*

The method of bisection is simple to visualize and guaranteed to find a zero of a continuous function f defined on an interval $[a, b]$ provided that $f(a)$ and $f(b)$ have opposite signs. A description of the method can be found in an Egyptian papyrus written circa 1700 B.C. by Ahmes. Let $I_0 = [a_0, b_0]$ be an interval included in the domain of a continuous function f, such that $f(a_0)$ and $f(b_0)$ have opposite signs. Since f is continuous, the intermediate value theorem guarantees that there is a number z in I_0 such that $f(z) = 0$. For example (5.1), let $I_0 = [0, 5]$, so that $f(a_0) = (e^{500} - 1) > 0$ while $f(b_0) = -5 < 0$. Now, set $c_0 = (a_0 + b_0)/2$. If $|f(c_0)|$ is less than a specified tolerance ϵ, the process terminates. Otherwise, let $I_1 = [a_1, b_1]$ be one of the intervals $[a_0, c_0]$ or $[c_0, b_0]$, such that $f(a_1)$ and $f(b_1)$ have opposite signs. In general, at the nth step choose I_{n+1} by setting $c_n = (a_n + b_n)/2$ and letting $I_{n+1} = [a_{n+1}, b_{n+1}]$ be one of the intervals $[a_n, c_n]$ or $[c_n, b_n]$, such that $f(a_{n+1})$ and $f(b_{n+1})$ have opposite signs. The sequence $c_0, c_1, c_2, \ldots$ is a bounded sequence and must therefore have a limit point z. Given that $|a_n - c_n| = \left(\frac{1}{2}\right)^n |b_0 - a_0|$, then $c_0, c_1, \ldots$ must also converge to z; the continuity of f implies that $f(z) = 0$. Notice that

$$|c_n - z| \leq 2^{-n} |b_0 - a_0| \tag{5.2}$$

The following criterion can be used to measure the rate or order of convergence of a sequence: if there exists a sequence $\epsilon_0, \epsilon_1, \epsilon_2, \ldots$ such that $|c_n - z| < \epsilon_n$ and there are numbers b and d such that $\epsilon_{n+1} < b(\epsilon_n)^d$, then $c_0, c_1, c_2 \ldots$ is said to converge to z with order d. If $d = 1$, the convergence is linear; if $d = 2$, the convergence rate is quadratic, and so on. For our example, $|c_n - z| < 2^{-n} \times 5 = \epsilon_n$, and since $\epsilon_{n+1} \leq \frac{1}{2}\epsilon_n$, the sequence converges to z with order one. The bisection method has the advantage that it always converges provided that f is continuous, and moreover the number of steps necessary to guarantee a given accuracy is known in advance. For example, if we desire a number y such that $|z - y| < 0.005$, then ten iterations would suffice since $2^{-10} \times 5 < 0.005$.

Regula falsi, a variation on the method of bisection, uses the idea of linear interpolation to enhance the rate of convergence. Suppose that the function f is continuous on the interval $[a_0, b_0]$ and that $f(a_0)$ and $f(b_0)$ are of opposite sign. Let L_0 denote the straight line containing the points $(a_0, f(a_0))$ and $(b_0, f(b_0))$. Let c_0 denote the abscissa of the intersection point of L_0 with the x-axis. If $|f(c_0)|$ is less than a specified tolerance ϵ, the algorithm terminates; otherwise, let $I_1 = [a_1, b_1]$ be one of the intervals $[a_0, c_0]$ and $[c_0, b_0]$ such that $f(a_1)$ and $f(b_1)$ have opposite signs. At the nth step, let L_n denote the straight line containing $(a_n, f(a_n))$ and $(b_n, f(b_n))$ and c_n be the abscissa of the point of intersection of L_n with the x-axis. If the convergence test is met, then the process is finished; if not, let $I_{n+1} = [a_{n+1}, b_{n+1}]$ be one of the intervals $[a_n, c_n]$ and $[c_n, b_n]$ such that $f(a_{n+1})$ and $f(b_{n+1})$ have opposite signs. $\{I_n\}_{n=0}^{\infty}$ is a nested sequence of intervals and therefore there is a number that is common to every one of the intervals. It can be shown that, neglecting round-off error, the sequence c_0, c_1, c_2, ... converges to a zero of f. A good choice of $[a_0, b_0]$ will clearly accelerate convergence. For example (5.1), $f(4.97) = 14.116$ and $f(5) = -5$, so by the intermediate value theorem, there must be a zero of f in $[a_0, b_0] = [4.97, 5]$.

Now, set $[a_0, b_0] = [0, 5]$ and notice that again $f(a_0)$ and $f(b_0)$ have opposite signs. The straight line L_0 is given by the equation $L_0(x) = (0.2e^{500} + 1)(e^{500} - 1)$ and the number c_0 is $5(1 - e^{500})/(e^{500} + 5)$. Notice that c_0 is between 0 and 5. However, in a machine with less than 217 decimal digits of accuracy, c_0 would be computed to be $5e^{500}/e^{500} = 5 = b_0$, and $I_0 = I_1 = I_2 \ldots$ This simple example demonstrates the very important role that round-off error plays in the analysis of numerical algorithms.

5.3 Newton's Method

For analysis of practical circuits, more sophisticated iterative methods such as Newton's method are required. Like bisection and *regula falsi*, Newton's method finds a zero z of a real valued function f defined on an interval $[a, b]$ by iteratively constructing a sequence which converges to z. Whereas the former methods require only continuity of f, Newton's method requires that f also be differentiable.

Choose an initial guess c_0 in $[a, b]$ such that $f'(c_0) \neq 0$ and let L_0 denote the line tangent to f at $(c_0, f(c_0))$. Since the slope of L_0 is not zero, the tangent must intersect the x-axis. Let c_1 denote the abscissa of the point of intersection. Suppose that approximate solutions $c_0, c_1, \cdots, c_n$ have been found on successively repeating this tangent construction. If $f(c_n)$ is smaller in magnitude than a user selected error tolerance ϵ and $|c_n - c_{n-1}| < \epsilon$, then c_n is said to be an ϵ-zero of f and the algorithm terminates. If not, and

$f'(c_n)$ is not zero, let L_n denote the line tangent to f at $(c_n, f(c_n))$,

$$y(x) = f(c_n) + f'(c_n)(x - c_n) \tag{5.3}$$

Let c_{n+1} be the abscissa of the point where L_{n+1} crosses the x-axis. Since $y(c_{n+1}) = 0$, (5.3) gives the recursion

$$c_{n+1} = c_n - \frac{f(c_n)}{f'(c_n)}, \quad f'(c_n) \neq 0 \tag{5.4}$$

Suppose that (5.4) is iterated for example (5.1) until $f(c_n)$ has magnitude less than 10^{-6}. Less than nine iterations are required for an initial guess of 4.97 or 5, but if $c_0 = 0$, the algorithm requires 502 iterations for convergence. The reason for this slow convergence is that if c_n is only slightly less than five, then $f(c_n)/f'(c_n)$ is approximately $\frac{-1}{100}$, so that c_{n+1} is roughly $c_n + \frac{1}{100}$. The efficiency of Newton's method depends very much on a good initial guess. Furthermore, Newton's method is most efficient at finding a zero of the function f when the second derivative of f is small and the first derivative of f is large.

More specifically, suppose that f has a continuous second derivative and that there exist positive numbers A and B such that $f'(x) > A$ and $f''(x) < B$. Then there exists an open interval S containing z such that, if c_0 is in S, the sequence $\{c_n\}$ defined recursively by (5.4) converges to z, and moreover for each positive integer n, $|c_{n+1} - z| < \frac{4B}{A}|c_n - z|^2$. Let us next consider the extension of this method to two dimensions. The algorithm may be conveniently described in strictly geometrical terms. This will make the proof of the theorem clear and will aid in certain discussions concerning the circuit simulation problem. Let g and h be two functions from $\mathbf{R}^2$ into $\mathbf{R}$ defined by $\boldsymbol{F}(x, y) = (g(x, y), h(x, y))$ and assume that the four partial derivatives $g_1 = \partial g/\partial x$, $g_2 = \partial g/\partial y$, $h_1 = \partial h/\partial x$, and $h_2 = \partial h/\partial y$ exist. Let Q denote the plane tangent to g at $\boldsymbol{p}_n = (x_n, y_n)$ and let R denote the plane tangent to h at (x_n, y_n). If either Q or R fails to intersect the xy-plane, then the method cannot continue. This would occur if both $g_1(\boldsymbol{p}_n)$ and $g_2(\boldsymbol{p}_n)$ were zero, causing Q to be parallel to the xy-plane or if both $h_1(\boldsymbol{p}_n)$ and $h_2(\boldsymbol{p}_n)$ were zero, in which case R would be parallel to the xy-plane. Suppose that this is not the case and let α denote the intersection of Q with the xy-plane and let β denote the intersection of R with the xy-plane. Notice that α and β are straight lines. If α and β are parallel, then Newton's method cannot continue. If this is not the case, the next iterate $\boldsymbol{p}_{n+1}$ is defined to be the intersection of α and β.

Notice that this approach is a natural generalization of the 1D construction. It will now be demonstrated that under suitable hypotheses, the sequence $\boldsymbol{p}_0, \boldsymbol{p}_1, \ldots$ will converge to a point $\boldsymbol{u}$ such that $\boldsymbol{F}(\boldsymbol{u}) = (0,0)$. The two-dimensional Newton method is most efficient when the tangent planes are good local approximators of the nonlinear functions (the second partial derivatives are small), the tangent planes intersect the xy-plane in a steep angle (at least one of the partial derivatives of each compononent function is large), and the two straight lines formed by the intersections with the tangent planes and the xy-plane intersect at an angle that is not too small. These statements are made more rigorous in the following theorem.

Theorem 5.1 *Let $\boldsymbol{F}$ denote a function from $\mathbf{R}^2$ to $\mathbf{R}^2$ such that $\boldsymbol{F}(\boldsymbol{u}) = (0,0)$. Write $\boldsymbol{F}(x,y) = (g(x,y), h(x,y))$ and assume that the second partial derivatives of g and h are continuous and bounded in magnitude by $B > 0$. Suppose moreover that there is a number A such that for each pair of points $\boldsymbol{p}$ and $\boldsymbol{q}$*

$$\|\boldsymbol{F}(\boldsymbol{p}) - \boldsymbol{F}(\boldsymbol{q})\| > A\|\boldsymbol{p} - \boldsymbol{q}\| \tag{5.5}$$

Then there is a neighborhood S of $\boldsymbol{u}$ such that if $\boldsymbol{p}_0$ is a point of S, then the Newton sequence $\boldsymbol{p}_0, \boldsymbol{p}_1, \ldots$ converges quadratically to $\boldsymbol{u}$. □

The inequality (5.5) in the two-dimensional theorem is a natural generalization of the one-dimensional case. The main idea of the proof is that if the first two terms of the Taylor series are used to approximate a function, then the error must be of higher order than linear and thus the convergence is quadratic. Let us examine a single step of the iteration from $\boldsymbol{p}_n$ to $\boldsymbol{p}_{n+1}$. This also serves to demonstrate certain aspects of circuit simulation involving the choice of integration timestep. The remainder of the argument necessary to complete a proof is left as an exercise.

Suppose S, $\boldsymbol{u}$, $\boldsymbol{F}$, g, h, A and B are prescribed as in the statement of the theorem. Set $\boldsymbol{p}_n = (a,b)$ and $\boldsymbol{p}_{n+1} = (c,d)$. Let $l(t)$ denote a parameterized representation of the line containing (a,b) and (c,d) defined by $l(t) = (x(t), y(t)) = (a(1-t)+ct, b(1-t)+dt)$. Notice that $l(0) = [a,b]$ and $l(1) = [c,d]$. Let Q denote the plane tangent to g at (a,b) and let R denote the plane tangent to h at (a,b). Let

$$\begin{aligned} \varepsilon_1 &= |Q(c,d) - g(c,d)| \\ \varepsilon_2 &= |R(c,d) - h(c,d)| \end{aligned} \tag{5.6}$$

so that

$$\boldsymbol{F}(c,d) = (Q(c,d), R(c,d)) + (\varepsilon_1, \varepsilon_2) \tag{5.7}$$

Substituting the equations for Q and R yields

$$\begin{aligned} \boldsymbol{F}(c,d) = (g(a,b), h(a,b)) + (&g_1(a,b)(c-a) + g_2(a,b)(d-b), \\ &h_1(a,b)(c-a) + h_2(a,b)(d-b)) + (\varepsilon_1, \varepsilon_2) \end{aligned} \tag{5.8}$$

The function $g(l(t))$ is a real valued function. The fundamental theorem of calculus gives

$$\int_0^1 (g(l(t)))'\,dt = g(l(t)) - g(l(0)) = g(c,d) - g(a,b) \tag{5.9}$$

Therefore,

$$\begin{aligned} g(c,d) &= g(a,b) + \int_0^1 (g(l(t)))'\,dt \\ &= g(a,b) + \int_0^1 g_1(l(t))l_1(t)\,dt + \int_0^1 g_2(l(t))l_2(t)\,dt \\ &= g(a,b) + (c-a)\int_0^1 g_1(l(t))\,dt + (d-b)\int_0^1 g_2(l(t))\,dt \end{aligned} \tag{5.10}$$

Now, using a technique employed in the proof of the Taylor remainder theorem, write

$$\begin{aligned} g(c,d) &= g(a,b) + (c-a)\int_0^1 (t-1)' g_1(l(t))\,dt \\ &\quad + (d-b)\int_0^1 (t-1)' g_2(l(t))\,dt \end{aligned} \tag{5.11}$$

Integrating (5.11) by parts

$$\begin{aligned} g(c,d) &= g(a,b) + (c-a)\Big(g_1(a,b) \\ &\quad - \int_0^1 (t-1)(g_{11}(l(t))l_1(t) + g_{12}(l(t))l_2(t))\,dt\Big) \\ &\quad + (d-b)\Big(g_2(a,b) \\ &\quad - \int_0^1 (t-1)(g_{21}(l(t))l_1(t) + g_{22}(l(t))l_2(t))\,dt\Big) \end{aligned} \tag{5.12}$$

Using the definition of the plane Q and recalling that the second partials of g are bounded by B, yields

$$\|g(c,d)\| < 2B(c-a)^2 + 2B(d-b)^2 \tag{5.13}$$

Similarly,

$$\|h(c,d)\| < 2B(c-a)^2 + 2B(d-b)^2 \tag{5.14}$$

Therefore

$$\|\boldsymbol{F}(c,d)\| < 4B\|\boldsymbol{p}_{n+1} - \boldsymbol{p}_n\|^2 \tag{5.15}$$

The definition of A implies that

$$\|\boldsymbol{p}_{n+1} - \boldsymbol{u}\| < \frac{1}{A}\|\boldsymbol{F}(c,d)\| \tag{5.16}$$

Therefore

$$\|\boldsymbol{p}_{n+1} - \boldsymbol{u}\| < \frac{4B}{A}\|\boldsymbol{p}_{n+1} - \boldsymbol{p}_n\|^2 \tag{5.17}$$

Inequality (5.17) is useful because the quantity $\|\boldsymbol{p}_{n+1} - \boldsymbol{p}_n\|$ is actually known whereas the quantity $\|\boldsymbol{u} - \boldsymbol{p}_n\|$ is not. The remainder of the proof is left as an exercise.

In two dimensions, Newton's method steps from $\boldsymbol{p}_n = (x_n, y_n)$ to $\boldsymbol{p}_{n+1} = (x_{n+1}, y_{n+1})$ by finding the point where the planes tangent to the component functions are both zero. In matrix form this is written

$$\begin{bmatrix} g(x_n, y_n) \\ h(x_n, y_n) \end{bmatrix} + \begin{bmatrix} g_1(x_n, y_n) & g_2(x_n, y_n) \\ h_1(x_n, y_n) & h_2(x_n, y_n) \end{bmatrix} \begin{bmatrix} x_{n+1} - x_n \\ y_{n+1} - y_n \end{bmatrix} = \begin{bmatrix} 0 \\ 0 \end{bmatrix} \tag{5.18}$$

This equation can be rewritten in the familiar form $\boldsymbol{A}\boldsymbol{x} = \boldsymbol{b}$ as follows

$$\begin{bmatrix} g_1(x_n, y_n) & g_2(x_n, y_n) \\ h_1(x_n, y_n) & h_2(x_n, y_n) \end{bmatrix} \begin{bmatrix} x_{n+1} - x_n \\ y_{n+1} - y_n \end{bmatrix} = \begin{bmatrix} -g(x_n, y_n) \\ -h(x_n, y_n) \end{bmatrix} \tag{5.19}$$

The matrix of partial derivatives in (5.19) is denoted by $\boldsymbol{F}'(\boldsymbol{p}_n)$ and is referred to as the Jacobian of $\boldsymbol{F}$. Note that there is a natural progression from dimension two to higher dimensions. Regardless of the dimension, (5.19) is of the form $\boldsymbol{A}\boldsymbol{x} = \boldsymbol{b}$ with

$$\boldsymbol{A} = \boldsymbol{F}'(\boldsymbol{p}_n), \quad \boldsymbol{x} = \boldsymbol{p}_{n+1} - \boldsymbol{p}_n, \quad \boldsymbol{b} = -\boldsymbol{F}(\boldsymbol{p}_n) \tag{5.20}$$

Multiplying both sides of (5.19) by the inverse of $\boldsymbol{F}'(\boldsymbol{p}_n)$ and transposing $\boldsymbol{p}_n$,

$$\boldsymbol{p}_{n+1} = \boldsymbol{p}_n - (\boldsymbol{F}'(\boldsymbol{p}_n))^{-1}(\boldsymbol{F}(\boldsymbol{p}_n)) \tag{5.21}$$

which has the same form as (5.4).

5.4 Further Examples of Nonlinear Circuits

In the case of a circuit consisting only of linear resistors, the system is linear so Newton's method always converges in one iteration independent of the initial guess x_0. For example, let us reconsider the circuit of Figure 5.1. Suppose that the voltages at the input nodes are given as functions of time and the problem is to find the voltages at the remaining nodes as functions of time. Specifically, given a sequence of times $t_0, t_1, \cdots t_k$, with $t_0 < t_1 < \cdots t_k$, construct the sequence $\boldsymbol{v}(t_0), \boldsymbol{v}(t_1), \cdots \boldsymbol{v}(t_k)$ where $\boldsymbol{v}(t) = (v_0(t), v_1(t), \cdots, v_{24}(t))$. Suppose that $\boldsymbol{v}(t_n)$ has been found and the problem is to find $\boldsymbol{v}(t_{n+1})$. We may employ Newton's method to accomplish this task. Let $\boldsymbol{x}_0$ be an initial guess for $\boldsymbol{v}(t_{n+1})$. If the two known voltage functions are continuous functions of time and $t_{n+1} - t_n$ is small, then $\|\boldsymbol{v}_{n+1} - \boldsymbol{v}_n\|$ will be small. Therefore an initial guess of $\boldsymbol{x}_0 = \boldsymbol{v}(t_n)$ would be close to the correct answer. Now proceed from $\boldsymbol{x}_0$ to $\boldsymbol{x}_1$ by using the Newton iteration (5.21). The value of $\boldsymbol{x}_1$ will be the exact solution regardless of the choice of $\boldsymbol{x}_0$ because, since the functions are linear, the tangent hyperplanes are just the component functions and therefore are independent of the choice of $\boldsymbol{x}_0$.

Now consider the circuit of Figure 5.2 that contains one linear resistor and two ideal diodes. Suppose that the resistor has a resistance of

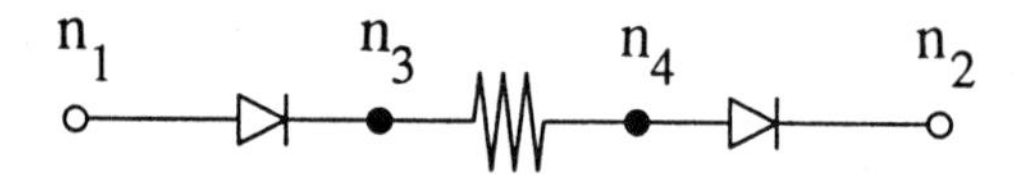

Figure 5.2: A circuit with two diodes and one resistor.

one ohm and the diodes are characterized by the parameters $I_0 = 1$ and $a = 100$. Let $v_i(t)$ denote the voltage at node n_i at time t and let $\boldsymbol{v}(t) = (v_0(t), v_1(t), v_2(t), v_3(t))$. Suppose that v_0 and v_1 are given continuous functions with domain $[0, 1]$, such that $v_0(0) = 0 = v_1(0) = 0$; $v_2(t)$ and $v_3(t)$ are to be determined for t in $[0, 1]$. For convenience, denote $v_2(t), v_3(t)$ by x, y, and let $g(x, y), h(x, y)$ denote the currents flowing at time t into nodes n_2, n_3, respectively. By Kirchoff's current law,

$$\begin{aligned} g(x,y) &= (e^{a(v_0(t)-x)} - 1) + (y - x) \\ h(x,y) &= (e^{a(y-v_0(t))} - 1) + (x - y) \end{aligned} \tag{5.22}$$

Set $\boldsymbol{F}(x, y) = (g(x, y), h(x, y))$; values for the voltages at n_2 and n_3 are found by solving the system $\boldsymbol{F}(x, y) = (0, 0)$ using Newton's method.

Let δ denote a bound on timestep size and ϵ, a bound on voltage error. The circuit simulator must construct a sequence of time points $\{t_j\}_{j=0}^{k}$ and a sequence of voltage vectors $\{r_j, s_j\}_{j=0}^{k}$ such that

$$0 = t_0 < t_1 < \cdots < t_k = 1, \quad t_{n+1} - t_n < \delta \tag{5.23}$$

and

$$|v_2(t_n) - r_n| < \epsilon, \quad |v_3(t_n) - s_n| < \epsilon \tag{5.24}$$

Since $v_0(0) = 0$ and $v_1(0) = 0$, $\boldsymbol{v}(t_0) = (0,0,0,0)$. Set $t_1 = t_0 + \Delta t$, and let (x_0, y_0) be an initial guess for (r_1, s_1); for example, $(x_0, y_0) = (r_0, s_0)$. Newton method constructs (x_1, y_1) from (x_0, y_0) by solving the linear system

$$\begin{bmatrix} ae^{a(v_0(t_1)-x_0)} - x_0 & y_0 \\ x_0 & -ae^{a(y_0-v_1(t_1))} - y_0 \end{bmatrix} \begin{bmatrix} x_1 - x_0 \\ y_1 - y_0 \end{bmatrix} = \begin{bmatrix} -(e^{a(v_0(t_1)-x_0)} - 1) + (y_0 - x_0) \\ -(e^{a(y_0-v_1(t_1))} - 1) + (x_0 - y_0) \end{bmatrix} \quad (5.25)$$

This process can be iterated to form a sequence $\{(x_j, y_j)\}$ which if it converges, has as its limit a solution of $\boldsymbol{F}(x, y) = \boldsymbol{0}$. Now set $t_2 = t_1 + \Delta t$ and continue the process. If the sequence does not converge, (x_0, y_0) is not in the region of attraction of (r_1, s_1). In the case of divergence, SPICE-type simulators try to proceed by taking a smaller time step, usually dividing the current time step by a factor of four. That is to say, set $t_1 = t_0 + \Delta t/4$, and repeat the above use of Newton's method. If the sequence x_j, y_j still does not converge, decrease the time step to $t_1 = t_0 + \Delta t/16$ and try again. For the circuit of Figure 5.2 this process will eventually succeed, because of the continuity of the system: g and h depend continuously on $v_0(t)$ and $v_1(t)$ which are themselves continuous functions of time. The radius of the region of attraction is continuous as a function of Δt. The detailed analysis guarantees that for sufficiently small Δt, Newton's method will converge to (r_1, s_1).

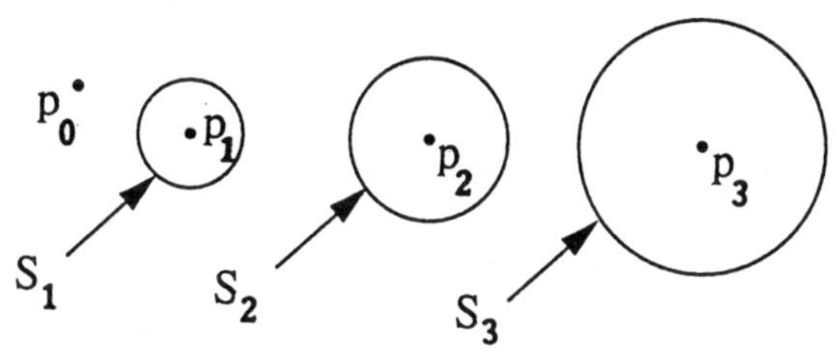

Figure 5.3: Failure in Newton's method.

The goal of a simulation is to produce an approximate solution that is accurate to within a predetermined error tolerance ϵ. In other words, for each discrete time t_i, $||(r_i, s_i) - (v_2(t_i), v_3(t_i))|| < \epsilon$. At time t_i, Newton's method produces a sequence $\{(x_j^i, y_j^i)\}$ converging to $(v_2(t_i), v_3(t_i))$. At what point is the iteration halted and (x_j^i, y_j^i) taken as (r_i, s_i)? It is tempting to require only $|x_{n+1} - x_n| < \epsilon$, $|y_{n+1} - y_n| < \epsilon$, since those quantities are immediately available. However, caution is warranted: e.g. recall that the harmonic series

$z_n = \sum_1^n \frac{1}{i}$ tends to infinity while the sequence $|z_n - z_{n+1}| = 1/(n(n+1))$ converges to zero. A more practical example is that of the circuit of Figure 5.2 in which $|x_{n+1} - x_n|$ is small precisely because x_n and x_{n+1} are far from correct.

Other quantities which are readily available are $g(x_n, y_n)$ and $h(x_n, y_n)$. Recall the previous condition stating that $\|\boldsymbol{F}(\boldsymbol{p}) - \boldsymbol{F}(\boldsymbol{q})\| > A\|\boldsymbol{p} - \boldsymbol{q}\|$ in (5.5). If $A = 1$, then $|g(x_n)| < \epsilon$ implies that $|x_n - v_2(t_i)| < \epsilon$, while if $A = \frac{1}{100}$, $|x_n - v_2(t_i)| < 100\epsilon$. This small value of A occurs in the circuit of Figure 5.2 if the voltage at node n_1 is much larger than that of n_0. Notice that if $(v_1, v_2, v_3, v_4) = (0, 10, 4, 4)$ then both g and h are nearly zero. The algorithm attempts to find the values of v_2 and v_3 based on the magnitude of g and h, but in this area of operation, as long as v_2 is nearly equal to v_3, g and h are not sensitive to a simultaneous change of v_2 and v_3.

There is yet another difficulty associated with simulating this circuit when $\boldsymbol{v} = (0, 10, 4, 4)$. The Jacobian matrix is

$$\begin{bmatrix} 100e^{-400} - 4 & 4 \\ 4 & -100e^{-600} - 4 \end{bmatrix} \approx \begin{bmatrix} -4 & 4 \\ 4 & -4 \end{bmatrix} \tag{5.26}$$

where $\approx$ denotes equality in the finite precision representation on the computer and the latter matrix is singular. Therefore, computer roundoff error will prevent Newton's method from successfully stepping from (x_0, y_0) to (x_1, y_1). The essentially singular structure of the Jacobian is not surprising given the extremely small value of A in $\|\boldsymbol{F}(\boldsymbol{p}) - \boldsymbol{F}(\boldsymbol{q})\| > A\|\boldsymbol{p} - \boldsymbol{q}\|$.

When the diode's ability to store charge is introduced into the diode model, yet another difficulty is encountered. The radius of convergence of Newton's Method becomes a function of timestep size. This is illustrated in Figure 5.3. In this figure, suppose that $\boldsymbol{p}_0$ is the numerical approximation to the solution of the differential equation at $t = 0$ and each $\boldsymbol{p}_i$ is the numerical solution at the next time step given that Δt is chosen to be $(1/2)^i$. The region S_i denotes the region of convergence of Newton's method at $\boldsymbol{p}_i$. Since the size of S_i is a function of Δt and the machine accuracy, there is no guarantee that $\boldsymbol{p}_0$ is contained in any of the regions S_i. This problem is solved using the techniques of ramping and damping discussed in the next section.

What can be said to summarize convergence and accuracy of Newton's method? If the known voltages at the known nodes are such that the voltages at the interior nodes depend in a reasonable way on the functions g and h from Kirchoff's current law, then the simulator will solve the nonlinear systems quickly and accurately. If however, the known voltages are such that the current law does not accurately predict the unknown voltages, the simulator will have difficulty.

The problem can be viewed in yet another way. When $v_0 = 0$ and $v_1 = 10$, the diodes serve as nearly perfect insulators, isolating nodes n_2 and n_3 from the rest of the circuit, and the voltages at these nodes cannot be determined. Such nodes are sometimes referred to as "floating".

5.5 Ramping and Damping

The simulation of the circuit in the last section is simplified by the assumption that $v_1(0) = v_2(0) = 0$. Therefore $v_3(0)$ and $v_4(0)$ are both known. Suppose now that $v_1(0) = a$, $v_2(0) = b$ and neither a nor b is zero. The problem can be reduced to the previous case by extending the domain of the functions v_1 and v_2. For t in $[-1, 0]$, set $v_1(t) = (t+1)a$ and $v_2(t) = (t+1)b$. Now the problem has been reduced to the known case since v_1 and v_2 are continuous functions on $[-1, 1]$ with $v_1(-1) = v_2(-1) = 0$. This method of extending the domain and increasing the applied voltages from zero to the given non-zero values is referred to as "ramping".

In the case that (x_0, y_0) is not in the region of attraction of (r_1, s_1), one can proceed by decreasing the size of the time step. Another method that can be employed in this case is referred to as "damping". The damped Newton method has been employed in the circuit simulator CAZM [14].

When stepping from (x_0, y_0) to (x_1, y_1), suppose that either $|g(x_1, y_1)| > |g(x_0, y_0)|$ or $|h(x_1, y_1)| > |h(x_0, y_0)|$. Construct the parameterized line connecting (x_0, y_0) to (x_1, y_1)

$$f(\eta) = (1 - \eta)(x_0, y_0) + u(x_1, y_1) \tag{5.27}$$

for η in $[0, 1]$. Notice that there exists some number a in $[0, 1]$ such that both $|g(f(a))| < |g(x_0, y_0)|$ and $|h(f(a))| < |h(x_0, y_0)|$ hold. Tentatively define (x_1, y_1) as the point given by Newton method, but if either g or h fails to decrease in going from (x_0, y_0) to (x_1, y_1), then redefine (x_1, y_1) as $f(\frac{1}{2})$. Continue this process until, for some positive integer n, both $|g(f(\frac{1}{2}^n))| < |g(x_0, y_0)|$ and $|h(f(\frac{1}{2}^n))| < |h(x_0, y_0)|$; then set $(x_1, y_1) = f(\frac{1}{2^n})$. By this method, a sequence $\{x_j, y_j\}$ is defined so that both sequences $\{g(x_j, y_j)\}$, and $\{h(x_j, y_j)\}$ are decreasing. Unless there is a point p where $|g(p)|$ and $|h(p)|$ are local minima of $|g|$ and $|h|$ respectively, then the damped Newton method will find $(v_2(t_1), v_3(t_1))$. This robust convergence comes at a price, because a damped Newton meth generally converges only linearly rather than quadratically. When total work is calculated there will be fewer attempted time steps, but more Newton iterations per step.

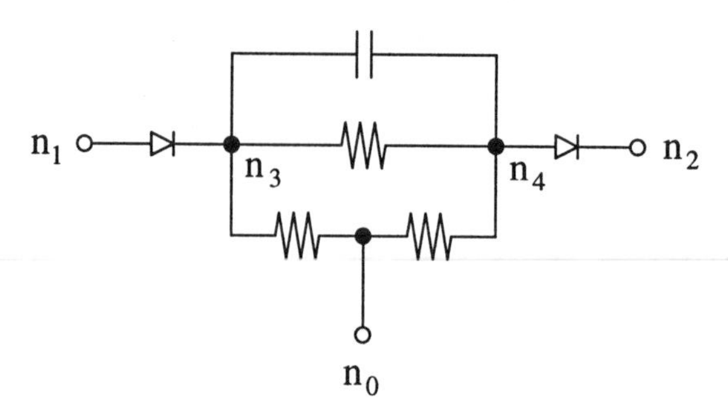

Figure 5.4: Another circuit illustrating Newton's Method.

The problem of floating nodes is at least partially solved in circuit simulation programs such as SPICE by connecting each interior node to ground through a branch that contains a small resistor. In the circuit simulation program, the admittance of these resistors is referred to as *gmin*. Experienced users of circuit simulation programs know that the simulators are more robust (at the expense of accuracy) if the value of *gmin* is made larger. For example, let us consider the circuit in Figure 5.4 with two interior nodes connected to the ground node n_0 by the two lower resistors each with resistance *gmin*. When the value of *gmin* is increased the geometric effect on the Jacobian system is to increase the angle between the lines that correspond to the intersection of the tangent planes and the xy plane. In practice the value of *gmin* can be increased until convergence is obtained. The solutions for the voltages using the large value of *gmin* can then be used as the input for an initial guess to the solution when *gmin* is reduced. Varying *gmin* in the inner loop can be thought of as being similar to the damping discussed previously. Both damping and varying *gmin* can be made automatic to form a robust simulator that need less tuning by the user.

5.6 Summary

The primary thrust of this chapter has been the introduction of Newton's method for solving nonlinear algebraic systems that arise in circuit simulation when using an implicit method such as the backward Euler integrator. The Newton recursion, $\boldsymbol{x}_{n+1} = \boldsymbol{x}_n - (\boldsymbol{F}'(\boldsymbol{x}_n))^{-1}\boldsymbol{F}(\boldsymbol{x}_n)$, requires the assembly and storage of a very large, sparse Jacobian matrix $\boldsymbol{F}'(\boldsymbol{x}_n)$, followed by its LU factorization. This expense is offset by a quadratic decrease in the error at each step, which gives the algorithm a clear advantage over other iterative methods for circuit simulation.

From a practical standpoint there are several difficulties which can occur when using Newton's method. First, convergence is only guaranteed when the initial guess is close enough to a zero $\boldsymbol{\xi}$ of $\boldsymbol{F}(\boldsymbol{x}) = \mathbf{0}$. Determining this domain of attraction is a delicate problem, and often the use of continuation methods are appropriate. In the case of circuit simulation this amounts to ramping the applied voltages from zero to their final values. Second, because of the exponential nonlinearities in the models for diodes and transistors, the Jacobian will often be nearly singular, causing the algorithm to break down. Recall the 2-D example in which the xy-intercepts of the tangent planes of the components of $\boldsymbol{F}$ are a pair of nearly parallel lines which intersect to give an $\boldsymbol{x}_{n+1}$ far from $\boldsymbol{x}_n$. Failure of the inner loop can in turn force the attempted timestep to be reduced in the outer loop. This can cause a vicious cycle which ultimately aborts the integration. Damping – taking some fraction of the Newton update – sometimes helps, but nothing entirely circumvents the difficulty caused by near singularity of a Jacobian. Similar issues arise in solving the nonlinear sparse systems for the discretized device and process problems described later.

5.7 Exercises

1. Modify the circuit of Figure 4.2 by adding a diode D on a branch from n_2 to n_3. Suppose that the current I through D is given by

$$I = e^{100v_{2,3}} - 1$$

   Set $v_1(t) = 1$ for $t \leq 0$ and $v_1(t) = \cos t$ for $t > 0$. Set $v_4(t) = 1$ for $t \leq 0$ and $v_4(t) = 1 + \sin t$ for $t > 0$. Simulate the circuit using the backward Euler technique and Newton's method.

2. Repeat Problem 5.1 but use the trapezoidal rule in place of the backward Euler scheme.

3. Repeat Problem 5.1 with $v_4(t) = 0$ for $t \leq 0$ and $v_4(t) = \sin t$ for $t > 0$ and compare results.

4. Simulate the circuit of Figure 4.2 using $c = 1$, $r_1 = r_2 = 0$ for $i = 1, 2$, $v_1(t) = 0$ if $t < 1$ and $v_1(t) = 1$ if $t \geq 1$ with $v_4(t) = c$ if $t \leq 0$ and $v_4(t) = \sin t$ if $t \geq 0$. Discuss your results.

5. Complete the proof of the quadratic convergence of Newton's method by finding a circle C with center $\boldsymbol{u}$ and a constant K such that if $\boldsymbol{p}_n$ is in C then p_{n+1} is in C and

$$||\boldsymbol{p}_{n+1} - \boldsymbol{u}|| < K||\boldsymbol{p}_n - \boldsymbol{u}||^2.$$

Chapter 6

Circuit Models and Parameter Extraction

6.1 Introduction

The extent to which a simulation of a given circuit reflects its actual operation depends upon several factors. The first is the numerical error of the solution of the differential-algebraic system of equations derived using Kirchhoff's laws, as discussed in the previous chapters. This numerical accuracy may be characterized mathematically and is usually well controlled by advanced integration methods. A second factor governing the overall reliability of a circuit simulation is the extent to which analytic device models accurately reflect the physics of the devices in the circuit. Finally, a very important factor affecting simulation quality is the extent to which the parameters in the device models fit the behavior of the actual fabricated devices. This chapter gives an overview of analytic device models for circuit simulation and discusses how the model parameters are obtained.

The behavior of devices such as resistors, capacitors, and transistors that make up an integrated circuit are modeled in the circuit simulator in terms of constitutive relations among the voltages, currents, and charges on the various terminals of the device. These constitutive relations are algebraic equations which attempt to capture macroscopically the physics of the electronic processes occurring within the device. The algebraic equations contain a number of parameters which must be adjusted to match the behavior of a particular device or class of devices. This adjustment process is known as parameter extraction or optimization. The standard method for extracting model parameters is to directly fit the parameters to measured electrical data by means of nonlinear least–squares optimization techniques. Elaborate strategies are often developed, requiring many individual sets of measurements, each aimed toward extraction of a certain subset of the parameters for the model.

An alternative to this traditional approach of using parametrized algebraic device models is to couple the circuit simulator directly to a device simulator. One simultaneously solves the time-dependent drift-diffusion partial differential equations and the differential-algebraic system resulting from Kirchhoff's laws. The currents and voltages at the terminals of each device are output from the device simulator and used as input to the circuit simulator. This approach, while potentially very accurate, is very expensive computationally and impractical for all but the smallest of circuits. This so-called mixed-mode simulation has been used for some special purpose investigations of device behavior within a circuit. An example is the simulation of the current induced in a memory cell due to an alpha particle hitting the cell [159, 365], which is important for the design of radiation-hardened circuits.

Still another approach is to completely by-pass the analytic device models, and instead construct tables of measured data for each device [114, 142, 156]. The circuit simulator then uses splines to interpolate between table values for a given set of bias conditions. There are several advantages to this method. First, splines guarantee smoothness and continuity, which helps minimize difficulties in numerical integration of the system of equations that govern the circuit. Second, the approximations and inaccuracies inherent in analytic device models are eliminated. Finally, table lookup models are usually much faster to evaluate than the corresponding analytic models, reducing the computational expense of circuit simulation.

On the other hand, there are some disadvantages to using tables. First, an enormous of data must be measured and stored for each different type of device. Another disadvantage of table models is the consequent lack of scaling. Usually an analytic model will allow some form of scaling, in that certain model parameters are functions of the device's geometry. This information is lost when using the table-based approach, in which a new table must be constructed for each new geometry. A third drawback is the absence of physical insight that might be provided by the analytic models, so that statistical variations due to fluctuations in the process conditions become more difficult to understand and predict.

6.2 Device Models for Circuit Simulation

The behavior of the devices in an integrated circuit are described by constitutive relations among the voltages and currents in the device. Chapters 2 and 3 give detailed derivations of these relations for several of the more common devices used in semiconductor circuits. The constitutive relations contain one or more parameters which must be adjusted to fit the measured behavior of a particular device. Together the parameters and the constitutive relations make up a model for a specific type of device.

Two-terminal Devices

A simple example of a device model is Ohm's law for the resistor. This model states that the current flowing through a resistor is $I = v_{i,j}/r$, where the resistor connects nodes i and j in the circuit, and $v_{i,j}$ is the difference between the voltages at those nodes. The resistance r is a model parameter for this device, and it depends on the physical properties and dimensions of the material from which the resistor is fabricated. For a particular device the resistance can be measured, assuming the validity of Ohm's law, by setting a voltage at each terminal of the resistor and measuring the current through the resistor. Alternatively, one could compute the current through a given resistor with a device simulator as described in Chapters 7 and 8, and thereby calculate the resistance, again assuming that Ohm's law is an accurate representation of the behavior of the device. Constitutive relations for several common devices are described in Chapters 2 and 3; for convenience, models for several two-terminal devices are summarized in Table 6.1.

MOSFET Models

Much more complicated models are required to describe adequately the behavior of MOSFET devices. There is a vast literature on the various models (see [529], for example), and some commercial circuit simulators contain a dozen or more MOSFET models from which to choose. Chapter 3 discussed the Shichman–Hodges MOSFET model [492], which is a simple empirical model that describes the general behavior of a MOSFET. More recent MOSFET models attempt to derive device behavior from semiconductor physics [318, 367, 429, 490, 529, 560]; these models are generally much more acceptable for accurate circuit simulation. However, for the purpose of illustrating parameter extraction later in this chapter, the simple Shichman–Hodges model will be used.

Device	Constitutive Equation	Parameters
Resistor	$I = \frac{v_{i,j}}{r}$	r
Capacitor	$I = c\frac{dv_{i,j}}{dt}$	c
Inductor	$\frac{dI}{dt} = Lv_{i,j}$	L
Diode	$I_d = I_s\left(\exp\left(\frac{qv_{i,j}}{NkT}\right) - 1\right)$	I_s, N

Table 6.1: Models for some of the common two-terminal devices.

The Shichman–Hodges MOSFET model is described in Chapter 3. For convenience we list the equations of the model here, with an emphasis on the parameters of the model. The constitutive equations for the current between the source and drain are divided into three regions, depending upon the values of the terminal voltages:

For $v_{g,s} \leq v_T$:

$$I_{ds} = 0 \tag{6.1}$$

For $0 < v_{d,s} < (v_{g,s} - v_T)$:

$$I_{ds} = \frac{K_P}{2}\frac{W}{L_{eff}}v_{d,s}(2(v_{g,s} - v_T) - v_{d,s})(1 + \lambda v_{d,s}) \tag{6.2}$$

For $0 < (v_{g,s} - v_T) \leq v_{d,s}$:

$$I_{ds} = \frac{K_P}{2}\frac{W}{L_{eff}}(v_{g,s} - v_T)^2(1 + \lambda v_{d,s}) \tag{6.3}$$

The threshold voltage is defined as

$$v_T = v_T^0 + \gamma(\sqrt{\phi - v_{b,s}} - \sqrt{\phi}) \tag{6.4}$$

and the effective channel length is defined as $L_{eff} = L - 2L_D$. In the above equations, W is the physical width of the gate of the MOSFET and L is the physical length of the gate of the MOSFET. The six parameters of the model are K_P, ϕ, λ, γ, v_T^0, and L_D.

Equations (6.1) – (6.3) model the MOSFET under DC conditions; that is, when the terminal voltages are constant. When the terminal voltages change with respect to time, the dynamic behavior of the MOSFET is governed by the charges in the gate, channel, source, and drain regions of the device. In the Shichman–Hodges MOSFET model, this dynamic behavior is modeled by placing voltage-dependent capacitors between the various terminals. In the three regions of operation, the corresponding capacitances are

$$\text{For} \quad v_{g,s} \leq v_T :$$
$$C_{gd} = 0; \quad C_{gs} = 0; \quad C_{gb} = C_{ox}WL \tag{6.5}$$
$$\text{For} \quad 0 < (v_{g,s} - v_T) \leq v_{d,s} :$$
$$C_{gd} = C_{gb} = 0; \quad C_{gs} = \frac{2}{3}C_{ox}WL \tag{6.6}$$
$$\text{For} \quad 0 \leq v_{d,s} < (v_{g,s} - v_T) :$$
$$C_{gd} = C_{gs} = \frac{1}{2}C_{ox}WL; \quad C_{gb} = 0 \tag{6.7}$$

where C_{ox} is a modeling parameter, equal to the parallel plate capacitance between the MOSFET gate and the substrate. The bulk and source terminals and the bulk and drain terminals are connected via a *pn*-junction diode, and their capacitances are modeled by

$$C_{bs} = \frac{C_{bs0}}{(1 - v_{b,s}/V_J)^m} \quad ; \quad C_{bd} = \frac{C_{bd0}}{(1 - v_{b,d}/V_J)^m} \tag{6.8}$$

where C_{bs0} and C_{bd0} are the capacitances of the junctions at zero bias, $v_{b,s}$ and $v_{b,d}$ are the voltages between the bulk and the source or drain, V_J is the junction built-in voltage, and m is the junction grading coefficient. Thus dynamic capacitance adds five additional parameters to the basic MOSFET model: $C_{ox}, C_{bs0}, C_{bd0}, V_J$, and m.

The Shichman–Hodges MOSFET model described above is one of the simplest models that captures the overall behavior of a MOSFET. However, since I_{ds} goes abruptly to zero at the threshold voltage, in contrast with the actual physics, this model is very poor for circuit designs that operate near v_T, such as many analog circuits. There is a large body of literature on more physically-based MOSFET models (see [318, 367, 429, 490, 529, 560] for a representative sample of these models). Tsividis and Suyama [530] describe a set of twelve requirements for a "good" MOSFET model from the point of view of the analog circuit designer. These twelve requirements are as follows.

1. The model should accurately fit the measured static (DC) behavior of the device, it should conserve charge, and it should predict with reasonable accuracy the switching speed. This is a necessary requirement for strictly digital circuit design.

2. The model should accurately predict the measured conductances g_{ds}, g_m, and g_{mb}, defined by

$$g_{ds} = \frac{dI_{ds}}{dv_{d,s}} \quad ; \quad g_m = \frac{dI_{ds}}{dv_{g,s}} \quad ; \quad g_{mb} = \frac{dI_{ds}}{dv_{b,s}} \tag{6.9}$$

   and the measured capacitances. All conductances and capacitances should be continuous with respect to any terminal voltage.

3. The model should give good results as the frequency of operation is increased.

4. The model should give accurate predictions of the noise generated by the device [529].

5. All the above criteria should be met over a wide range of terminal voltages.

6. All the above criteria should be met over the range of temperature of interest.

7. All the above criteria should be met for any combination of device width and length.

8. One set of model parameters should suffice for all dimensions (width and length) of MOSFETS of a given type; that is, the model should scale with geometric dimensions.

9. When implemented in a circuit simulator, the model should provide a warning if it is used outside its region of validity.

10. The model should have as few parameters as possible, and the parameters should be linked as closely as possible to the device structure and the process parameters, for example, thickness of the gate oxide.

11. The model should be linked to a parameter extraction methodology.

12. The model should provide links to process and device simulators.

The twelve Tsividis–Suyama requirements for a MOSFET model may also be extended to other devices. For example, the diode model in Table 6.1 should also meet all twelve requirements (with appropriate adjustments in the conductance definitions in item 2). Tsividis and Suyama also present a set of benchmark tests for MOSFET devices models.

Circuit simulation models that describe the constitutive relations among terminal voltages and currents of other types of devices can be developed in a similar manner. Models have been developed for the bipolar junction transistor [151, 229, 305], the MESFET [123, 369, 507] and a variety of other devices [7].

Behavioral Models

All of the models discussed above are "behavioral" models in that they attempt to reflect the physics of the devices in a form that is suitable for numerical circuit simulation. In a similar way, behavioral models of more elaborate combinations of circuit elements can also be created. For example, consider an operational amplifier circuit. This circuit is designed to have high gain, high input impedance, and low output resistance. Implementation of such a circuit requires dozens of semiconductor devices. Suppose, however, that the opamp is itself only a small piece of a much larger electrical system, and that one wants to model the overall qualitative behavior of the system. In the simplest approximation, an ideal voltage source can replace the semiconductor devices of this opamp during the circuit simulation of the entire system. A more elaborate and accurate behavioral macromodel of the opamp can also be constructed using combinations of simple devices [61].

Many circuit simulators include ideal current source and voltage source elements, and the value of each source can be a function of voltages at other nodes or currents through other elements in the circuit. Using these ideal sources, a large number of behavioral models can be constructed. For example, a voltage-controlled oscillator (VCO) is a circuit that generates a sinusoidal output with frequency controlled by the value of the voltage applied to the input of the circuit. A VCO requires many semiconductor devices to implement and the circuit simulation can be computationally expensive. However, the behavior of the VCO can be modeled with an ideal voltage source in the following way. Assume the VCO frequency is controlled by a voltage $V_c(t)$ as $f_{osc} = f_0(1 + V_c(t))$. Then the instantaneous output of the VCO is the sin of the time integral of f_{osc}. An ideal current source with value f_{osc} in series with a capacitor generates a voltage V_{int} equal to the time integral of f_{osc}. An ideal voltage source whose value is equal to the sin of V_{int} produces the final output of the VCO [542].

It is relatively straightforward to build ideal models of other circuits, such as integrators, voltage comparators, logic gates, and phase-locked loops using ideal sources [109, 328, 498]. One can further abstract this idea, and treat the circuit simulator as a general solver for a system of differential-algebraic equations. For example, SPICE3 [439] has been used to simulate an accelerometer sensor together with the controlling electronics that trigger deployment of an automobile airbag system [424]. The accelerometer was modeled by an ODE based on Newton's law, and the electronic components were modeled by typical circuit elements. A number of languages to describe such mixed systems for simulation are available [346, 471]. These abstract models and their parameters require much the same level of care to implement as the models which are built into the simulator for semiconductor devices. In fact, given the generality of behavioral simulation, the modeling and parameter issues that arise are often exacerbated.

6.3 Parameter Extraction

Given an analytic device model such as those described in the previous sections, it becomes necessary to adjust the parameters of the model to fit measurements from actual physical devices. The first step in parameter extraction is the collection of data from the target manufacturing process. The data may be actual electrical or physical measurements of manufactured devices, or derived values from process and device simulations, or a combination of the two. A good understanding of the model equations for a particular device is necessary in order to ascertain what data are relevant for the model parameters. For example, consider the parameters of the Shichman–Hodges MOSFET model (6.1) – (6.3). If the device is biased with $v_{d,s}$ equal to a small value, say 0.1 V, then the current becomes

$$I_{ds} \approx \frac{K_P}{2} \frac{W}{L_{eff}} 2 v_{d,s}(v_{g,s} - v_T) \tag{6.10}$$

Therefore, measurements of I_{ds} versus $v_{g,s}$ should yield a straight line whose x-intercept is equal to v_T. Other bias points are used for the remaining model parameters. Thus the selection of what data to gather from experiment and what extraction methodology to use are intimately connected with the model equations and parameters themselves.

Data Collection

Usually, the most accurate method for collecting data is to measure directly the electrical and physical characteristics of devices that have been manufactured. Often several discrete device structures suitable for such measurements are included on the silicon wafer along with the actual integrated circuits being manufactured. These discrete devices are used both for monitoring the process itself and for device parameter extraction. For example, an MOS integrated circuit might include several individual MOSFETS of various geometries, with interconnects to enable electrical probing of each individual device. In addition, there might be several special structures to allow external measurements of such quantities as the parasitic capacitances between adjacent structures or the resistances of interconnect lines.

Detailed extraction of parameters for device models is usually done on only a small number of devices. It is extremely important that these devices be representative of the manufacturing process; that is, that they are in some sense "average" devices. The manufacturing line usually samples a few key electrical properties of individual devices at the completion of the manufacturing process. In a MOS process, for example, the source to drain current when $v_{g,s} = v_{d,s} = 3$ V might be measured at five different sites on a wafer, for five different wafers in each manufacturing lot. Over time, statistical information regarding the value of this current would be collected in a database, and any wafer that deviated substantially from the average would signal a potential processing problem. This same data can be used to help establish statistical limits for circuit simulation and to verify the extracted parameters. In selecting particular wafers and devices for detailed parameter characterization, it is essential that they be chosen using statistical sampling techniques.

Once a set of devices has been selected, electrical measurements are made. Depending on the particular device model, data are collected under various voltage or current bias conditions. For example, in the Shichman–Hodges MOSFET model described in the previous section, the quantity of primary interest is the drain current under various gate, source, and drain voltage biases. When the gate–source voltage is less than the threshold voltage, the drain current is modeled as zero, so it is not necessary to measure such data. Instead, one biases the device in the linear and saturation regions, and measures drain current as a function of the terminal voltages.

It is often not possible to measure all the relevant parameters needed for circuit simulation models directly from physical structures. It might be necessary to begin circuit design work using a new process before any devices have actually been produced, as is often the case for a new generation of process technology. Also, even when fabricated devices are available for measurement, certain data may not be physically measurable. For example, the overlap capacitance between the gate and the source diffusion region of a MOSFET may be too small to measure unless a special test structure is provided on the wafer.

With the use of process and device simulators, the manufacture and operation of a device structure can be simulated numerically and "data" can be measured and extracted from the output. Circuit simulation parameters can be estimated before device fabrication, and properties that are difficult or impossible to measure directly can be determined. In practice, process and device simulation is usually used to complement direct measurements rather than supplant them. As discussed in Chapter 1, this use of process and device simulation is part of an iterative cycle of technology and design.

Even if sufficiently many device structures exist for complete characterization of device model parameters, simulations are often of interest in generating statistical data based on best and worst case fluctuations in the various processing steps. It might be very difficult, for example, to find and measure an actual MOS wafer that had experienced lower than normal temperature in the gate oxide growth, as well as smaller than usual gate line width. It is, however, straightforward to simulate such a device numerically, and thus determine the final effect of such process fluctuations on electrical behavior. Data from simulations can be used in conjunction with measured data and statistical information from the process line itself to generate best and worst case envelopes for the device model parameters.

Fitting data to the device model equations

Once data have been obtained for a particular device, it is used to determine the parameters for the analytic models used to represent the device's behavior in a circuit simulator, a process known as parameter extraction. A nonlinear least–squares algorithm is usually used to fit the parameters to the data. To obtain the "best" set of parameters $(a_1, a_2, ..., a_M)$ fitting a given function $y(\boldsymbol{x}; \boldsymbol{a})$ to a set of M data points (x_i, y_i), define the error functional

$$\chi^2(\boldsymbol{a}) = \sum_{i=1}^{M} \left(\frac{y_i - y(x_i; \boldsymbol{a})}{\sigma_i} \right)^2 \tag{6.11}$$

where σ_i is the standard deviation, or measurement error tolerance, of the ith data point, and y_i is the specified data at x_i. The function $y(\boldsymbol{x}; \boldsymbol{a})$

is defined by the device model equation, and the parameters $\boldsymbol{a}$ are to be determined such that $\chi^2(\boldsymbol{a})$ is minimized. For example, in the equation for the diode current in Table 6.1, $y(\boldsymbol{x}; \boldsymbol{a})$ is the equation for I_d, with one independent variable, $x = v_d$, and two parameters, $(a_1, a_2) = (I_s, N)$. The data values y_i would be measurements or simulated values of the diode current for successive values of applied voltage v_d.

A standard method for minimizing (6.11) with respect to the set of parameters (a_i) is the Levenberg–Marquardt algorithm [319, 351, 437], which is briefly outlined here. The method combines the steepest descent algorithm and Newton's method (see Chapter 13). Steepest descent minimizes a functional by proceeding in the direction of the negative gradient. That is, for iterate $k + 1$,

$$\boldsymbol{a}^{(k+1)} = \boldsymbol{a}^{(k)} - \delta \boldsymbol{\nabla} \chi^2(\boldsymbol{a}^{(k)}) \tag{6.12}$$

If the functional being minimized is quadratic, an optimal choice of δ is available, as discussed in Chapter 13, but in general this is not the case, and a choice for δ must be made on heuristic grounds. The Levenberg–Marquardt method modifies (6.12) by choosing a different δ for each component:

$$a_i^{(k+1)} = a_i^{(k)} - \frac{1}{\lambda \alpha_{ii}} \frac{\partial \chi^2(\boldsymbol{a}^{(k)})}{\partial a_i} \tag{6.13}$$

where λ is a unitless constant and α_{ii} is the ith diagonal element of the Hessian matrix

$$H_{ij} = \frac{\partial^2 \chi^2}{\partial a_i \partial a_j} \tag{6.14}$$

The reason for using such a step length is far from arbitrary and it becomes clear when one considers another method for minimizing (6.11), namely Newton's method.

Note that a local minimum $\boldsymbol{a}$ of (6.11) is characterized by the fact that $\boldsymbol{\nabla} \chi^2(\boldsymbol{a}) = 0$, so that our minimization problem can be recast as finding the zeros of the gradient function. Applying Newton's method yields

$$H(\boldsymbol{a}^{(k+1)} - \boldsymbol{a}^{(k)}) = -\boldsymbol{\nabla} \chi^2(\boldsymbol{a}^{(k)}) \tag{6.15}$$

where H is the derivative of $\boldsymbol{\nabla} \chi^2$, that is the Hessian of χ^2. Note that considerable expense is involved in solving the linear system (6.15) as compared with solving (6.12) which only requires computing the gradient. This expense is usually more than justified by the quadratic convergence property of Newton's method, at least near a zero of $-\boldsymbol{\nabla} \chi^2$.

In the Levenberg–Marquardt algorithm, the two separate iteration formulas, (6.13) and (6.15), are combined to yield the system of equations

$$\sum_{j=1}^{M} \beta_{ij}(a_j^{(k+1)} - a_j^{(k)}) = -\frac{\partial \chi^2(a^{(k)})}{\partial a_i} \qquad i = 1, ..., M \tag{6.16}$$

where $\beta_{ij} = H_{ij}$ for $i \neq j$ and $\beta_{ii} = (1 + \lambda)H_{ii}$. As the unitless parameter λ becomes very large, the matrix β becomes diagonally dominant, and (6.16) approaches the steepest descent iteration (6.13). When $\boldsymbol{a}$ is far from a minimumof χ^2, this is the preferred algorithm. On the other hand, when $\boldsymbol{a}$ is near a minimum, Newton's method becomes more efficient than steepest descent, and the Newton iteration formula (6.15) is preferable. By letting λ approach zero in (6.16), the Newton formula is recovered. After each step produces a new estimate of the parameter vector $\boldsymbol{a}$, the Levenberg–Marquardt algorithm increases the value of λ if χ^2 increases or reduces λ if χ^2 decreases. Thus the algorithm switches between a steepest descent and a Newton optimization in a smooth and controllable manner. When χ^2 decreases by a negligible amount between iterations, a local minimum and hence a best fit of the data to the parameters has been found. This algorithm is robust and works very well in practice. The final parameters must be analyzed carefully, however, since their accuracy depends heavily on the extent to which the model equations faithfully describe the physics of the device, as well as upon the reliability of the measured data.

Note that the Levenberg–Marquardt method can be considered conceptually as a modified or quasi-Newton method. The size of the linear system is quite modest (< 100) compared with the sizes encountered in circuit, device, or process simulation (10^4 to 10^6), and hence can be solved using a direct method such as LU decomposition. Standard caveats for steepest descent and Newton's method apply here as well. Near a zero of the derivative of $\nabla\chi^2$, that is a singular point of the Hessian of χ^2, the linear system (6.16) will become ill-conditioned with attendant numerical difficulties. Far from a minimum, steepest descent may undergo "hemstitching" and repeatedly overshoot the domain of attraction of a local minimum. Another difficulty in minimizing (6.11) is the fact that there are often many local minima, and the one produced by the algorithm will depend on the initial guess $\boldsymbol{a}_0$ used to start the iteration. The value of $\boldsymbol{a}_0$ can often be supplied based upon physical reasoning or previous parameter extractions.

One could simply input the model equations, the unknown parameters, and the raw data to a nonlinear least–squares routine, which would then find an optimized set of parameters with which to represent the data. This

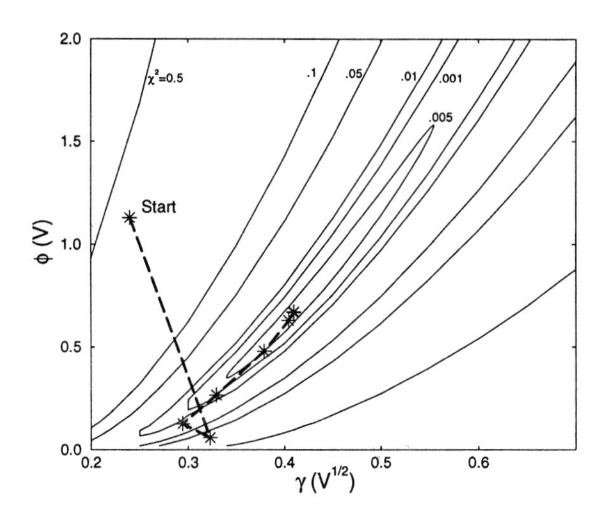

Figure 6.1: Optimization trajectory of the Levenberg–Marquardt algorithm applied to the parameters for the threshold voltage of a MOSFET. Solid lines are contours of χ^2 in the $\phi - \gamma$ plane, and the dashed line traces the steps taken by the algorithm.

is usually not the preferred approach, however. A typical MOSFET model has several dozen parameters to be fit, and there may in fact be many local minima satisfying a least–squares condition. Approaching the task purely from the viewpoint of statistical estimation, while neglecting the physical origins of the problem, can give disappointing results. A better technique is to extract the parameters in a series of well-controlled steps, each of which exploits the physics of the device model. For example, one might make measurements specifically to extract the threshold voltage of a MOSFET, and then with a different set of measurements extract the channel length parameter. With these parameters fixed or strongly constrained, the global optimizer could then extract the remaining parameters with greater confidence. Developing a parameter extraction strategy for a given set of analytic device equations remains very much an art.

As an example of the Levenberg–Marquardt algorithm, consider the optimization of the parameters of (6.4) for the threshold voltage of a MOSFET. Here the threshold voltage v_T is a function of the bulk to source voltage $v_{b,s}$, with parameters v_T^0, γ, and ϕ. Assume that measurements of the v_T for various values of $v_{b,s}$ have been made. Note that by definition, the parameter v_T^0 is the threshold voltage when $v_{b,s}$ is zero, so its value is fixed. The Levenberg–Marquardt algorithm is then invoked to adjust ϕ and γ so as to minimize χ^2. Figure 6.1 shows contour lines of χ^2 in the $\phi - \gamma$ plane. With an initial guess of $\phi = 1.2$ and $\gamma = 0.25$, the algorithm minimizes χ^2 in seven iterations, with final best-fit parameter values of $\phi = 0.673$ and $\gamma = 0.41$. In this example, the algorithm was implemented so that the value of the dimensionless parameter λ in (6.16) is decreased (increased) by one order of magnitude whenever an iteration reduces (increases) the value of χ^2. In this example, the value of λ was initially set to 1, and it had decreased to 10^{-7} by the final iteration.

Parameter Verification

The final step in parameter extraction is also the most critical: verification of the extracted parameters using circuit simulations. The obvious first step in verification is checking that each parameter for a device model is present in the set and is within reasonable bounds, based on physical reasoning. The threshold voltage of a MOSFET, for example, might be expected to fall between 0.5 V and 0.8 V; values generated by least–squares which fall outside this range would indicate a problem with the extraction methodology, the original device data, or the model equations themselves. The second step is to perform consistency checks between the various parameters which are known to be correlated. For example, the oxide thickness parameter should normally be the same for both p-channel and n-channel MOSFETs. Once a set of parameters has passed the completeness, bounds, and consistency checks, extensive circuit simulations using the parameter set can be run. These simulations allow for further consistency checks, such as verification of the expected device behavior as a function of temperature. If a database of previously extracted parameter sets is available, sets from similar processes can be identified and compared to the current one. Likewise, if actual measurements from a mature process are available from the manufacturing line, the parameter set can be used in simulations compared directly with the measured data. If best and worst case parameter sets have been extracted (see below), the simulated statistical variations can also be compared to the measured variations. Finally, it is desirable to compare various operating characteristics of a circuit to those simulated using the extracted parameters. A simple circuit such as a ring oscillator (discussed in Chapter 17) is often used for this purpose.

6.4 Statistical Variations of Model Parameters

Variations in the manufacturing process, such as small changes in the temperature of a diffusion furnace or the pressure in a reactive ion etch, inevitably lead to statistical variations in the electrical behavior of the devices. The circuit designer must take into account these variations to ensure the ultimate manufacturability of the circuit. Some of the methodologies that have been developed to help simulate these variations in a circuit design include best case/worst case modeling, Monte Carlo analysis, and response surface methodology.

The best case/worst case methodology attempts to set absolute bounds on the variations of a manufacturing process. Suppose that measurements of a "typical" MOSFET have been made, parameters based on the Shichman–Hodges model have been extracted for the device, and the parameters have been verified according to the previous sections. What should the parameters be if, instead of the "typical" process conditions, the device had been fabricated under a combination of the most unfavorable conditions? For example, suppose that the furnace which grows the gate oxide were a few degrees hotter than normal, resulting in an increase in the oxide thickness, that the MOSFET gate patterning produced a physical gate length 0.2 μm longer than usual, and that the implant and subsequent diffusions produced a higher than normal channel doping and smaller than normal source/drain underdiffusion. All these conditions would combine to produce a MOSFET which exhibits a much lower drain current than normal. When all such possibilities occur simultaneously, the "worst-case" device results. Conversely, processing conditions can also occur to create a MOSFET which exhibits much higher drain current than usual, resulting in a "best case" device. One could in principle deliberately skew each critical step in the manufacturing process to produce best and worst case devices, measure the devices, and then extract parameters from measurements as outlined in the previous section. In practice, however, this is usually not feasible due to time and cost constraints. Process simulation can play a critical role in simulating variations at each step, and the results can be input to a device simulator to produce simulated electrical data. This is a particularly potent application of process and device simulation, since the device parameters used in circuit simulation are often not easily related to specific physical process conditions. Figure 6.2 shows the drain current of a MOSFET assuming best case, worst case, and typical parameters.

When designing a circuit, one would simulate the behavior of the circuit with both the best and worst case parameters, helping ensure that the

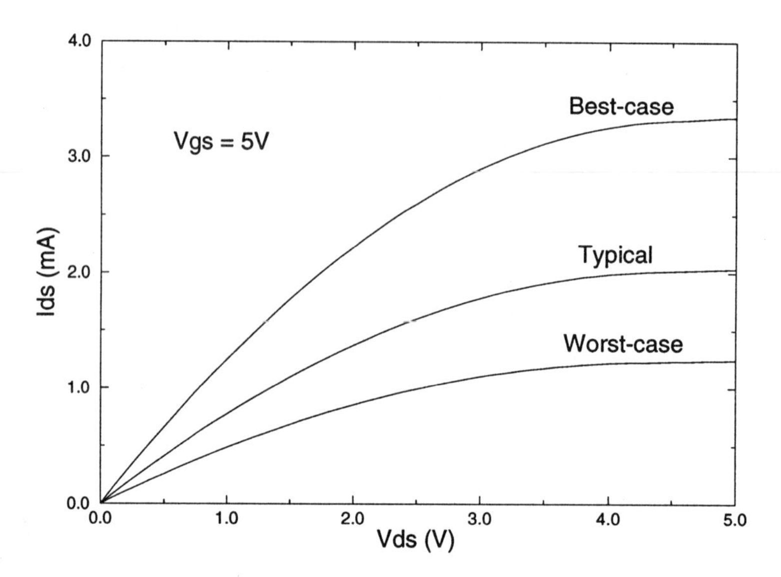

Figure 6.2: Illustration of best case, worst case, and typical device currents for a MOSFET model.

circuit will behave as expected over the range of manufacturing conditions. Consider, for example, a simple circuit consisting of a chain of five inverters. Using the typical, best case, and worst case process parameters as illustrated in Figure 6.2 (along with a similar set of parameters for the p-channel MOSFET), simulation of the delay from the input to the output of the chain yields 0.61 ns, 1.1 ns, and 1.9 ns for best case, typical, and worst case model parameters, respectively.

One of the problems with the best case/worst case methodology is that it is often overly conservative, and the circuit designer is forced to make tradeoffs in order to satisfy processing variations that may never occur in practice. Monte Carlo techniques can be used to sample statistically the distribution of model parameters [60, 437] and hence judge the applicability of the best case/worst case parameter values for a particular design. As a simple illustration, assume that the best case/worst case conditions in Figure 6.2 have been determined by varying each parameter by $\pm 3\sigma$ from the typical case. Now assuming that each parameter is independent (which is often not justified), a series of simulations can be run varying each parameter p_i by

$$p_i = p_i^T + R_g \sigma_i \tag{6.17}$$

where p_i^T is the typical value of the parameter, R_g is a random variable with Gaussian distribution, and σ_i is the standard deviation of p_i. The number of independent simulations needed in the Monte Carlo analysis can be determined given a particular confidence level and precision [60]. Figure 6.3

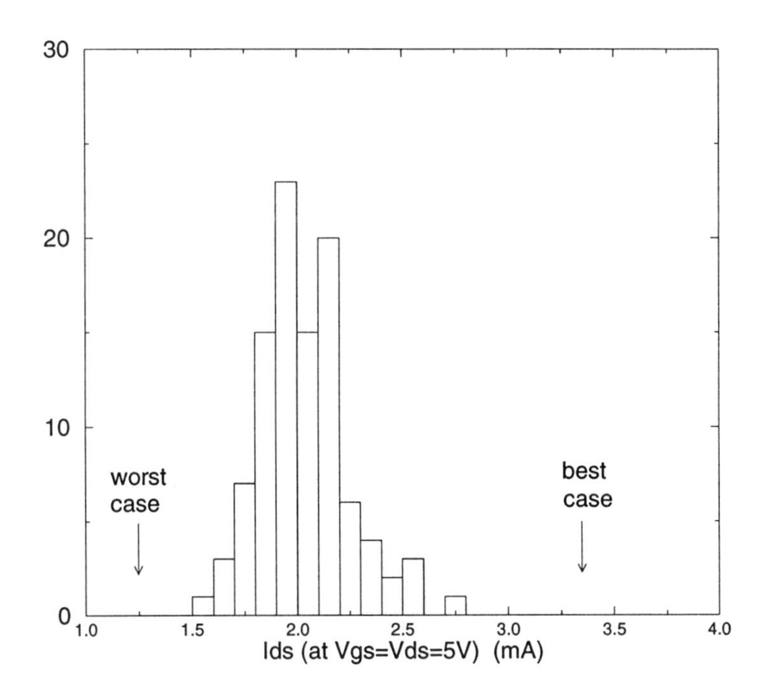

Figure 6.3: Monte Carlo calculation of the distribution of maximum drain current (when $v_{g,s} = v_{d,s} = 5\text{V}$) for the MOSFET shown in the previous figure.

shows the distribution of maximum drain current for the same MOSFET of Figure 6.2, obtained from a series of Monte Carlo simulations with each model parameter independently adjusted according to (6.17). The runs were repeated to a confidence level of 95 percent. The distribution is somewhat tighter than the best case/worst case limits of I_{ds}. To see the consequences of the Monte Carlo variation of model parameters in an actual circuit, a simple delay chain consisting of five identical inverters was constructed using the n-channel MOSFET of Figure 6.2 and a complementary p-channel

MOSFET with similar parameter variation. The distribution of the signal delay from the input to the output of the chain, as calculated by Monte Carlo simulation, is shown in Figure 6.4. Here the distribution is decisively skewed toward the best case set of parameters. Much more elaborate sta-

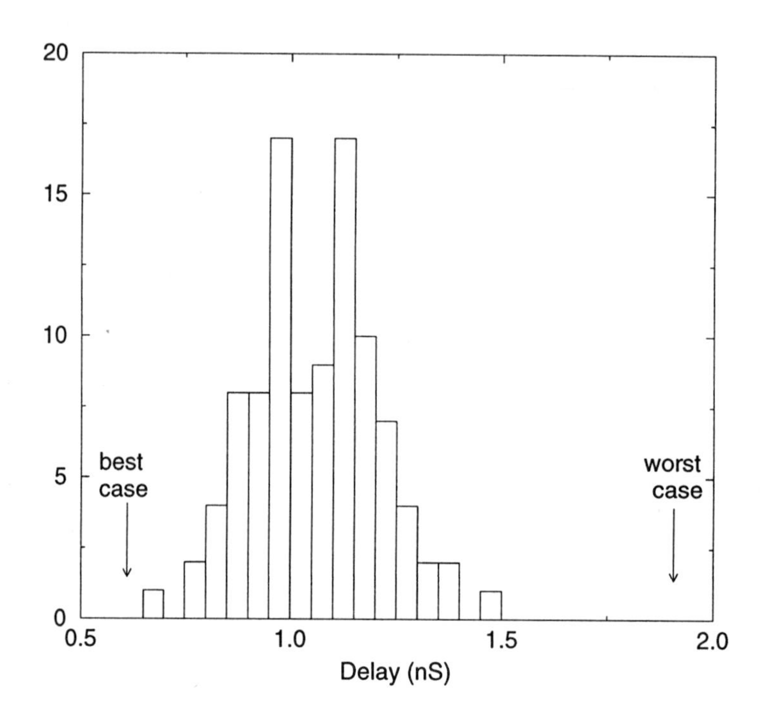

Figure 6.4: Monte Carlo calculation of the distribution of the delay through a chain of five inverters.

tistical simulation schemes have been developed, such as sensitivity-based optimization schemes and response surface methodology [3]. It is important in all statistical simulation to establish correlations among the various model parameters. This requires careful design of the set of experiments, both those conducted during the actual fabrication and those run during process and device simulations.

6.5 Summary

To a circuit designer, the device model equations and model parameters represent "the truth" in terms of the physics of the devices from which a circuit must be built. Therefore great care must be exercised in developing models and in extracting parameters for the model. In addition to accurately representing the physics of the device, a model must be continuous through the first derivative with respect to the terminal voltages for proper behavior within a circuit simulator. Since many of the more complex semiconductor devices such as MOSFETS are modeled in a piecewise fashion over various regions of operation, special attention must be paid to ensure continuity at the boundaries between these regions. Fitting the parameters of the model to experimental or simulated (via process and device simulation) data is a multi-dimensional optimization problem. The Levenberg–Marquardt algorithm, which combines the techniques of steepest descent and Newton's method, is a general and robust method for these problems. Once the optimal set of parameters has been obtained, careful verification using the circuit simulator itself must be performed in order to test the accuracy of the parameters as well as the applicability of the device model. The effect of the inevitable variations in the semiconductor manufacturing process on the model parameters and subsequent circuit behavior can be estimated by simple best and worst case analysis or by more sophisticated statistical techniques.

6.6 Exercises

1. Plot the current I_d through a diode as a function of the voltage drop $v_{i,j}$ across the diode using the model in Table 6.1. For parameters, assume that $I_s = 10^{-14}$ amps and $N = 1$. Also, at room temperature, the thermal voltage kT/q is approximately 0.025 V. Assuming that a computer will generate a floating point overflow error for $\exp(x)$ when $x > 700$, at what value of $v_{i,j}$ will the diode model exhibit numerical overflow? Since it is possible for such values to be reached during the Newton iteration in a circuit simulation, modify the diode model to prevent such an overflow by introducing a linear extrapolation for large values of $qv_{i,j}/NkT$, also ensuring that the model is continuous in the first derivative.

2. Show that the Shichman–Hodges MOSFET model, (6.1)–(6.3), is continuous and that the derivatives with respect to the terminal voltages

are continuous, for $v_{g,s} > 0$ and $v_{d,s} > 0$.

3. Examine (6.1)–(6.5) from the point of view of the Tsividis and Suyama criteria for a "good" MOSFET model. Calculate g_{ds}, g_m, and g_{mb}, and plot these values for $v_{g,s}$ near v_T.

4. In the Shichman–Hodges model, I_{ds} is zero when $v_{g,s} < v_T$. Experimentally, the current does not go abruptly to zero at threshold, but rather falls exponentially as $\exp[-Av_{g,s}]$ for $0 < v_{g,s} < v_T$, where A is a constant. Extend the model to include this behavior in such a way that I_{ds} is continuous as a function of $v_{g,s}$.

5. Consider the parameters of the Shichman–Hodges model, K_P, ϕ, λ, γ, v_T^0, and L_D. In which direction would one vary each parameter to increase the drain current? To decrease the drain current? Compute the sensitivity of the drain current to variations in each parameter (for example, $\lambda \partial I_{ds}/\partial\lambda$).

6. Implement the Levenberg–Marquardt algorithm and use it to extract the parameters of the Shichman–Hodges model: K_P, ϕ, λ, γ, v_T^0, and L_D. Assume that Figure 6.2 contains experimental data of I_{ds} versus $v_{d,s}$ with $L = 2$ μm and $W = 10$ μm. Extract three sets of parameters representing best case, worst case, and typical processing conditions, as indicated in the figure.

7. Suppose that for a particular process, the threshold voltage v_T^0 of n-channel MOSFETS were measured for different values of the gate length L: $(L, v_T^0) = ((10\ \mu\text{m}, 0.75\ \text{V}), (2\ \mu\text{m}, 0.75\text{V}), (1\ \mu\text{m}, 0.74\text{V}), (0.8\ \mu\text{m}, 0.70\text{V}), (0.6\ \mu\text{m}, 0.63\text{V}))$. What effect does this correlation have upon best case and worst case parameter sets?

Chapter 7

Semiconductor Device Modeling

7.1 Introduction

New semiconductor device fabrication technologies have radically transformed the field of microelectronics. Until relatively recently, the development of new devices was guided primarily by experiment, aided by the use of simple analytical models. In many instances, one-dimensional analytical models sufficed to give a viable characterization of current–voltage (I–V) curves for the device. With the advent of submicron devices, these simple models have become of limited, and in some cases questionable, use. More sophisticated physical and mathematical models in higher dimensions may be necessary, and their solution involves more sophisticated analysis and numerical simulation techniques.

Specifically, an effective design process now requires a combination of experimental studies with numerical simulation for solution of the associated partial differential equations describing the device transport problem. This dual strategy is important since the use of simulation can drastically reduce the number of admissible designs, and experimentation can be limited to the most relevant configurations. Experiments too are important in validating models and in establishing their limitations. One-dimensional numerical models still play a vital role in ascertaining many of the qualitative characteristics of a device. With this knowledge, more detailed two-dimensional and three-dimensional simulations can then be made to explore the effects of reduced feature size and complicated geometries.

Under the application of a bias voltage, conduction band electrons and valence band electrons move through a device. In one of the simpler models applicable to certain large devices, the transport of valence band electrons may be neglected and the problem reduces to solving a transport PDE for a single "carrier". As the device size is reduced, this model is less adequate and a two-carrier model for "electrons" (conduction band electrons) and "holes" (valence band electrons) may be applied. This model can be described by a coupled set of partial differential equations for the electrostatic potential ψ, electron concentration n and hole concentration p. Since the device must operate under a variety of conditions, solutions to these governing nonlinear partial differential equations are required for a range of applied voltages. The electrostatic potential and carrier concentrations exhibit large gradients in the neighborhood of pn-junctions and also near the gate oxide–silicon interface for MOS transistors. Finely graded computational meshes are required in these regions to resolve the solution adequately. The following sections present some fundamentals of transport theory, followed by the drift-diffusion equations and their properties. Finally, we develop the approximate analysis and solution schemes, including special features such as adaptive grid strategies.

7.2 Transport Equation

Prior to developing the fundamental drift-diffusion equations for the transport of carriers in semiconductor devices, it is instructive to review briefly some basic theory related to reference frames and the material derivative that lead to the transport equation. These ideas also provide a general framework to describe transport phenomena, including those associated with diffusion processes considered later.

There are two main reference frames – lagrangian and eulerian. In the lagrangian frame the state of the system or process is described with reference to the initial configuration at $t = 0$ and proves most useful in applications such as elasticity problems involving small deformations. On the other hand, in an eulerian frame the state is described with reference to the present configuration and this is appropriate for modeling transport processes, particularly when convective effects (drift) are important. To demonstrate the latter point, let us introduce the material derivative D/Dt of a scalar field ϕ such as carrier concentration or temperature. Then in the lagrangian frame we have simply $D\phi/Dt = \partial\phi/\partial t$, but the transformation relating the present configuration at time t to the initial state must be specified. In the eulerian frame, let $\boldsymbol{x}(t)$ denote the position vector for the present state. Then, since

$\phi = \phi(\boldsymbol{x}(t), t)$ the material derivative is defined as

$$\frac{D\phi}{Dt} = \frac{\partial \phi}{\partial t} + \sum_{i=1}^{n} \frac{\partial \phi}{\partial x_i} \frac{\partial x_i}{\partial t} \tag{7.1}$$

where $n = 1, 2, 3$ is the dimension and the chain rule has been used. Since the velocity $\boldsymbol{u}$ is the time rate of change of position, then $u_i = \partial x_i / \partial t$ and we can write (7.1) compactly as

$$\frac{D\phi}{Dt} = \frac{\partial \phi}{\partial t} + \boldsymbol{u} \cdot \boldsymbol{\nabla} \phi \tag{7.2}$$

so that the velocity now enters explicitly in the drift term $\boldsymbol{u} \cdot \boldsymbol{\nabla}\phi$.

The transport equation for a scalar field ϕ can be derived by appealing to the associated conservation law. For example, let ρ be the constant mass density, ϕ the species concentration, $\boldsymbol{q}$ the species flux vector and f the source or sink density function (possibly associated with reactions or recombination effects). Consider an arbitrary volume V in the transport domain and let $\boldsymbol{n}$ denote the outward unit normal to boundary surface S of V. The conservation or balance law equating the time rate of change of species in V to flow through the boundary surface S and production/consumption by f is then given by

$$-\int_V \frac{D(\rho\phi)}{Dt} dV = \int_S \rho \boldsymbol{q} \cdot \boldsymbol{n} dS + \int_V \rho f dV \tag{7.3}$$

By applying Gauss divergence theorem to the surface integral and simplifying, we obtain

$$\int_V \left(\frac{D\phi}{Dt} + \boldsymbol{\nabla} \cdot \boldsymbol{q} + f \right) dV = 0 \tag{7.4}$$

Since (7.4) holds for arbitrary volume V, it follows that the integrand must vanish and we have

$$\frac{D\phi}{Dt} + \boldsymbol{\nabla} \cdot \boldsymbol{q} + f = 0 \tag{7.5}$$

Substituting (7.2) in (7.5) and introducing a Fourier-type constitutive relation $\boldsymbol{q} = -k\boldsymbol{\nabla}\phi$, with material property k, the transport equation is

$$\frac{\partial \phi}{\partial t} + \boldsymbol{u} \cdot \boldsymbol{\nabla}\phi - \boldsymbol{\nabla} \cdot (k\boldsymbol{\nabla}\phi) = -f \tag{7.6}$$

Then (7.6) constitutes the fundamental equation describing drift ($\boldsymbol{u} \cdot \boldsymbol{\nabla}\phi$), diffusion ($\boldsymbol{\nabla} \cdot (k\boldsymbol{\nabla}\phi)$) and reaction ($f$) processes. Each of these respective contributions must be adequately characterized if the resulting discretized numerical schemes are to approximate the system reliably. This fundamental scalar transport problem therefore provides a convenient vehicle to test numerical schemes, as seen later for both device and process problems. In the development of the process theory we also model diffusion as a random walk and present an alternative deterministic formulation. In the next section we focus on the drift-diffusion equations for carrier transport in devices.

7.3 Drift-Diffusion Equations

The transport of electrons and holes in semiconductor devices may be mathematically modeled using the Boltzmann transport equation, under appropriate assumptions on the motion of the carriers. The carrier motion within the semiconductor lattice is assumed to take place through a series of accelerations due to applied external fields, and through a series of scattering events. For silicon devices with active dimensions of 0.1 μm or larger, this semiclassical approach remains valid. Hence, direct solution of the Boltzmann transport equations is one possible approach to analyze semiconductor devices. Although this approach is appropriate for theoretical studies, it is not practical for device design since the solution of the Boltzmann system is computationally prohibitive even on supercomputers, and design optimization requires many simulations.

Additional simplifying assumptions are required to produce a tractable mathematical model that lends itself to practical design studies. A quasi-static local field approximation may be used to reduce the problem from one involving spatial and momentum coordinates to one involving spatial coordinates alone. The basic premise is that the carriers respond instantaneously to a change in the electric field. It follows that the product of the particle velocity and relaxation time is small compared with the active device dimensions. This simplifying assumption in the Boltzmann transport equation leads to the basic semiconductor transport equations [460, 494].

The drift-diffusion system can also be developed from Maxwell's equations in electromagnetic theory

$$\nabla \cdot \boldsymbol{D} = \rho, \quad \frac{\partial \boldsymbol{D}}{\partial t} + \boldsymbol{J} = \nabla \times \boldsymbol{H} \tag{7.7}$$

$$\nabla \cdot \boldsymbol{B} = 0, \quad \frac{\partial \boldsymbol{B}}{\partial t} = -\nabla \times \boldsymbol{E} \tag{7.8}$$

where $\boldsymbol{D}$, $\boldsymbol{B}$ and $\boldsymbol{J}$ represent the electric displacement, magnetic induction and conduction current density vectors, respectively, and ρ is the electric charge density. The electric displacement $\boldsymbol{D}$ and magnetic induction $\boldsymbol{B}$ can be related to the electric field $\boldsymbol{E}$ and magnetic field $\boldsymbol{H}$ by constitutive relations

$$\boldsymbol{D} = \varepsilon \boldsymbol{E}, \qquad \boldsymbol{B} = \mu \boldsymbol{H} \tag{7.9}$$

where ε and μ are the material dielectric permittivity tensor and magnetic permeability tensor, respectively. For semiconductor modeling, these tensor quantities may be taken as scalars.

Next introduce an electrostatic potential ψ with

$$\boldsymbol{E} = -\boldsymbol{\nabla}\psi \tag{7.10}$$

so that using (7.9) and (7.10) in the first equation of (7.7), the elliptic partial differential equation for electrostatic potential is

$$-\boldsymbol{\nabla} \cdot (\varepsilon \boldsymbol{\nabla}\psi) = \rho \tag{7.11}$$

If the permittivity in a region is constant, the problem reduces to one of solving Poisson's equation for specified charge density ρ. The charge density is related to the concentration of electrons and holes in the material. Assuming complete ionization of impurities, this implies

$$\rho = q(p - n + C) \tag{7.12}$$

where q is the unit electronic charge and C is the electrically active net impurity concentration. Substituting in (7.11), the equation for electrostatic potential becomes

$$-\boldsymbol{\nabla} \cdot (\varepsilon \boldsymbol{\nabla}\psi) = q\,(p - n + C) \tag{7.13}$$

This equation is coupled to the transport equations for carrier concentrations p and n, which enter as forcing functions on the right hand side of (7.13). Taking the divergence of both sides of the second equation in (7.7) yields

$$\frac{\partial \rho}{\partial t} + \boldsymbol{\nabla} \cdot \boldsymbol{J} = 0 \tag{7.14}$$

Separating the current density $\boldsymbol{J}$ into electron and hole components, $\boldsymbol{J} = \boldsymbol{J}_n + \boldsymbol{J}_p$, and rearranging terms

$$\frac{\partial p}{\partial t} + \frac{1}{q}\boldsymbol{\nabla} \cdot \boldsymbol{J}_p = \frac{\partial n}{\partial t} - \frac{1}{q}\boldsymbol{\nabla} \cdot \boldsymbol{J}_n = -R(\psi, n, p) \tag{7.15}$$

where R is the net generation–recombination rate of electron–hole pairs per unit volume and it is assumed that C is essentially independent of t.

In device design, the steady-state response to an applied voltage is of primary interest. The coupled set of equations for electrostatic potential and carrier concentrations in the device domain Ω is then

$$-\nabla \cdot (\varepsilon \nabla \psi) = q(p - n + C) \quad (7.16)$$

$$\nabla \cdot \boldsymbol{J}_n = qR(\psi, n, p) \quad (7.17)$$

$$\nabla \cdot \boldsymbol{J}_p = -qR(\psi, n, p) \quad (7.18)$$

The electron current density $\boldsymbol{J}_n$ and hole current density $\boldsymbol{J}_p$ may be related to the electric field and carrier concentration gradient by the drift-diffusion constitutive relations

$$\boldsymbol{J}_n = qn\mu_n \boldsymbol{E} + qD_n \nabla n \quad (7.19)$$

$$\boldsymbol{J}_p = qp\mu_p \boldsymbol{E} - qD_p \nabla p \quad (7.20)$$

where μ_n, μ_p are the electron and hole mobilities and D_n, D_p are the corresponding diffusivities.

Substituting these constitutive relations into the transport equations (7.17) and (7.18) leads to a coupled set of three partial differential equations for device modeling. The PDE system can be interpreted as a diffusion equation for electrostatic potential coupled to a pair of convection–diffusion equations for carrier transport. This system is commonly referred to as the drift-diffusion system or model.

The mobilities and diffusivities in equations (7.19) and (7.20) are actually functions of the local electric field. Moreover, $R(\psi, n, p)$ and $qn\mu_n\boldsymbol{E}$ make the problem nonlinear even if μ_n and D_n are constant. Also, it is far more common to have a field-dependent mobility than diffusivity. The system (7.16)–(7.20) cannot model velocity overshoot and, in particular, hot carrier effects since the local models of mobility are no longer valid and equations describing conservation of carrier momentum and energy are required instead of the drift-diffusion constitutive relations. This hot carrier problem and associated "hydrodynamic equations" are treated in the next chapter.

Mathematical representations for the carrier mobilities, the diffusivities and the recombination rate R are required to complete the drift-diffusion model. Various relations have been proposed for mobilities and recombination, based on experimental studies. For instance, a field-dependent mobility model which accounts for velocity saturation is described for silicon by Sze [514] and has the form

$$\mu(E) = \mu_0 \left(1 + \left(\frac{\mu_0 E}{v_{sat}}\right)^{\beta}\right)^{-1/\beta} \quad (7.21)$$

where v_{sat} is the saturation drift velocity, β is a constant, μ_0 is a low field mobility and the scalar E is the component of the electric field in the direction of current flow. The low field mobility includes the effects of carrier scattering due to ionized impurities in the silicon lattice. Expression (7.21) gives a mobility which is approximately constant for low applied electric fields, while at higher electric fields the drift-velocity saturation causes a reduction in the carrier mobility. The diffusion coefficients and mobilities are related through the Einstein relation

$$D_n = \mu_n \frac{kT}{q}, \qquad D_p = \mu_p \frac{kT}{q} \tag{7.22}$$

where k is the Boltzmann constant and T is the carrier temperature.

Recombination effects are small in MOS devices, but in bipolar devices they strongly influence the current gain. The recombination term R can be written as

$$R = R^{SRH} + R^{AU} \tag{7.23}$$

where R^{SRH} corresponds to indirect recombination and generation through trap centers in the energy band gap, and R^{AU} represents net recombination due to Auger processes. In Auger recombination an electron–hole pair recombine without the aid of a trapping center whereby energy is transferred to another electron or hole. For indirect Shockley–Read–Hall recombination

$$R^{SRH} = \frac{np - n_I^2}{\tau_p (n + n_T) + \tau_n (p + p_T)} \tag{7.24}$$

with τ_n and τ_p as the electron–hole recombination lifetimes, n_I the intrinsic carrier concentration, and n_T and p_T depending on the position and occupancy of the traps. For the Auger process

$$R^{AU} = \left(np - n_I^2\right) (c_n n + c_p p) \tag{7.25}$$

where c_n and c_p are the Auger coefficients. This contribution is significant if the concentration of electrons and holes is high in a given spatial region when compared with the density of the background doping atoms. The effect of impact ionization is sometimes included in the recombination term — impact ionization is dependent on the electric field rather than the carrier concentrations and becomes significant only for high electric fields.

At the ohmic contact of a MOSFET, Dirichlet boundary conditions for electrostatic potential ψ and carrier concentrations n and p drive the device, as indicated in Figure 7.1. The side and lower boundaries are assumed to be sufficiently remote from the ohmic contacts that the assumption of no electron or hole current flow across these boundaries is reasonable. This implies that the normal component of the electric field is approximately zero, and therefore,

$$\boldsymbol{E} \cdot \boldsymbol{n} = 0, \qquad \boldsymbol{J}_n \cdot \boldsymbol{n} = 0\,, \qquad \boldsymbol{J}_p \cdot \boldsymbol{n} = 0 \tag{7.26}$$

where $\boldsymbol{n}$ is the outward unit normal. These Neumann boundary conditions for zero flux also apply along the remaining gate oxide edges. Flux balance is required at the oxide–semiconductor interface.

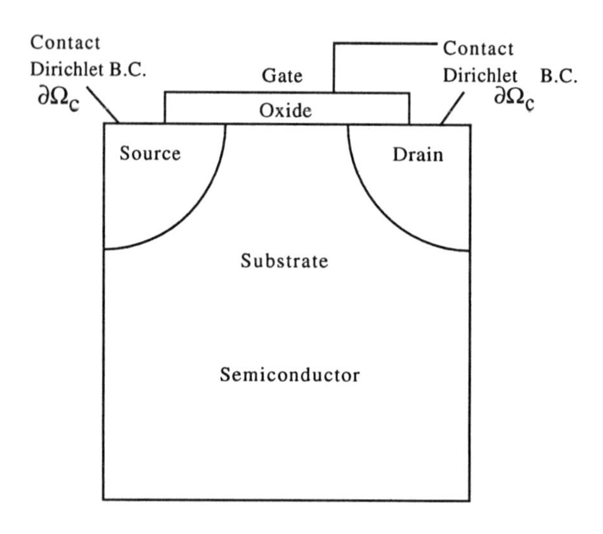

Figure 7.1: Typical Dirichlet boundary conditions at the contacts $\partial\Omega_c$ and Neumann conditions (zero flux) elsewhere for MOSFET example.

Assuming space charge neutrality and thermal equilibrium,

$$\rho = q(p - n + C) = 0, \qquad np = n_I^2 \tag{7.27}$$

hold at the ohmic contacts. The Dirichlet boundary data for the carrier concentrations on these surfaces $\partial\Omega_c$ then become

$$n = \tfrac{1}{2}\left(C + (C^2 + 4n_I^2)^{1/2}\right), \quad p = \tfrac{1}{2}\left(-C + (C^2 + 4n_I^2)^{1/2}\right) \tag{7.28}$$

The electrostatic potential at the contacts can be related to the applied voltage V by setting the value to that of a semiconductor at thermal equilibrium,

$$\psi = V + \sinh^{-1}(C/2n_I) \tag{7.29}$$

In the MOSFET example of Figure 7.1, the oxide is an insulator and hence there are no mobile charge carriers and negligible amounts of impurities in this subregion. That is, $n = p = C = 0$ in the gate oxide, so the electrostatic potential equation in this region simplifies to Laplace's equation, $\Delta\psi = 0$. At the semiconductor–insulator interface, flux conservation implies

$$\left[\!\left[\varepsilon \frac{\partial \psi}{\partial n}\right]\!\right] = Q_{ss} \tag{7.30}$$

where $[\![\cdot]\!]$ denotes the jump and Q_{ss} is the charge per unit surface area on the interface.

Other types of boundary conditions are appropriate for devices such as MESFETS and Schottky diodes. The present discussion of boundary conditions assumes an isolated semiconductor device, which actually forms part of a more complex integrated circuit. A more accurate model would include neighboring devices and passive circuit elements. Moreover, the applied terminal voltages may not be known *a priori* and, ideally, would be determined by solving the external circuit equations in conjunction with the local governing partial differential equations that apply within the device.

The governing equations can also be recast in terms of quasi-Fermi potentials ϕ_n and ϕ_p, which are defined by [380]

$$n = n_I \exp\left(\frac{q(\psi - \phi_n)}{kT}\right), \quad p = n_I \exp\left(\frac{q(\phi_p - \psi)}{kT}\right) \tag{7.31}$$

Substituting these expressions into the previous system of equations (7.16)–(7.18) yields

$$-\nabla \cdot (\varepsilon \nabla \psi) = q\left(n_I \exp\left(\frac{q(\phi_p - \psi)}{kT}\right) - n_I \exp\left(\frac{q(\psi - \phi_n)}{kT}\right) + C\right) \tag{7.32}$$

$$\nabla \cdot \boldsymbol{J}_n = qR(\psi, \phi_n, \phi_p) \tag{7.33}$$

$$\nabla \cdot \boldsymbol{J}_p = -qR(\psi, \phi_n, \phi_p) \tag{7.34}$$

The drift-diffusion constitutive relations for electron and hole current densities similarly transform to

$$\boldsymbol{J}_n = -q\mu_n n_I \exp\left(\frac{q(\psi - \phi_n)}{kT}\right) \nabla \phi_n \tag{7.35}$$

$$\boldsymbol{J}_p = -q\mu_p n_I \exp\left(\frac{q(\phi_p - \psi)}{kT}\right) \nabla \phi_p \tag{7.36}$$

The discrete approximations of the current continuity equations (7.33) and (7.34) may be poorly conditioned, however, when using this set of variables.

Exponentials of the quasi-Fermi potentials have also been employed as variables[502]. Defining

$$\nu = n_I \exp\left(-\frac{q\phi_n}{kT}\right), \quad \omega = n_I \exp\left(\frac{q\phi_p}{kT}\right) \tag{7.37}$$

the current density relations in terms of the Slotboom variables are transformed to

$$\boldsymbol{J}_n = qD_n \exp\left(\frac{q\psi}{kT}\right) \nabla\nu \tag{7.38}$$

$$\boldsymbol{J}_p = -qD_p \exp\left(\frac{-q\psi}{kT}\right) \nabla\omega \tag{7.39}$$

The numerical range for ν and ω is greater than that for n and p, which can lead to scaling problems for the numerical solution of the discretized problem. Aside from these difficulties, the basic strategy for developing an approximate model for the coupled equations is similar to the previous treatment irrespective of the choice of variables. The set of variables ψ, n and p is more common, and will be used in the following approximate analysis.

7.4 Approximate Formulation

In this section we consider the numerical discretization of the drift-diffusion system of partial differential equations in standard ψ, n, and p variables. This is the most common formulation and it forms the basis of most semiconductor design software currently in use. Accordingly, the governing equations and constitutive relations in (7.16)–(7.20) are taken and first rescaled to a dimensionally convenient form. Introducing the scaling of Markowich *et al.* [349], the coupled system of partial differential equations become

$$\lambda^2\nabla^2\psi = n - p + C \tag{7.40}$$

$$\nabla \cdot (-n\mu_n \nabla\psi + D_n \nabla n) = R(\psi, n, p) \tag{7.41}$$

$$\nabla \cdot (p\mu_p \nabla\psi + D_p \nabla p) = R(\psi, n, p) \tag{7.42}$$

where the variables are nondimensionalized and

$$\lambda^2 = \frac{\varepsilon\psi_0}{qC_0 x_0^2} \tag{7.43}$$

represents the minimum Debye length for the device. The scaling factor x_0 is the maximum length scale, C_0 corresponds to the maximum value of $|C|$ and ψ_0 is the thermal voltage kT/q. The Debye parameter is small for most device configurations and an asymptotic analysis using singular perturbation theory can be applied for one-dimensional *pn* junctions. This one-dimensional solution can be utilized in two-dimensional numerical simulations to provide a good starting guess.

The earliest numerical studies of semiconductor devices considered one-dimensional models of bipolar transistors [228]. Here, the governing equations for electrostatic potential and current continuity were decoupled in a solution algorithm that is now commonly referred to as a Gummel iteration. Convergence properties of the iteration have been analyzed [380].

Expanding the divergence expressions in (7.41), we easily identify a diffusion term involving $D_n \Delta n$, where "Δ" denotes the Laplacian, and a convective term $-\mu_n \nabla\psi \cdot \nabla n$, where $-\mu_n \nabla\psi$ is the convective flux. These two transport mechanisms of diffusion and convection "compete" since convection dominates in high field regions where $\nabla\psi$ is large, whereas in layer regions Δn is large so diffusion is significant. It is this competition between directional drift and diffusive transport processes that complicates the numerical problem. The current continuity equations are convection–diffusion equations and central difference schemes tend to generate oscillations when the convection (drift) term is strong. This difficulty can be circumvented by introducing sufficiently fine grids that are graded into the layers or by introducing additional dissipation, either directly as an added artificial diffusion term, or indirectly through upstream differencing or upwind elements as an added numerical diffusion. The effect of upwinding or artifical dissipation can also be achieved by other means. In particular, certain simplifying assumptions at the cell or element level can be used to construct a local solution and develop a difference scheme oriented with the electric field. This is the basis of the Scharfetter–Gummel approach.

Scharfetter–Gummel Scheme

Before analyzing the full system, first consider the steady one-dimensional current continuity equation for electrons with zero recombination. Equation (7.17) becomes

$$\frac{dJ_n}{dx} = 0 \tag{7.44}$$

By assuming that the electron mobility and diffusivity are constants related by the Einstein relation, so that after scaling $\mu_n = D_n$, the drift-diffusion constitutive relation (7.19) becomes

$$J_n = D_n \left(\frac{dn}{dx} + nE \right) \tag{7.45}$$

with $E = -d\psi/dx$. Substituting (7.45) in (7.44) and expanding yields the one-dimensional drift-diffusion equation for electron concentration n with drift velocity E. Clearly (7.44) implies J_n is constant and using (7.45),

$$D_n \left(\frac{dn}{dx} + nE \right) = \text{constant} \tag{7.46}$$

In the Scharfetter–Gummel discrete formulation, the electrostatic potential ψ is assumed to vary linearly over an element of the discretization. Differentiating ψ on the element, the electric field $E = -\psi'$ is constant on an element joining two nodes. Introducing the local integrating factor $\exp(Ex)$ in (7.46) and integrating on element Ω_i from left node x_i to any interior point x, the electron concentration within the element is

$$n(x) = n_i e^{(\psi - \psi_i)} + \left(\frac{J_n}{D_n E} \right)_{i+\frac{1}{2}} \left(1 - e^{(\psi - \psi_i)} \right) \tag{7.47}$$

where n_i, ψ_i denote the grid point values at node i, subscript $i + \frac{1}{2}$ denotes the (constant) value at the midpoint of element i, and we have set $E = (\psi - \psi_i)/(x - x_i)$. Using (7.47) to solve for the current density in the element

$$J_n|_{i+\frac{1}{2}} = D_n \left(\frac{2\alpha_{i+\frac{1}{2}}}{h} \right) \left(\frac{n_{i+1} e^{\alpha_{i+\frac{1}{2}}} - n_i e^{-\alpha_{i+\frac{1}{2}}}}{e^{\alpha_{i+\frac{1}{2}}} - e^{-\alpha_{i+\frac{1}{2}}}} \right) \tag{7.48}$$

where $\alpha_{i+\frac{1}{2}} = \frac{1}{2} E_{i+\frac{1}{2}} h_i = -\frac{1}{2} (\psi_{i+1} - \psi_i)$ and $h_i = x_{i+1} - x_i$ is the element length. The cell Peclet number $\alpha_{i+\frac{1}{2}}$ represents the ratio of convection to diffusion in the cell.

The conservation equation for current density J_n implies a flux balance at grid point x_i between adjacent cells in the discrete model, so that

$$[\![J_n]\!] = 0 \text{ at } x = x_i \tag{7.49}$$

where $[\![\cdot]\!]$ denotes the jump. Substituting the expression (7.48) for adjacent cells at grid point i into (7.49) yields the classical Scharfetter–Gummel discretization for the one-dimensional current continuity equation,

$$\left(\alpha_{i-\frac{1}{2}} \left(\coth \alpha_{i-\frac{1}{2}} - 1 \right) \right) n_{i-1}$$

$$- \left(\alpha_{i-\frac{1}{2}} \left(\coth \alpha_{i-\frac{1}{2}} + 1\right) + \alpha_{i+\frac{1}{2}} \left(\coth \alpha_{i+\frac{1}{2}} - 1\right)\right) n_i$$

$$+ \left(\alpha_{i+\frac{1}{2}} \left(\coth \alpha_{i+\frac{1}{2}} + 1\right)\right) n_{i+1} = 0 \qquad (7.50)$$

This scheme provides accurate solutions without the numerical oscillations frequently encountered with more standard central difference approximation of the first derivative.

Numerical Oscillations

It is instructive to examine the drift-diffusion equation with the assumption of a constant field E, and analyze the approximate solutions given by standard central differences or Galerkin finite elements on a uniform grid. For convenience and consistency with the transport literature on the convection–diffusion equation, let the electron velocity ψ' be denoted by v, the diffusivity D_n by D, and the transport variable(n or p) be denoted by u. The corresponding model convection–diffusion equation can be conveniently expressed as

$$vu' - Du'' = 0 \qquad (7.51)$$

By introducing a uniform discretization with grid points x_i and mesh size h, the finite difference equation at interior grid point i becomes

$$v\,\frac{(u_{i+1} - u_{i-1})}{2h} - D\,\frac{(u_{i+1} - 2u_i + u_{i-1})}{h^2} = 0 \qquad (7.52)$$

This central difference equation is identical to that obtained using a finite element Galerkin formulation with linear basis functions and is second-order accurate; that is, the truncation error obtained by Taylor series expansion of (7.52) is $O(h^2)$. Moreover, the linear interpolant of the nodal solution converges asymptotically to the exact solution with an L^2 error (mean square integral error) that is also $O(h^2)$. Writing

$$s_{i-\frac{1}{2}} = \frac{(u_i - u_{i-1})}{h}, \qquad s_{i+\frac{1}{2}} = \frac{(u_{i+1} - u_i)}{h} \qquad (7.53)$$

for the respective slopes of the approximation on cells i and $i+1$, the difference equation (7.52) can be rewritten

$$\frac{v}{2}\left(s_{i+\frac{1}{2}} + s_{i-\frac{1}{2}}\right) = \frac{D}{h}\left(s_{i+\frac{1}{2}} - s_{i-\frac{1}{2}}\right) \qquad (7.54)$$

Regrouping terms

$$\left(\frac{v}{2} - \frac{D}{h}\right) s_{i+\frac{1}{2}} = -\left(\frac{v}{2} + \frac{D}{h}\right) s_{i-\frac{1}{2}} \qquad (7.55)$$

from which it is seen that the signs of the slopes between adjacent elements must be opposite if the cell Peclet condition

$$\frac{vh}{D} \leq 2 \tag{7.56}$$

is violated. Therefore, the numerical solution must oscillate when this condition fails to hold. This result implies that the central difference scheme for the drift-diffusion equation is oscillatory if the grid is too coarse or if the drift velocity is too large relative to diffusion. Note that the cell requirement (7.56) also implies that the coefficient matrix defined by (7.52) is diagonally dominant.

It is easy to construct a Galerkin finite element scheme for this model problem[38]. Simply multiplying (7.51) by a weight (test) function w and integrating by parts we get: find u satisfying the end conditions and such that

$$\int_0^1 (Du'w' + vu'w)\,dx = 0 \tag{7.57}$$

holds for all admissible weight functions w with $w(0) = w(1) = 0$. Introducing a discretization of E elements this implies

$$\sum_{e=1}^{E} \int_{\Omega_e} (Du_e'w_e' + vu_e'w_e)\,dx = 0 \tag{7.58}$$

The element contributions follow on introducing the element basis functions. For example, the linear Lagrange basis yields the same tridiagonal system (7.52) as before.

It is instructive to examine the transient drift problem in one dimension. The governing transport equation becomes

$$\frac{\partial u}{\partial t} + v\frac{\partial u}{\partial x} = 0 \tag{7.59}$$

Recalling the definition of the material derivative, it follows from (7.59) that $Du/Dt = 0$ when $\frac{\partial x}{\partial t} = v$. That is, the carrier concentration u is constant along the characteristic lines $x - vt = \xi$ =constant. Clearly (7.59) is a first-order hyperbolic equation describing signal propagation due to drift velocity v in the absence of diffusion. It follows that, for $v > 0$, signals propagate from left to right. This suggests that a backward or upwind difference scheme might be more appropriate.

We also remark that the central difference scheme for the stationary problem $v\frac{\partial u}{\partial x} = 0$ implies that $u_{i+1} = u_{i-1}$ which certainly admits the correct solution u=constant. However, since alternate grid points are decoupled an oscillatory solution $u_i = (-1)^i$ is also admissible. More specifically, let us simply set $u_i = p^i$ in the above equation. Then $u_{i+1} - u_{i-1} = p^{i+1} - p^{i-1} = 0$ implies $(p^2 - 1) = 0$ or $p = \pm 1$ and we have the general solution $u_i = A + B(-1)^i$. The two solution modes $(\pm 1)^i$ are a consequence of approximating a first-order differential operator by a second-order difference operator.

When a backward difference approximation from grid point i to $i-1$ is used for the convective term in (7.51), the finite difference equation becomes

$$v\,\frac{(u_i - u_{i-1})}{h} - D\,\frac{(u_{i+1} - 2u_i + u_{i-1})}{h^2} = 0 \tag{7.60}$$

and the same analysis reveals that this scheme will not oscillate, irrespective of the size of the Peclet number. Introducing Taylor series expansions about grid point i for the difference operator on the left in (7.60)

$$L_h u|_i = (vu_x - Du_{xx})\,|_i - \frac{vh}{2} u_{xx}|_i + O(h^2) \tag{7.61}$$

Hence, this difference scheme $L_h u|_i = 0$ can be interpreted as a first-order accurate approximation of the original drift-diffusion equation, or as a second-order accurate approximation to a modified convection–diffusion equation

$$vu_x - \left(D + \frac{vh}{2}\right) u_{xx} = 0 \tag{7.62}$$

The contribution $\frac{1}{2} vhu_{xx}$ from the truncation error of (7.60) now acts as a numerical diffusion to stabilize the oscillations. The additional numerical diffusivity $\frac{vh}{2}$ scales linearly with the convective flux coefficient v, acting to smear steep gradients and suppress oscillations.

Finally, as an alternative formulation, multiply the convection–diffusion equation (7.51) by an integrating factor $\exp(-vx/D)$, to rewrite the equation in self-adjoint form as

$$\left(e^{-vx/D} u'\right)' = 0 \tag{7.63}$$

which can be centrally differenced or used to construct a Galerkin finite element scheme. Let us consider the latter approach. First, take the weighted integral projection of (7.63) with test function w. Integrating by parts in the associated weighted residual statement, the weak variational form becomes: find u satisfying the essential boundary conditions and such that

$$\int_\Omega e^{-vx/D} u' w' dx = 0 \tag{7.64}$$

holds for all admissible test functions w. Note that the exponential integrating factor enters in the corresponding integral expression.

A piecewise-linear finite element basis is introduced on a discretization of E elements and the discrete model for (7.64) is constructed by accumulating element integral contributions

$$\sum_{e=1}^{E} \int_{\Omega_e} e^{-\frac{v}{D}x} u_e' w_e' dx = 0 \tag{7.65}$$

where u_e, w_e are the element trial and test functions. The resulting algebraic system is symmetric and numerical experiments reveal that the approximation with linear elements is nonoscillatory for all Peclet numbers. Similar results are obtained using a central finite difference model for (7.63). Furthermore, the approximate solution is exact at the grid points; that is, it interpolates the exact solution. One distinction between this formulation and the standard Galerkin approach for (7.51) is the presence of the exponential weighting factor. If the exponential function is grouped with the test function, the method can be interpreted as a Petrov–Galerkin method, in which the basis functions for the trial space and test space differ [80]. This formulation is similar to that obtained in the Scharfetter–Gummel scheme described earlier, since both involve, in effect, the use of an exponential integrating factor.

Higher Dimensions

The Scharfetter–Gummel finite difference strategy can be extended to higher dimensions to produce analogous nonoscillatory discrete schemes. Simply applying the Scharfetter–Gummel approach in the respective x and y directions for the two-dimensional problem may lead, however, to a scheme that is too dissipative. This implies that sharp layers in the concentration variables may be excessively smeared by numerical dissipation in the scheme. Ideally, it is desirable to add dissipation in the direction of the flux vector to control the oscillations, but limit the artificial dissipation in the transverse (crosswind) direction. This observation has led to the use of directed flux-upwind schemes similar to the streamline-upwind techniques of computational fluid dynamics.

In practice, efficient solutions will be obtained if nonuniform grids are employed with the mesh graded in regions where sharp layers occur. Mapping techniques can be applied to take a nonuniform finite difference grid in the physical domain to a uniform grid in the computational domain. Finite volume and finite element schemes, however, are more generally applicable, particularly when local adaptive grid refinement is considered. This topic is discussed further in Chapter 9.

A weighted residual form of the current continuity equations can be introduced to construct a Petrov–Galerkin formulation of the drift-diffusion problem. As noted previously, this formulation introduces different trial and test bases in the integral statement. The effect of using the modified test basis is to add dissipation in the direction of the carrier drift velocity, thereby suppressing numerical oscillations. This is similar to a multi-dimensional Scharfetter–Gummel scheme, with the added flexibility of allowing irregular, nonuniformly graded meshes, as well as controlling the added dissipation.

Consider the weighted residual form of the electron continuity equation

$$\int_\Omega (\nabla \cdot \boldsymbol{J}_n)\, w^* dx dy = \int_\Omega R w^* dx dy \tag{7.66}$$

for admissible test functions $w^* \in G$, the space of test functions. For convenience, write the test function w^* as $w + q^*$ where w is the standard Galerkin test function and q^* is an upwind-biased test function, to be specified. Substituting in the weighted residual statement, separating contributions, and integrating by parts

$$\int_\Omega \boldsymbol{J}_n \cdot \nabla w \, dx dy + \int_\Omega (-\nabla \cdot \boldsymbol{J}_n + R)\, q^* dx dy = -\int_\Omega R w \, dx dy \tag{7.67}$$

The boundary integral arising out of the integration by parts procedure is zero, since $w = 0$ on the ohmic contacts where Dirichlet conditions are specified, and the normal component of the current density satisfies $\boldsymbol{J}_n \cdot \boldsymbol{n} = 0$ on the remaining boundary, where $\boldsymbol{n}$ is the outward unit normal (recall the MOSFET example in Figure 7.1).

For the part of the test function biased in the direction of the electron drift velocity, take

$$q^* = \frac{D_n^*}{|\boldsymbol{v}_n|^2} \boldsymbol{v}_n \cdot \nabla w \tag{7.68}$$

where $\boldsymbol{v}_n$ is the electron drift velocity and $D_n^* \geq 0$ is a parameter controlling the numerical dissipation. Substituting in the weighted residual statement,

$$\begin{aligned} &\int_\Omega (D_n \nabla n \cdot \nabla w - n \boldsymbol{v}_n \cdot \nabla w)\, dx dy \\ &+ \int_\Omega (-\nabla \cdot \boldsymbol{J}_n + R)\, \frac{D_n^*}{|\boldsymbol{v}_n|^2} \boldsymbol{v}_n \cdot \nabla w \, dx dy = -\int_\Omega R(\psi, n, p) w \, dx dy \end{aligned} \tag{7.69}$$

constitutes a Petrov–Galerkin formulation for the electron continuity equation. The next step is to construct a finite element partition of the domain and replace the integrals in (7.69) by a sum of element contributions

$$\sum_e \int_{\Omega_e} (D_n \nabla n \cdot \nabla w - n\boldsymbol{v}_n \cdot \nabla w)\, dxdy$$

$$+\sum_e \int_{\Omega_e} (-\nabla \cdot \boldsymbol{J}_n + R) \frac{D_n^*}{|\boldsymbol{v}_n|^2} \boldsymbol{v}_n \cdot \nabla w dxdy = -\sum_e \int_{\Omega_e} Rw dxdy \qquad (7.70)$$

Substituting the finite element expansions for the electrostatic potential and electron concentrations in (7.70), and setting w_h as the finite element test basis, leads to the approximate problem. When a piecewise-polynomial basis of linear triangles or bilinear rectangles is chosen, $\nabla \cdot \boldsymbol{J}_n$ reduces in the interior of each element to $-\boldsymbol{v}_n \cdot \nabla n$, so that the finite element statement of the variational problem for the device domain Ω becomes: find $n_h \in H^h$ satisfying $n_h = n_h^D$ on the contact surface $\partial\Omega_c$ and such that

$$\sum_e \int_{\Omega_e} \left(D_n \nabla n_h \cdot \nabla w_h + \frac{D_n^*}{|\boldsymbol{v}_n|^2} (\boldsymbol{v}_n \cdot \nabla n_h)(\boldsymbol{v}_n \cdot \nabla w_h) \right) dxdy$$

$$-\sum_e \int_{\Omega_e} (n_h \boldsymbol{v}_n \cdot \nabla w_h)\, dxdy = -\sum_e \int_{\Omega_e} R \left(w_h + \frac{D_n^*}{|\boldsymbol{v}_n|^2} (\boldsymbol{v}_n \cdot \nabla w_h) \right) dxdy \qquad (7.71)$$

for all $w_h \in H^h$ with $w_h = 0$ on $\partial\Omega_c$. Here the electron mobility and diffusion coefficient are taken to be approximately constant on each element. A similar weighted-residual statement can be developed for the hole current continuity equation (7.18) or (7.41). The normal flux conditions (7.26) are automatically approximated as natural boundary conditions on the sides of $\partial\Omega_s = \partial\Omega - \partial\Omega_c$ by the variational statement. From equation (7.71) it is clear that this Petrov–Galerkin approach introduces a numerical dissipation tensor of the form

$$\boldsymbol{D}^* = \frac{D_n^*}{|\boldsymbol{v}_h|^2} (\boldsymbol{v}_n \otimes \boldsymbol{v}_n) \qquad (7.72)$$

where $\boldsymbol{v}_n$ is the electron drift velocity and $\otimes$ denotes the tensor product. Thus, the numerical dissipation is added in the direction of the carrier drift velocity.

In the following numerical studies, $D_n^* = (\bar{K}_\xi + \bar{K}_\eta)D_n$ where $\bar{K}_\xi = \alpha_\xi(\coth \alpha_\xi - \alpha_\xi^{-1})$, $\bar{K}_\eta = \alpha_\eta(\coth \alpha_\eta - \alpha_\eta^{-1})$ and $\alpha_\xi = (\boldsymbol{E} \cdot \hat{\boldsymbol{e}}_\xi)h_\xi/2$, $\alpha_\eta = (\boldsymbol{E} \cdot \hat{\boldsymbol{e}}_\eta)h_\eta/2$. This particular choice is motivated by the optimal one-dimensional result at the nodes using exponential weighting discussed previously. Mesh parameters h_ξ, h_η correspond to the lengths of the bisectors of the opposite sides of a quadrilateral and $\hat{\mathbf{e}}_\xi, \hat{\mathbf{e}}_\eta$ are unit vectors aligned in these respective directions. Quantities α_ξ, α_η resemble directional Peclet numbers and this choice of numerical dissipation should be superior to the standard approach of simply adding numerical dissipation in the x and y directions based on the one-dimensional Scharfetter–Gummel discretization. The present scheme has been termed a Flux-Upwind Petrov–Galerkin method (FUPG) because of its close similarity to Streamline Upwind (SUPG) schemes in computational fluid dynamics.

7.5 Grid Refinement

Mesh grading and grid refinement offer the possibility of computing accurate, reliable, inexpensive solutions in regions by tailoring the mesh to fit the solution. The basic idea behind adaptive refinement is to solve the problem on a coarse grid and then refine the mesh in regions where the solution gradients are large. Various *a posteriori* error analyses for post-processing the solution to guide refinement have been developed, and sophisticated algorithms for refinement of the grid have been implemented [82]. Some aspects of this subject are explored further in Chapter 9. Attention is now focused on the use of adaptive refinement in the context of device simulation, but the procedures also apply to the process modeling problems described in later chapters.

Recall the model drift-diffusion problem $vu' - Du'' = 0$ in (7.51) with $u(0) = 0$, $u(1) = 1$ and $v/D \gg 1$. The solution has a boundary layer near $x = 1$. As demonstrated previously, when this problem is solved on a coarse uniform grid with central differences and the cell condition $vh/D \leq 2$ is violated, then the inter-element solution must oscillate. The oscillations may be interpreted as resulting from the inability of the method to resolve the layer using a coarse grid. In this sense the failure of the method to produce a smooth solution provides valuable information to the numerical analyst. Simply grading the mesh into the layer permits the layer to be accurately resolved (Figure 7.2) and the oscillations suppressed. Thus, grid refinement or mesh grading is an appropriate strategy for alleviating the problem of numerical oscillations associated with the drift-diffusion equation.

The adaptive refinement algorithm begins with a solution on a coarse grid which is post-processed to identify the layer region. The grid is locally refined in this subregion and the solution process is repeated. This solution-local refinement iteration is repeated until the final graded mesh

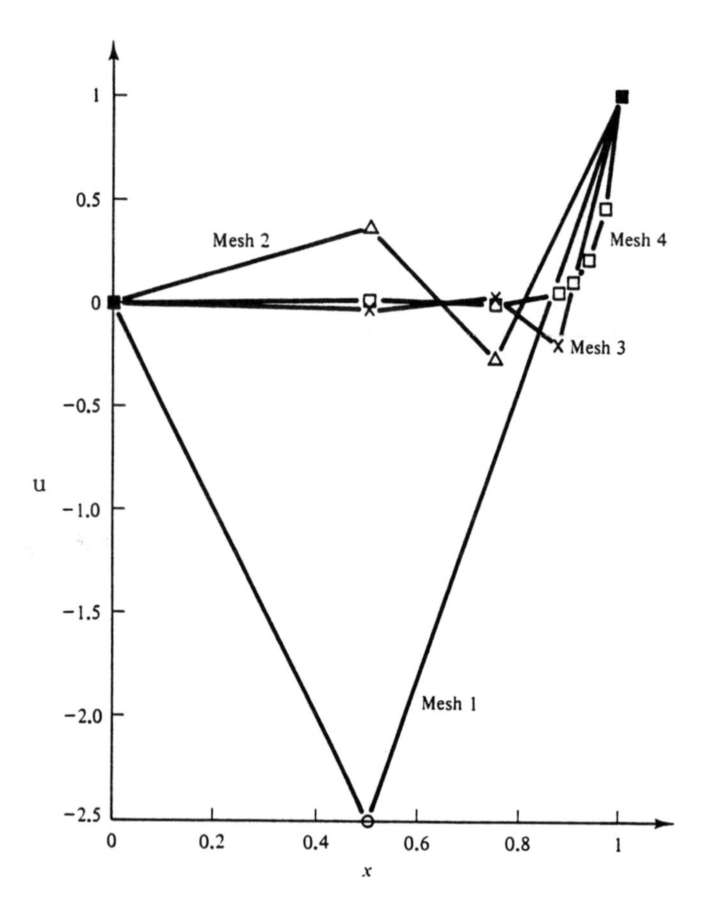

Figure 7.2: Numerical solutions to the drift-diffusion equation using adaptive refinement.

and non-oscillatory solution is obtained. The approach can be extended to higher dimensions and necessitates an appropriate data structure to support refinement. Several adaptive refinement schemes with associated data structures have been used for the solution of boundary-value and evolution problems[26, 34, 95]. An example of a data structure for 2D adaptive refinement and some test calculations are given next.

Refinement Tree Structure

Consider the problem of adaptive refinement for two-dimensional device modeling. The natural refinement of the triangular or quadrilateral element is to a quartet of subelements as indicated in Figure 7.3. Reference "parent" triangle and square are shown in the lower part of the figure together with

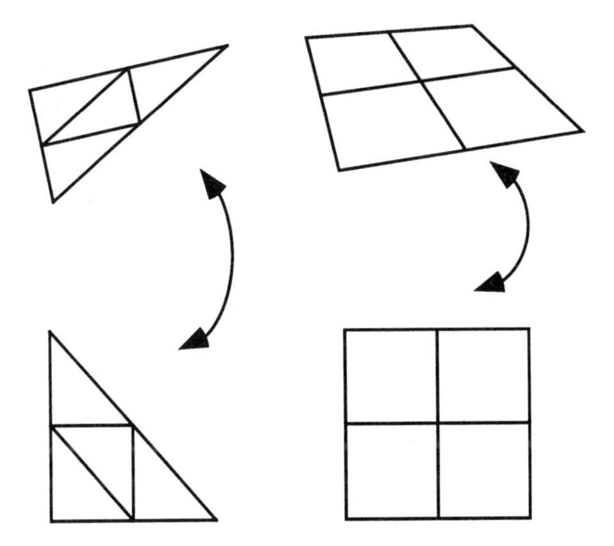

Figure 7.3: Refinement of reference elements and associated map to physical plane.

the quartet of congruent subelements for the regular reference domain. The isoparametrically mapped (linear and bilinear) partitions for a general triangle and quadrilateral are indicated in the upper part of the figure. Each of the four new subelements generated is a "child" of the original "parent" element. The net result is a quadtree data structure in which the nodes in the tree correspond to elements in the mesh and branches extend downward from elements to their repeatedly refined subelements (Figure 7.4). The "leaf" nodes correspond to unrefined elements in the finest mesh generated.

It is desirable that the transition from large to small elements be gradual to avoid ill-conditioning in the associated algebraic system. One strategy is to require that elements sharing a common side be no more than one level apart in the quadtree, since the grid transition should not be too abrupt. Moreover, continuity constraints on the approximation should be applied at the semi-detached constraint nodes. This implies that a single element cannot be refined more than once without refinement of adjacent elements (see Figure 7.5). Therefore, before an element can be refined, its neighbors must be checked; if a neighbor is at a lower level, then it must first be refined, and so on, recursively. This rule requires that the data structure contains information about an element's neighbors and their level in the quadtree. The corresponding octree for 3D computations is considered in Chapter 9.

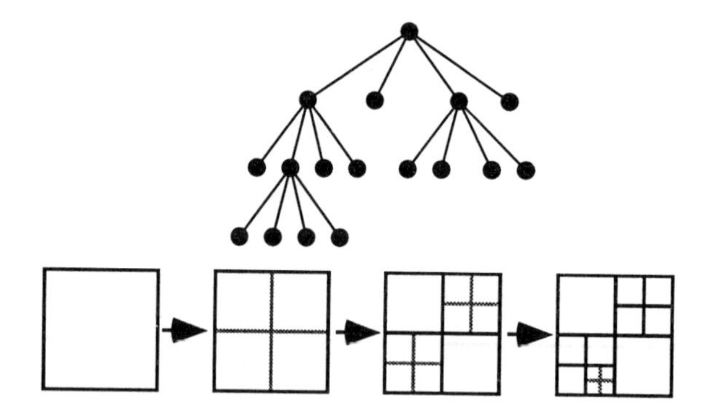

Figure 7.4: Quadtree data structure for refinement scheme and corresponding mesh sequence.

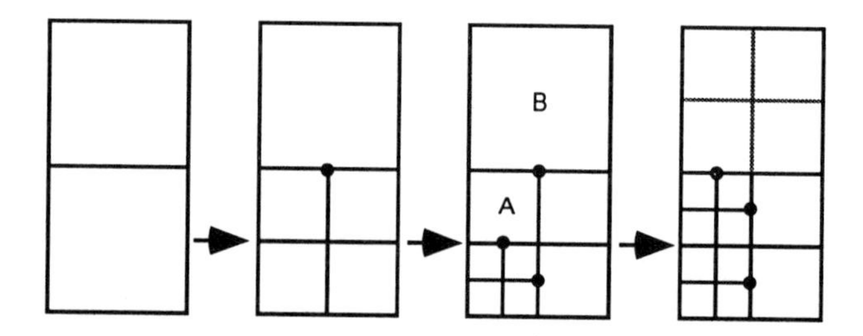

Figure 7.5: One-level rule for adjacent element refinement: element A cannot be refined until element B is refined. Constrained nodes are marked by symbol ".".

Following each successive refinement of the grid, a finite element or finite volume system is solved on the current mesh. The finite element calculation involves computation of the element matrix contributions, assembly of the global matrix, and sparse linear system solution. Element number, global node number, nodal coordinate information, material properties, etc. are required for these calculations. In the interests of efficiency, the algorithm should generate the necessary information rapidly from a minimal data base. It is easy to design "greedy" algorithms in which large quantities of data are generated at each refinement step. This wastes storage and prohibits practical use on engineering workstations and most mainframe computers. Instead, by knowing the current refinement level for a given element, the tree can be traversed in the reverse direction to find the original background element. The nodal coordinate information from this macro-element can then be used to compute nodal coordinates for all of its subelements at any level in the tree. Material properties and other data can be similarly transferred from the macro-element to the active elements of the tree. The quadtree data structure uses small arrays of fixed size and still permits efficient computation of the node and element data. The required calculations are fast and the use of storage is very efficient.

When an element is refined it becomes inactive and is replaced by the quartet of active subelements. There are several strategies for storing the relevant data. For example, an initial level-zero coarse discretization of the domain may be defined, consisting of background elements with their associated nodal coordinates and material data. Refinement level, element and node information are prescribed in two arrays for each refined element, and the active data set for the computation scheme can be regenerated in a constant multiple of the time for accessing data in these arrays. During refinement, the position of a subelement on the quadtree is needed, together with certain neighbor, node and edge information. An integer array containing tree pointers and nodal numbers defining the connectivity of elements may be constructed for refinement. This array specifies level l, parent p, macro-element m and child pointer c_j of element j; node numbers of the parent element, edge node numbers, center node number and other node numbers depending on the element type. The second array contains information on the types of nodes, i.e. an original interior node, a boundary node, a constrained node between elements at different levels, etc.

As seen in Figure 7.5, refinement with the advocated scheme will introduce new nodes on the boundary of the refined element. Consider the case where the adjacent element is not refined. The value of the solution at the new edge node on the interface between refined and unrefined elements should be constrained to guarantee continuity of the global approximation across the interface. For example, if a linear finite element basis is used on triangles or a bilinear basis on quadrilaterals, then the new midpoint node is constrained to be the average of the values of the two end nodes on that edge.

Adaptive refinement can also be used for transient problems. First, the grid is refined based on the initial data, and then it is adaptively refined at subsequent time intervals so as to follow the evolving solution. Note that, for such a procedure to be effective, the scheme should also permit coarsening of the grid. This can be easily accommodated in the quadtree data structure, so that the grid is automatically refined in front of an advancing layer and coarsened behind it. Adaptive refinement impacts on the 3D problem most significantly. We elaborate on the associated octree data structure and give an example in Chapter 9.

Refinement with Iteration

Recall that the Gummel iteration for the decoupled system of device equations requires repeated successive solution of the discrete equations for the electrostatic potential and for the electron and hole concentrations. These decoupled linear systems can be solved by sparse elimination or by iterative linear solvers. Iterative solution schemes are attractive for several reasons. First, the solution at the previous Gummel iteration provides a good starting iterate for the next linear system solution. Second, the electrostatic potential system is symmetric and positive definite. Therefore, it is well suited to solution by iterative methods such as successive over-relaxation or conjugate gradient schemes as discussed in Chapter 13. The carrier transport system is nonsymmetric and can be treated using a Lanczos type of scheme such as the biconjugate gradient method. Finally, with adaptive refinement, the solution iterate on a coarse grid can be easily interpolated to a fine grid. This interpolation is a necessary part of multigrid algorithms, which can also be incorporated in the scheme to accelerate convergence.

Since new nodes are introduced when a parent element is subdivided, the nodal numbering can vary dramatically across the subelements in a quartet. The bandwidth of the resulting sparse matrix depends upon the local difference in node numbering on a patch. Therefore, the adaptive refinement scheme may severely degrade the band structure of the matrix and standard band solvers will perform very inefficiently. The grid can be renumbered at each refinement step using re-ordering schemes, but this is not efficient if numerous refinement steps are taken. Similar difficulties arise with frontal elimination solvers, since they rely on close element numbering within a patch if the frontwidth is to be small. e.g. Subdividing a parent element may introduce new elements with large element numbers adjacent to old elements that have low element numbers. Once again, renumbering can be introduced to alleviate the problem and efficient frontal elimination can then be carried out. Alternatively, sparse iterative solvers can be introduced but sparsity structure is still an issue. However, several sparse matrix formats are applicable for iterative solution using packages such as ITPACK or NSPCG [230]. Pointers identify the non-zeros in each row of the matrix so the global assembled matrix is stored in a compact form. The matrix-vector products in the conjugate gradient method, for instance, can then be written compactly using only the non-zero entries in a given row.

As noted above, the current continuity equations are not self-adjoint and therefore the resulting linearized system is not symmetric. In the numerical results shown later, a biconjugate gradient scheme is used. For finite volume and finite element implementations, these "gradient" type iterative methods can be recast to an element-by-element form that does not require assembly of the global matrix and allows the use of local element matrices directly in the solution algorithm. There is, in general, no guarantee of convergence with iterative schemes for nonsymmetric problems. Numerical experience nevertheless indicates that breakdown of iterative methods is rare and can sometimes be treated by restarting if the iteration stagnates or divergence is detected. Certain "look ahead" strategies to detect breakdown may also be introduced.

To illustrate the element-by-element calculation, consider the conjugate gradient method applied to the electrostatic potential system $\boldsymbol{A}\boldsymbol{\psi} = \boldsymbol{b}$ where $\boldsymbol{A}$ corresponds to the discretized Laplacian and $\boldsymbol{b}$ is computed from the previous Gummel iterate of the decoupled PDE system. One of the major computations in the gradient iterative solver involves repeated evaluation of the residual

$$\boldsymbol{r}^{(n)} = \boldsymbol{b} - \boldsymbol{A}\boldsymbol{\psi}^{(n)} \tag{7.73}$$

where $\boldsymbol{\psi}^{(n)}$ is the iterate. Writing (7.73) as a sum of element contributions,

$$\boldsymbol{r}^{(n)} = \sum_e \hat{\boldsymbol{b}}_e - \left(\sum_e \hat{\boldsymbol{A}}_e\right) \boldsymbol{\psi}^{(n)} \tag{7.74}$$

where $\hat{\boldsymbol{A}}_e$ and $\hat{\boldsymbol{b}}_e$ are the contributions of element e to the global matrix $\boldsymbol{A}$ and the right-hand side $\boldsymbol{b}$, expanded to full system size. That is, $\hat{\boldsymbol{b}}_e = \boldsymbol{B}_e^T \boldsymbol{b}_e$ and $\hat{\boldsymbol{A}}_e = \boldsymbol{B}_e^T \boldsymbol{A}_e \boldsymbol{B}_e$ where $\boldsymbol{B}_e$ is the Boolean or adjacency matrix for the map between local and global numbering systems and $\boldsymbol{b}_e$, $\boldsymbol{A}_e$ are the local element vector and matrix contributions in a local numbering system for element e. The local element contributions $\boldsymbol{b}_e$, $\boldsymbol{A}_e$ can be evaluated in parallel for all elements in the mesh and accumulated into the global residual vector as

$$\boldsymbol{r}^{(n)} = \sum_e \hat{\boldsymbol{r}}_e = \sum_e \boldsymbol{B}_e^T \boldsymbol{r}_e \quad \text{with} \quad \boldsymbol{r}_e = \boldsymbol{b}_e - \boldsymbol{A}_e \boldsymbol{\psi}_e \tag{7.75}$$

where $\boldsymbol{\psi}_e = \boldsymbol{B}_e \boldsymbol{\psi}$ extracts the element values from the global vector $\boldsymbol{\psi}$. It is important to note that the matrix–vector products are the computationally intensive part of these calculations and can be vectorized or parallelized over the elements.

A fundamental problem still remains: under even moderate applied voltages, the decoupled Gummel iteration may not converge on coarse meshes with standard non-dissipative finite difference or finite element schemes.

This is largely due to the feedback of oscillatory instabilities from the discretized current continuity equations. The problem can be circumvented by the introduction of numerical dissipation via the Scharfetter–Gummel approach or similar upwind strategies. This stabilizes the problem, but at the expense of introducing artificial dissipation that degrades the accuracy of the solution and the resolution of layers. The approach can, however, be combined with the adaptive refinement procedure to considerable advantage. As the mesh is selectively refined to resolve steep fronts, the amount of numerical dissipation, which depends on the local grid size, is reduced proportionally. At the conclusion of a sequence of solution and adaptive refinement steps, the amount of numerical dissipation remaining will be small and the layers will be well approximated. Continuation techniques in which the applied voltage is incrementally increased to the desired level can also be introduced to enhance convergence, as discussed in the next section.

7.6 Numerical Modeling Results

A long-channel MOSFET (recall Figure 7.1) is considered and the solution is computed using adaptive refinement with the flux-upwind Petrov–Galerkin (FUPG) scheme. The flux-upwind scheme introduces less cross-wind diffusion into the solution than does the standard upwind scheme. A surface plot of the final electrostatic potential solution is given in Figure 7.6.

When large voltages are applied directly to a device, the Gummel iteration may fail. Convergence is sensitive to the strength of the nonlinear coupling between the electrostatic potential and the current continuity equations, as well as the applied boundary data. A general strategy for making nonlinear iterations more robust is to apply continuation in a parameter. The parameter is first set to a value for which a solution can be easily obtained and is then incrementally adjusted or continued towards the final desired value. The previous MOSFET results are obtained using incremental continuation in applied voltage. The initial coarse grid solution is first obtained for a lower applied voltage. This provides a starting iterate for the solution at the next voltage, and so on, until the actual applied voltage is reached. The mesh refinement process can be carried out conveniently in conjunction with this continuation scheme. For example, the grid can be adaptively refined to produce an accurate solution at a given voltage and then the voltage is increased and refinement continued. This produces a more efficient overall scheme when several solutions at increasing applied voltages are needed to trace out an I–V curve for a device.

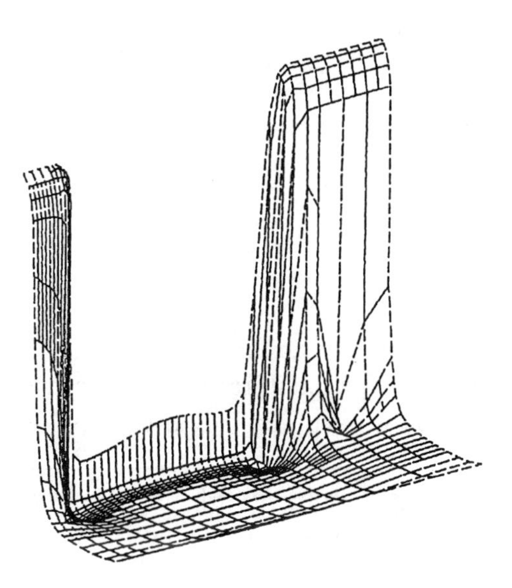

Figure 7.6: Electrostatic potential surface plot using adaptive refinement and flux-upwinding (from [484]).

In this incremental continuation scheme, the potential solution at the previous voltage is the starting iterate for the next voltage level. An improved starting iterate can be generated by use of Euler–Newton continuation along a tangent to the curve $\psi(V)$ as follows. Let the discretized nonlinear system of equations for potential vector $\boldsymbol{\psi}(V)$ at applied voltage V be expressed parametrically in terms of voltage by

$$\boldsymbol{F}\left(\boldsymbol{\psi}(V), V\right) = \mathbf{0} \tag{7.76}$$

Differentiating with respect to the voltage parameter,

$$\frac{\partial \boldsymbol{F}}{\partial \boldsymbol{\psi}} \frac{\partial \boldsymbol{\psi}}{\partial V} + \frac{\partial \boldsymbol{F}}{\partial V} = \mathbf{0} \tag{7.77}$$

so that

$$\frac{\partial \boldsymbol{F}}{\partial \boldsymbol{\psi}} \frac{\partial \boldsymbol{\psi}}{\partial V} = -\frac{\partial \boldsymbol{F}}{\partial V} \tag{7.78}$$

This can be interpreted as an ODE in the continuation parameter and may be integrated approximately using schemes such as the forward Euler method discussed in the circuit and process simulation chapters. Introducing the forward Euler integrator for (7.78), the continuation step for specified voltage increment δV approximately satisfies

$$\frac{\partial F}{\partial \psi}\delta\psi = -\delta F \tag{7.79}$$

The starting iterate for electrostatic potential at the new voltage level is now obtained from the solution $\delta\psi$ of (7.79) as

$$\psi^0(V + \delta V) = \psi(V) + \delta\psi \tag{7.80}$$

Using this starting iterate, the nonlinear potential equation (7.76) is now solved using, for instance, Newton iteration. Successive approximation using the electrostatic solution is then applied to the nonlinear carrier equations in the second part of the Gummel iteration. The current–voltage drain characteristics of the long-channel MOSFET are given in Figure 7.7, based on simulation using this continuation scheme.

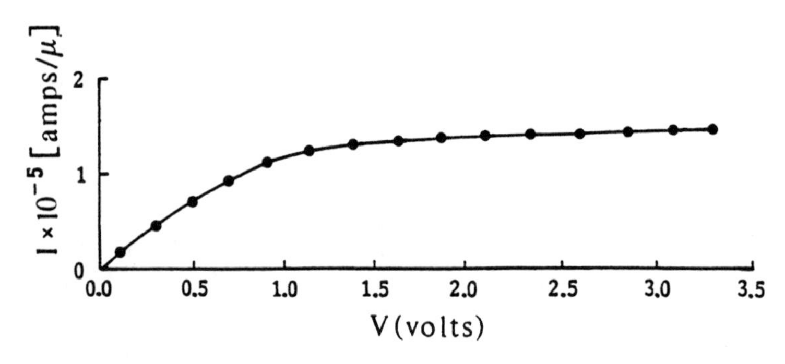

Figure 7.7: *I–V* characteristics for long-channel MOSFET using incremental continuation in applied voltage V.

Three-Dimensional Device Simulation

By reducing the device size, circuit density can be increased and therefore overall circuit speed can be improved and cost can be decreased. Three-dimensional effects, however, become increasingly important as the size of the device decreases and as novel structures are introduced in VLSI technology. Moreover, 3D effects become more significant due to the proximity of adjacent devices and layering of devices in 3D. It is important to determine how 3D effects scale as the device becomes smaller. For example, short-channel effects cause the threshold voltage to depend in a complex 3D manner on device dimensions. The 3D nature of internal device potentials and concentrations can also strongly influence sub-threshold effects such as "punch-through" and "leakage". Current–voltage relations, parasitic capacitance and resistances around contacts may also be influenced significantly by 3D effects. Device sensitivity to process parameters is increased substantially due to the tighter interaction of field, channel and other dopings as indicated in the chapters on process modeling. Similarly, in small scale bipolar junction transistor (BJT) devices, the average base transit time depends on the emitter and base geometries. Finally, we remark that three-dimensional effects are obviously important in interconnects.

Novel structures such as the trench capacitor discussed in the chapter on silicon oxidation also exhibit 3D effects. The purpose of the trench is to electrically isolate adjacent MOSFETs or BJTs, providing better protection from leakage and latch-up. The voltages on nearby devices, however, may lead to depletion or inversion along the trench walls and the extent of these depletion regions will depend upon the 3D geometry of the trench as well as 3D doping profiles. Leakage will be strongly influenced by the 3D potentials in the vicinity of the trench especially near the corners. As noted in the discussion of oxidation in Chapter 15, local thinning of the oxide layer in corners may promote leakage. Similar effects are observed in trench capacitors for DRAMs.

Simulation can provide important information on 3D scaling effects. Present generation simulators vary in the complexity of their solution techniques, ranging from solution of the single nonlinear Poisson equation to the full set of drift-diffusion equations, with some limited to modeling a particular device and others to certain geometries. Both finite difference and finite element or finite volume methods are employed. Generalization of the present two-dimensional capability to include adaptive refinement in general 3D geometries is a topic of current research, as is the extension to more complex physical models such as the hydrodynamic model discussed in the next chapter.

Device simulation in three dimensions is formally similar to that described for the two-dimensional case, with the obvious complications associated with describing the geometry and discretization, as well as the exponential increase in the computational work required. For realistic applications, the problem is both computationally and storage intensive, so that special strategies such as adaptive refinement now become essential to developing a viable simulator. The 3D problem may be approached with this strategy in mind: an initial grid of quadrilateral brick (hexahedral) elements is constructed using mappings of uniform grids from the reference cube. Alternatively, one can generate tetrahedral triangulations of the 3D device using extensions of the Delaunay idea. The quadtree data structure for the triangle and quadrilateral elements generalizes to an octree data structure for the quadrilateral brick element. The 3D brick is subdivided into an octet of subelements, and pointers at each level of the octree relate the "parent" elements to the subelement "children." Constraints are now applied on both edges and faces between adjacent refined and unrefined elements. Since there are 6 face neighbors and 12 edge neighbors for a quadrilateral brick element, the neighbor information is more extensive and the data structure more complex. These ideas are discussed further in Chapter 9.

Nonlinear Potential Model

To introduce the main ideas for $3D$ simulation, let us consider the nonlinear Poisson equation. This will permit analysis of an interesting subclass of 3D device problems which can be solved efficiently in a workstation environment. The extension to the fully coupled system with Gummel iteration follows as in the 2-D case, but is computationally more intensive.

The nonlinear equation for electrostatic potential is

$$-\nabla \cdot (\varepsilon \nabla \psi) = \rho \tag{7.81}$$

where ε is permittivity, ψ is the electrostatic potential and ρ is the charge density. For convenience, define the electrostatic potential relative to a reference vacuum potential ψ_V as

$$\psi = \psi_V - \phi_{si}/q \tag{7.82}$$

where ϕ_{si} is the work function of intrinsic silicon and q is the magnitude of charge for an electron. In insulating regions, the charge density is zero so (7.81) reduces to Laplace's equation $\Delta\psi = 0$. In the semiconductor regions

$$\rho = q\,(p - n + N_D - N_A) + \delta(r - r_0)Q_{ss} \tag{7.83}$$

where N_D is donor concentration, N_A is acceptor concentration, δ is the Dirac delta, and Q_{ss} is the density of fixed charge on a 3D surface at radial distance $r = r_0$. Using Boltzmann statistics to model electron and hole concentrations

$$n = n_I e^{q(\psi - \phi_n)/kT}, \quad p = n_I e^{q(\phi_p - \psi)/kT} \tag{7.84}$$

where n_I is intrinsic electron concentration, k is the Boltzmann constant, T is the temperature in degrees Kelvin, ϕ_n is the electron quasi-Fermi potential and ϕ_p is the hole quasi-Fermi potential. Dirichlet boundary conditions for the electrostatic potential on the ohmic contact are computed from

$$\psi = V - \phi_{ms} \tag{7.85}$$

where V is the applied voltage and ϕ_{ms} is the contact metal-to-semiconductor work function. The jump in the normal electric displacement across an interface is equal to the surface charge density on the interface.

After normalization, the nonlinear potential equation (7.81) has the form

$$-\boldsymbol{\nabla} \cdot (\varepsilon \boldsymbol{\nabla} \psi) = e^{(\phi_p - \psi)} - e^{(\psi - \phi_n)} + C \tag{7.86}$$

A finite element scheme can now be constructed as in the two-dimensional case described earlier. For example, after integrating by parts in a weighted-residual expression, we get the weak statement: find $\psi \in H$ satisfying the essential boundary conditions on the ohmic contacts and such that

$$\int_\Omega \varepsilon \boldsymbol{\nabla} \psi \cdot \boldsymbol{\nabla} w \, dxdydz = \int_\Omega \left(e^{(\phi_p - \psi)} - e^{(\psi - \phi_n)} + C \right) w \, dxdydz \tag{7.87}$$

for all admissible test functions w vanishing on the contacts. Introducing a tetrahedral finite element discretization and basis in (7.87), the resulting nonlinear algebraic system has the form

$$\boldsymbol{g}(\boldsymbol{\psi}) = \boldsymbol{A}\boldsymbol{\psi} - \boldsymbol{b}(\boldsymbol{\psi}) = \boldsymbol{0} \tag{7.88}$$

Equation (7.88) can again be solved using either successive approximation or quasi-Newton schemes to obtain the potential field.

A finite volume scheme on non-uniform rectangular grids has been applied [327] to generate discrete systems of equations similar to (7.88) and to analyze a series of electrically erasable programmable read-only memory (EEPROM) structures, a MOSFET and a trench structure. For example, a floating gate tunnel oxide transistor is considered in Figure 7.8. The floating gate is insulated from the silicon substrate by a standard gate oxide, except over a portion of the drain where a thin dielectric exists. The equivalent circuit is also shown in Figure 7.8 and consists of a word line, a select line and a ground line with a metal bit line which contacts the drain of each select transistor. To write the device, the source, drain and substrate are held at ground and the control gate is pulsed to a high positive voltage. A primary design concern for EEPROM devices is write speed, and the nonlinear potential solver is suitable for this particular problem since the currents that occur during writing are small.

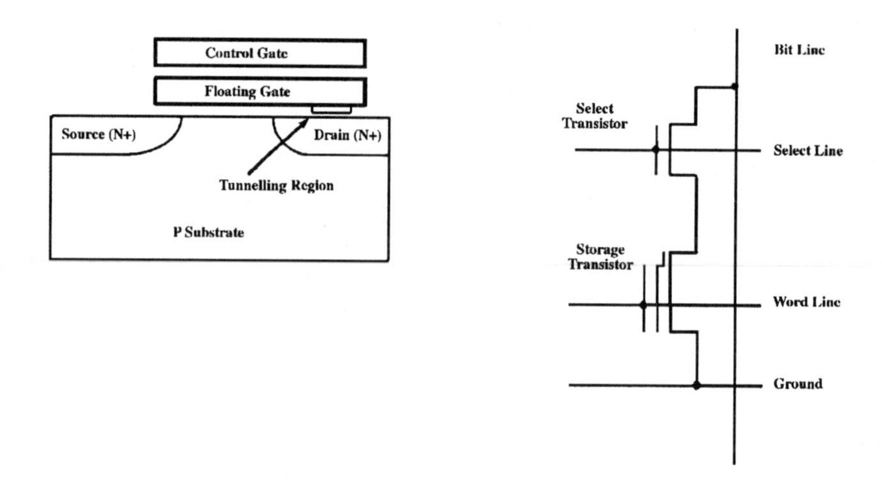

Figure 7.8: Model for EEPROM and equivalent circuit [327].

Equivalent circuit models have been extensively utilized for the design of EEPROM cells. These models utilize lumped capacitance to describe coupling of external voltages to the floating gate, as well as a Fowler–Nordheim expression for tunneling current. The capacitances are estimated with a flat plate capacitor formula. During writing and erasing, charge is transported through the thin dielectric between the floating gate and the drain, and the potential model provides a means of analyzing this charge transport.

The adaptive refinement ideas and quadtree data structure discussed previously for 2D problems can be directly extended to an octree data structure for 3D applications such as the nonlinear potential problem discussed above. Such an octree data structure and adaptive refinement scheme has been implemented in a prototype program for 3D device simulation [95]. Once again, the element-by-element conjugate gradient method is applicable, and permits the scheme to be utilized for realistic applications on engineering workstations, as well as vector and parallel mainframes.

7.7 Augmented Drift-Diffusion Models

In the next chapter we develop the hydrodynamic model and related advanced transport models that are designed to include hot carrier effects in ultra-small devices. These advanced transport models are more complex and sensitive than the established drift-diffusion model. Moreover, these hydrodynamic and Boltzmann systems that are capable of modeling non-local effects are more expensive to solve. An intermediate approach that can be readily incorporated into existing drift-diffusion software, is to augment the electron mobility to include the influence of the rate of change of the electric field. In this way the range of applicability of drift-diffusion models and software may be enlarged. This model may also be used to generate an improved starting iterate for the hydrodynamic model.

In the augmented model a term proportional to the total rate of change of the electric field is added to the current density relation. In one dimension we then have

$$J_n = \mu_n \left(\frac{\partial n}{\partial x} - n \frac{\partial \psi}{\partial x} \right) - n\gamma \frac{dE}{dt} \tag{7.89}$$

where γ is the proportionality constant. From (7.89) the velocity can be expressed as

$$v_n = \mu_n \frac{\partial \psi}{\partial x} - \frac{\mu_n}{n} \frac{\partial n}{\partial x} + \gamma \frac{dE}{dt} \tag{7.90}$$

In the steady state, the total rate of change (material derivative) satisfies

$$\frac{dE}{dt} = v_n \frac{\partial E}{\partial x} \tag{7.91}$$

so that (7.90) implies

$$v_n = \frac{\mu_n}{1 - \gamma \frac{\partial E}{\partial x}} \left(\frac{\partial \psi}{\partial x} - \frac{1}{n} \frac{\partial n}{\partial x} \right) \tag{7.92}$$

Using (7.90) and (7.92) in (7.89) and collecting terms,

$$J_n = \hat{\mu}_n \left(n \frac{\partial \psi}{\partial x} - \frac{\partial n}{\partial x} \right) \tag{7.93}$$

where $\hat{\mu}_n = \mu_n/(1 - \gamma \partial E/\partial x)$ is the augmented mobility. Thus, by simply modifying the mobility model in the drift-diffusion formulation, some aspects of hot carrier effects can be treated. The proportionality coefficient γ has to be determined from experiment or Monte Carlo simulation, and it is a function of E and dE/dt. This identifies one of the main limitations of this approach. Another difficulty is the extension of the idea to higher dimensions. One strategy is to assume that γ depends only on the component of the electric field parallel to the drift-velocity. Note, however, that the additional nonlinearity and field gradient contributions may adversely affect convergence of the Gummel iteration. Moreover, we need $\gamma E' > 1$ for the problem to remain physically well posed.

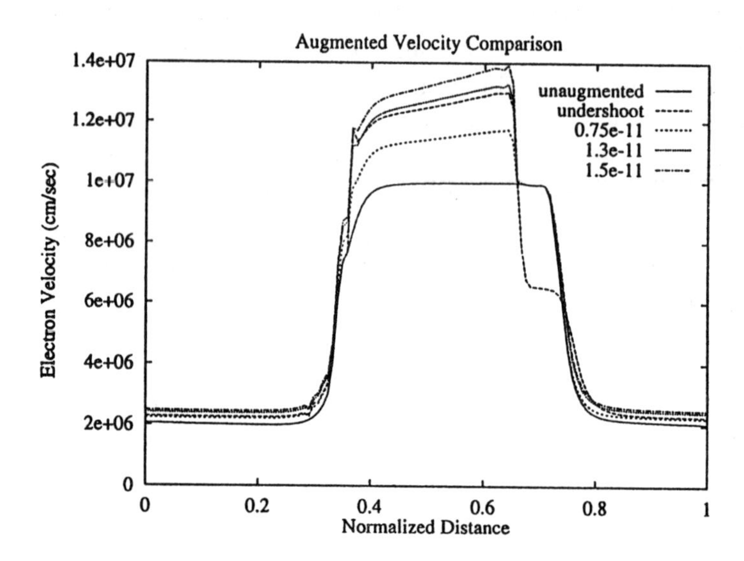

Figure 7.9: Comparison of standard drift-diffusion model and augmented drift-diffusion for simple diode problem with different choices of γ (from [127]).

The finite difference and finite element formulations of the augmented drift-diffusion system are completely analogous to the classical case and therefore need not be developed here. As a numerical test problem, we again consider the simple one-dimensional $n_+ - n - n_+$ diode with $n_+ = 5 \times 10^{17}$ and $n = 2 \times 10^{17}$. The total device length is 0.3 μm and the active length for comparison is 0.1 μm. Solution for an applied voltage of 1 volt was obtained in two continuation steps, each of 0.5 volt. Drift-diffusion and augmented results for the velocity profile are shown in Figure 7.9 for computation on a uniform grid with mesh size $h = 1/64$ and several values of γ. In the active region, the augmented model exhibits velocity overshoot above the saturation velocity. Further details are given in [127]. The above results were obtained efficiently using a novel multigrid/multilevel solution scheme which we now briefly consider.

7.8 Multigrid and Multilevel Schemes

As the grid for the drift-diffusion calculations is refined, the size of the sparse algebraic systems grows as $h^{-\alpha}$ where h is the mesh size and α is the spatial dimension. For many practical applications in two and three dimensions, elimination solvers will be prohibitive due to both storage and CPU limitations. For this reason, sparse iterative solvers are appealing. However, the matrix conditioning of the systems deteriorates as the grid is refined and this implies that the convergence of these iterative methods is severely degraded. Preconditioning to remove this dependence on the grid size is needed to produce reliable, efficient solvers. As an example, let us consider the electrostatic potential equation in the first part of the Gummel decoupled iteration

$$-\Delta\psi = f \tag{7.94}$$

where f depends on the carrier concentrations obtained from the solution of the discretized carrier transport equations. Hence f is effectively a function of spatial position for this solution step.

The differential operator in (7.94) is self-adjoint and discretizing with finite differences or finite elements leads to a sparse algebraic system of the form

$$\boldsymbol{A}\psi = \boldsymbol{b} \tag{7.95}$$

that is symmetric and positive definite (real positive eigenvalues). Moreover, for the standard central difference scheme or linear finite elements in two dimensions, the conditioning of matrix $\boldsymbol{A}$ is proportional to h^{-2} where h is the (average) mesh size. As an example, let the central difference scheme be used with Jacobi iteration on a uniform grid with (i, j) natural ordering. Then

$$\psi_{ij}^{(k+1)} = \frac{1}{4}\left(\psi_{i-1\,j}^{(k)} + \psi_{ij-1}^{(k)} + \psi_{i+1\,j}^{(k)} + \psi_{ij+1}^{(k)}\right) + \frac{h^2}{4} f_{ij} \tag{7.96}$$

where k is the Jacobi iteration index. This iteration converges in $O(h^{-2})$ iterations. Hence as the mesh is refined and h decreases both the amount of work per iteration and the number of iterations increase.

The multigrid idea can be conveniently interpreted as a "filtering" process in which nested coarser grids are used to accelerate convergence. Let ψ_{ij} be the exact solution of (7.96) and let $\psi_{ij}^{(k)}$ be the computed iterate on a fine grid of size h. Then the error $e_{ij} = \psi_{ij} - \psi_{ij}^{(k)}$ satisfies

$$\boldsymbol{A}_h \boldsymbol{e}_h = \boldsymbol{r}_h \tag{7.97}$$

where $\boldsymbol{r}_h = \boldsymbol{b}_h - \boldsymbol{A}_h \boldsymbol{\psi}_h^{(k)}$ is the algebraic residual.

Now the error $\boldsymbol{e}_h$ can be expressed as a linear combination of eigenvectors (modes) of $\boldsymbol{A}_h$ and under-relaxing (7.96) produces a scheme that will preferentially damp the highest frequency modes for the grid. This implies that this iterative "smoother" rapidly removes high frequency errors but the lowest error modes decay very slowly indeed. This suggests that coarser grids be used to filter error modes as follows:

1. First project the residual $\boldsymbol{r}_h$ in (7.97) to the next coarse level grid of size $2h$

$$\boldsymbol{r}_{2h} = \boldsymbol{P}\boldsymbol{r}_h \tag{7.98}$$

2. Next define the associated problem for the coarse grid error modes

$$\boldsymbol{A}_{2h}\boldsymbol{e}_{2h} = \boldsymbol{r}_{2h} \tag{7.99}$$

3. In a simple two-level scheme this coarse grid system is solved (usually by elimination) to obtain $\boldsymbol{e}_{2h}$.

4. The coarse grid correction $\boldsymbol{e}_{2h}$ is projected back to the fine grid to obtain

$$\boldsymbol{e}_h^* = \boldsymbol{P}^T\boldsymbol{e}_{2h} \tag{7.100}$$

5. The fine grid solution is updated to

$$\boldsymbol{\psi}_h^{(k)} = \boldsymbol{\psi}_h^{(k)} + \boldsymbol{e}_h^* \tag{7.101}$$

   and a few smoothing iterations are carried out on the fine grid.

6. The above steps are repeated to meet specified convergence criteria on the solution iterates and residual.

In practice, this can be directly extended to additional coarser mesh levels by replacing the direct solve in step 3 by smoothing iterations followed by a residual projection to the next grid of size $4h$ and so on. The cycle from fine grid down to the coarsest grid and back to fine grid is termed a V-cycle.

The relationship of this approach to preconditioning the fine grid matrix can easily be seen as follows. From step 3, $\boldsymbol{e}_{2h} = \boldsymbol{A}_{2h}^{-1}\boldsymbol{r}_{2h}$. Substituting in (7.100) and using (7.98), we see that

$$\boldsymbol{e}_h^* = \boldsymbol{P}^T\boldsymbol{A}_{2h}^{-1}\boldsymbol{P}\boldsymbol{r}_h \tag{7.102}$$

is the computed approximation to $e_h = A_h^{-1} r_h$. That is, the composition $P^T A_{2h}^{-1} P$ is an approximation to A_h^{-1} or, equivalently, $Q_h = P^T A_{2h}^{-1} P$ is a preconditioner for A_h^{-1} such that the system $Q_h A_h e_h = Q_h r_h$ is well conditioned.

The carrier transport equations also require the solution of linear sparse systems similar in form to (7.95) with the noteworthy exception that the systems are no longer symmetric. For the fine grid, the system may be diagonally dominant and symmetrizable. However, as the grid size is coarsened through the multigrid process this diagonal dominance is not preserved. This implies that the iterative "smoother" on these coarser grids will not be effective and the multigrid scheme will fail. There is, therefore, a need for improved iterative smoothers to circumvent this difficulty. One approach is to change the discretization scheme so that stronger upwinding is used on the coarser grids and the cell Peclet or Reynolds number requirement is met. This can be accomplished by adding numerical dissipation or artificial dissipation to the coarse grid problem, as described earlier in this chapter. Both strategies imply that diagonal dominance is restored, and they effectively generate a new smoothing iteration.

7.9 p-methods and Multilevel Schemes

Accuracy can be improved by either refining the mesh or by increasing the order of the method. Most numerical schemes rely on the mesh refinement approach. However, there is an increasing interest in both higher-order finite difference methods and higher-degree (p) finite element schemes.

In the standard higher-order difference scheme, additional surrounding grid point values are included in the stencil to annihilate the leading terms in the truncation error. For example, the standard five-point difference stencil for the Laplacian in (7.94) produces the discrete operator in (7.96) and Taylor series analysis of the linear problem reveals that the truncation error is $O(h^2)$. If, however, the adjacent grid points $(i-2, j)$, $(i+2, j)$ and $(i, j-2)$, $(i, j+2)$ are included then the resulting nine-point stencil gives a scheme with $O(h^4)$ accuracy. Of course, the bandwidth is significantly increased and there are now nine non-zeros per row in A rather than five so the computational effort increases. This nine-point scheme is not "compact" in the sense that the stencil involves points outside the patch of cells immediately adjacent to (i, j). The boundary conditions also require more care.

Another strategy that has proven effective is to develop higher-order compact schemes that use the differential equation to annihilate the leading truncation error terms. The approach is most simply described by developing it for the one-dimensional model drift-diffusion equation in (7.50)

$$vu' - Du'' = 0 \tag{7.103}$$

where the drift-velocity v and diffusivity D are taken to be constants. By central differencing about grid point i in a uniform mesh of size h,

$$\begin{aligned}(vu' - Du'')\,|_{x_i} &= v\frac{(u_{i+1} - u_{i-1})}{2h} - D\frac{(u_{i+1} - 2u_i + u_{i-1})}{h^2} \\ &\quad -\frac{vh^2}{6}u_i''' + \frac{Dh^2}{12}u_i^{(4)} + O(h^4)\end{aligned} \tag{7.104}$$

Now the third and fourth derivatives in the leading truncation error term of (7.104) can be expressed in terms of lower derivatives by means of (7.103). (A related idea is used in the Lax–Wendoff technique described in the next chapter.) By differentiating (7.103), we have $u''' = (v/D)u''$ and $u^{(4)} = (v/D)^2u''$ so the leading truncation error term in (7.104) simplifies to $(-v^2h^2/12D)u_i''$ which is dissipative. Differencing u_i'' centrally, the new three-point difference approximation is

$$\frac{v}{2h}(u_{i+1} - u_{i-1}) - \left(\frac{D}{h^2} + \frac{v^2h^2}{12D}\right)(u_{i+1} - 2u_i + u_{i-1}) = 0 \tag{7.105}$$

which is $O(h^4)$ accurate and still compact. Moreover, a simple test following (7.53) verifies that the scheme is nonoscillatory. This implies that (7.105) would be not only more accurate but could be used as a better coarse grid smoothing iteration in a multigrid scheme. These high-order compact schemes can be developed for higher-dimensional problems and more complicated differential equations, but the algebraic complexity in deriving the schemes becomes prohibitive for the coupled nonlinear problems of interest here.

Higher-degree (p) finite element schemes provide a better approach to improving the accuracy on a fixed mesh. Recall that the standard L^2 error estimates are asymptotically $O(h^{p+1})$ for regular elliptic problems, so that increasing polynomial degree p dramatically reduces the error. Once again the bandwidth is increased and there are more non-zeros per row so computational cost increases.

Considering again the potential problem (7.94) in the first step of the Gummel decoupling, the weak integral statement becomes: find $\psi \in H$

satisfying the essential (Dirichlet) boundary conditions on $\partial\Omega_1$, and such that

$$\int_\Omega \nabla\psi \cdot \nabla v \, dxdy = \int_\Omega fv \, dxdy \tag{7.106}$$

for all admissible test functions $v \in H$ with $v = 0$ on $\partial\Omega_1$. Zero flux conditions are automatically enforced as natural boundary conditions on the remainder of the boundary.

Discretizing the domain as a union of E finite elements and introducing a finite element basis of degree p, we obtain

$$\sum_{e=1}^{E}\sum_{j=1}^{N_e}\left(\int_{\Omega_e} \nabla\chi_i \cdot \nabla\chi_j \, dxdy\right)\psi_j^e = \int_{\Omega_e} f\chi_i \, dxdy \tag{7.107}$$

where χ_i denote the element basis functions and $\{\psi_i^e\}$ are local element nodal values. Assembling the element contributions, we again have a system of the form (7.95). Note that the element matrix in (7.107) is of size $N_e \times N_e$, where N_e is the number of nodal unknowns for the element. If a tensor-product basis of degree p is used, then $N_e = (p+1)^d$ where $d = 2$ is the dimension.

The standard finite element basis functions are constructed using Lagrange polynomials. For p methods in which the degree of the basis is increased it is useful to introduce hierarchic bases such as those defined by the Legendre or Chebyshev polynomials. These bases have the important property that they are nested. That is, the basis of degree p explicitly contains all lower degree bases. This implies that the element matrices in (7.107) are similarly nested if p is increased. Therefore only the new rows and columns of the element matrices need to be generated. Likewise, if p is reduced the related rows and columns of the element matrix are discarded.

These features are appealing for multilevel p schemes. The ideas are similar to those of multigrid methods:

1. First a few solution iterates are constructed for the high p system using, for instance, an element-by-element iterative smoother.

2. Next the residual is projected to the level $(p-1)$ approximation space using the $(p-1)$ degree basis.

3. The level $(p-1)$ element matrices are then used to solve the coarse level system, the error correction is projected back up to level p, and so on as before.

This scheme has been applied successfully to the augmented drift-diffusion problem [127]. There it was observed in exploratory numerical studies that, for a problem with a discontinuous doping profile, the element boundary should coincide with the discontinuity line or the Gummel iteration exhibits convergence problems. For elements of degree greater than four, matrix conditioning is poor (despite the multilevel strategy) and more fine grid smoothing iterations are needed. This can be circumvented by combining multilevel and multigrid strategies or some other domain decomposition preconditioner. We remark that the multilevel scheme is amenable to parallel element-by-element solution [126].

7.10 Summary

Here we begin with the idea of the material derivative to construct the model transport problem and show how the numerical behavior of the drift-diffusion problem depends critically on the treatment of the drift term. The close connection between "upwind" techniques, the Scharfetter–Gummel discretization and exponentially-upwinded finite elements is analysed. Some grid-related issues such as the need for local adaptive refinement in the solution layers are demonstrated for drift-diffusion equations and a MOSFET example. The associated questions of nested grid iteration lead naturally to a brief introduction for multigrid schemes and spectral finite elements.

7.11 Exercises

1. Carry out the detailed steps to derive the relation (7.47) for $n(x)$ and the resulting one-dimensional Sharfetter–Gummel difference scheme (7.50) for the current continuity equation. Implement the scheme in a one-dimensional finite difference program for a single carrier and test it for a simple diode problem.

2. Test the central scheme in (7.52) for the model drift-diffusion problem with $v/D = 20$ for uniform grids of size $h = 1/5$, $h = 1/10$ and $h = 1/20$. Graph the results. Modify the scheme to use a backward-difference approximation of the drift term as in (7.60) and compare the results with the coarse grid calculation.

3. Construct a central finite difference scheme for the self-adjoint form of the model problem in (7.63). Apply this to the test problem on a grid

with $h = \frac{1}{2}$. Compare the computed nodal solution at $x = 0.5$ to the exact solution and discuss your findings.

4. Construct the element matrix for linear elements in the bilinear functional in (7.65) and compare this with the element matrix for the standard Galerkin treatment of (7.51). Study the behavior as $h \to 0$ in each case.

5. Verify that the cell Peclet condition for the test problem (7.51) also corresponds to the condition that the tridiagonal system (7.52) be diagonally dominant. Use this diagonal dominance condition for the corresponding two-dimensional problem with constant drift velocity vector $\boldsymbol{v}$ and diffusivity D to determine a similar restriction on the mesh. Apply this result in a two-dimensional slope analysis to test if it is sufficient. Comment on necessity.

6. Verify for the linear triangle that $\nabla \cdot J_n$ reduces in the interior of each element to $-\boldsymbol{v}_n \cdot \nabla n$. Show, however, that this implies an additional approximation in (7.71) and modify this formulation to include the "neglected" interface contribution on the element boundary.

7. Expand the dissipation tensor in (7.72) to a representation in a cartesian (x, y) frame. Compare this form with that obtained if upwinding is applied directly in the respective x and y directions.

8. Using Taylor series analysis, construct a finite difference scheme analogous to (7.52) but for a non-uniform grid. Give the Taylor series truncation error for this scheme and confirm that it reverts to the previous $O(h^2)$ scheme when the mesh is uniform. Develop an *a posteriori* error estimator based on the truncation error. Implement this in a program and test it on the one-dimensional model drift-diffusion equation in (7.51) with $u(0) = 0$, $u(1) = 1$. Compare your results with those in Figure 7.2.

9. Sketch the mesh and the corresponding quadtree for a triangle that is repeatedly refined towards a corner. Determine the number of active elements in the grid at any level of refinement and compare this with the number if the grid is uniformly refined to the same level.

10. The continuation scheme in Section 7.6 breaks down with a singular Jacobian matrix as a limit point is approached. Introduce the arc-length s as a new parameter to regularize the problem by removing

such singular points. Construct the modified Jacobian system for the solution pair $u(s)$, $\lambda(s)$ and give the revised continuation algorithm.

11. Construct a Galerkin finite element formulation for the 3D nonlinear potential problem in (7.81) using tetrahedral elements. Introduce volume coordinates to define the linear basis on the tetrahedron and give the detailed form of the 4×4 element matrix contribution. Sketch out an algorithm to solve (7.81) using successive approximation with this finite element formulation. Comment on the anticipated convergence behavior of this algorithm.

12. Develop a decoupled Gummel scheme for the augmented drift-diffusion model in Section 7.6. Describe the modifications to the previous finite difference and finite element schemes arising from augmentation.

13. Write a program to compute the solution to (7.95) using the Jacobi iteration in (7.96). Compare the relative decay rates for the amplitude of three modes corresponding to initial "disturbances" of $\sin(\pi x)\sin(\pi y)$, $\sin(4\pi x)\sin(4\pi y)$ and $\sin(8\pi x)\sin(8\pi y)$ on a 32×32 grid. Relate your observations to the "filter" argument for error modes and multigrid schemes.

14. Verify that the higher-order compact scheme in (7.105) is not oscillatory.

15. Develop a p finite element scheme with the Lagrange basis for the drift-diffusion problem in one dimension. Confirm that the conditioning of the systems deteriorates as p increases.

Chapter 8

Hydrodynamic Device Equations

8.1 Introduction

The drift-diffusion models described in the previous chapter assume a local correspondence between the electric field and the average carrier energy and velocity. These models employ a field-dependent mobility and are unable to characterize non-local behavior, such as velocity overshoot, that is observed in some submicron devices. The basic scattering processes implicit in the device physics are energy-dependent. This suggests that including the energy transport equation with, say the average carrier energy or carrier temperature as a new dependent variable, may be appropriate in modeling devices with high electric fields and very large field gradients. For example, in a submicron MOSFET the drain voltage does not scale linearly with channel length and the lateral electric field can amount to several hundred kV cm^{-1}. Since hot carriers may significantly degrade the performance and reliability of a device, it is important that these effects be adequatelly modeled in numerical simulations. Moreover, an effective energy transport model can be used for practical device design, since this approach is computationally efficient relative to Monte Carlo methods which require simulation of the energy and velocity distributions of carriers, rather than their averages.

8.2 Hot Carriers

Inclusion of hot carrier transport via an energy transport equation leads to an expanded set of partial differential equations that closely resembles the systems modeling gas dynamics and in which the electrons may be interpreted as an "electron gas." This analogy has led to the terminology the hydrodynamic equations to distinguish this model from the previous drift-diffusion equations. There are, however, some important features of the hydrodynamic equations that distinguish them from the Euler equations of gas dynamics and make the device problem numerically more difficult. These concern primarily the non-zero source term and coupling to the potential equation. Examination of the hydrodynamic operator reveals that its mathematical properties, like those of the gas dynamic equations, may vary with the local velocity. That is, the governing system of equations may exhibit a different type (elliptic, hyperbolic, or parabolic) in different regions of the device (See Chapter 15 for a brief discussion of classification of PDEs). Conceptually, the motion of electrons in the device can be viewed as similar to that of molecules in a gas flow. For the latter there is a critical speed, the speed of sound, below which the flow is subsonic and the equation is elliptic. At flow velocities above this critical value, the flow is locally supersonic and the equation is hyperbolic. A similar situation exists for the dependence of hydrodynamic equation type on carrier velocity, which implies that both subcritical and supercritical velocities may occur.

The hydrodynamic equations comprise a set of transport PDEs describing conservation of particle number, momentum and energy, together with the Poisson equation for the electrostatic potential. Here we consider the formulation for a single carrier. The starting point for the hydrodynamic model is the Boltzmann transport equation

$$\frac{\partial f}{\partial t} + \frac{\boldsymbol{p}}{m} \cdot \nabla f + c\boldsymbol{E} \cdot \frac{\partial f}{\partial \boldsymbol{p}} = \left(\frac{\partial f}{\partial t}\right)_c \tag{8.1}$$

for a complete single-particle phase space distribution $f(\boldsymbol{x}, \boldsymbol{p}, t)$, where $(\frac{\partial f}{\partial t})_c$ indicates collision contributions and $\boldsymbol{x}, \boldsymbol{p}$ denote spatial and momentum coordinates. The momentum and energy conservation equations are obtained by taking the first three moments of the Boltzmann transport equation. The zero-order moment corresponds to the particle number, the first-order moment to the average momentum of the ensemble, and the second-order moment to the average energy. The resulting system of PDEs for electron density n, momentum density $\boldsymbol{p}$ and energy density W has the form:

$$\frac{\partial n}{\partial t} + \nabla \cdot (n\boldsymbol{v}) = \left(\frac{\partial n}{\partial t}\right)_c \tag{8.2}$$

$$\frac{\partial \boldsymbol{p}}{\partial t} + \boldsymbol{v}\nabla \cdot \boldsymbol{p} + \boldsymbol{p} \cdot \nabla \boldsymbol{v} = -en\boldsymbol{E} - \nabla(nT) + \left(\frac{\partial \boldsymbol{p}}{\partial t}\right)_c \tag{8.3}$$

$$\frac{\partial W}{\partial t} + \nabla \cdot (\boldsymbol{v}W) = -en\boldsymbol{v} \cdot \boldsymbol{E} - \boldsymbol{\nabla} \cdot (\boldsymbol{v}nT) - \boldsymbol{\nabla} \cdot \boldsymbol{q} + \left(\frac{\partial W}{\partial t}\right)_c \tag{8.4}$$

where $\boldsymbol{E}$ is the electric field, T is the temperature in energy units, e is the electronic charge, $\boldsymbol{q}$ is the heat flux, and the subscript c again indicates collision contributions. (The Boltzmann coefficient has been normalized to unity for convenience.) Note that this system is not closed. In particular, $\boldsymbol{q}$ appears in (8.4) as an unknown "heat flux". Additional moments could be taken to yield additional equations but this would introduce further unknowns and a still larger, more complex system of PDEs. Instead a constitutive equation relating $\boldsymbol{q}$ to T is usually introduced. Similarly, p and W can be related to other variables.

As constitutive relations, a standard approach is to adopt a Fourier relation and parabolic energy bands so that

$$\boldsymbol{q} = -\kappa \boldsymbol{\nabla} T\,, \quad \boldsymbol{p} = mn\boldsymbol{v}\,, \quad W = \frac{3}{2}nT + \frac{m}{2}n|\boldsymbol{v}|^2 \tag{8.5}$$

where κ is "thermal" conductivity and m is the effective electron mass. The thermal conductivity of the electron gas can be specified using a Weidemann–Franz relation and further simplified to the form $\kappa = R\gamma n$, where R is a parameter of order unity and $\gamma = 7.2 \times 10^{-26}$ watt m^2 K^{-1}. If $\kappa = 0$, then (8.2)–(8.4) correspond to the Euler equations of gas dynamics with gas constant 5/3 and with source terms on the right due to the collision terms and the electric field. It follows that these equations are hyperbolic with equivalent sound speed $a = (\gamma T/m)^{1/2}$. There are five possible nonlinear waves possessing discontinuity surfaces [199]. The equations permit shock discontinuities in v, n, T and contact discontinuities in v_t and T, where v is the particle velocity normal to the wave and v_t is the particle velocity tangential to the wave. If instead, κ is non-zero, only four nonlinear waves are possible and physical diffusion in the temperature equation excludes discontinuities in T. However, steep gradients in T may still be present. Finally, as in the drift-diffusion model, the potential ψ satisfies the Poisson equation

$$-\boldsymbol{\nabla} \cdot (\varepsilon \boldsymbol{E}) = e(N_D - N_A - n) \tag{8.6}$$

where $\boldsymbol{E} = \boldsymbol{\nabla}\psi$, and N_D and N_A are, respectively, the concentrations of donor and acceptor impurities. Equations (8.2)–(8.6), constitute the hydrodynamic model. Using (8.5), equations (8.3) and (8.4) become

$$\frac{\partial \boldsymbol{v}}{\partial t} + \boldsymbol{v} \cdot \boldsymbol{\nabla}\boldsymbol{v} = -\frac{e}{m}\boldsymbol{E} - \frac{1}{mn}\boldsymbol{\nabla}(nT) + \left(\frac{\partial \boldsymbol{v}}{\partial t}\right)_c \tag{8.7}$$

$$\frac{\partial T}{\partial t} + \boldsymbol{v} \cdot \boldsymbol{\nabla}T = -\frac{2}{3}(\boldsymbol{\nabla} \cdot \boldsymbol{v})T + \frac{2}{3n}\boldsymbol{\nabla} \cdot (\kappa \boldsymbol{\nabla} T) + \left(\frac{\partial T}{\partial t}\right)_c \tag{8.8}$$

where

$$\left(\frac{\partial \boldsymbol{v}}{\partial t}\right)_c = \frac{1}{mn}\left(\frac{\partial \boldsymbol{p}}{\partial t}\right)_c - \frac{\boldsymbol{v}}{n}\left(\frac{\partial n}{\partial t}\right)_c$$

$$\frac{3}{2}\left(\frac{\partial T}{\partial t}\right)_c = \frac{1}{n}\left(\frac{\partial W}{\partial t}\right)_c - \frac{\boldsymbol{v}}{n}\left(\frac{\partial \boldsymbol{p}}{\partial t}\right)_c + \frac{1}{n}\left(\frac{m}{n}v^2 - \frac{3}{2}T\right)\left(\frac{\partial n}{\partial t}\right)_c$$

so that the governing PDE system is now given by (8.2) and (8.6)–(8.8).

Several major assumptions have been made in the choice of constitutive relations. For example, the Fourier relation between heat flux and temperature has the form $\boldsymbol{q} = -\kappa \nabla T$ encountered in heat conduction. Furthermore, relaxation times for the model are obtained from uniform field Monte Carlo simulations and are parameterized by the average energy. It is not known how reliable these assumptions are for nonuniform fields in which the distribution function is distorted and there are related gradient effects. The parabolic energy band relation $\boldsymbol{p} = mn\boldsymbol{v}$ has also been utilized to simplify the model. Relaxation times derived from Monte Carlo simulation with nonparabolic bands may be employed in this model in an attempt to include nonparabolic effects. From a theoretical standpoint, nonparabolicity introduces additional terms into the hydrodynamic model. These terms depend on higher-order moments of the distribution function and result in further unknowns and a more complex system. The collision source terms account for dissipation of energy and momentum due to a combination of many scattering processes, and the forms given here are approximate since they depend on average values rather than the actual nonequilibrium distributions. Hence, there are still a number of open questions regarding the physical model and choice of parameters. Nevertheless, the basic structure of the equations is essentially fixed and a modular approximate model can be developed to make predictive computations and to compare different physical models or parameter ranges. With this view in mind, we now develop an approximate scheme in 1D.

8.3 1D Steady-state Problem

The one-dimensional steady-state problem is a convenient starting point for an investigation of numerical models and simulation techniques. The governing equations (8.2) and (8.6)–(8.8) simplify to

$$\frac{d}{dx}(nv) = 0 \tag{8.9}$$

$$v\frac{dv}{dx} + \frac{e}{m}E + \frac{1}{mn}\frac{d}{dx}(nT) = \frac{1}{m}\left(\frac{dp}{dt}\right)_c = -\frac{v}{\tau_p(w)} \tag{8.10}$$

$$v\frac{dw}{dx} + \frac{1}{n}\frac{d}{dx}(nTv + q) + evE = \left(\frac{dw}{dt}\right)_c = -\frac{w - w_0}{\tau_w(w)} \tag{8.11}$$

$$\varepsilon\psi_{xx} + e(N - n) = 0 \tag{8.12}$$

Setting

$$w = \frac{3}{2}T + \frac{1}{2}mv^2, \quad \tau_p = m\frac{\mu_{no}}{e}\frac{T_0}{T}, \quad \kappa = \frac{3}{2}\frac{\mu_{n0}}{e}nT_0$$
$$w_0 = \frac{3}{2}T_L \quad \tau_w = \frac{m}{2}\frac{\mu_{no}}{e}\frac{T_0}{T} + \frac{3}{2}\frac{\mu_{no}}{ev_s^2}\frac{TT_0}{T + T_0}$$

where μ_{no} is the low-field electron mobility and v_s is the saturation velocity and substituting in (8.11)

$$\frac{3}{2}v\frac{dT}{dx} + T\frac{dv}{dx} + \frac{1}{n}\frac{d}{dx}\left(\kappa\frac{dT}{dx}\right) + \frac{\frac{m}{2}v^2 + \frac{3}{2}(T - T_0)}{\tau_w} - \frac{mv^2}{\tau_p} = 0 \tag{8.13}$$

Equations (8.9), (8.10), (8.12) and (8.13) constitute the form of the 1D steady hydrodynamic equations to be considered here.

This first-order coupled system of nonlinear ordinary differential equations can be discretized and solved numerically in a variety of ways. For example, the spatial derivatives can be approximated by finite differences on a uniform grid with central differences for the diffusion contributions such as $d/dx(\kappa dT/dx)$ and upstream differences for drift terms vdv/dx, vdT/dx, etc. These ideas are similar to those described for the drift-diffusion problem. A weak integral statement can also be constructed and a Petrov–Galerkin finite element method developed.

Let $\boldsymbol{g}(\boldsymbol{u}) = \boldsymbol{0}$ denote the resulting nonlinear algebraic system where $\boldsymbol{u}$ is the vector of grid point unknowns $(\boldsymbol{n}, \boldsymbol{v}, \boldsymbol{T}, \boldsymbol{\psi})$. This system can be solved using Successive approximation, Newton's method or similar strategies. For example, the Newton iteration proceeds as follows: from initial iterate $\boldsymbol{u}_0$ compute

$$\boldsymbol{u}_{r+1} = \boldsymbol{u}_r - \boldsymbol{J}_r^{-1}\boldsymbol{g}_r \qquad r = 0, 1, 2, \ldots \tag{8.14}$$

where $\boldsymbol{J}_r$ is the Jacobian $\boldsymbol{J}_r = \boldsymbol{J}(\boldsymbol{u}_r) = (\partial g_i/\partial u_j)$ evaluated at $\boldsymbol{u}_r$ and $\boldsymbol{g}_r = \boldsymbol{g}(\boldsymbol{u}_r)$. Equivalently, (8.14) can be expressed as the system

$$\boldsymbol{J}_r\boldsymbol{\delta}_{r+1} = -\boldsymbol{g}_r \tag{8.15}$$

where $\boldsymbol{\delta}_{r+1} = \boldsymbol{u}_{r+1} - \boldsymbol{u}_r$ is the correction to the current iterate. The contraction properties and other aspects of Newton's method are discussed in Section 13.5.

Since different combinations of primary field variables n, $\boldsymbol{v}$, T and ψ appear in the respective governing equations, it follows that the Jacobian $\boldsymbol{J}$ will have a sparse block structure that can be exploited in the Newton solution step. Let $\boldsymbol{g}_n$, $\boldsymbol{g}_v$, $\boldsymbol{g}_T$ and $\boldsymbol{g}_\psi$ denote the contributions to the nonlinear discretized equations resulting from (8.9), (8.10), (8.13), and (8.12), respectively. Then the Jacobian has the block form

$$\boldsymbol{J} = \begin{bmatrix} \frac{\partial \boldsymbol{g}_n}{\partial \boldsymbol{n}} & \frac{\partial \boldsymbol{g}_n}{\partial \boldsymbol{v}} & 0 & 0 \\ \frac{\partial \boldsymbol{g}_v}{\partial \boldsymbol{n}} & \frac{\partial \boldsymbol{g}_v}{\partial \boldsymbol{v}} & \frac{\partial \boldsymbol{g}_v}{\partial \boldsymbol{T}} & \frac{\partial \boldsymbol{g}_v}{\partial \boldsymbol{\psi}} \\ \frac{\partial \boldsymbol{g}_T}{\partial \boldsymbol{n}} & \frac{\partial \boldsymbol{g}_T}{\partial \boldsymbol{v}} & \frac{\partial \boldsymbol{g}_T}{\partial \boldsymbol{T}} & 0 \\ \frac{\partial \boldsymbol{g}_\psi}{\partial \boldsymbol{n}} & 0 & 0 & \frac{\partial \boldsymbol{g}_\psi}{\partial \boldsymbol{\psi}} \end{bmatrix} \tag{8.16}$$

where each block matrix has a regular diagonal, bidiagonal or tridiagonal structure if standard low-order finite difference or finite element schemes are used. As noted previously for the drift-diffusion model, the nonlinear systems may be difficult to solve by standard Newton procedures. Incremental continuation in applied voltage will again alleviate some of these difficulties.

Numerical results from Gardner *et al* [199] are given in Figure 8.1 for a submicron n^+-n-n^+ silicon diode which models the channel in a MOSFET. The n^+ regions are each 0.1 μm with doping density $N = 5 \times 10^{17}$ cm^{-3} and the central region is 0.4 μm with $N = 2 \times 10^{15}$ cm^{-3}. The ambient device temperature is $T_0 = 300°\text{K} = 0.025$ eV. The parameters are those chosen by Odeh *et al* [411] who also show that the results compare favorably with Monte Carlo simulations for this device. The curves in Figures 8.1(a) and 8.1(b) correspond to electron temperature and velocity profiles respectively for voltage biases $V = 0.5, 1.0, 1.5, 2.0$ as marked. As V increases, the layers in temperature and velocity become more pronounced. The temperature increases as the electrons flow through the central n region and the peak temperature for $V = 2$ is an order of magnitude higher than the ambient temperature $T_0 = 0.025$ eV. This can be contrasted with the drift-diffusion assumption that the process is isothermal with $T = T_0$ constant. Velocity overshoot in much of the n region is evident in Figure 8.1(b).

The Jacobian system in (8.15) can be solved by Gaussian elimination or block iterative methods. The structure of $\boldsymbol{J}$ in (8.16) suggests that iterative decoupling may be fruitful. This is similar in spirit to the Gummel decou-

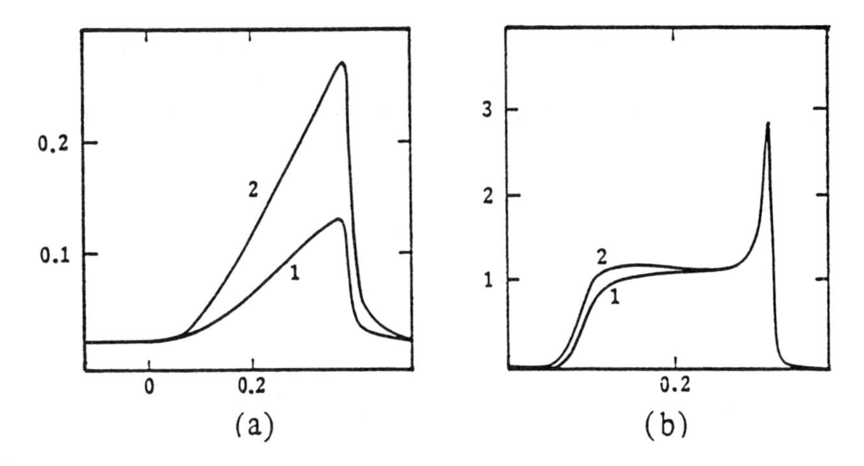

Figure 8.1: (a) Electron temperatures and (b) velocity profiles for a submicron n_+-n-n_+ silicon diode at voltage biases 0.5, 1.0, 1.5, 2.0 (from [199]).

pling for drift-diffusion. For example, lagging $\boldsymbol{v}$ in the first block of (8.16) yields the smaller block system at iterate $r+1$

$$\frac{\partial \boldsymbol{g}_n}{\partial \boldsymbol{n}} \delta \boldsymbol{n} = -\boldsymbol{g}_n(\boldsymbol{u}_r) - \frac{\partial \boldsymbol{g}_n}{\partial \boldsymbol{v}} \delta \boldsymbol{v} \tag{8.17}$$

which can be solved to produce an approximation for $\delta \boldsymbol{n}$. Then the final block system for ψ in (8.16) can be solved using the result for $\boldsymbol{n}_{r+1}$ from (8.17). Finally, the block subsystems for $\delta \boldsymbol{v}$ and $\delta \boldsymbol{T}$ can be solved to complete one step of the block-decoupled algorithm. The results from this block-decoupled step can now be used to update (8.17) and the subsystem solutions continued. The expected quadratic convergence of Newton's method for this problem under appropriate assumptions on the choice of initial iterate and strength of the nonlinearity has been demonstrated[199]. However, such nonlinear problems may admit multiple solutions and the solution paths or bifurcation diagrams for certain parameter ranges may be of practical interest. General discussions of path-following techniques such as arc-length continuation are given in [440].

A series of numerical experiments for the 1D steady equations has been carried out under the further non-physical restriction that the electric field E is a known function of position [53]. These parametric studies explore some of the limitations of the hydrodynamic model and associated constitutive assumptions for slowly and rapidly varying fields. For example, the results suggest that premature electron cooling in an abruptly decreasing electric field encourages velocity overshoot, while delayed cooling yields a higher drift velocity in the low-field region. This raises further questions regarding validity of the Fourier heat transfer assumption. An extended study that models non-parabolic energy band effects is also warranted to characterize average electron energy more accurately. Finally, improved models for energy dissipation are needed.

8.4 1D Time-dependent Problem

The extension to a discrete formulation of the transient 1D equations is straightforward. Perhaps the most direct strategy is to discretize in space using finite differencing or finite elements as in the steady-state problem. This yields a system of ordinary differential equations to be integrated with respect to time from specified initial data. Such ODE system integrators are developed in the chapter on multiple-species diffusion models. However, since sharp solution gradients will develop in the device problem, the high frequency content of the solution is important and stiff integrators will be required. The effect of the source term will also be significant. Such implicit schemes will require solution of a nonlinear algebraic system at each timestep. If Newton iteration is introduced, the problem reverts to one similar to (8.15), (8.16). Once again, the block submatrix structure of the Jacobian can be exploited to decouple iteratively the various field variables and thereby reduce the size of the systems that are solved. Since the solution at the previous time provides a good starting value for the new time, rapid convergence is possible even for relatively large timesteps.

For non-stiff or moderately-stiff problems, high-order Runge–Kutta methods with extended ranges of stability are appealing explicit methods. They offer reasonable stability properties compared with other explicit schemes. Moreover, these schemes are readily vectorized or parallelized and this becomes important for 3D hot carrier simulations. As in the steady-state case, special care must be taken to discretize the drift term using Scharfetter–Gummel upwinding or a similar strategy. Steep layers in the solution may also evolve and the large gradients in these regions can lead to numerical oscillations which cause the Newton iteration to diverge. Special techniques developed for shock capturing in gas dynamics can be applied, including various numerical dissipation schemes such as Lax–Wendroff methods, higher-order upwind schemes, flux-corrected transport (FCT), flux-splitting and flux-limiting. In essence, these methods attempt to add sufficient dissipation to control oscillations without smearing the layers or adversely effecting the phase (wave speeds).

As an illustrative case, consider the time evolution of the temperature and velocity for an $n^+ - n - n^+$ diode of length 1 μm [174]. The doping profile is $N_D(x) = 5 \times 10^{17}$ cm^{-3} for $0 \leq x \leq 0.25$ and $0.75 \leq x \leq 1$ with $N_D(x) = 2 \times 10^{15}$ cm^{-3} for $0.35 \leq x \leq 0.65$. The two regions are connected by a polynomial $Q(x) = -5x^7 + 21x^5 - 35x^3 + 35x + 16$. The grid is uniform with mesh size $\Delta x = 0.01$ and a simple forward Euler integrator has been used with stepsize $\Delta t = 0.1(\Delta x)$. In this test problem, bias voltage is applied as a "ramp" and the electric field and temperature evolve as indicated in Figure 8.2.

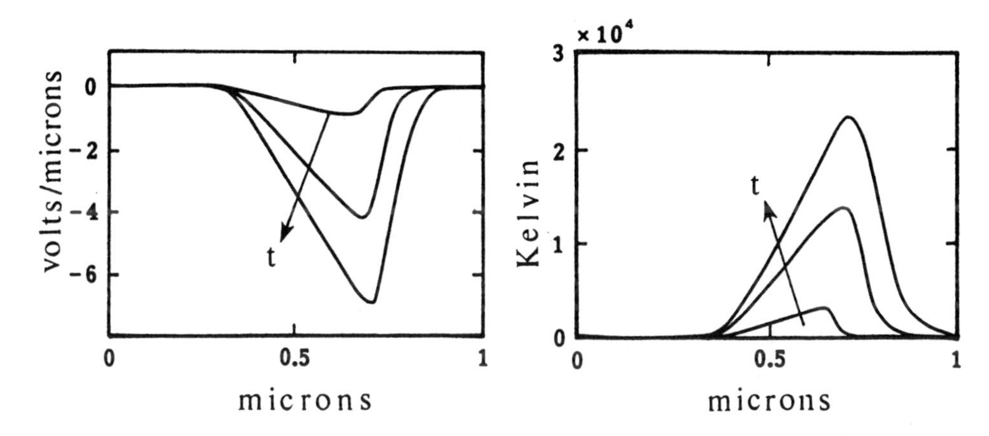

Figure 8.2: Evolution of electric field and temperature solutions for $n^+ - n - n^+$ diode (from [174]).

8.5 Lax–Wendroff and Taylor–Galerkin Schemes

The Taylor–Galerkin finite element method has been an effective scheme for solution of hyperbolic problems and convection-dominated transport processes. The basic idea is related to the Lax–Wendroff approach for dissipative finite difference schemes so we first describe this procedure for the model drift equation

$$\frac{\partial u}{\partial t} + \alpha \frac{\partial u}{\partial x} = 0 \tag{8.18}$$

where the drift coefficient α is constant. Introducing a forward difference approximation in time, this implies

$$\frac{u^{n+1}(x) - u^n(x)}{\Delta t} = -\alpha \frac{\partial u}{\partial x}(x, t_n) + \frac{\Delta t}{2}\frac{\partial^2 u}{\partial t^2}(x, t_n) + \ldots \tag{8.19}$$

Now the time truncation error term of order $O(\Delta t)$ can be represented in terms of spatial derivatives by differentiating (8.18) with respect to t and using $\frac{\partial u}{\partial t} = -\alpha \frac{\partial u}{\partial x}$ to write

$$\frac{\partial^2 u}{\partial t^2} = -\alpha \frac{\partial^2 u}{\partial x \partial t} = -\alpha \frac{\partial}{\partial x}\left(\frac{\partial u}{\partial t}\right) = \alpha^2 \frac{\partial^2 u}{\partial x^2} \tag{8.20}$$

Substituting into (8.19),

$$\frac{u^{n+1}(x) - u^n(x)}{\Delta t} = -\alpha \frac{\partial u}{\partial x}(x, t_n) + \frac{\alpha^2(\Delta t)}{2}\frac{\partial^2 u}{\partial x^2}(x, t_n) + O(\Delta t^2) \tag{8.21}$$

This equation can now be centrally differenced with respect to x to obtain a dissipative scheme that is second-order accurate in both space and time. The Taylor–Galerkin variant of this scheme is constructed by forming the weak Galerkin projection of (8.21) instead of spatial differencing. That is, the weighted-residual statement is constructed and the dissipative term is integrated by parts to get

$$\int_\Omega \frac{(u^{n+1} - u^n)}{\Delta t} v(x)\, dx = -\int_\Omega \alpha \frac{\partial u}{\partial x} v(x)\, dx - \int_\Omega \frac{\alpha^2 \Delta t}{2} \frac{\partial u}{\partial x}\frac{\partial v}{\partial x}\, dx \tag{8.22}$$

for all admissible test functions $v(x)$ and where we have assumed $u(0,t)$ is given at the inlet (source) and $\frac{\partial u}{\partial x} = 0$ is enforced as a natural condition at the outflow boundary (drain). As noted previously, this scheme is dissipative with dissipation proportional to the stepsize Δt.

Next, let us consider the development of this strategy for the hydrodynamic transport equations. Following the basic approach in [58], the non-dimensional transport equations can be expressed in the form

$$\frac{\partial \boldsymbol{U}}{\partial t} + \frac{\partial \boldsymbol{F}(\boldsymbol{U})}{\partial x} = \boldsymbol{S}(\boldsymbol{U}, \psi) - \nabla \cdot \boldsymbol{Q}(\boldsymbol{U}) \tag{8.23}$$

where $\boldsymbol{U} = (n, nu, n\omega)$, flux vector $\boldsymbol{F}(\boldsymbol{U}) = (nu, nu^2 + P, u(n\omega + P))$, source vector $\boldsymbol{S}(\boldsymbol{U}, \psi) = (0, \theta n\psi' - nu/\tau_p, \theta nu\psi' - n(\omega - \omega_0)/\tau_\omega)$ and heat flux $\boldsymbol{Q}(\boldsymbol{U}) = (0, 0, -\kappa(\boldsymbol{U})T')$. The "primes" denote differentiation with respect to x. Here n is the electron concentration, u is the drift-velocity, ω is the average energy, T is carrier temperature, P is thermodynamic pressure, ψ is electrostatic potential and $\theta = eL^2 n_\infty / \varepsilon\psi_\infty$ is a nondimensional group where L is the length scale and n_∞, ψ_∞ are the concentration and potential scales. The important constitutive parameters are the thermal conductivity $\kappa(\boldsymbol{U})$ and relaxation time scales τ_p and τ_ω for momentum and energy respectively. For example, the Baccarani–Wordeman models express the relaxation times in terms of carrier temperature as

$$\tau_p = \tau_{p0}/T\,, \quad \tau_w = \tau_p/2 + \alpha\tau_{p0}/(\tau_{p0} + \tau_p) \tag{8.24}$$

with $\alpha = (3\tau_{p_0} kT_0)/(mv_s^2)$ and the low field relaxation time τ_{p_0} is related to the low field mobility μ_0 by $\tau_{p_0} = m\mu_0/e$, m is the effective mass, v_s is the drift-velocity scale, k is the Boltzmann constant and T_0 is the lattice temperature scale. The thermal conductivity is specified by the Weidemann–Franz relation

$$\kappa = n\kappa_0 T_0 \tau_{p_0} k^2/m^2 \tag{8.25}$$

The electrostatic potential equation (8.12) transforms similarly to $\Delta\psi = \theta(n - N_D)$. The potential and transport equations may be iteratively decoupled using the Gummel approach in the same manner as for the drift-diffusion equations in the previous chapter. The weak Galerkin statement for the potential problem becomes: find $\psi \in H$ satisfying the essential end conditions $\psi(x_l) = 0$, $\psi(x_r) = V$ and such that

$$\int_{x_l}^{x_r} \psi' v'\, dx = -\int_{x_l}^{x_r} \theta(n - N_D) v\, dx \tag{8.26}$$

for all admissible test functions v satisfying $v(x_l) = v(x_r) = 0$.

By introducing a finite element partition and linear Lagrange basis, we have the linear system

$$\sum_{j=1}^{N} \left(\int_{x_l}^{x_r} \chi_i' \chi_j'\, dx \right) \psi_j = -\int_{x_l}^{x_r} \theta(n - N_D)\chi_i\, dx \tag{8.27}$$

where n is given from the previous iteration on the transport system. The system (8.27) is tridiagonal and symmetric positive definite and therefore is efficiently solved by tridiagonal symmetric Gaussian elimination.

Turning now to the transport system (8.23), forward differencing in time and replacing the time truncation error term by a corresponding spatial derivative using the Lax–Wendroff idea yields a regularized PDE problem with a spatial dissipative term. Introducing the weighted residual integral and integrating by parts, we have the forward Euler–Taylor–Galerkin (FETG) representation

$$\int_{x_l}^{x_r} \boldsymbol{W}^t(\boldsymbol{U}^{n+1} - \boldsymbol{U}^n)\,dx = \Delta t\left(\int_{x_l}^{x_r} (\boldsymbol{W}_t\boldsymbol{S} + \frac{d\boldsymbol{W}^t}{dx}\boldsymbol{F})\,dx + \int_{x_l}^{x_r} \frac{\Delta t}{2}(\boldsymbol{W}^t\boldsymbol{B} + \frac{d\boldsymbol{W}^t}{dx}\boldsymbol{A})\frac{\partial \boldsymbol{U}}{\partial t}\,dx - \boldsymbol{W}^t\boldsymbol{F}|_{x_l}^{x_r} - \frac{\Delta t}{2}\,\boldsymbol{W}^t\boldsymbol{A}\frac{\partial \boldsymbol{U}}{\partial t}|_{x_l}^{x_r}\right) \tag{8.28}$$

where $\frac{\partial \boldsymbol{U}}{\partial t} = S(\boldsymbol{U}) - \boldsymbol{A}\frac{\partial \boldsymbol{U}}{\partial x}$ and the flux and source Jacobians are respectively,

$$\boldsymbol{A} = \begin{bmatrix} 0 & 1 & 0 \\ (\gamma-3)u^2/2 & -(\gamma-3)u & \hat{\gamma} \\ u(-\gamma\omega + \hat{\gamma}u^2) & \gamma\omega - 3\hat{\gamma}u^2/2 & \gamma u \end{bmatrix},$$

$$\boldsymbol{B} = \begin{bmatrix} 0 & 0 & 0 \\ D\psi' + u/\tau_p & \tau_p^{-1} & \frac{-u}{\tau_p\omega} \\ \omega_l/\tau_\omega & D\psi' & \tau_\omega^{-1} \end{bmatrix} \tag{8.29}$$

The FETG finite element approximation follows on introducing finite element expansions $\boldsymbol{U}_h$ for $\boldsymbol{U}$ and test functions $\boldsymbol{W}_h$ for $\boldsymbol{W}$ in (8.28) with potential ψ_h specified from the solution of (8.27). The resulting system has the form

$$\boldsymbol{M}(\boldsymbol{U}^{n+1} - \boldsymbol{U}^n) = \boldsymbol{f} \tag{8.30}$$

where, in practice, the coefficient matrix $\boldsymbol{M}$ is approximated as a diagonal "lumped" matrix by using trapezoidal integration on the left in (8.28). The solution $\boldsymbol{U}^{n+1}$ can then be explicitly advanced through timestep Δt at each node i of the grid using

$$\boldsymbol{U}_i^{n+1} = \boldsymbol{U}_i^n + \boldsymbol{M}_{ii}^{-1}\boldsymbol{f}_i \tag{8.31}$$

where $\boldsymbol{U}_i$ denotes the (3×1) vector of nodal unknowns at grid point i.

For problems exhibiting strong "shocked" solutions, the steep solution gradient at the shock may lead to nonlinear instabilities and divergence. To circumvent this difficulty, an additional artificial dissipative term may be added on the right in (8.28) of the form

$$d(\boldsymbol{U},\boldsymbol{W}) = -\int_{x_l}^{x_r} \mu \frac{\partial \boldsymbol{W}^t}{\partial x}\frac{\partial \boldsymbol{U}}{\partial x}\, dx \tag{8.32}$$

where the artificial diffusivity μ depends on the magnitude of the solution gradient. This implies that this dissipation will be negligible away from shock-like layers and will act to stabilize the local gradients.

Boundary conditions for the system are specified based on the properties of hyperbolic problems and the characteristic directions. More specifically, the eigenvalues of the flux Jacobian $\boldsymbol{A}$ in (8.30) determine the incoming modes. At the source contact there are two incoming modes and at the drain there is only one incoming mode. Hence two source boundary conditions and one drain boundary condition are specified. The eigenvectors of the flux Jacobian can be used to construct a transformation matrix $\boldsymbol{P}$ for a characteristic projection to update the boundary conditions at each timestep. Further details are given in [58].

The FETG scheme described here can also be used for time-iterative recursion to a steady-state solution. In this case local timestepping can be introduced to accelerate convergence. That is, the timestep Δt can be taken independently as Δt_i at node i, provided, of course, that the local stability condition or CFL condition is still satisfied. The resulting scheme obviously is not time accurate and may be interpreted as a form of point-iterative solver for the steady-state solution in which local adaptive relaxation is applied. The curves in Figure 8.3 compare the steady-state residuals for the algebraic system with this time-iterative solver using both the maximum uniform step at each time level and also the local timestep relaxation for a 0.4 μm diode. The uniform step scheme requires 3500 steps to meet a specified tolerance that the local scheme attains in only 2500 steps.

As a simple illustrative example, we show computed results for a 0.4 μm silicon diode with a bias of 3.0 volts. The doping is given by the piecewise-linear profile in Figure 8.4(a) with doping concentration 2×10^{18} cm^{-3} at source and drain and 2×10^{15} cm^{-3} in the channel. The steady-state solution was computed first on a uniform grid of 100 elements and it gave good results for carrier concentration, average energy and drift velocity (see Figures 8.4(a)–(c)). However, there are significant oscillations in the drift-velocity near the drain junction due to the coarse grid. An adaptive local refinement scheme was applied and the final graded mesh of 150 elements produced a non-oscillatory drift-velocity profile Improvements in the other solution variables are also realized through this local refinement process.

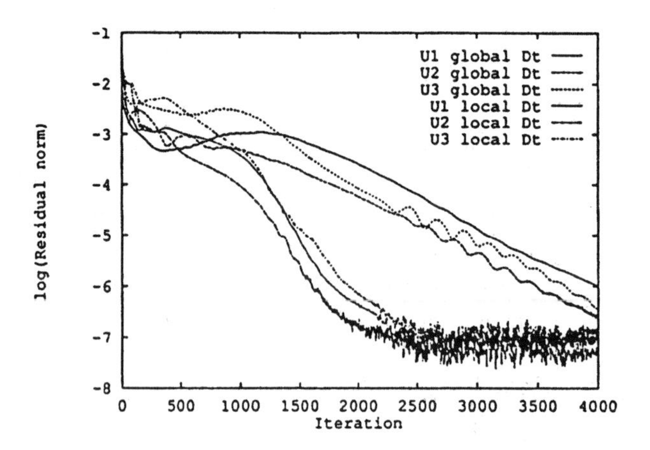

Figure 8.3: Convergence plot for timestep recursion [58].

8.6 Quantum Hydrodynamics

The hydrodynamic model can also be modified to include quantum effects that may be important in extremely small devices. By introducing the momentum displaced Wigner distribution function in the general moment equations, the collisionless quantum hydrodynamic equations are obtained. Under further simplifying assumptions, a quantum-corrected collisionless nonparabolic hydrodynamic transport system can be formulated.

A full 3D quantum hydrodynamic (QHD) model has been constructed [197] from a moment expansion of the Wigner distribution function. In the one-dimensional case this reduces to the QHD model in [225]. The QHD equations can be classified using an eigenvalue analysis of the flux Jacobians as mentioned previously for the standard HD model: the 3D quantum model has two Schrodinger modes, two hyperbolic modes, one parabolic mode and one elliptic mode. Ideas from gas dynamics can be applied to infer qualitative properties of the QHD system and solution. For example, associated with the two hyperbolic modes are two contact discontinuities in the tangential velocity across the wave. The two hyperbolic modes associated with shocks in the standard HD model are transformed into Schrodinger modes (waves) by the addition of the quantum terms.

We now summarize the upwind scheme from [197] for the steady-state one-dimensional quantum hydrodynamic system written in the form

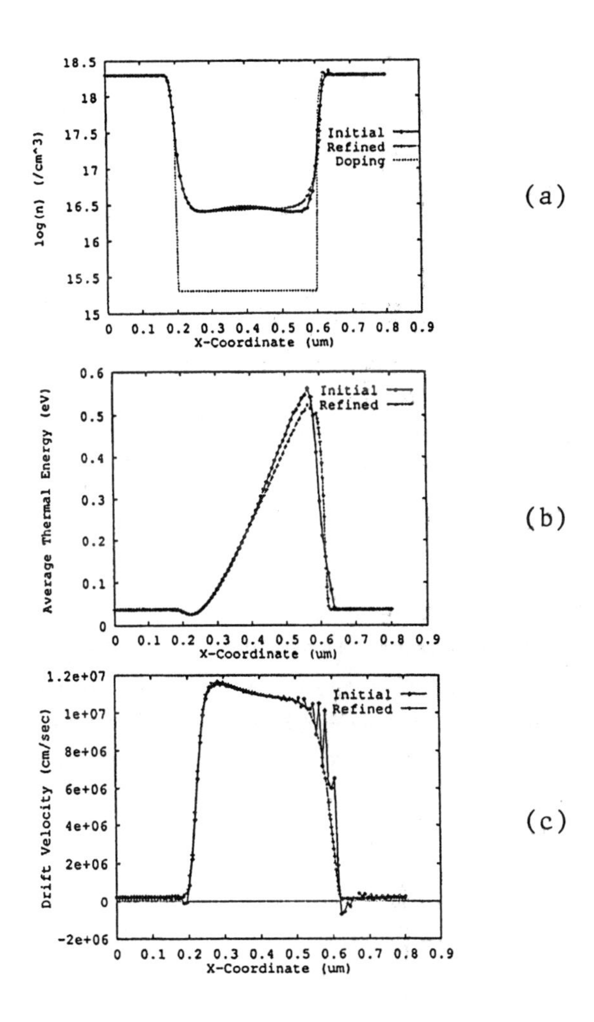

Figure 8.4: FETG results for (a) carrier concentration, (b) average energy, and (c) drift velocity on adaptively-refined grid (from [58]).

$$\frac{d\boldsymbol{U}}{dx} = -\boldsymbol{S} - \boldsymbol{H} \tag{8.33}$$

with $\boldsymbol{U} = (ug_n, ug_u, ug_T, 0)$, $\boldsymbol{S} = (0, s_u, s_T, s_V)$, $\boldsymbol{H} = (0, H_n, H_T, H_V)$ where

$$\begin{aligned}
&g_n = n, \quad g_u = mnu \\
&g_T = \tfrac{5}{2}nT + \tfrac{1}{2}mnu^2 - \tfrac{\bar{h}^2 n}{8m}\tfrac{d^2}{dx^2}(\log n) + nV \\
&H_n = \tfrac{d}{dx}(nT) - \tfrac{d}{dx}\left(\tfrac{\bar{h}^2 n}{12m}\tfrac{d^2}{dx^2}\log n\right) + n\tfrac{dV}{dx} \\
&H_T = -\tfrac{d}{dx}\left(\kappa\tfrac{dT}{dx}\right), \quad H_V = \epsilon\tfrac{d^2V}{dx} \\
&s_u = \tfrac{mnu}{\tau_p}, \quad s_T = \left(\tfrac{3}{2}nT + \tfrac{1}{2}mnu^2 - \tfrac{\bar{h}^2 n}{24m}\tfrac{d^2}{dx^2}\log n - \tfrac{3}{2}nT_0\right)/\tau_\omega \\
&s_V = e^2(N_D - N_A - n)
\end{aligned} \tag{8.34}$$

Here the basic variables are again those of the hydrodynamic model with the Fourier heat conduction assumption and the new quantum contributions being identified with the terms involving Planck's constant $\bar{h}$. In (8.34), $V = -\varepsilon\phi$ is the potential energy, $e > 0$ is the electronic charge, T_0 is the lattice temperature and N_D, N_A are the densities of donors and acceptors.

The backward difference scheme employs a staggered grid with velocity values $u_{i+\frac{1}{2}}$ at the centers of the elements and n, T and V at the end nodes of the elements. Accordingly, the second equation in the system is differenced at the element midpoints $\{i+\frac{1}{2}\}$ and the other equations at the end nodes $\{i\}$. The convective terms are differenced upstream as in

$$\frac{d}{dx}(ug)_i \approx (u_{i+\frac{1}{2}}g_R - u_{i-\frac{1}{2}}g_L)/\Delta x \tag{8.35}$$

where

$$g_R = \begin{cases} g_i, (u_{i+\frac{1}{2}} > 0) \\ g_{i+1}, (u_{i+\frac{1}{2}} < 0) \end{cases}, \quad g_L = \begin{cases} g_{i-1}, (u_{i-\frac{1}{2}} > 0) \\ g_i, (u_{i-\frac{1}{2}} < 0) \end{cases} \tag{8.36}$$

and second-order central differences are used for H_n, H_T, H_V. A damped Newton method with incremental continuation in voltage is applied to solve the resulting nonlinear system.

Resonant Tunneling Diode

The above scheme is applied to simulate a GaAs resonant tunneling diode with double $Al_xGa_{1-x}As$ barriers. The barrier height $\beta = 0.209$ eV (for Al mole fraction 0.3). The n_+ source and drain regions have doping density 10^{18} cm^{-3} with n channel of doping density 5×10^{15} cm^{-3} and length 250 Å as shown in Figure 8.5. The barriers are 50 Å wide as indicated by the shaded bands in the figure. The quantum well between barriers is also 50 Å wide. The device has 50 Å spacers between the barriers and the source or drain contacts. The barrier height β is incorporated in the QHD equations by replacing V by $V + \beta$.

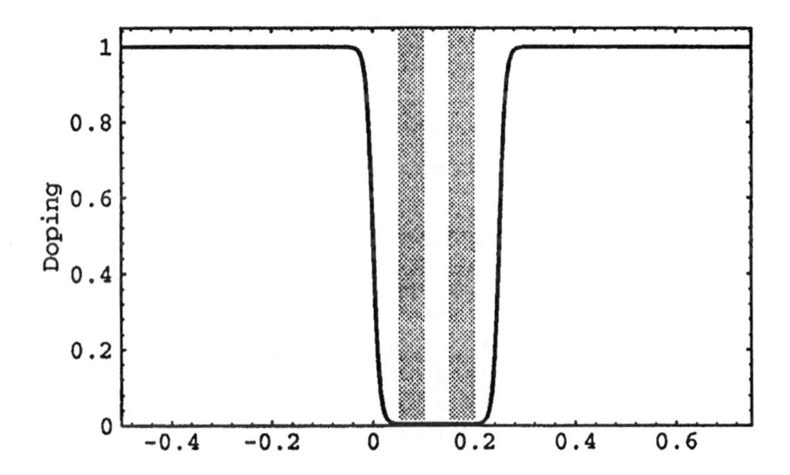

Figure 8.5: Doping profile for *RT* diode showing quantum well barriers as shaded regions. (Horizontal x scale in tenths of μm).

Numerical results for electron density, velocity and temperature for the RT diode are presented in Figure 8.6 for $\kappa_0 = 0.4$ in the Weidemann–Franz relation for thermal conductivity $\kappa = \kappa_0 \tau_{p_0} n T_0 / m$. Note the charge enhancement in the quantum well. The electron density at the center of the quantum well is more than two orders of magnitude greater than the background doping density. The velocity profile confirms that the electrons spend significantly more time in the well than in the barriers, as would be anticipated. The temperature plot indicates that the electrons cool as they pass through the first barrier, then heat up as they accelerate near the channel drain junction and finally cool down again in the drain. Conduction band bending in the well and barriers is evident in Figure 8.7. The above computations required incremental continuation in voltage to obtain convergence. The current–voltage curve for the device is plotted in Figure 8.8 and it displays a region of negative differential resistance from the peak to the valley of the curve. Further details are given in [197].

8.7 Some Extensions

Hydrodynamic models based on non-parabolic energy band assumptions have been suggested as more appropriate for modeling hot carrier effects [518]. A discussion of the system structure and scaling choice are given in [426]. In this latter treatment we also develop and implement a modified Scharfetter–Gummel (S-G) strategy for curvilinear graded meshes. Let us consider the two-dimensional case. The basic idea is to introduce a transformation between the physical (x, y) domain and a reference (ξ, η) domain. The hydrodynamic system transforms accordingly with the metric coefficients of the transformation now entering in the transformed system on the reference domain. This system can now be differenced on a cartesian grid in the reference domain and corresponds to solving the problem on the mapped, graded, (and possibly curvilinear) grid in the physical (x, y) domain.

Recalling the fundamental S-G approach developed in Section 7.3, the (scaled) current density $\boldsymbol{J}$ in the transformed (ξ, η) domain can be written as

$$\begin{aligned} \boldsymbol{J} &= \left(-\frac{2}{3}M_1\tau_p\frac{\partial(Bnw)}{\partial\xi} + M_2\tau_p n\frac{\partial\psi}{\partial\xi}\right) \mid \nabla\xi \mid \boldsymbol{k}_1 \\ &+ \left(-\frac{2}{3}M_1\tau_p\frac{\partial(Bnw)}{\partial\eta} + M_2\tau_p n\frac{\partial\psi}{\partial\eta}\right) \mid \nabla\eta \mid \boldsymbol{k}_2 \end{aligned} \tag{8.37}$$

where $\boldsymbol{k}_1$ and $\boldsymbol{k}_2$ are unit vectors in the respective ξ and η directions, n is the electron density, w is the total energy, τ_p is the momentum relaxation time, $B(w) = (1 + \alpha w/q)/(1 + 2\alpha w/q)$ for charge q, $M_1 = t_s w_s/m v_s x_s$, $M_2 = q_s t_s \psi_s/m v_s x_s$ for reference timescale t_s, length scale x_s, potential ψ_s, velocity v_s and m is the effective mass.

A rectangular grid is introduced for the (ξ, η) domain with $\boldsymbol{J}, B$ assumed locally constant and ψ, w locally linear on each rectangular cell. We also use the empirical relation $\tau_p \propto 1/w$. Next, $\boldsymbol{J}$ is projected in the respective ξ, η directions and the components J_1, J_2 integrated along the edges of each rectangular cell using the unknown local values of n, w and ψ for boundary

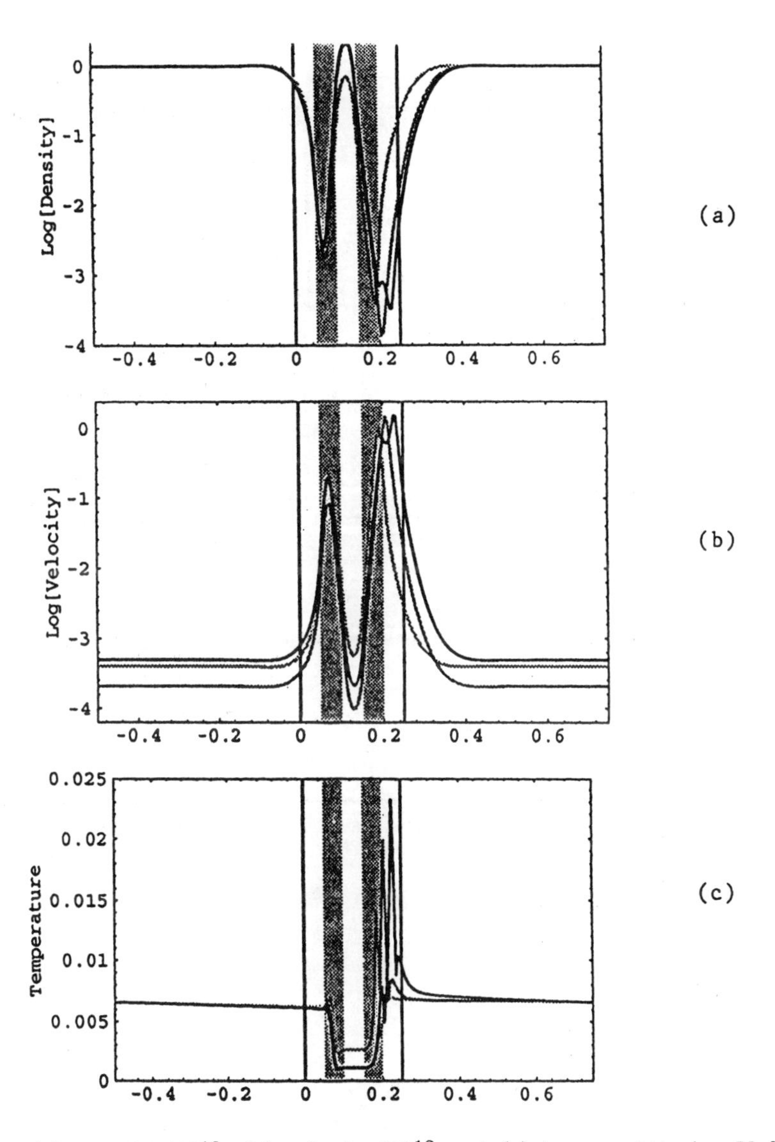

Figure 8.6: (a) Density/10^{18}, (b) velocity/10^{18} and (c) temperature in eV for ΔV=0.097, 0.191, and 0.22 volts. The horizontal scale x is in tenths of a μm (from [197]).

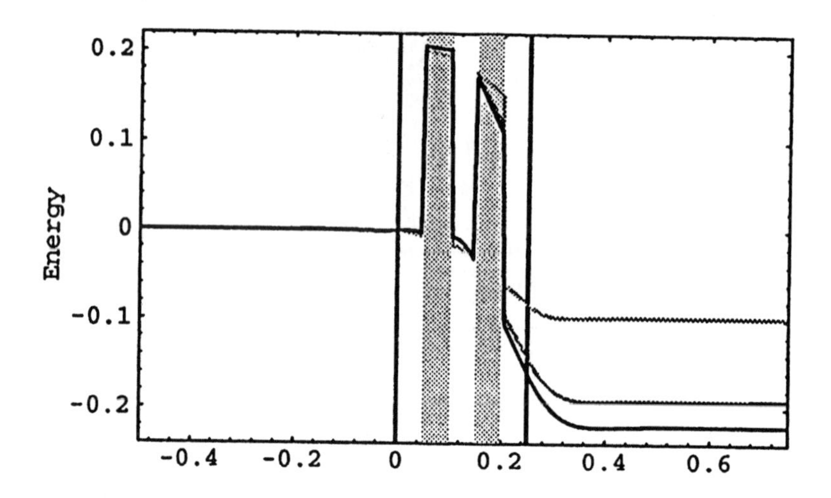

Figure 8.7: Conduction band bending in the well and barriers.

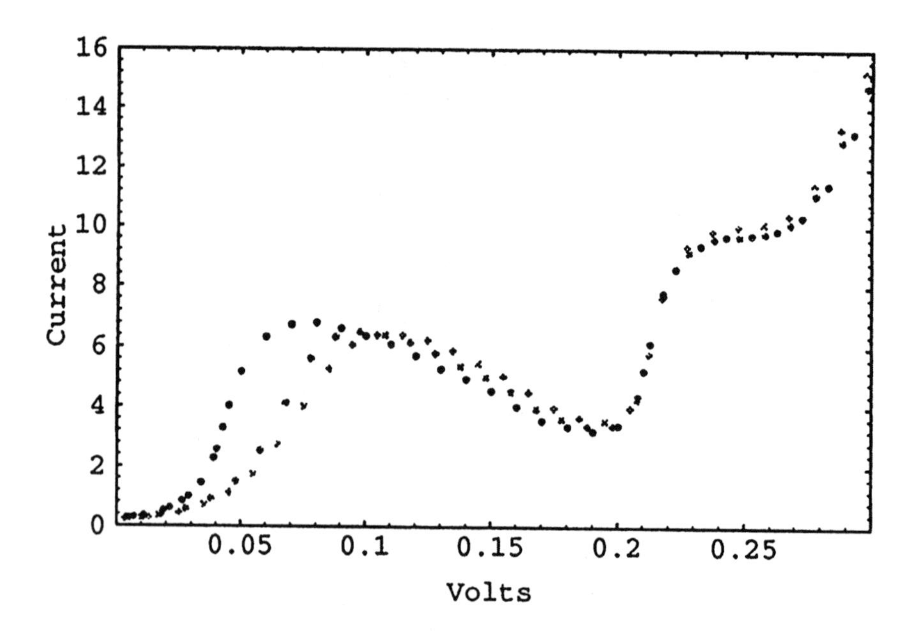

Figure 8.8: Current–voltage curve for the device.

conditions to obtain

$$\frac{J_1}{\mid \nabla \xi \mid} = -\frac{2}{3} M_1 C_r B(w_i^{av}) \left[\frac{n_{i+1}}{w_{i+1}} \beta(-X) - \frac{n_i}{w_i} \beta(X)\right] \frac{(w_{i+1} - w_i)}{ln(w_{i+1}/w_i)\Delta \xi} \quad (8.38)$$

where

$$X = \left[2 - \frac{3}{2} \frac{M_2}{M_1 B(w_i^{av}} \frac{(\psi_{i+1} - \psi_i)}{(w_{i+1} - w_i)}\right] ln(w_{i+1}/w_i) \quad (8.39)$$

and $\beta(x) = x/(e^x - 1)$ is the Bernoulli function. A similar relation applies for J_2 on replacing i by j and ξ by η. In (8.38) and (8.39), C_r is a scaled coefficient of τ_p, $\Delta\xi = \xi_{i+1} - \xi_i$ and w_i^{av} is average energy along interval $\Delta\xi$.

Substituting these relations for $\boldsymbol{J}$ into $\nabla \cdot \boldsymbol{J}$ and the transport equations yields the final semidiscrete system of ODE's for the carrier continuity equations. A similar extension of the S-G approach and ODE systems can be constructed for the energy equation in precisely the same way. Finally, the electrostatic potential equation is centrally differenced to complete the system. The complete fully-coupled ODE system then has the form

$$\frac{d\boldsymbol{U}}{dt} = \boldsymbol{F}(\boldsymbol{U}) \quad (8.40)$$

where $\boldsymbol{U}$ the vector of nodal unknowns (potential, carrier concentration, velocity and energy). This ODE system can then be integrated numerically using the schemes described for semidiscrete ODE systems described previously to determine the transient response. The system can also be integrated to steady state as described in the following example.

As a test case we consider a MOSFET-like structure with gate length approximately 0.6 μm and simple rectangular source and drain doping regions as shown in Figure 8.9. The solution was computed on a 61×41 mesh graded strongly into the junction regions of the device. A backward Euler integration scheme was utilized for the ODE system with sparse nonsymmetric biconjugate gradient squared (BCGS) solver to enhance efficiency. Steady state surface plots for the energy and electrostatic potential are shown in Figure 8.10. Further details are given in [426].

A variety of modeling capabilities now exist in commercial, industrial and university device simulators. These range from simple one-dimensional drift-diffusion models to three-dimensional advanced transport models and Monte Carlo schemes. The push towards the deep submicron regime has made the simpler drift-diffusion physical models and one-dimensional analysis less effective. More complex models and higher-dimensional capability

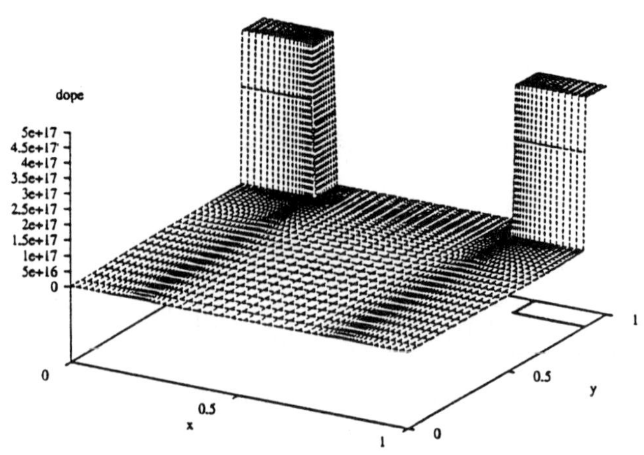

Figure 8.9: Doping regions and graded 61×41 mesh for test problem.

are then needed for reliable analysis and design studies. At the same time it should be recognized that most simulation work for device design is carried out with simpler models in one dimension and to some extent in two dimensions. An argument can be made for developing a hierarchy of models and capabilities. The initial design studies can then be made with simpler models to estimate threshold voltage, drain current, etc. Successively more refined models and higher-dimensional simulations are then made for substrate current predictions, analysis of hot carrier effects and device reliability calculations for the final or near-final design. This implies that the overall analysis and product cycle time will be more efficiently handled. It also implies that full 3D simulations on fine grids with complex models will be kept to the minimum.

The 3D problem mentioned above remains a formidable obstacle: the steep gradients in the solution fields necessitate fine grids so grid generation and grid refinement are key bottlenecks, inhibiting effective solution and the 3D discretization leads to algebraic systems with many unknowns and the computation and storage requirements scale nonlinearly. These topics remain areas of current research interest and are an important component in the next generation simulators. Several key issues concerning grid generation and adaptive grid refinement are taken up in the next chapter.

8.8 Summary

Hot carrier effects become important when the gate length of the device is reduced to the deep submicron scale. The need to model energetic carriers then implies that the energy equation should be included in the mathemat-

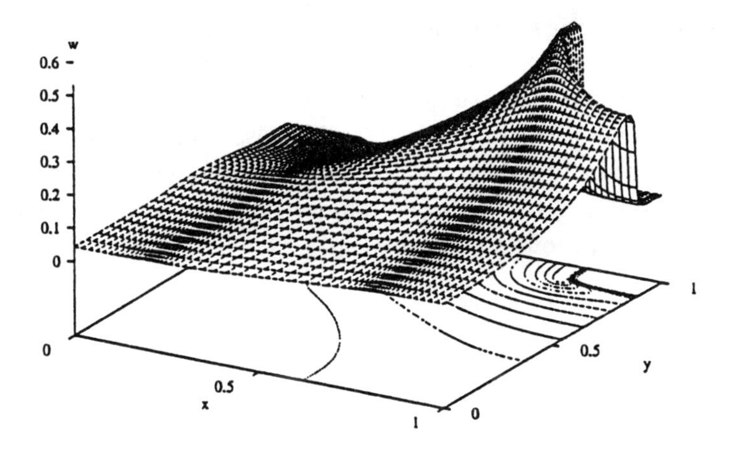

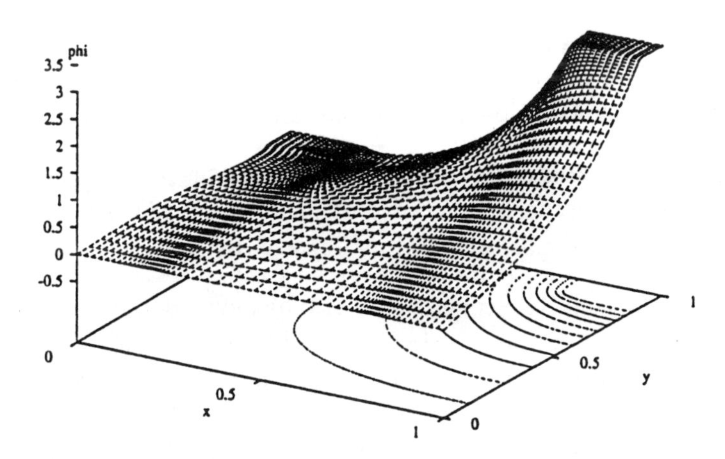

Figure 8.10: Surface plots of energy and electrostatic potential from non-parabolic hydrodynamic solution(from [426]).

ical model. The resulting transport operator closely resembles the euler system operator in gas dynamics and hence the terminology hydrodynamic system. It is not surprising then that techniques from computational fluid dynamics can be successfully extended to the hot carrier transport problem. Here we describe the Lax–Wendroff and related Taylor–Galerkin schemes for this system and give sample results. We also include some results from the literature that include quantum effects in the model. This area is a topic of present research interest and new numerical models are needed to treat the various physical models efficiently.

8.9 Exercises

1. Develop a finite difference scheme for the steady one-dimensional hydrodynamic equations (8.9)–(8.12) using backward differencing of the convective terms in the transport system and the Gummel iteration. Briefly describe how this scheme would be extended to two and three dimensions.

2. Investigate the sensitivity of the solution to the one-dimensional diode test problem corresponding to Figure 8.1 by perturbing the conductivity κ and relaxation time parameters by 10 percent.

3. Complete the finite difference derivation of the Lax–Wendroff scheme for the model hyperbolic transport problem (8.18). Develop a nonlinear extension of this scheme for the case $\alpha = u$.

4. Least-squares finite elements may also be developed for transport problems. In this approach (8.18) is first differenced with respect to time and then the least-squares residual is constructed. Minimizing the residual yields a weak statement similar to a Petrov–Galerkin form. Construct the weak statement for this test problem and then use integration by parts to recover a corresponding Euler–Lagrange differential equation. Compare the dissipative contribution with that of the Taylor–Galerkin scheme.

5. Determine the eigenvalues of the flux Jacobian $\boldsymbol{A}$ in (8.29) and comment on the nature of the associated modes. What are the implications regarding source (inflow) and drain (outflow) boundary conditions?

Chapter 9

Grid Generation and Refinement

9.1 Introduction

Semiconductor process and device problems typically involve PDE systems to be solved on regions of irregular geometric shape with several material layers. This implies that unstructured irregular grids are useful to discretize the domain efficiently. Secondly, the associated transport problems frequently involve solutions that may exhibit layer structures or steep gradients. Thus a basic requirement is an efficient grid generation capability for unstructured grids in irregular domains. The most efficient approach to obtain accurate, reliable solutions is to grade the underlying mesh into the layer regions. Since these regions and the necessary mesh resolution are often not known in advance, an adaptive mesh refinement strategy is useful. In the case of steady-state problems the solution on a coarse background grid can be computed and the grid then locally refined in regions where the errors in the computed solution are ascertained to be large. Hence, computable error indicators are needed in this step to determine which subregions, patches, or elements should be refined. These error indicators should ideally be based on a rigorous mathematical foundation centered on the mathematical statement of the problem and the approximation properties of the discretization method. *A posteriori* error analysis provides such a framework for developing error indicators. Similar ideas apply for evolution problems. In this case the grid may require refining in some part of the domain and coarsening in another part of the domain with adaptive time-stepping and integration order selection as the solution evolves. This implies, of course, that the computational complexity increases significantly.

In the case of large-scale two-dimensional and three-dimensional simulations, mesh refinement can bring within our scope problems that previously were too large for practical calculation. It can also permit a greater variety of parameter ranges to be investigated for the same reasons. This is particularly true of nonlinear problems where many solutions may be necessary, solution bifurcation branches may need to be traced out, or a range of parametric studies must be computed. It is also well known that the ability to obtain iterative convergence in many nonlinear applications may depend on the choice of starting iterate and the quality of the mesh. That is, the nonlinearity may be such that a convergent iteration is not possible on a given mesh, whereas a near-optimal mesh of the same size may be successful. Finally, adaptive refinement through a sequence of meshes provides a natural framework for a continuation approach in treating the nonlinearity – the nonlinearity is gradually increased as the mesh and solution are progressively improved. The solution on an intermediate mesh provides a good starting iterate for the nonlinear solution and mesh refinement at the next continuation step. Incremental continuation in a parameter or arc length continuation, is a common strategy for addressing this difficulty. Similarly, in time-stepping algorithms the solution at the previous timestep and grid will be a good starting iterate for the next solution step.

Some basic ideas for adaptive refinement were introduced in Chapter 7 and applied to the device problem. Here we consider some extensions of these techniques and associated error analysis. We begin with point insertion schemes and associated Delaunay triangulations since this approach is very flexible both for grid generation and local adaptive refinement. Next we return to the quadtree and octree data structures introduced in Chapter 7 and describe some associated element subdivision strategies. The mathematical ideas related to local error estimators and error indicators are discussed in Section 9.4. The benefits of interweaving solution iteration with mesh refinement are then summarized in the next section. Some brief comments on mesh redistribution and moving meshes conclude the chapter. These topics are discussed at length in [82] which also provides an extensive bibliography.

9.2 Point Insertion Strategies

As noted in the introduction, for most practical applications, unstructured grids are needed even if adaptive refinement is not applied. Thus, adaptive grid techniques based on local mesh refinement fit conveniently in this set-

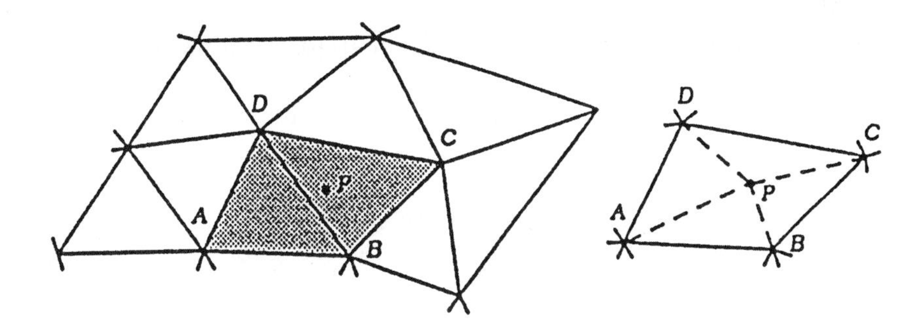

Figure 9.1: Grid enrichment by point insertion. Point P is added and the circumcircle test determines the enclosing "cavity" for vertex reconnection.

ting. However, local refinement will necessitate more careful consideration of the supporting data structures, as we see later.

To fix ideas, let us begin by summarizing the main steps in a calculation involving layers adaptive refinement for a steady state problem:

1. An initial background grid is generated.
2. The discretized problem is solved on the current mesh.
3. Quit if converged or at maximum allowable grid.
4. The mesh is locally refined and data structure updated.
5. Return to Step 2.

Now there are several related ideas here that merit elaboration: first, the initial background grid may be generated in an unstructured way from an initial set of grid points determined by the definition of the boundary and the material layers. Alternatively, a marching front strategy can be used that works in from the boundary and out from interfaces. Finally, a point-insertion strategy can be developed (Figure 9.1). For example, unstructured grids of two-dimensional triangulations and three-dimensional tetrahedral tesselations can be developed in this manner. For brevity we outline the point insertion approach for unstructured two-dimensional triangulations. This scheme is very efficient for practical applications, and there is public domain and industrial software available.

Assume that we are given a set of ordered points defining the boundary and material interfaces. The problem is to construct a "good" unstructured triangulation of the set of points. The main steps in the algorithm become:

1. First normalize the coordinates of the points so that the domain is contained in the unit square.

2. Next, define a supertriangle containing the unit square and such that the vertices of the supertriangle are remote (more than several units from the square boundary).

3. Now insert the first point (x_1, y_1) and connect it to the vertices of the supertriangle. Label the new triangles and nodes and construct element vertex and neighbor arrays defining the connectivity and data structure.

4. Insert the next point; locate the containing triangle and perform circumcircle tests for this point with adjacent triangles to carry out Delaunay swaps; this will maximize the minimal angle to avoid slender triangles. For example, the added point D in Figure 9.2 lies inside the circumcircle of triangle ABC so connect D to vertices B and C and exchange diagonal DA for diagonal BC.

5. Step 4 is continued recursively for all points in the grid.

6. Finally, the boundary and material interface contours are reconstructed by diagonal swaps and triangles exterior to the domain are deleted.

The above procedure yields an optimal Delaunay triangulation of the domain. The computational complexity of this scheme is only $O(N \log N)$ where N is the number of mesh points. Hence the algorithm is very fast and can be conveniently implemented on a graphics workstation. The scheme can be further improved by including a bin sort. For example, a triangulation with several thousand elements can be generated in this fashion in a few seconds on a medium-performance workstation. One of the important features of the Delaunay procedure is that it automatically involves neighbor element information. This is important in adaptive refinement as we now see.

The general approach described above can be extended easily to adaptive refinement using point-insertion ideas. That is, assume that we have generated the grid in this manner and now computed a solution and constructed error indicators. Those elements that require refinement are flagged and the

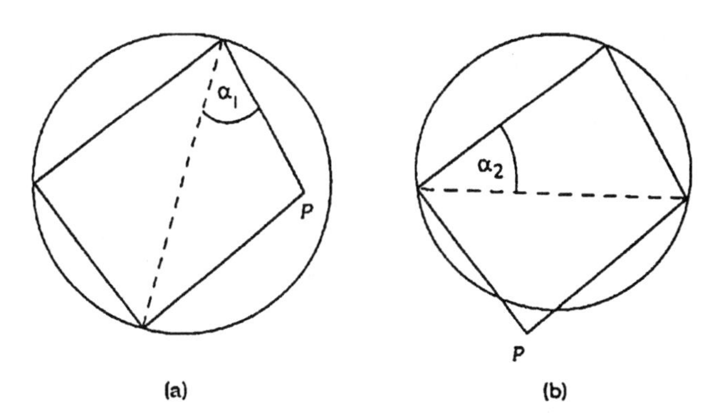

Figure 9.2: Max-min angle or circumcircle test for diagonal swap.

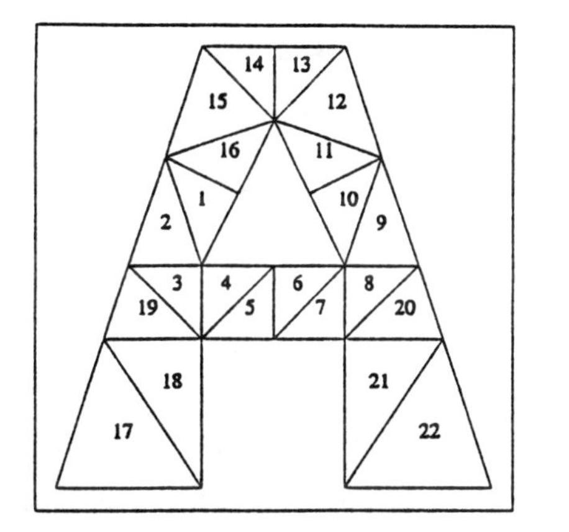

Figure 9.3: Example of refined triangulation obtained by point insertion based on specified "attractors"(from [57]).

point insertion algorithm can obviously be applied recursively to update the grid. The solution/refinement steps are then repeated as indicated previously. An example of a grid generated using a variant of this approach is given in Figure 9.3. Here, the points are inserted based on certain attractor zones that may be associated with error indicators and this leads to the graded final mesh shown. We remark that there is considerable flexibility in the point insertion process for adaptive refinement. That is, the scheme may be based on inserting a mid-point in the longest side of a triangle, at the centroid or elsewhere. An example employing this approach to refine into a layer defined by the doping profile is shown in Figure 9.4. Only a detail of the grid near the doping layer is shown here.

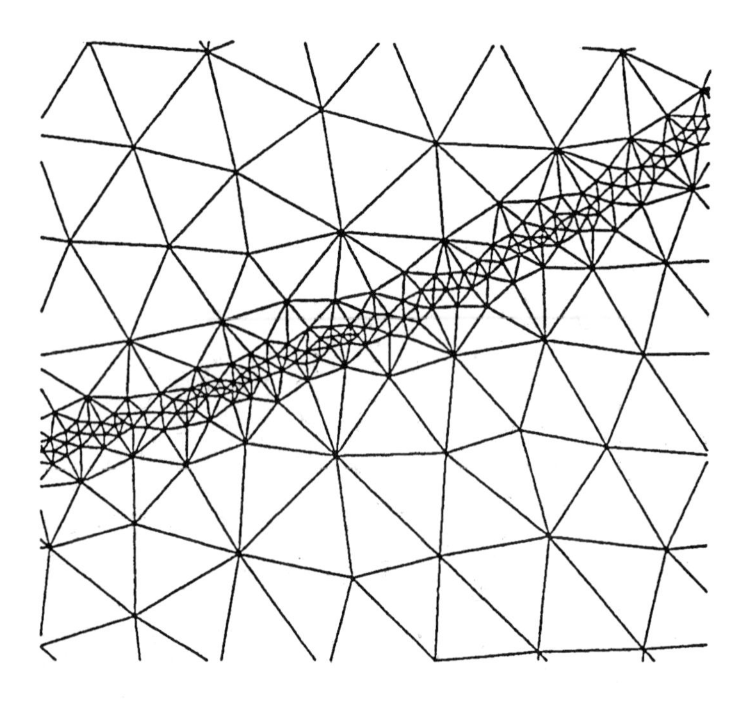

Figure 9.4: Detail of refined grid in "source" region.

The above strategy extends trivially to prismatic discretization of certain three-dimensional regions. Here the spatial (x, y) domain is triangulated as above and the triangulation translated vertically in a marching front strategy to define the prism elements. For true three-dimensional applications the grid generation and adaptive refinement schemes are more complex but are very important to an efficient solution. This is particularly true for problems where fine scale details of the solution are to be resolved. The basic grid generation strategy for tetrahedral tesselations proceeds in a manner similar to that for the two-dimensional triangulation: a super-tetrahedron containing the domain is first constructed; then points are inserted sequentially and triangle face swaps carried out by testing if the inserted point lies within the circumsphere of the adjacent tetrahedra. Finally, exterior tetrahedra are "pruned" from the mesh. Not surprisingly, this scheme has higher computational complexity than its two-dimensional counterpart and practical estimates for moderate grids are $O(N^{3/2})$. There are some other important distinctions from the two-dimensional case: first of all the circumcircle test guarantees that the triangles are well-shaped in the sense that it performs a diagonal swap to maximize the minimum angle. This property is not preserved for the circumsphere test. Hence, tetrahedra with small solid angles may be generated. In practice, a post-processing step can be used to smooth the grid and alleviate this problem. Another important distinction is that a face swap for a pair of adjacent tetrahedra can lead to a configuration of three tetrahedra.

9.3 Quadtree and Octree Data Structures

An alternative approach to point insertion is to use the ideas of quadtree and octree data structures introduced in Chapter 7 for two-dimensional and three-dimensional refinement, respectively. Here the initial background grid is generated as before and provides the level 0 "macroelement" discretization. In the subsequent refinement step, elements are subdivided, typically to a quartet in two dimensions or an octet in three dimensions, although other variations on these partitions are also useful in certain applications; e.g. in the device problem directional refinement is essential in layer regions. This quartet subdivision is continued recursively to the subelements and further levels in the quadtree (recall Figure 7.4). As a given element is refined it becomes "inactive" and the data structure is updated to remove this element and add the new sub-elements so obtained. The active leaf nodes of the tree define the current mesh at any stage of the algorithm.

This strategy can be applied both for grid generation and for refinement. In the former case the coarse unstructured background grid may be uniformly refined through several levels. If the initial coarse grid is appropriately graded, this uniform unstructured refinement strategy can be effective. However, it is also obvious that the number of elements increases at an exponential rate with $N_0 2^{dl}$ elements at level l for dimension d, where N_0 is the number of elements in the initial grid. A better strategy is to locally refine the elements in an adaptive manner to yield an "unbalanced" tree with active elements at several different levels. The local gradient and curvature can be calculated as a feature indicator, or a more analytical interpolation error estimate can be introduced to accomplish this step, as seen in the next section.

If a triangle is sub-divided to a quartet of sub-triangles by connecting mid-side points, then the new sub-triangles are congruent and each is similar to the parent triangle. This implies that the angles remain well behaved under refinement. However, some care must be exercised in handling the transition to adjacent elements that are not refined. This can be made by connecting the new node to an opposite vertex, by defining special transition elements or by enforcing constraints on the solution variation across this mesh interface. In general, it is desirable that no two adjacent elements lie more than one or two levels apart in the quadtree. This recursion implies that coding in C or a similar language is recommended.

In the case of tetrahedral refinement, connecting the mid-edge points yields four corner sub-tetrahedra each similar to the original tetrahedron, and in addition an interior octahedron which can be sub-divided by any diagonal to define four further sub-tetrahedra. However, these interior sub-tetrahedra are not similar to the vertex tetrahedra. Refining a cube to an octet of sub-cubes proceeds similarly. Note that an interior cube has 6 face neighbors and 12 edge neighbors, so the neighbor pointers and data structures are significantly more complicated than in the two-dimensional case.

A modified octree approach was developed in [244] for three-dimensional semiconductor device structures. The device is first partitioned into a set of macro-elements consisting of cubes, rectangular prisms and rectangular pyramids. The grid generation and refinement process then proceeds by sub-dividing these macro-elements in a manner similar to that described previously. The scheme allows directional refinement towards boundary surfaces so that cubes may be divided, for instance, into octets, quartets or halves as needed (see Figure 9.5). The added edge nodes are handled by refining the adjacent elements appropriately to accommodate the transition and achieve a valid finite element mesh.

In this way a modified octree is generated. As an example, part of the mesh for a MOS-controlled thyristor is shown in Figure 9.6. The device structure is actually 10 μm $\times$ 10 μm $\times$ 100 μm and the mesh shown corresponds to part of the top. The final mesh contains 41,716 elements with 31,846 nodes and required 860 CPU seconds to generate on a SPARC station 1+ with 20 MB main memory.

9.4 Error Indicators

Let us now turn to the issue of error indicators and the *a posteriori* error estimators that underlie them. The simplest computable indicators are based on solution features such as the magnitude of local solution gradients. For problems with certain types of singularities and layers, this approach may be effective but generally it will not be adequate. Instead, we seek an error indicator that is more closely correlated with the unknown error behavior.

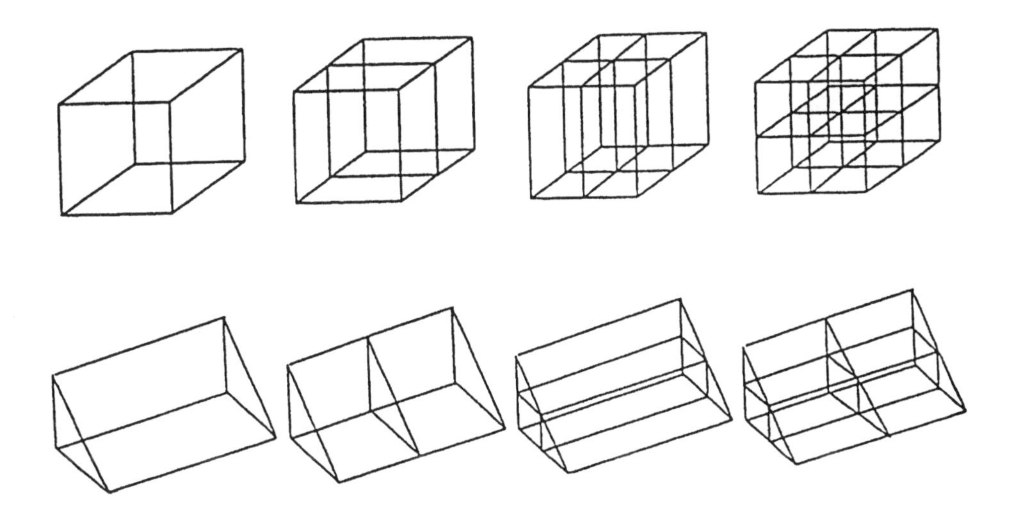

Figure 9.5: Refinement of cube and prism elements.

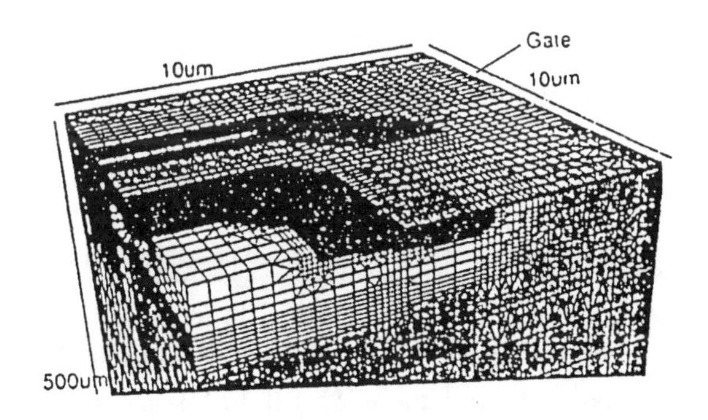

Figure 9.6: Adaptive 3D grid in top section of thyristor (from [244]).

There is a wealth of literature dealing with *a priori* error estimates for finite difference and finite element solution of a broad class of linear and nonlinear monotone problems. The finite difference estimates basically rely on Taylor series and truncation error analyis. *A priori* finite element estimates use the observation that the Galerkin finite element approximation to a linear elliptic boundary-value problem is the best possible approximation in the associated energy norm. Hence the approximation error $E = u - u_h$ can be bounded by the interpolation error $E_I = u - \tilde{u}_h$ where $\tilde{u}_h$ is the finite element interpolant of exact solution u on the given mesh. One can then appeal to interpolation theory to obtain an interpolation estimate on element

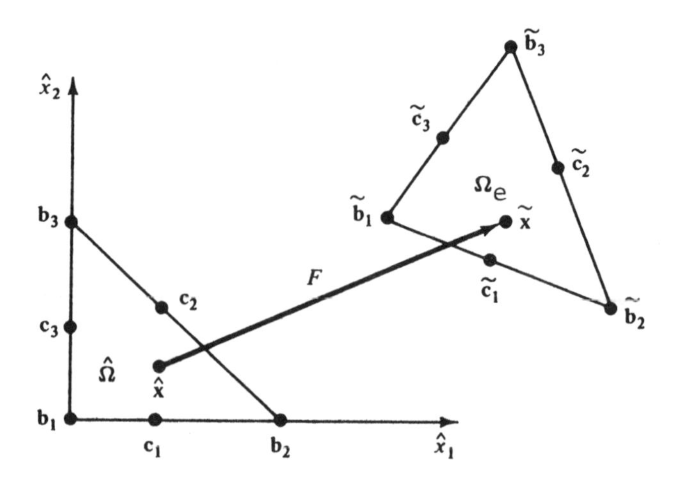

Figure 9.7: Affine map from reference triangle $\hat{\Omega}$ to representative triangle $\hat{\Omega}_e$ in the discretization.

Ω_e (Figure 9.7) of the general form

$$\|u - \tilde{u}_h\|_{m,q,\Omega_e} \leq C\frac{h_e^{k+1}}{\rho_e^m}\text{meas}(\Omega_e)^{\frac{1}{q}-\frac{1}{p}} \mid u \mid_{k+1,p,\Omega_e} \tag{9.1}$$

where $\|\cdot\|$ denotes the $W^{m,q}$ norm on Ω_e, $|\cdot|$ similarly denotes a seminorm, $p = q = 2$ corresponds to the standard Hilbert space $H^m = W^{m,2}$, h_e is the element circumcircle diameter, ρ_e is the element incircle diameter, k is the degree of the element, and meas(Ω_e) denotes the area or volume of the element in two dimensions or three dimensions respectively.

Now if the quantity on the right in (9.1) can be approximated and the result computed, this will form the foundation for an error indicator based on this interpolatory estimate. For example, let us assume that the generic constant C is slowly varying over the elements and can be calibrated approximately from an auxiliary problem with known solution. Furthermore, let us smooth the computed finite element solution on a patch containing element Ω_e to define u_h^*. Then we can compute an approximation $|u_h^*|_{k+1,p,\Omega_e}$ to the final factor on the right in (9.1) and this will yield a computable local error indicator,

$$\varepsilon_e = (h_e^{k+1}/\rho_e^m)\text{meas}(\Omega_e)^{\frac{1}{q}-\frac{1}{p}} \mid u_h^* \mid_{k+1,p,\Omega_e} \tag{9.2}$$

This set of element error indicators $\{\varepsilon_e\}$ can now be rank-ordered in magnitude. The subset of elements with indicators of large magnitude defines which elements are to be refined. As an example, if linear triangles or tetrahedra are used, then we can replace (h_e^{k+1}/ρ_e^m) by $\tilde{h}_e^{k+1-m}$ with element degree $k = 1$ and $m = 0$ or 1, where $\tilde{h}_e$ is the maximum side of the element. A smooth approximation u_h^* on a local patch of elements can be used to approximate the *second* derivatives on Ω_e. Alternatively, we can use the interface normal flux jumps between adjacent elements to compute approximations to second derivatives for Ω_e.

There is a tacit assumption in the above approach that the *local* interpolation estimate is an appropriate estimator for the local Galerkin error but this is clearly not rigorous. That is, the fact that the interpolation error provides a global bound does not imply a similar local property. Nevertheless the local approximation to the interpolation error based on using the current finite element Galerkin approximation is a rational strategy for determining an error indicator and has proven useful in practice.

Another class of estimators and indicators can be derived using residuals. Here the basic idea is to tie the refinement criteria more closely to the error in the approximation to the governing equations rather than the error in the approximate solution. For example, the truncation error for a finite difference approximation of a differential equation can be used to assess the adequacy of the mesh resolution in finite difference computations. Since most standard finite element schemes are derived from some form of weighted-residual statement, the residual of the PDE is the natural analog of the truncation error and we will now examine this approach.

To illustrate the ideas, let us take the linear operator equation (or system) written as

$$Lu = f \tag{9.3}$$

in domain Ω with appropriate boundary conditions specified on $\partial\Omega$. Let $\tilde{u}$ be an admissible trial function, $\tilde{u} \in H$ where H is the class of admissible functions satisfying the essential boundary conditions. Substituting $\tilde{u}$ for u in (9.3), we define the residual r by

$$r = f - L\tilde{u} \tag{9.4}$$

Now setting $f = Lu$ from (9.3) in (9.4), and using the linearity of L,

$$r = L(u - \tilde{u}) \tag{9.5}$$

That is, the error $E = u - \tilde{u}$ satisfies the differential equation $LE = r$, where the residual r is the forcing function. Formally, we may introduce the inverse operator L^{-1} and write $E = L^{-1}r$. Taking norms, it follows that

$$\|E\| \leq \|L^{-1}\|\|r\| \tag{9.6}$$

so that the error is bounded by an associated norm of the residual. This implies that refining the grid to reduce the residual will reduce the bounding term on the right and the solution error will be reduced accordingly. Moreover, the above argument applies on any subdomain of the problem and hence can be used to construct a residual-based error indicator for an adaptive refinement algorithm. For example, a smooth approximation u_h^* can again be interpolated on a patch and then used in (9.4) on the element to approximate the element residual as an error indicator for refinement. In an analogous manner, the truncation error in a finite difference scheme can be approximated on a patch to determine a corresponding local truncation error indicator.

In a collocation finite element model, the residual is set to zero at a specified number of interior collocation points $\boldsymbol{x}_c$ within each element. That is, the algebraic system is obtained by setting $r(x_c) = 0$, $c = 1, 2, \ldots, N$. For example, in solving a second-order elliptic boundary-value problem, a C^1 finite element basis is used. The L^2 norm of the residual can be computed on each element using the resulting solution to obtain the residual indicators. This can be compared with a Galerkin finite element scheme where the interface flux jumps, arising from the Gauss formula, are also included.

There have also been several attempts to derive indicators that are based on certain local superconvergence properties or superconvergent post-processing strategies. These schemes seek to exploit the observation that the solution, flux or a derived quantity may be exceptionally accurate at certain points such as the nodes or Gauss points. For example, the fluxes $\boldsymbol{\sigma}_h$ on an element can be obtained by differentiating the local approximation u_h. Let us assume that a superior local flux approximation $\boldsymbol{\sigma}_h^*$ can be constructed using some form of patch averaging or a similar projection. Then $\|\boldsymbol{\sigma}_h^* - \boldsymbol{\sigma}_h\|$ on Ω_e can be used as a local error indicator. Consider the case of the linear basis on the triangle with nodes at the vertices. The flux vector $\boldsymbol{\sigma}_e$ is constant on the element. Next take the continuous piecewise-linear least-squares approximation to this piecewise-constant field on a patch P_i of elements surrounding vertex node i using a three-point nodal quadrature on each element of the patch. This yields the area-weighting formula at central node i of the patch

$$\boldsymbol{\sigma}_i^* = \sum_{e \in P_i} \frac{A_e}{A_P} \boldsymbol{\sigma}_e \tag{9.7}$$

where A_e is the area of element e and A_P is the area of the entire patch (Figure 9.8). Now the vertex values $\boldsymbol{\sigma}_i^*$ can be interpolated on each triangle to obtain $\boldsymbol{\sigma}_h^*$ and the error indicator is computed as $\|\boldsymbol{\sigma}_h - \boldsymbol{\sigma}_h^*\|$ on Ω_e.

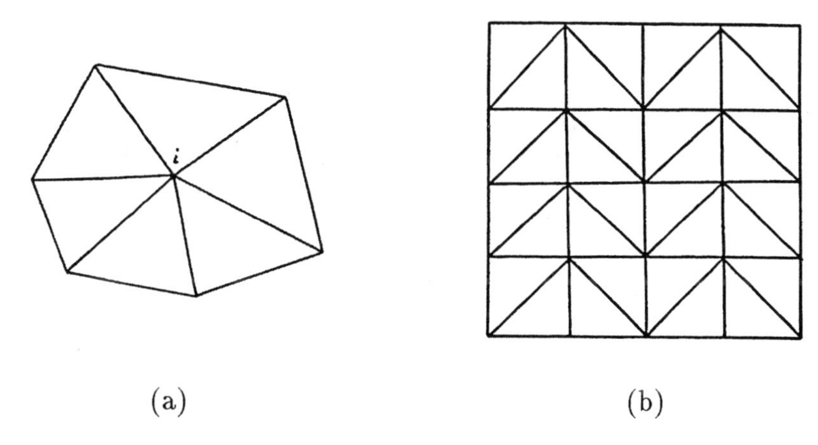

Figure 9.8: (a) Patch of triangles surrounding vertex node i and (b) chevron grid for asymptotic test.

As a final category, we may consider solving a local boundary-value problem on each element to develop an inexpensive accurate local solution u_h^* and then directly evaluate $\|u_h^* - u_h\|$ on Ω_e as an error indicator. The most direct approach is to interpolate boundary data on an element using the existing approximate solution and then simply apply local refinement or increase the local basis degree and approximate the solution to $Lu = f$ on Ω_e. Other variations of this approach involve constructing a local Neumann problem, constructing a related auxiliary problem on the element, or a similar strategy.

Certain asymptotic properties of the estimator are desirable. Let ε_i be an estimate of the true error E_i on subregion Ω_i. Then, as the mesh is refined, $E_i \to 0$ and ε_i should behave similarly. More specifically, we want $\varepsilon_i \approx E_i$ as the mesh is refined and we refer to this property as "asymptotically exact". Therefore, the effectivity index $\rho_i = \|\varepsilon_i\|/\|E_i\|$ is frequently introduced to evaluate and calibrate an estimator for a problem with known solution, usually with Ω_i taken as the full domain Ω. The asymptotic exactness property then implies that

$$\lim_{h \to 0} \rho_i = 1 \tag{9.8}$$

Note that if $E_i \to 0$ faster than ε_i as $h \to 0$, then ρ will grow as some power of $\frac{1}{h}$. If E_i and ε_i approach zero with h and at the same rate then $\rho_i \rightarrow C$, constant. However if $C \neq 1$ then the estimator is not asymptotically exact. This will imply over-refining or under-refining the mesh in practical applications. As an example, consider the Dirichlet problem $-\Delta u = 0$ on the unit square with exact solution $u = xy$ using a uniform rectangular mesh subdivided by diagonals to right isosceles triangles. The approximate solution with linear elements is exact at the vertices (independent of the diagonal alignment). Now apply the flux estimator described previously and consider a sequence of uniform refinements for the particular case of the "chevron" or "herringbone" triangulation (Figure 9.8(b)) where the diagonals in alternate grid columns are respectively of positive and negative slopes. Then $\|\sigma - \sigma_h\| = h^2/\sqrt{6}$ but $\epsilon = (\sqrt{7}/\sqrt{3})h^2/6$ on Ω_e so $\rho = \sqrt{7}/\sqrt{3}$ on each grid, and the estimator is not asymptotically exact.

9.5 Iterative Solution with Refinement

As suggested in Chapter 7, much is to be gained by using iterative solution strategies in conjunction with adaptive refinement. In particular, the solution on the current grid can be interpolated to provide a good starting iterate on the next grid, and so on. Multigrid acceleration strategies on the hierarchy of grids can also be exploited to significant advantage. This can also be used in conjunction with domain decomposition. Finally, in nonlinear problems continuation can be applied both in the nonlinear parameters and in the grid.

Since the refinement process leads naturally to unstructured grids and the node point numbering will be irregular, the sparsity pattern of the resulting algebraic system will be quite general. Similar difficulties arise with respect to element numbering. Hence standard sparse elimination schemes and frontal elimination solvers will perform poorly. This difficulty can be treated by using node or element renumbering algorithms. These are fast and will alleviate the problem, but the approach will generally still be inferior to iterative solution for large-scale problems.

Generalized gradient iterative methods such as conjugate gradient, biconjugate gradient, generalized minimum residual (GMRES) and quasi-minimum residual (QMR) algorithms are being increasingly adapted for practical applications. In Section 7.4, we briefly described how conjugate gradient and biconjugate gradient can be applied within a Gummel iteration for the drift-diffusion equations. We also indicated how these schemes can be recast at the element level so that system assembly is avoided.

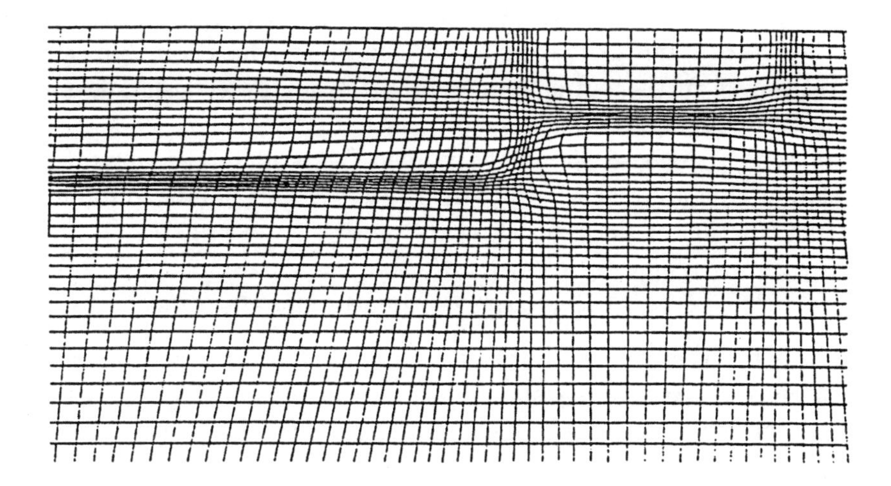

Figure 9.9: Enlarged view of redistributed structured grid(source region).

As stated in the introduction, semiconductor process and device computations are known from experience to be grid sensitive, especially for submicron devices. This is because there are extreme layers in the solution and strong nonlinearities. If the grid is not appropriately graded into the layers then perturbations in the grid have been seen to produce significant changes in the solution. As device sizes continue to shrink into the "deep-submicron" range these numerical difficulties become more pronounced and the calculations less reliable. The problem is aggravated by the need to use grids with as few node points as possible so the calculations are still economical. This has led device engineers to rely on grids with slender triangles where the aspect ratio may exceed 1000 : 1. The numerical conditioning of the resulting algebraic systems is then also suspect. Adaptive grid refinement and redistribution strategies may be introduced to circumvent these difficulties. Directional refinement and Delaunay procedures may also ameliorate the aspect-ratio problem.

9.6 Redistribution

Redistribution techniques provide an alternative strategy whereby the topology of the grid remains fixed, but the grid points move into the layers. For example, the doping profile for a device may provide a starting point for clustering the grid, as indicated by the subregion near a MOSFET source in

Figure 9.9. Here an initial rectangular grid has been redistributed to minimize the error in approximating a specified doping function. Subsequently, the grid can be further adjusted based on the computed solution and error indicators for the approximate solution.

In practice the grid should be graded but still retain cells of reasonable shape. One approach is to redistribute the grid points to minimize an objective function of the general form

$$F(\boldsymbol{v}) = \alpha E(\boldsymbol{v}) + \beta S(\boldsymbol{v}) + \gamma O(\boldsymbol{v}) \tag{9.9}$$

where $\boldsymbol{v}$ is the vector of node point coordinates, $S(\boldsymbol{v})$ is a measure of grid "smoothness", $O(\boldsymbol{v})$ is a measure of local orthogonality of the grid lines and $E(\boldsymbol{v})$ is a measure of the solution error. The factors α, β, γ may be varied to weight any one or more of the respective contributions. As an example, consider a representative patch centered at grid point (i, j) in a topologically rectangular structured grid. Let the local radial vectors connecting (i, j) to the adjacent grid points in the "curvilinear" five-point stencil be denoted $\boldsymbol{r}_1$, $\boldsymbol{r}_2$, $\boldsymbol{r}_3$ and $\boldsymbol{r}_4$. Then, the departure from a smooth orthogonal grid can be defined, for instance, by using the local measures

$$S_{ij} = \sum |\boldsymbol{r}_i|^2, \quad O_{ij} = (r_1 \cdot r_2)^2 + (r_2 \cdot r_3)^2 + (r_3 \cdot r_4)^2 + (r_4 \cdot r_1)^2 \tag{9.10}$$

In closing, we remark that a grid generated for a similar device may be particularly useful as an initial grid in this redistribution procedure for a new device. This is particularly appropriate for updating a design or as part of an automated design procedure so a grid "archive" should be considered for such applications.

9.7 Moving Grids

In the oxidation process discussed in Chapter 15, the grid deforms with the moving surfaces and this usually leads at some stage to poorly shaped cells. For two-dimensional oxidation simulation with triangular elements, it may suffice to employ local diagonal "swaps" in a Delaunay procedure. Since the Delaunay process is very fast, this is also economical. Local smoothing (redistribution of the grid points) can be combined with the Delaunay idea.

Other strategies may also be applied to treat this problem. For example, in [469] a quadtree-based grid generator is used in conjunction with a geometry movement algorithm for the oxidation problem. The flow chart

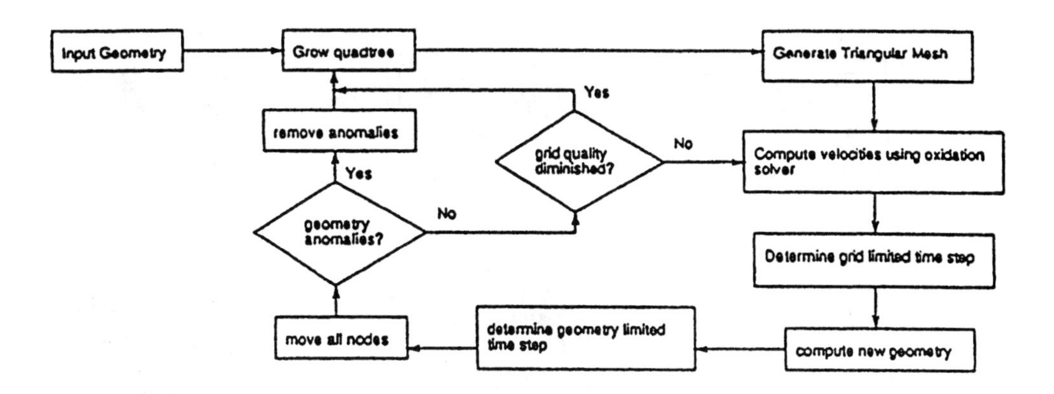

Figure 9.10: Flowchart for grid evolution during oxidation simulation (from [469].

for this combined algorithm is shown in Figure 9.10. First the initial geometry is specified using geometric modeling software, and a quadtree mesh of quadrilaterals is constructed and subsequently triangulated. The oxidation computation is initiated on this using SUPREM IV and the grid nodes are adjusted, based on the computed oxide velocity field. Remeshing is carried out if the elements become too slender or the mesh grading and accuracy deteriorate. The level of recursive refinement during remeshing can be varied to include local refinement at the moving boundaries and in interior layer regions. "Loops" in the evolving moving boundary can also be detected by the geometric modeller, and the boundary surface intersection similarly monitored. A simple example showing the geometry and meshes before and after several oxidation steps is given in Figure 9.11.

9.8 Summary

Efficient, reliable and robust unstructured irregular grids are needed for semiconductor process and device problems. In the present work we begin by summarizing the main steps in unstructured grid generation and adaptive refinement to enhance the grid. *A posteriori* error estimates and associated error indicators provide a mathematical framework for local use in adaptive mesh refinement and for quantitative reliability assessment of the computed solution. Several different indicators are constructed and their relative merits discussed. The role of an effectivity index and properties such as asymptotic exactness of the *a posteriori* estimator are examined. Local mesh refinement strategies that use these indicators are briefly described,

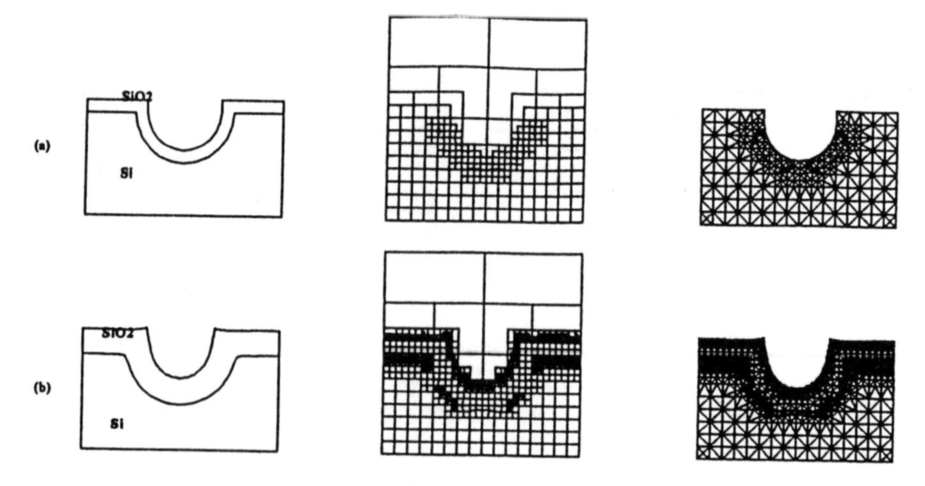

Figure 9.11: Geometry, quadtree mesh and triangular meshes before (a) and after several oxidation steps (b) (from [469]).

together with the supporting data structures and software consideration. Finally, we consider some aspects of refinement and iterative solution that are advantageous for unstructured adaptive grids. The objective here is to provide a brief overview of some of the main aspects of this technology. A more extensive treatment with detailed algorithms and analysis is available in [82] and in the references.

9.9 Exercises

1. Take a small set of 7 points generated randomly in the unit square and sketch the sequence of triangulations generated by the point insertion strategy with supertriangle vertices at (5,0), (0,3) and (-3,-3).

2. Show the equivalence of the circumcircle test and the max-min angle test for a pair of triangles defining a convex quadrilateral. Comment on the implications for the tests if the quadrilateral is not convex.

3. Assume that an adaptive refinement scheme refines 25 percent of the active elements at each refinement step. Determine the number of active elements at any later stage for both the two-dimensional quadtree and three-dimensional octree cases.

4. Consider a one-dimensional problem with non-uniform grid $\{x_i\}$ of linear elements. Use the interpolation bound in (9.1) to construct a computable error indicator for $m = 0$, $p = q = 2$. Apply this indicator

to the model drift-diffusion problem $vu' - Du'' = 0$, $u(0) = 0$, $u(1) = 1$ with $v/D = 50$.

5. Use the one-dimensional form of the flux averaging formula (9.7) in the test problem of the previous exercise and compare the resulting approximate solutions and grids. Test if the estimator is asymptotically exact.

6. Show how the quadtree and octree data structures can be used to allow simultaneous refinement and coarsening of the grid in different subregions of the domain. Discuss the implications for the data structures and, in particular, the element arrays and nodal arrays.

7. Sketch an example for a circumsphere face swap applied to a pair of adjacent tetrahedra.

8. Introduce a partition of a right tetrahedron to 8 subtetrahedra by joining mid-edge nodes. Compare the shapes of the resulting vertex and interior sub-tetrahedra and discuss the implications of repeated refinement.

9. Assume a triangle is subdivided to a quartet of subtriangles and that a linear approximation is made on each subtriangle. Verify that the approximation is discontinuous across the interface to an unrefined neighbor triangle. Investigate the implications of splitting the neighbor triangle or constraining the node on the interface. Suggest and evaluate an alternative strategy. Comment on the analogous situation in three dimensions.

Chapter 10

Ion Implantation

10.1 Introduction

Ion implantation is the principal means of introducing dopant impurities into semiconductors during the device manufacturing process. Dopant impurity ions are accelerated to high energies, usually in the range of tens or hundreds of keV, before impacting on the semiconductor target, coming to rest below the surface of the wafer. This technique allows very precise control over the amount of dopant deposited into the material. In addition, the purity of the implanted species is guaranteed, because only those ions with the desired charge and mass will survive the flight to the wafer surface. In a manufacturing environment, ion implantation is the most reliable and repeatable means of doping a semiconductor.

Once the proper amount of dopant has been added to the wafer, a high temperature anneal is necessary to allow the impurity ions to form chemical bonds with the semiconductor atoms, making the impurity a donor or acceptor in the semiconductor. From the process modeling viewpoint, ion implantation therefore provides the initial condition for subsequent diffusion modeling. In submicron technologies, the diffusion temperature and time are usually kept to a minimum, so it becomes extremely important to ensure that the initial profile resulting from the ion implantation is modeled very accurately. In addition, the implantation process damages the crystalline semiconductor, producing vacancies, self-interstitials, and extended defects. As discussed in Chapter 12, these defects play a major role in subsequent high temperature diffusion.

Computer aided design models for ion implantation fall into two broad categories: analytic distribution functions and particle, or Monte Carlo, methods. The distribution function approach is statistical in nature, relying upon fits to experimental data to reproduce the observed profiles of dopant ions. It is computationally very inexpensive and works well for simple geometries in one dimension. The particle methods attempt a first-principles calculation based upon two-body scattering theory. Although computationally expensive, they easily handle the most complicated structures and, in addition, can provide the initial defect profiles required for the advanced diffusion models discussed in Chapter 12. The two methods are complementary; each plays an important role in the modeling of ion implantation.

10.2 Analytic Distribution Functions

Analytic distribution function models for ion implantation profiles are the simplest to implement and the fastest methods to execute. When the models are used in conjunction with experimental data for the particular implant conditions, they can be an extremely effective means of modeling profiles in the vertical dimension (normal to the wafer surface). The analytic models are based on distribution functions which represent the concentration of the implanted impurity as a function of x, the depth into the wafer. The particular distribution function chosen should have the properties that: (1) the function has a unique maximum, and (2) the integral of the distribution function from the surface of the wafer ($x = 0$) to the back of the wafer must equal the total dose of the implant, that is, the total number of ions implanted per unit area. Several distribution functions satisfy these properties. The ones most frequently used to model ion implantation are the Gaussian, the Pearson Type IV, and the double Pearson Type IV.

Gaussian Distribution

The concentration of implanted impurities as a function of depth x normal to the wafer surface is modeled with the Gaussian distribution by

$$C(x) = \frac{D_T}{\sqrt{2\pi}\sigma} \exp\left(\frac{-(x-R)^2}{2\sigma^2}\right) \tag{10.1}$$

where C is the number of ions per unit volume, and D_T is the number of ions per unit area impacting on the wafer surface. There are two parameters of the Gaussian distribution: R, the range, which defines the x position of the maximum of the distribution, and σ, the standard deviation, which are used to fit the distribution function to measured data. Tables of distribution function parameters for common ion/target combinations have been calculated based upon stopping theory [208], and are widely available in process simulation programs.

A slight variation of (10.1) is the two-sided Gaussian distribution, which uses a different value for the standard deviation to the left and to the right of the implant profile peak:

$$C(x) = \begin{cases} \dfrac{2D_T}{\sqrt{2\pi}(\sigma_1 + \sigma_2)} \exp\left(\dfrac{-(x-R)^2}{2\sigma_1^2}\right) & \text{for } x < R \\ C(x) = \dfrac{2D_T}{\sqrt{2\pi}(\sigma_1 + \sigma_2)} \exp\left(\dfrac{-(x-R)^2}{2\sigma_2^2}\right) & \text{for } x > R \end{cases} \tag{10.2}$$

This allows more accurate fitting to asymmetric impurity profiles.

Pearson Distribution

The Pearson Type IV distribution uses four parameters to model the implant profile [250, 465, 556]. This distribution is defined as

$$C(x) \;=\; D_T f(x) \tag{10.3}$$

where $f(x)$ is given by

$$\begin{aligned} f(x) \;=\; & K|b_2(x-R)^2 \;+\; b_1(x-R) \;+\; b_0|^{\frac{1}{2b_2}} \;\times \\ & \exp\left(-\frac{b_1/b_2 + 2b_1}{\sqrt{4b_2b_0 - b_1^2}} \arctan\left(\frac{2b_2(x-R) + b_1}{\sqrt{4b_2b_0 - b_1^2}}\right)\right) \end{aligned} \tag{10.4}$$

K is a normalization factor, chosen such that

$$\int_{-\infty}^{+\infty} f(x)\, dx \;=\; 1 \tag{10.5}$$

and the constants are given by

$$b_0 \;=\; -\frac{\sigma^2(4\beta - 3\gamma^2)}{10\beta - 12\gamma^2 - 18} \tag{10.6}$$

$$b_1 \;=\; -\frac{\sigma\gamma(\beta + 3)}{10\beta - 12\gamma^2 - 18} \tag{10.7}$$

$$b_2 \;=\; -\frac{2\beta - 3\gamma^2 - 6}{10\beta - 12\gamma^2 - 18} \tag{10.8}$$

The four parameters which determine the concentration profile are the range R, the standard deviation σ, the skewness γ, and the kurtosis β, all of which are related to the first four moments of the distribution function in (10.4). For a Type IV Pearson distribution, the skewness and kurtosis must satisfy the relation:

$$\beta > \frac{48 + 39\gamma^2 + 6(\gamma^2 + 4)^{3/2}}{32 - \gamma^2} \tag{10.9}$$

In the limit as $\gamma \to 0$ and $\beta \to 3$, the Pearson IV distribution reduces to the Gaussian distribution.

As in the case of the Gaussian distribution, parameters for several ion and target combinations are available in table form [208]. They can also be calculated for amorphous target materials using stopping theory [580]; however optimized fits to experimental data are generally more reliable [59, 465]. The Pearson distribution function has been used very effectively in fitting some experimental implant conditions, especially where the target material is non-crystalline and ion channeling effects are not significant.

Double Pearson Distribution

Channeling along crystal planes causes an observed tail of implant profiles. This is especially pronounced for light ions such as boron, whose profiles cannot be accurately described by the standard Pearson Type IV distribution function. However, by using the sum of two Pearson distributions, one modeling the peak of the profile, and the other modeling the tail, accurate fits can be obtained [427, 519].

Using a double Pearson distribution, the ion implant concentration is modeled by

$$C(x) = D_1 f_1(x) + D_2 f_2(x) \tag{10.10}$$

where f_1 and f_2 are Pearson distribution functions defined in (10.4), and $D_1 + D_2$ is equal to the total dose of the implant. Thus the double Pearson distribution has nine fitting parameters, four for each of the individual Pearson distributions, and D_1. Tables of double Pearson distribution parameters for several boron and boron difloride implants into crystalline silicon have been published [519].

Comparison of the Analytic Distribution Models

Typical plots of the Gaussian, Pearson, and double Pearson distribution functions discussed above are shown in Figure 10.1. The symmetry of the Gaussian distribution is not usually observed in practical situations, so its use is limited to situations where accuracy of the initial distribution is not important. This can be the case if the implantation is to be followed by a long, high temperature anneal.

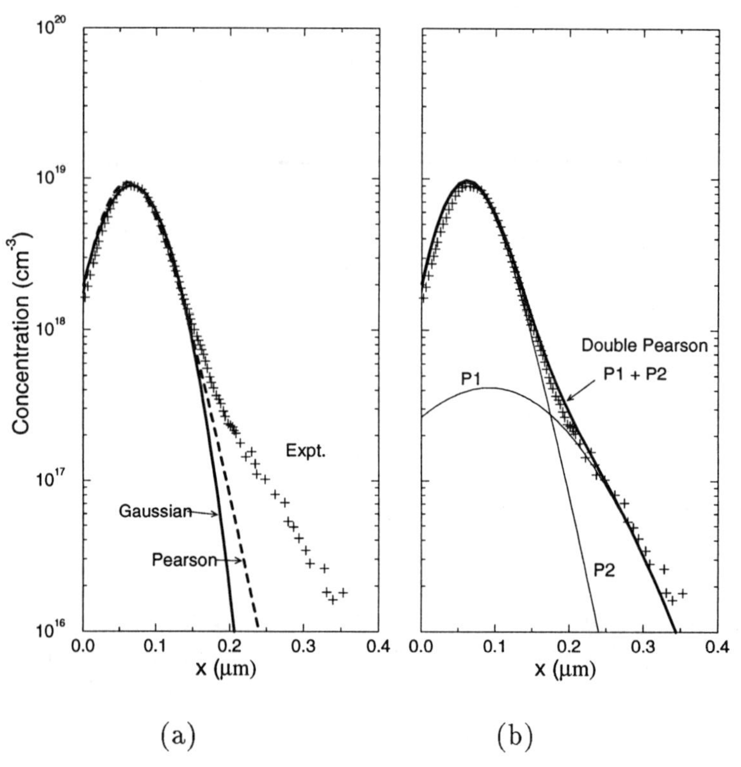

Figure 10.1: Typical shapes of the (a) Gaussian and Pearson Type IV distributions, and (b) the double Pearson distribution. The symbols are experimental data.

The Pearson Type IV distribution has gained wide acceptance in the semiconductor industry. It is the standard distribution function in many process simulation programs. The Pearson distribution is only accurate, however, for those situations where crystalline channeling is not important. This is the case, of course, whenever the target material is amorphous, and it also holds when a relatively heavy ion, such as arsenic, is implanted. A heavy ion imparts so much damage to the crystal during the initial stages of the implant that the surface layer of the target quickly becomes amorphous and channeling is minimized. The double Pearson distribution overcomes most of the limitations of the Pearson distribution in situations where channeling is significant.

Multi-layer Targets

Up to this point, the discussion of analytic distributions has assumed an implant into a single, homogeneous target material. In semiconductor processing, it is often the case that the impurity ions are implanted through several different layers of material. A typical situation is the implantation through a thin layer of SiO_2, say 20 nm thick, into an underlying layer of crystalline silicon. Since the material properties of the various layers are different, the penetration of the ions into each layer will be different, and a simple distribution function is usually not capable of modeling the final profile.

One straightforward method to model implantation into a multi-layer target is to compute each layer independently, using one of the distribution functions discussed above, and then merge the solutions [465]. Suppose that the target has N layers of various materials, and that the boundary between layer i and layer $i+1$ is at depth x_i. First, the implant profile in the top layer is computed using one of the analytic distribution functions, assuming that the layer extends to infinity. The profile is then truncated at the boundary x_1 between the top layer and the next layer, and the ion dose within that layer is calculated by integrating the doping profile over the thickness of the layer. Next, the implant profile is calculated in the second layer, as if that layer extended from zero to infinity. This profile is integrated from zero to a depth x_0, where x_0 is chosen such that the integral is equal to the dose calculated in the previous layer. Then the entire second layer profile is shifted so that x_0 aligns with the layer boundary, x_1. The process is then repeated for all the layers in the target. This algorithm guarantees that the total ion dose in the target is equal to the correct value implanted; however, the overall profile will be discontinuous at each layer boundary.

10.3 Energy Loss and Scattering

The analytic distribution functions described in the previous section give excellent results, provided that experimental data are available to fit the moments of the distribution. However, such data are not always available for a given implant condition, in which case it is necessary to use a more physically-based approach to model ion implantation. Particle methods (also known as Monte Carlo methods) can incorporate much of the physics underlying the scattering of individual ions as they travel through the target. The particle methods provide full three-dimensional information on the resulting profile, which is extremely difficult to obtain experimentally. In addition, these techniques can model implantation into arbitrary geometries, such as trench structures, and multiple layers of materials. They accurately simulate channeling in crystalline materials, and they can provide insight into defect generation due to ion impact. Particle methods are, however, much more computationally expensive than using distribution functions. The remainder of this section presents the basic theory behind particle methods for modeling ion implantation.

As an energetic ion enters a solid target, it loses energy by two basic mechanisms. The first is nuclear scattering, in which the nucleus of the ion elastically collides with the nucleus of an atom in the target. Each scattering event causes the ion to lose energy, and also to change direction. The second mechanism is electron energy loss, which occurs when an ion interacts with the electrons of the target atoms and slows in a manner analogous to frictional drag. This mechanism does not alter the direction of the ion's trajectory, only its energy.

Nuclear Scattering

The basic assumption in modeling energy loss due to nuclear collision is that the ion will interact with only one target atom at a time. This simplification allows the use of binary scattering theory from classical mechanics [215]. Consider the two particles shown in Figure 10.2. Particle 1 has mass m_1 and kinetic energy E_0, and approaches particle 2, an initially stationary particle with mass m_2.

The distance p in the figure is called the impact parameter. After the collision, particle 1 will deviate from its original course by an angle θ. By applying the laws of conservation of energy and conservation of momentum, it can be shown [215, 580] that after the collision, particle 1 will lose kinetic energy

$$\frac{\Delta E}{E_0} = \frac{4m_1 m_2}{(m_1 + m_2)^2} \cos^2(pI) \tag{10.11}$$

where ΔE is the energy lost, E_0 is the kinetic energy before collision, and I is the integral

$$I = \int_0^{s_{max}} \left(1 - \frac{V(s)}{E_r} - p^2 s^2\right)^{-1/2} ds \tag{10.12}$$

In (10.12), s is the inverse separation between the two particles ($s = 1/r$), $V(s)$ is the potential between the two particles, assumed to be repulsive, and

$$E_r = \frac{E_0}{1 + m_1/m_2} \tag{10.13}$$

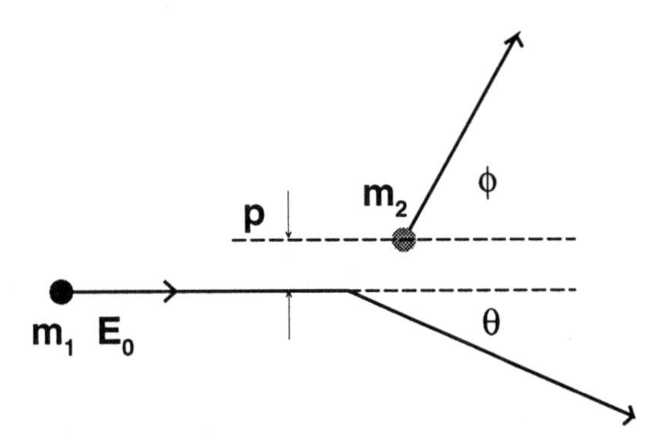

Figure 10.2: Particle 1 with mass m_1 and energy E_0 scatters off particle 2 of mass m_2. After collision, particle 1 is deflected by angle θ.

is the reduced energy in the center-of-mass coordinates. The upper limit of the integral, s_{max}, is the inverse distance of closest approach of the two particles, and is given by the solution to the equation

$$1 - \frac{V(s_{max})}{E_r} - p^2 s_{max}^2 = 0 \tag{10.14}$$

The angle θ by which the incoming particle has been deflected is given by

$$\cos\theta = \frac{1 - 0.5(1 + m_2/m_1)\Delta E/E_0}{\sqrt{1 - \Delta E/E_0}} \tag{10.15}$$

Equations (10.11)–(10.15) are the basic equations for classical two-body scattering.

It is convenient to express the scattering integral, (10.12), in a dimensionless form. Let a be a screening length defined by [580]

$$a = \frac{0.8853 a_0}{Z_1^{0.23} + Z_2^{0.23}} \tag{10.16}$$

where a_0 is the Bohr radius (0.529×10^{-10}m), Z_1 is the atomic number of particle 1, and Z_2 is the atomic number of particle 2. The potential in (10.12) can be written as

$$V(s) = Z_1 Z_2 k s f(\tilde{s}) \tag{10.17}$$

where $k = q^2/(4\pi\epsilon_0)$ and $\tilde{s} = as$. Now by introducing a dimensionless impact parameter

$$b = p/a \tag{10.18}$$

and a dimensionless energy

$$\epsilon = \frac{aE_r}{Z_1 Z_2 k} \tag{10.19}$$

the scattering integral (10.12) becomes

$$I = \frac{1}{a}\int_0^{\tilde{s}_{max}} \left(1 - \tilde{s}\frac{f(\tilde{s})}{\epsilon} - b^2\tilde{s}^2\right)^{-1/2} d\tilde{s} \tag{10.20}$$

In order to calculate the energy lost in a given collision, and the angle at which the ion is scattered, we evaluate $\cos^2(pI)$, which becomes

$$\cos^2(pI) = \cos^2\left(b\int_0^{\tilde{s}_{max}} \left(1 - \tilde{s}\frac{f(\tilde{s})}{\epsilon} - b^2\tilde{s}^2\right)^{-1/2} d\tilde{s}\right) \tag{10.21}$$

The quantity $\cos^2(pI)$ is thus a function of two dimensionless parameters, b and ϵ, which for a given collision between two particles can be obtained from (10.18) and (10.19). The energy lost by particle 1 due to the collision is given by (10.11), while the angle at which particle 1 leaves the collision can be found from (10.15).

As an example, consider the Coulomb potential between two particles, $V(r) = Z_1Z_2k/r$, or $V(s) = Z_1Z_2ks$. In this case, $f(\tilde{s}) = 1$, and from (10.21)

$$\cos^2(pI) = \cos^2\left(b\int_0^{\tilde{s}_{max}} \left(1 - \frac{\tilde{s}}{\epsilon} - b^2\tilde{s}^2\right)^{-1/2} d\tilde{s}\right) \tag{10.22}$$

with

$$\tilde{s}_{max} = \frac{-1}{2\epsilon b^2} + \frac{1}{2\epsilon b^2}\sqrt{1 + 4b^2\epsilon^2} \tag{10.23}$$

the solution of (10.14). The integral can be evaluated exactly to yield

$$\cos^2(pI) = \frac{1}{1 + 4b^2\epsilon^2} \tag{10.24}$$

Equation (10.11) gives the energy loss for the Coulomb potential as

$$\Delta E = E_0\frac{4m_1m_2}{(m_1 + m_2)^2}\left(\frac{1}{1 + 4b^2\epsilon^2}\right) \tag{10.25}$$

and (10.15) gives the scattering angle θ for the collision. Figure 10.3 plots the energy loss and the scattering angle as a function of the dimensionless impact parameter for the special case of the Coulomb potential with $m_1 = m_2$. In a "head-on" collision, $b = 0$ and $\Delta E/E_0 = 1$, indicating that all the energy is transferred from particle 1 to particle 2 in the collision. For a positive value of b, the higher the incident particle energy, the smaller the incremental energy loss, $\Delta E/E_0$, and the smaller the scattering angle.

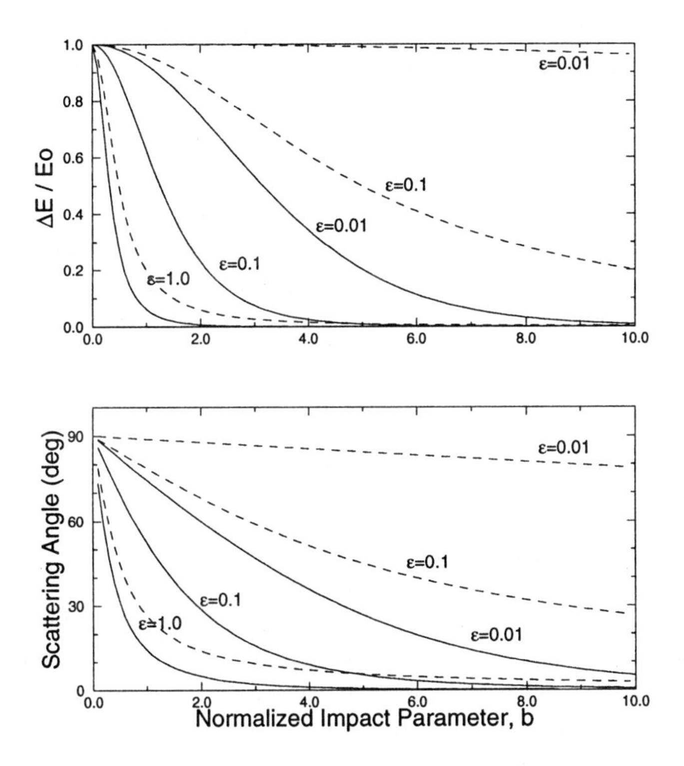

Figure 10.3: Energy loss and scattering angle as a function of the dimensionless impact parameter for two particles of equal mass interacting via the Coulomb potential, shown in dashed lines, and via the universal potential of Equation (10.26), shown in solid lines.

Of course the Coulomb potential is suitable only for two isolated charged particles. For more realistic inter-atomic potentials, the Coulomb potential is screened by the electrons that surround the two nuclei. Ziegler, Biersack and Littmark [580] found that the following potential gives excellent results for a wide variety of atom combinations (and hence they named it the "universal" potential):

$$V(r) = \frac{Z_1 Z_2 k}{r}\Big(0.1818e^{-3.2r/a} + 0.5099e^{-0.9423r/a} + 0.2802e^{-0.4029r/a} + 0.02817e^{-0.2016r/a}\Big) \quad (10.26)$$

Figure 10.3 shows the energy loss and scattering angle as a function of dimensionless impact parameter for this potential. It is clear that the energy loss is short-ranged as compared with the Coulomb potential; that is, for larger values of the normalized impact parameter, the scattering is negligible. With this universal potential, equation (10.14) for s_{max} is transcendental, and a closed-form solution cannot be found. However, it is straightforward to obtain an accurate approximation to s_{max} numerically by using Newton's method, as discussed in Chapter 5. Once s_{max} is known, the scattering integral in (10.21) can be approximated, and various schemes have been proposed in the literature for this computation [43, 455, 580]. However, since there are only two parameters in the dimensionless integral, it is computationally efficient to use numerical quadrature to obtain highly accurate values of the integral, and to construct tables of $\cos^2(pI)$ as a function of b and ϵ. Values of the scattering integral may then be obtained by table lookup [248, 393].

Electron Energy Loss

In addition to energy loss by nuclear scattering, the ion loses energy through inelastic interactions with electrons in the solid. The physics of this process are quite complex [580], but a relatively simple model that works reasonably well for typical semiconductor processing implant conditions assumes that the energy loss is proportional to the velocity of the ion, in analogy with macroscopic friction. Since $v_i = \sqrt{2E_0/m_1}$, the energy loss due to interactions between the ion and the electrons can be modeled as

$$\Delta E_e = -(k_0/L^2)\sqrt{E_0} \quad (10.27)$$

Here L is the length of the path that the ion travels through the solid, E_0 is the kinetic energy of the ion, and k_0 is the constant

$$k_0 = \frac{1.212\ Z_1^{7/6} Z_2}{\left(Z_1^{2/3} + Z_2^{2/3}\right)^{3/2} m_1^{1/2}} \quad (10.28)$$

in units of $\text{Å}^2\text{eV}^{1/2}$ [326]. There is no change in the ion's direction due to this process. Several authors have suggested modifications to the electron energy loss formula for particular ion/target combinations and for high energy ions [153, 556, 558, 580]. A more detailed analysis which takes into account the electron density as a function of position within the crystal lattice has also been implemented [298].

10.4 Ion Trajectories in Amorphous Targets

Using the nuclear and electronic energy loss formulas discussed above, one can calculate the trajectory of an ion through an amorphous solid target [43]. Assume that an ion with kinetic energy E_0 hits an amorphous target at an angle θ_0 to the target normal. The surface of the target is assumed to be at $x = 0$, with x increasing into the target. The horizontal direction, along the surface of the target, is measured by the variable y. The number of atoms per unit volume in the target material is defined by

$$N = N_a \frac{\rho}{m_2} \tag{10.29}$$

where N_a is Avogadro's number, ρ is the mass density, and m_2 is the mass of the target, in atomic mass units. If the target is composed of more than one type of atom, them m_2 is an average atomic mass, given by

$$m_2 = \frac{i \times m_A \ + j \times m_B \ + k \times m_C + ...}{i + j + k + ...} \tag{10.30}$$

for a material with chemical composition $A_i B_j C_k$... For example, silicon nitride, with composition Si_3N_4, has an average atomic mass

$$m_2 = \frac{3 \times 28.086 + 4 \times 14.007}{3 + 4} = 20.04 \text{ a.m.u.} \tag{10.31}$$

The mean atomic separation between atoms in the target is $N^{-1/3}$, so the ion will on average travel a distance $L = N^{-1/3}$ between each scattering event. As the ion enters the target material, it will approach the first target atom with impact parameter p, as defined in the previous section. The probability of finding a target atom between p and $p + \delta p$ is given by [43]

$$w(p)\delta p = 2\pi N^{2/3} p \delta p \tag{10.32}$$

for $p < \pi^{-1/2} N^{-1/3}$. If R_n is a uniformly distributed random number in [0,1], then this probability distribution gives

$$p \ = \ \sqrt{\frac{R_n}{\pi N^{2/3}}} \tag{10.33}$$

The algorithm for calculating an ion trajectory in an amorphous target proceeds as follows.

Algorithm 10.1 (2D ion trajectory in an amorphous target)

1. *Set the initial ion energy (E) and angle of impact (α) with respect to the target normal. Use N and m_2 to calculate the initial target material properties.*

2. *Select a random number in [0,1], and calculate the normalized impact parameter and normalized energy:*

$$b = \frac{1}{a}\sqrt{\frac{R_n}{\pi N^{2/3}}} \; ; \; \epsilon = \frac{aE}{(1 + m_1/m_2)Z_1 Z_2 k} \tag{10.34}$$

3. *Look up the value of $\cos^2(pI)$ as a function of b and ϵ in a previously calculated table, and calculate the nuclear stopping energy*

$$\Delta E_n = E\frac{4m_1 m_2}{(m_1 + m_2)^2}\cos^2(pI) \tag{10.35}$$

4. *Calculate the energy loss due to electronic processes, ΔE_e from (10.27).*

5. *Update the ion energy after the collision, $E \leftarrow E - \Delta E_n - \Delta E_e$.*

6. *Update the angle of travel of the ion, $\alpha \leftarrow \alpha + \theta$ where θ is given by*

$$\theta \;=\; \pm\ \arccos\left(\frac{1 - 0.5(1 + m_2/m_1)\Delta E_n/E_0}{\sqrt{1 - \Delta E_n/E_0}}\right) \tag{10.36}$$

   The $\pm$ factor is selected by another random number, and indicates whether the ion scatters to the left or to the right in this particular collision.

7. *Calculate the new ion position*

$$x \leftarrow x + L\cos\alpha \; ; \; y \leftarrow y + L\sin\alpha \tag{10.37}$$

8. *If $x < 0$, the ion has reflected out of the target and we exit the algorithm. If the ion has crossed a boundary into a different material, new values for the material constants N and m_2 are calculated.*

9. *If the ion energy E has decreased below some pre-defined value (say 5 eV), then the final (x, y) position is recorded, and we exit the algorithm. Otherwise, return to step 2 and repeat.*

Figure 10.4 shows some typical trajectories for phosphorus ions implanted into amorphous silicon calculated using this algorithm. Each ion suffered on average 250 collisions before finally stopping.

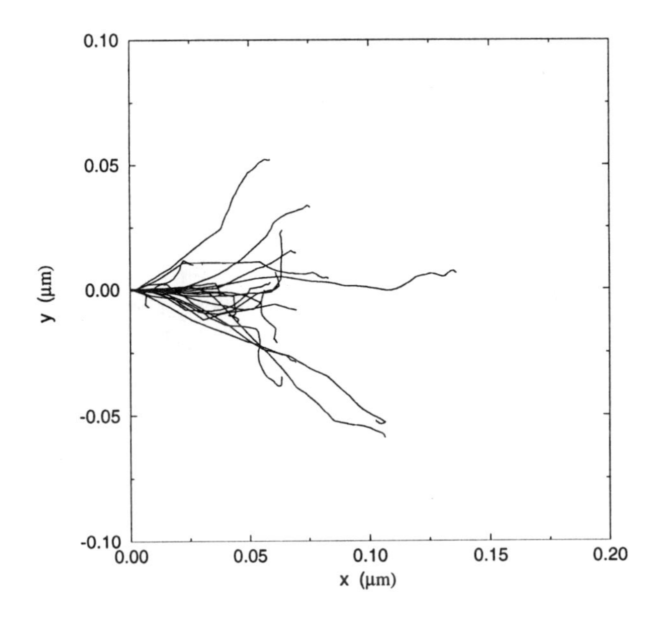

Figure 10.4: Typical calculated trajectories of a 50 keV phosphorus implant into amorphous silicon. The y-axis represents the surface of the target, and the ions enter the target normal to the surface at $(x, y) = (0, 0)$.

In order to get meaningful statistical information on the final distribution of implanted ions, one must calculate hundreds or thousands of individual trajectories, and store the final stopped position of each ion. Then the target is discretized, and the number of stopped ions in each cell is counted. In one dimension this produces the familiar histogram. Final concentration values (number of ions per unit volume) for each discrete cell are obtained by scaling to match the implanted dose. Figure 10.5 shows the final stopped positions of phosphorus ions implanted into amorphous silicon, for various numbers of incident ions. The final concentration profiles obtained by mapping the raw data onto a one-dimensional grid are shown in Figure 10.6. For this implant, 10^4 to 10^5 ions are needed to obtain a relatively smooth profile.

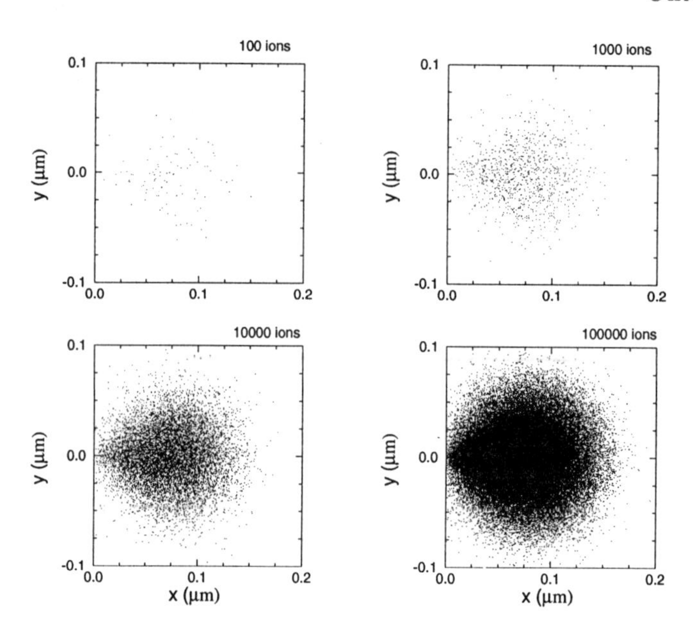

Figure 10.5: Final positions of 10^2, 10^3, 10^4, and 10^5 phosphorus ions implanted at 50 keV into amorphous Si.

10.5 Ion Trajectories in Crystalline Targets

When the target material is crystalline, the target atoms occupy positions on a well-defined lattice, and the algorithm of the previous section must be altered in several ways. The selection of the normalized impact parameter for a particular scattering event is no longer a random process; the geometry of the lattice and the trajectory of the ion now determine the impact parameter for scattering. Furthermore, the distance of travel between collisions is obtained directly from the geometry of the lattice, rather than from the average separation between atoms in the target. With crystalline targets, scattering in materials containing different types of atoms on the lattice, such as GaAs, can be treated directly, rather than by using an average value for atomic number and mass of the target atoms.

Atoms in crystals arrange themselves into regular arrays with various well-defined geometries, so as to minimize the overall energy of the crystal [18]. Silicon crystallizes into the diamond lattice structure, which is illustrated in Figure 10.7. GaAs also crystallizes into this same structure, with the Ga and As atoms occupying alternate positions on the lattice. It is clear from Figure 10.7 that an ion could travel a relatively long distance along a

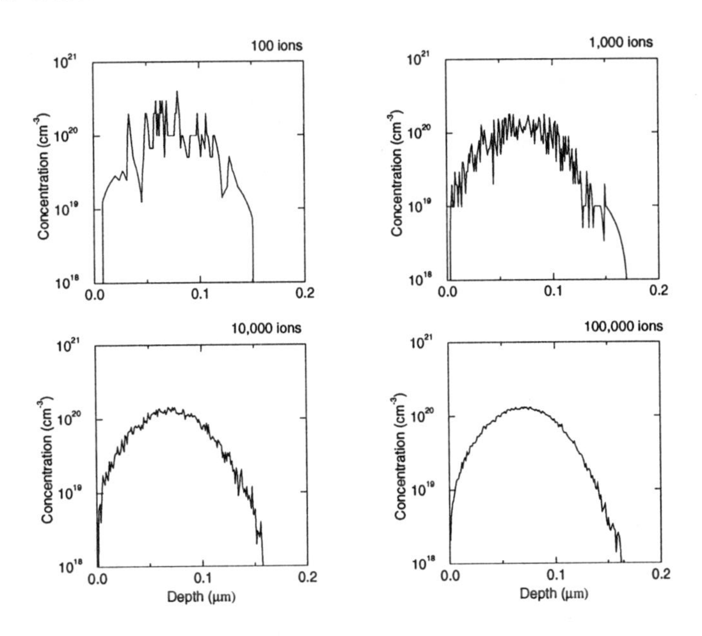

Figure 10.6: Concentration profiles for phosphorus implanted into amorphous silicon at 50 keV with a dose of 10^{15} cm^{-2}, calculated from the values of the final positions shown in Figure 10.5.

primary axis of the crystal without a significant scattering event occurring with one of the lattice atoms. This is the phenomenon of channeling. The net result is a somewhat deeper ion penetration than would be expected in an amorphous target of the same material.

The algorithm for calculating the trajectory of an ion through a crystal lattice begins by determining the initial velocity vector of the ion as it reaches the target surface. The magnitude of the velocity vector is $\sqrt{2E_0/m_1}$ where m_1 is the mass and E_0 is the initial energy of the ion. The components of the initial velocity vector are specified as initial conditions. The ion will impact on the target surface at a random point. Once the initial position with respect to the lattice and the initial ion velocity have been determined, the trajectory can be computed from the geometry of the lattice. First, the positions of the target lattice atoms in the vicinity of the ion are generated by translating the basic cell of the lattice in the x, y, and z directions. Let the vector $\boldsymbol{X}_n$ identify the position of the nth atom in the lattice. The projected distance of closest approach of the ion to lattice site n is given by

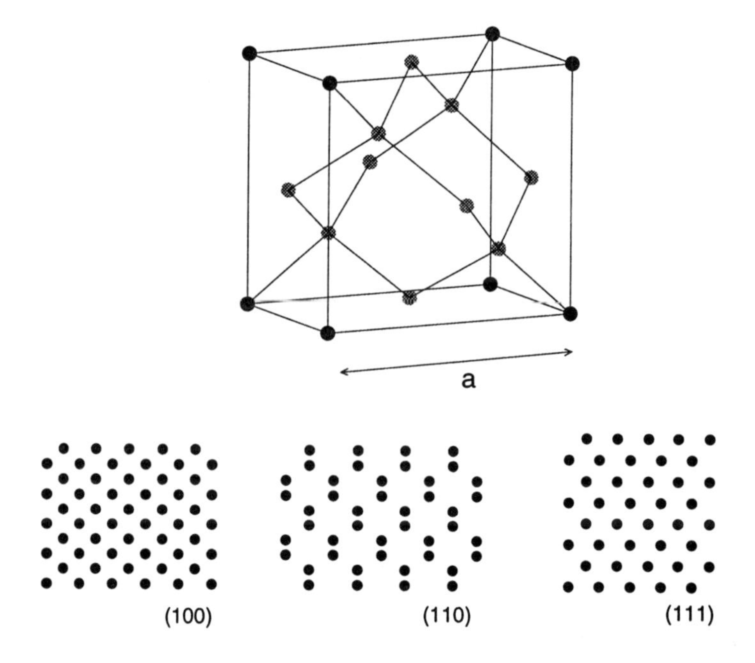

Figure 10.7: Illustration of a crystalline lattice for the diamond structure. For silicon, the distance a is 5.43×10^{-10} m. When viewed from various directions (for example, (100), (110), and (111) in the illustration), opportunities for ion channeling become evident.

the dot product

$$t_n = \frac{\boldsymbol{v}}{|\boldsymbol{v}|} \cdot (\boldsymbol{X}_n - \boldsymbol{x}) \tag{10.38}$$

where $\boldsymbol{x}$ is the current position of the ion. The impact parameter for scattering between the ion and lattice site n is then given by the magnitude of the vector

$$\boldsymbol{p}_n = \boldsymbol{X}_n - \boldsymbol{x} - t_n \frac{\boldsymbol{v}}{|\boldsymbol{v}|} \tag{10.39}$$

The lattice atom with the minimum distance of closest approach among those in the vicinity of the ion is the one chosen for the scattering event. The ion is advanced to this position of closest approach, losing energy via electronic interactions according to (10.27), where L is now the actual distance traveled. Then the ion scatters by nuclear interaction with this lattice atom according to the usual binary scattering formulas from the previous section. A new ion velocity is calculated, and the next scattering event is chosen. This process is repeated until the ion comes to rest.

This is the basic algorithm for ion implantation in a crystalline target; however, a number of refinements are necessary to model accurately the effects of attendant physical processes. These processes include position-dependent electronic stopping, lattice vibrations, multiple target scattering, and the build up of damage to the lattice.

The electronic stopping relation in (10.27) assumed a uniform distribution of electrons in the solid, but in a crystal lattice the electron density is strongly dependent upon position. Near the lattice sites, the electron density is much greater than in the channel region midway between atomic sites. A relatively simple way to model this effect empirically is to use one electronic stopping coefficient in the channel region, and another coefficient near the lattice sites [116]. Let a_0 be the distance at which crossover to the channel occurs. The coefficient k_0 in (10.27) is modified to

$$k_0 = k_ch + (k_1 - k_ch)(\frac{a_0 - p}{a_0})^2 \tag{10.40}$$

for impact parameter p less than a_0. For $p \geq a_0$, k_0 is equal to the constant k_ch. A somewhat different empirical formula has been used by Hobler for boron implantation into silicon [247]. A more precise method for calculating the position dependence of electronic stopping is to calculate explicitly the electron density for the particular crystalline lattice from energy band theory, and then integrate over the actual ion path [297].

Another refinement to the trajectory algorithm for crystalline targets is to allow thermal vibrations of the lattice atoms about their equilibrium lattice positions. The root mean square displacement from equilibrium is dependent upon the temperature and the properties of the material, and is determined by Debye theory [18]. In this case the impact parameter for a given collision is adjusted by a Gaussian random variable multiplied by the root mean square displacement. Physically, the main effect of the thermal vibrations is to increase the likelihood that a well-channeled ion will scatter out of its channel.

A third refinement to the algorithm concerns the binary collision approximation for nuclear scattering. The algorithm outlined above selects a single target atom for nuclear scattering, when in fact the ion interacts with all nearby lattice atoms to some extent. Consider the case in which the ion travels midway between, and perpendicular to, two atoms in the lattice. The magnitude of the scattering angle will be exactly the same, but in opposite directions, due to interactions with each lattice atom, so the net effect will be no change in the direction of the ion. There will, of course, be energy loss due to interactions between the ion and the two atoms. The theory described previously is valid only for binary scattering, where one particle scatters off another. The more general case of multi-particle interactions is an unsolved problem of classical mechanics. A reasonable approach is to independently calculate the contribution of atoms in the immediate vicinity of the ion, and then sum these contributions to obtain a single energy loss and scattering angle.

The final modification that we discuss is to consider the effect of damage to the lattice as the implant proceeds. A single scattering event can impart enough energy to the target atom to dislodge the atom from its lattice site, introducing a vacancy defect at that position. The dislodged atom then travels through the lattice, eventually losing enough energy to stop. It will come to rest in an otherwise perfect lattice, and therefore exist as a self-interstitial defect. Thus each primary scattering event of the incoming ion with a lattice atom has the potential to create a vacancy at the site of the initial collision, and a self-interstitial at the location where the recoiling lattice atom comes to rest. Furthermore, the recoil can dislodge other lattice atoms along its path, creating more point defects. The refined algorithm calculates the initial velocity of each recoil, and then follows the trajectory of the recoil until it stops. Any additional recoils created along this trajectory are likewise followed in a recursive manner. The end result is a "damage tree" for each incoming ion. Figure 10.8 shows the damage trees calculated for a single boron ion and for a single arsenic ion implanted into silicon. Since boron (atomic mass 10.81 a.m.u.) is much lighter than silicon (28.086 a.m.u.), the boron ion produces relatively little damage to the lattice. On the other hand, since arsenic (74.9 a.m.u.) is much heavier than silicon, a single arsenic ion creates extensive damage to the lattice.

Another effect of recoil generation is the gradual accumulation of damage as the ion implantation proceeds. The second ion will see a less perfect lattice than the first, and so forth as more ions are implanted. Indeed, for a relatively heavy ion such as arsenic, the crystalline lattice will quickly become amorphous, suppressing channeling completely. Figure 10.9 illustrates this effect: out of a total of 10^4 ions implanted, the final position is shown for the first 100 and the last 100 ions. The average penetration into the target of the former group is much greater than the latter group, because damage to the lattice has gradually accumulated, suppressing channeling.

Particle simulations illustrate that the ion implantation process is inherently destructive since it damages the crystalline structure of the target.

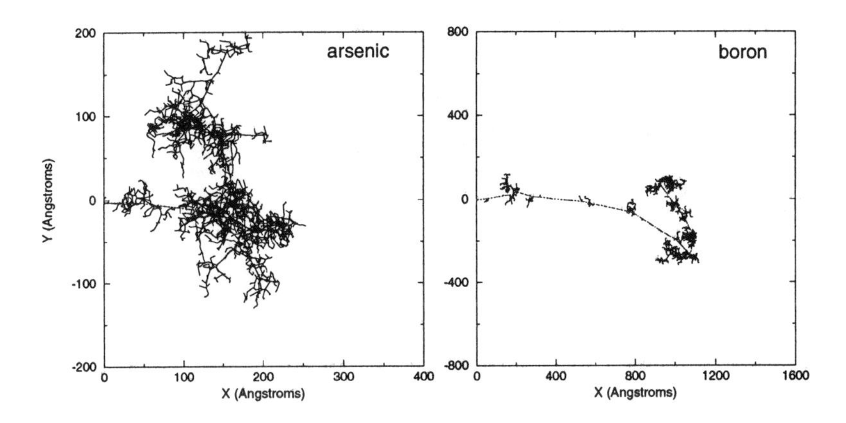

Figure 10.8: Damage tree generated by a single ion of arsenic and boron implanted at 50 keV into crystalline silicon. The final stopping position of the arsenic ion is $(x, y) = (208, -36)$ and that of the boron ion is $(x, y) = (903, 43)$.

Therefore, in processing the wafer, it is always necessary to follow an ion implantation with a high temperature diffusion (or anneal). As the temperature is raised, the atoms migrate to a position of lowest energy, gradually rebuilding the original lattice. Now each implanted impurity ion will come to occupy a lattice site, substituting for one of the original target atoms. Once incorporated into the lattice, the impurity ion may donate or accept an electron, depending upon its atomic properties, and thus change the electrical behavior of the semiconductor in its vicinity.

The high temperature diffusion that must follow ion implantation also redistributes the impurity ions, as discussed in the next two chapters. It is shown there that the point defects (vacancies and interstitials) play a major role in the diffusion of the impurity ions. Particle simulations of the ion implantion process can provide the initial distributions of vacancies and interstitials which are required for subsequent diffusion modeling.

10.6 Summary

Ion implantation plays a major role in semiconductor device processing. With precise control over processing conditions, ion implantation is the most

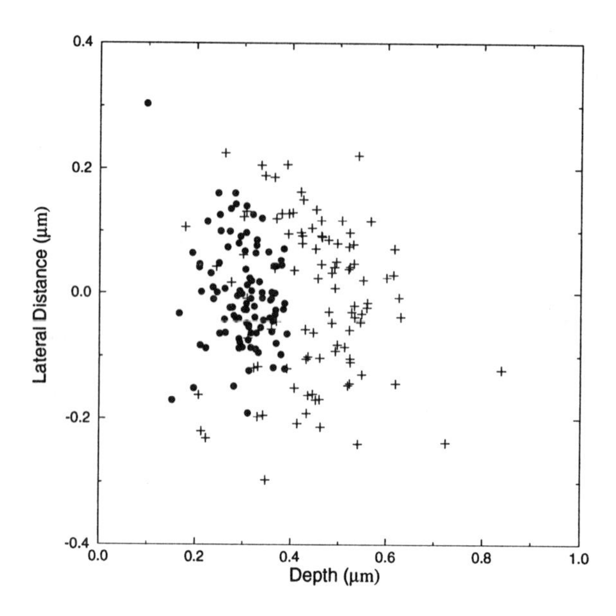

Figure 10.9: Final stopping position of the first 100 ions (plus signs) and the last 100 ions (circles), from a total of 10^4 implanted ions, illustrating the effect of damage buildup during implantation.

reliable and repeatable means of doping a semiconductor. Accurate computer modeling of this process is critical because the distribution of dopant ions determines to a large extent the electrical performance of the semiconductor device.

Process models for ion implantation fall into two broad classes: analytic distribution function models and particle (or Monte Carlo) models. The distribution functions can provide very accurate fits to measurements of ion concentration versus depth in one dimension. However, the distribution functions lack predictive capability for situations where experimental data are not readily obtained, such as ion concentration profiles in the lateral dimensions, or for complex multi-layer two- and three-dimensional targets. In contrast, particle methods attempt to model the implantation process at the atomic level. They are inherently three-dimensional, and can predict profiles in situations where experimental data is not available. Particle methods can also provide information on distributions of point defects for use in subsequent diffusion simulations. However, these methods are very expensive computationally, since they track one ion at a time at the atomic level. Analytic distribution functions and particle methods play complementary roles in modeling ion implantation.

10.7 Exercises

1. Assuming a Gaussian distribution, plot the concentration profiles of boron, phosphorus, and arsenic implanted into silicon at 100 keV and a dose of 10^{15} cm^{-2}. For distribution parameters, use (R, σ) = $(0.3\ \mu m, 0.07\ \mu m)$, $(0.14\ \mu m, 0.05\ \mu m)$, $(0.06\ \mu m, 0.02\ \mu m)$ for the three species, respectively.

2. The double Pearson distribution can give excellent fits to experimental data. Create a "double Gaussian" distribution with five parameters, R_1, σ_1, R_2, σ_2, and D_1, and fit it to the experimental data in Figure 10.1.

3. Assume that a target has an alternating sequence of layers of Si and SiO_2, each 100 nm thick. Calculate the Gaussian profile of boron implanted into this structure at 100 keV and 10^{15} cm^{-2}.

4. In Figure 10.2, particle 2 is initially at rest. After the collision, it moves in the direction indicated by the angle ϕ, and the magnitude of the velocity is $\sqrt{2\Delta E_0/m_2}$. Particle 1 moves in a direction indicated by the angle θ, and the magnitude of its velocity is $\sqrt{2(E_0 - \Delta E_0)/m_1}$. Using conservation of momentum, derive an expression for the angle ϕ.

5. Equations (10.22)–(10.25) give the solution of the scattering integral for the case of the Coulomb potential. Evaluate the scattering integral (10.12), for the case of the collision of two infinitely hard spheres, i.e. a "billiard ball" potential: $V(r) = 0$ for $r > 2r_0$ and $V(r) \to \infty$ for $r \leq 2r_0$, where r_0 is the radius of the billiard balls. Note that for this potential, $s_{max} = 1/(2r_0)$, and therefore $V(\tilde{s}) = 0$ over the entire range of the scattering integral. Then, assuming that the masses of the two balls are equal, write an expression for the energy loss of the incident ball, and the scattering angles of the two balls as a function of p and E_0.

6. The binary scattering theory of Section 10.3 assumes that the ion is traveling at non-relativistic velocity, that is, a velocity much less than the speed of light. Is this assumption reasonable for arsenic implants at 100 keV, 1000 keV, and 1 MeV? At what energy is the velocity

approximately 10 percent of the speed of light for arsenic and for boron?

7. Algorithm 1 outlines a procedure for calculating of the trajectory of an ion through an amorphous solid. Modify the algorithm for the case of an ion traveling through a crystalline solid to include the following effects: (a) deterministic, rather than random, selection of impact parameter for each scattering event (see (10.39)); (b) the position-dependent electronic energy loss (e.g. (10.40)); (c) trajectories of secondary atoms.

Chapter 11

Single Species Diffusion

11.1 Introduction

This chapter is concerned with the transport of an impurity during the high temperature processes of semiconductor fabrication. In general, transport processes involve diffusion, convection due to fluid motion, and reaction, all of which occur simultaneously. In many instances, however, one of these three is the dominant factor causing transport and for this reason, as well as tractability, it is appropriate to study each separately. Diffusion characterizes the movement of a species or of a quantity such as heat through a medium, from regions of high concentration to low. Beginning with the work of Fourier [185] in 1822 on the conduction of heat in a solid, a vast literature exists that spans many disciplines: mathematics, physics, materials science, and solid-state electronics, to name a few. Crank's book [118] is a classic introduction to general diffusion, with many closed form solutions for 1D problems and an overview of diffusion in a heterogeneous medium (see also [99]).

Modeling diffusion in semiconductors has become quite specialized, with an expert in silicon diffusion focusing on different issues than would a researcher modeling gallium arsenide diffusion. The text [49] is an early effort to present a systematic account of diffusion in semiconductors. Articles on the calculation and measurement of diffusion coefficients in silicon, III–V compounds, and oxides can be found in [485], while the reference [532] provides an excellent introduction to the physical mechanisms, such as point defects, responsible for semiconductor diffusion. Other treatments of diffusion in device and processing technology include [224, 464, 559].

The present chapter covers the basic diffusion processes, later chapters describe several kinetic models for diffusion in silicon, the numerical methods that can be used to implement these models in three dimensions, and several specialized topics such as rapid thermal annealing. We emphasize that these physical models, mathematical formulations, and numerical algorithms are representative of other diffusion processes of interest in semiconductor technology, such as diffusion of oxygen during oxidation (Chapter 15) or diffusion during chemical vapor deposition.

11.2 Diffusion as a Random Walk

One of the most important processing steps in semiconductor fabrication is diffusion. It is used to introduce impurities into silicon for the purpose of forming *pn*-junctions and to anneal damage after ion implantation. Also, some diffusion of impurities inevitably occurs during other high temperature treatments such as oxidation or epitaxy. Diffusion is perhaps best understood in terms of a random walk through a regular Cartesian lattice. Although the actual situation in silicon is considerably more complicated, this special case allows a straightforward derivation of the diffusion equation, which models low dopant concentration diffusion in silicon very well. Figure 10.7 illustrated the actual silicon lattice; Figure 11.1 shows a potential well representing its idealized counterpart in 1D. In this simple model, an impurity atom located at a lattice site remains there until it acquires enough energy (W) to overcome the potential barrier and exchange places with a neighboring silicon atom.

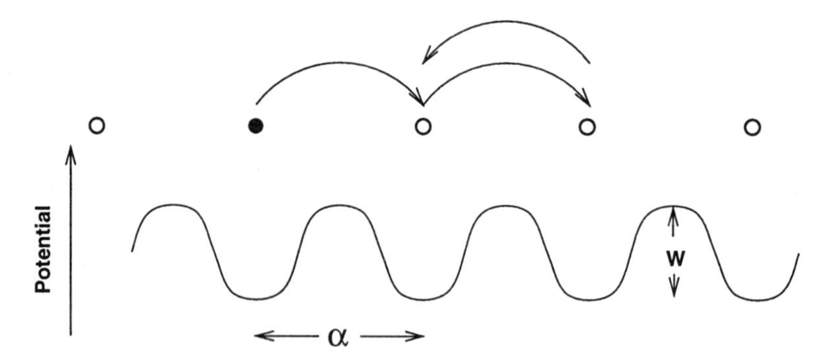

Figure 11.1: Idealized silicon lattice in 1D represented as a potential well.

Consider the unrestricted one-dimensional random walk, which corresponds to 1D diffusion on an unbounded domain. Let α be the lattice constant, i.e. the mean distance between lattice points. An impurity atom initially located at $x = 0$ executes random jumps of length α to the right or left every Δt seconds. Let X_i be a random variable that assumes the value α if the atom moves to the right by changing places with a neighboring silicon atom, and $-\alpha$ if it moves to the left; the associated probabilities are

$$p = P(X_i = \alpha), \quad q = P(X_i = -\alpha) \text{ with } p + q = 1 \tag{11.1}$$

If $p > 1/2$, there is a tendency for the particle to drift to the right during the course of the random walk. The position of the atom at the nth step is given by the random variable $S_n = X_1 + X_2 + \cdots + X_n$. Assuming independence of the random variables, the probability that the particle is located at a given point after n steps is given by the binomial distribution, defined by the coefficients in the expansion of $(p + q)^n$. The expectation and variance of S_n are

$$E(S_n) = (p - q)n\alpha, \qquad V(S_n) = 4pqn\alpha^2 \tag{11.2}$$

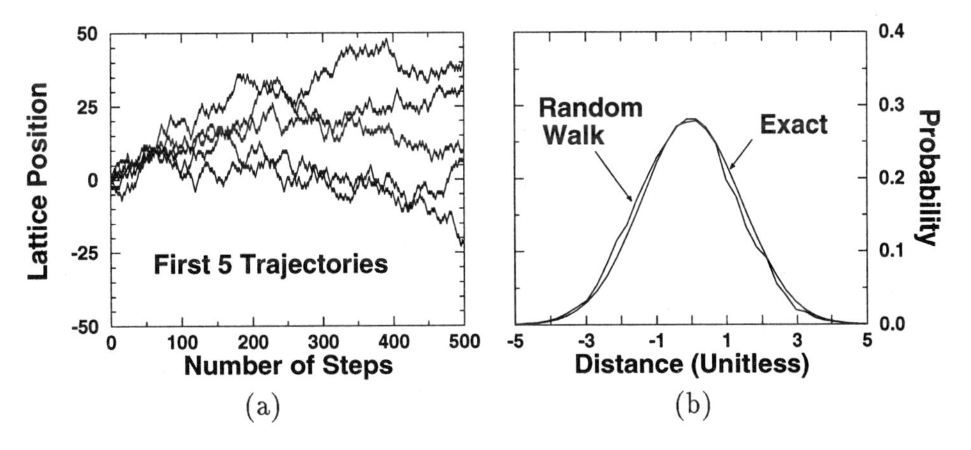

Figure 11.2: (a) Sample trajectories of a particle executing a random walk with probability 0.5 of moving to the right at each step. (b) Comparison of the exact and approximate solutions to the 1D diffusion equation with the Dirac delta as initial condition.

We may ask: what happens in the continuum limit as the stepsize α approaches zero and the number of steps n per unit time becomes unbounded? Experimentally, the mean displacement of the particle per unit time and the variance are found to have values $\mathcal{C}$ and $\mathcal{D}$, respectively. For very large n we expect E and V to have values extremely close to $\mathcal{C}$ and $\mathcal{D}$, i.e. $\mathcal{C} = (p-q)n\alpha$ and $\mathcal{D} = 4pqn\alpha^2$. In this continuum limit model, let $v(x, t)$ be the probability that at time t the particle is located at the point x. Then for timestep

$\Delta t = 1/n$,

$$v(x, t + \frac{1}{n}) = p\, v(x - \alpha, t) + q\, v(x + \alpha, t) \tag{11.3}$$

because to have reached x at the current step, at the previous step either it was at $x - \alpha$ and jumped to the right, or it jumped to the left from $x + \alpha$. Next, introduce the Taylor series expansions

$$v(x, t + \frac{1}{n}) = v(x, t) + \frac{1}{n}\frac{\partial v}{\partial t}(x, t) + O\left(\frac{1}{n^2}\right) \tag{11.4}$$

$$v(x \pm \alpha, t) = v(x, t) \pm \alpha \frac{\partial v}{\partial x}(x, t) + \frac{1}{2}\alpha^2 \frac{\partial^2 v}{\partial x^2}(x, t) + O\left(\alpha^3\right) \tag{11.5}$$

Substituting (11.4) and (11.5) into (11.3) and simplifying,

$$\frac{\partial v}{\partial t}(t, x) = n\alpha(q - p)\frac{\partial v}{\partial x}(t, x) + \frac{1}{2}n\alpha^2 \frac{\partial^2 v}{\partial x^2}(t, x) + O\left(\frac{1}{n}\right) + O\left(n\alpha^3\right) \tag{11.6}$$

In the limit as $\alpha \to 0$ and $n \to \infty$ with $n\alpha^2$ constant, the Fokker–Planck drift-diffusion equation is obtained

$$\frac{\partial v}{\partial t} = -\mathcal{C}\frac{\partial v}{\partial x} + \frac{\mathcal{D}}{8pq}\frac{\partial^2 v}{\partial x^2} \tag{11.7}$$

If $p \neq q$, $\mathcal{D} = 4pq\alpha\mathcal{C}/(p - q)$ which approaches 0 as $\alpha \to 0$. To ensure a nonzero variance, take $p = q = 1/2$, so that $\mathcal{C} = /$, and one obtains the diffusion equation

$$\frac{\partial v}{\partial t} = D\frac{\partial^2 v}{\partial x^2} \tag{11.8}$$

For convenience, a macroscopic diffusivity D is defined as $\mathcal{D}/2$, and this is the "diffusion coefficent" discussed in the following sections.

Mathematically, this is an initial-boundary value problem. The values of the impurity concentration in a region Ω at time zero and the values of the concentration or its normal flux on the boundary $\partial\Omega$ for $t > 0$ are typical initial and boundary conditions to complete the statement of a well-posed problem. A more sophisticated derivation can be given that uses a correlated random walk in which the random variables $X_1, \ldots, X_n$ are no longer assumed to be independent. In this case p need not be equal to q for a finite speed $\mathcal{C}$. There are other features of interest in the derivation given above. Note that if $\mathcal{D} = 0$, the problem reduces to the first-order wave equation $v_t + \mathcal{C}v_x = 0$. The quantity on the left hand side of this equation is just the directional derivative of v in the direction $(\mathcal{C}, 1)$, hence v must be constant on lines parallel to $(\mathcal{C}, 1)$. For further details see the discussion of the material derivative and the transport theorem in Chapter 7.

In recent years, the deep connections between differential equations and probability theory have begun to be appreciated. Historically, the simplified random walk considered in this section was known as the gambler's ruin problem, because at each throw of the die a gambler wins or loses a dollar to an adversary with probabilities p and q, respectively. In fact, this is a special instance of a Markov chain, and more generally, a stochastic process [177].

11.3 Diffusion in a Continuum

The diffusion equation can also be derived at the macroscopic level assuming a continuum with conservation laws and constitutive theory. In 1855 Fick formulated a constitutive relation for diffusion by considering flux, which is defined physically as the number of atoms or molecules passing through a unit area in a unit time. Fick's first law states that the instantaneous flux vector is proportional to the concentration gradient. That is,

$$\boldsymbol{F} = -D\,\nabla C \tag{11.9}$$

where C is the concentration of the diffusing substance and the constant of proportionality D is called the diffusion coefficient or diffusivity. In some cases, such as diffusion at low dopant concentration in silicon, D is approximately constant, while in others D shows a marked dependence on position or concentration. The flux through a surface S is $\int_S \boldsymbol{F} \cdot \boldsymbol{n}\, ds$, where $\boldsymbol{n}$ is the outward unit normal function defined on S. If the concentration inside S is higher than it is outside S, then the flux through S is negative .

Consider an arbitrary region Ω in $\mathbf{R}^3$ with boundary surface $\partial\Omega$, as shown in Figure 11.3. Suppose that a flow field exists, specified by a velocity function $\boldsymbol{v}(x,t)$, so that some transport occurs due to the advection of the impurity. The time rate of change of species mass number in Ω is given by

$$\frac{d}{dt}\int_\Omega C(\boldsymbol{x},t)\,dxdydz = \int_\Omega \frac{\partial C}{\partial t}(\boldsymbol{x},t)\,dxdydz. \tag{11.10}$$

Assuming conservation of mass, a change in the amount of C in Ω can only take place in two ways. The first is by movement through the boundary of the region. This contributes two terms, one due to Fick's law and the other to advection, which together give the total flux through the boundary,

$$\int_{\partial\Omega} -D\nabla C\cdot \boldsymbol{n}\, ds + \int_{\partial\Omega} -C\boldsymbol{v}\cdot\boldsymbol{n}\, ds = -\int_{\Omega} \nabla\cdot(D\nabla C + C\boldsymbol{v})\, dxdydz \quad (11.11)$$

where the divergence theorem has been applied.

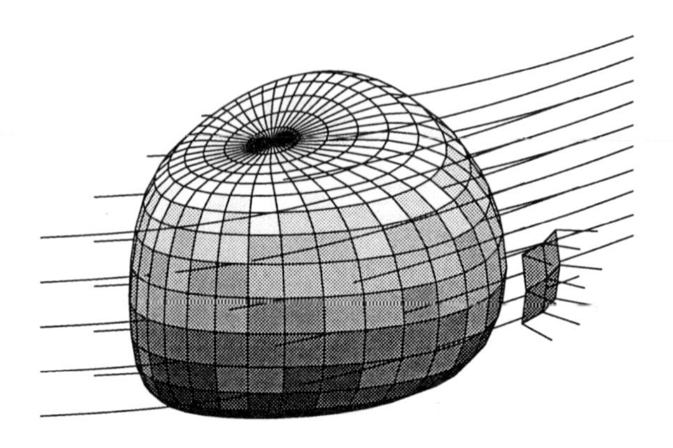

Figure 11.3: Fick's second law can be derived by applying conservation of mass and the divergence theorem to an arbitrary region Ω. A small section of $\partial\Omega$ and its outward unit normals are shown at the right.

Secondly, a change in C can occur if there are sources or sinks within Ω, where the species of interest is either created or destroyed. This is the case when chemical reactions are occurring simultaneously with the transport. For example, in the advanced diffusion models discussed in the following chapter, species such as silicon interstitials and vacancies can annihilate each other, while at the same time they are constantly being produced because it is thermodynamically favorable to have some disorder in the lattice at any temperature above absolute zero. If G denotes the generation of the quantity C per unit volume and R the recombination or elimination, then this gives a contribution to the overall balance of $\int_\Omega (G(\boldsymbol{x},t) - R(\boldsymbol{x},t))\, dxdydz$. Summing these various terms gives the identity (recall that a net outward flux implies concentration is decreasing)

$$\int_\Omega \left(\frac{\partial C}{\partial t} - \nabla\cdot(D\nabla C + C\boldsymbol{v}) - (G-R)\right) dxdydz = 0 \qquad (11.12)$$

Since the integral is assumed to be continuous and the region Ω is arbitrary, the integrand must be zero everywhere in $\mathbf{R}^3$ so the general drift-diffusion-reaction equation follows

$$\frac{\partial C}{\partial t} = \nabla\cdot(D\nabla C + C\boldsymbol{v}) + G - R \qquad (11.13)$$

Recall from Chapter 7 that the current density of electrons is given by $\boldsymbol{J} = D_n \nabla n + \mu_n n \boldsymbol{E}$, so that (11.13) corresponds to the current continuity equation, $\frac{\partial n}{\partial t} + \nabla \cdot \boldsymbol{J} = 0$. Some materials including crystals, textile fibers, and polymer films exhibit anisotropic diffusion: they have different diffusion properties in different directions. In this case the scalar diffusivity must be replaced by a tensor, a 3×3 matrix acting on the vector ∇C. To a good approximation, impurity diffusion in silicon is isotropic. It is also possible to derive (11.13) directly from the Boltzmann transport equation by taking moments.

11.4 Intrinsic (Low Concentration) Diffusion

Consider the problem of diffusing an impurity species into silicon at temperature T. When the concentration of an impurity is below the intrinsic electron concentration of approximately 10^{18} atoms cm^{-3} at 1000°C, the linear equation $\partial C/\partial t = D\Delta C$ suffices to model dopant diffusion. From experimental studies, the temperature dependence of the diffusivity can be approximated by an Arrenhius expression

$$D(T) = D_0 e^{-E_a/kT} \tag{11.14}$$

where D_0 is the pre-exponential factor, k is the Boltzmann constant, T is the temperature in degrees Kelvin, and E_a is the activation energy. Experimentally measured diffusivities of various impurities in silicon as a function of temperature are shown in Figure 11.4(a).

Analytical solutions for the linear diffusion equation were used in some of the very early process simulators and are still of value [464]. A knowledge of these solutions allows the process engineer to develop a feel for what general doping profiles are possible when forming *pn*-junctions and serves as a consistency check on simulator output. As discussed in Chapter 17, for a statistical simulator, many runs must be made to relate a physical parameter such as mean diffusivity to the input bias current for, say, an operational amplifier. The loss of accuracy in using a simple analytical model over the more precise physically-based models discussed later is offset by the greatly reduced execution time and the requirement to determine, not the exact value of a quantity, but a statistical confidence interval.

Two important cases of intrinsic diffusion occur in older technologies and have closed-form solutions in one dimension. The first is predeposition, in which the surface concentration is held constant and the impurity diffuses into the region $x > 0$. Because the wafer thickness ($\approx 2 - 3$ mm) is much

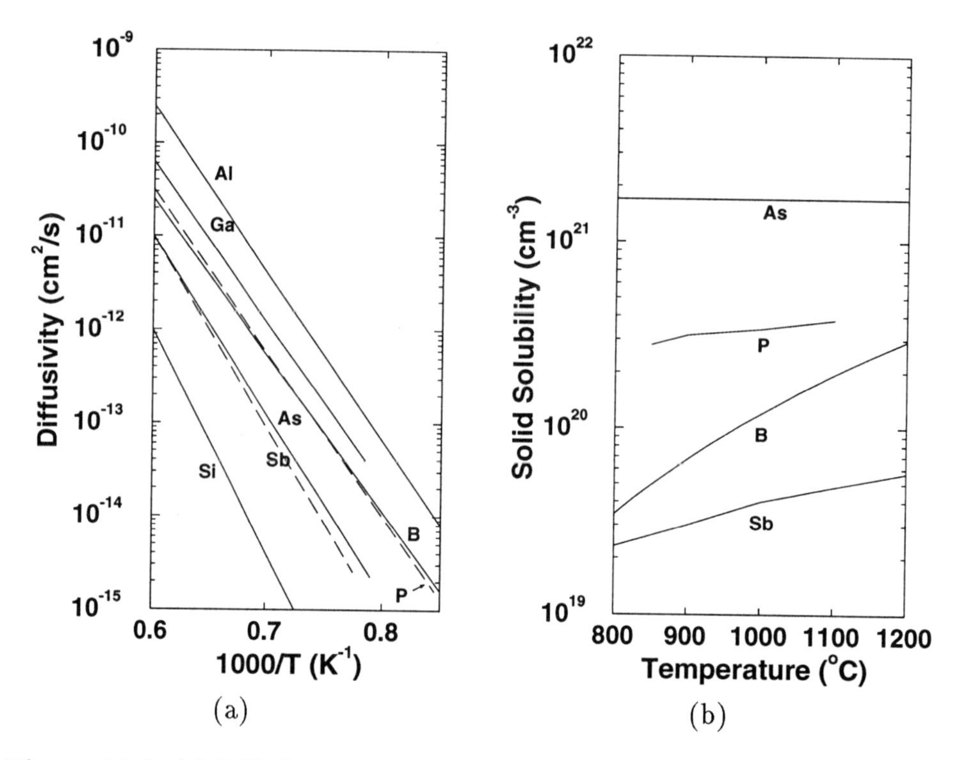

Figure 11.4: (a) Diffusion coefficient (diffusivity) as a function of the reciprocal of temperature for several common impurities in silicon. (b) Solid solubilities in silicon of four common dopants (from [14] (B), [407] (P), and [526] (Sb)).

greater than typical junction depths ($0.5-2\mu$m), a reasonable approximation is that the far boundary is at infinity. Equation (11.8) is then solved for $t > 0$ with the Dirichlet boundary conditions $C(0,t) = C_s$, $C(\infty, t) = 0$ and the initial concentration, $C(x,0) = 0$ for $x > 0$. Judicious use of Laplace transforms yields the predeposition solution

$$C(x,t) = C_s \text{ erfc}\left(\frac{x}{2\sqrt{Dt}}\right) \tag{11.15}$$

where erfc is the complementary error function. It is more difficult to achieve a precisely controlled profile with predeposition than with ion implantation, and the latter has largely supplanted predeposition for introducing impurities into the wafer.

The other case with a closed-form solution is a "drive-in" that follows a predeposition or that is used to anneal the damage caused during a shallow ion implant. To a first approximation the total amount of dopant in the wafer, Q, is conserved. Assuming an initial complementary error function profile given by (11.15), which is so shallow that it approximates a Dirac delta distribution, together with boundary conditions $\frac{\partial C}{\partial x}(0,t) = 0$ and $C(\infty, t) = 0$, the exact solution for the drive-in is

$$C(x,t) = \frac{Q}{\sqrt{\pi D t}}\, e^{-x^2/(4Dt)} \tag{11.16}$$

Figure 11.5 shows the evolution in the (x,t)–plane of an impurity profile which at time $t = 0$ is given by a very shallow implant – for all practical purposes a Dirac delta.

11.5 Extrinsic (High Concentration) Diffusion

As the impurity concentration is increased so that the electron concentration n exceeds its intrinsic value n_i, measured diffusivities show a marked concentration dependence. This is because diffusion is more than a simple random-walk nearest-neighbor interchange mechanism — it is assisted by point defects. Far from having a perfect crystal structure stretching in all directions, even the purest crystal ingot contains a multitude of defects: thermal (phonons) and electronic (excitons) defects; vacancies, lattice sites not occupied by a silicon atom; divacancies, two adjacent unoccupied sites; self-interstitials, silicon atoms not residing on a lattice site; and impurity atoms. At any temperature above absolute zero, the crystal achieves an overall lower energy state by allowing a measure of disorder via these irregularities. During typical processing conditions, background concentrations of vacancies and interstitials can vary in the range 10^{10}–10^{16} cm^{-3}. This is much less than typical doping concentrations of 10^{16}–10^{21} cm^{-3} and the number of lattice sites 10^{22} cm^{-3}, but the defects are nevertheless extremely important because they provide a mechanism for impurity diffusion. There are other irregularities in the lattice - called extended defects - such as defect complexes that extend over 10–20 atoms, dislocation loops, stacking faults, and the crystal surface itself. The effects of these irregularities are quite significant in some applications and progress is being made in modeling them numerically [54, 480]. In what follows, the term *point defect* will refer to either a vacancy or silicon self-interstitial.

Early papers suggested that phosphorus, for example, diffuses by first bonding with a vacancy to form a pair called an E-center [572, 565, 566, 570]. It is thermodynamically more favorable for the pair to move through the lattice than for phosphorus to interchange positions directly with an adjacent silicon atom. Vacancies are known to exist in several charge states, V^0, V^-, $V^=$, V^+, with a concentration that is temperature and Fermi level dependent

$$[V^-] = [V^-]_i \; e^{(E_F - E_i)/(kT)} \tag{11.17}$$

where brackets denote concentration, i an intrinsic value, and E_F the Fermi level.

In process modeling, it is generally assumed that charge neutrality exists during the diffusion process. If the densities of states in the conduction and valence bands are calculated quantum mechanically, and if the probability of an electron occupying these states is accounted for, the following expressions for a semiconductor in equilibrium are obtained:

$$n = N_C \; e^{-(E_C - E_F)/(kT)} \; , \quad p = N_V \; e^{-(E_F - E_V)/(kT)} \tag{11.18}$$

where E_C is the conduction band level, E_V is the valence band level, N_C and N_V are the density of states, and n and p are the concentrations of electrons and holes, respectively. The product of n and p is independent of the Fermi level and its square root is defined to be the intrinsic concentration n_i of electrons. That is,

$$n\,p = n_i^2 = N_C N_V \; e^{-(E_C - E_V)/(kT)} \tag{11.19}$$

This relationship can also be derived on the basis of the law of mass action from the reaction terms in Shockley–Read–Hall [495] recombination theory, as discussed in Chapter 7. Now, assume that the net charge density in any volume element is zero; then the electric field is zero, which gives

$$0 = n - p + N_A - N_D \tag{11.20}$$

where N_A is the concentration of acceptor impurities and N_D that of donors. Combining (11.19) and (11.20), the algebraic condition for electron density in terms of the net impurity concentration $N_{net} = N_D - N_A$ is

$$n(N) = \frac{1}{2}\left((N_D - N_A) + \sqrt{(N_D - N_A)^2 + 4n_i^2}\right) \tag{11.21}$$

The assumption of vanishing space charge is perhaps questionable when considering coupled diffusion of dopants at very high concentrations in a structure with pn-junctions. It is, however, much easier to assume (11.21) than to add the continuity equation for electrons and Poisson's equation to the solution process (see (11.33)–(11.34)), as done in Chapter 7. Simulations [433] show that this simplification is usually justified.

As the number of charged dopant atoms increases, the electron concentration is forced above its intrinsic level, which in turn produces more charged vacancies and enhances diffusion. That is

$$n \gg n_i \quad \Rightarrow \quad [V^-] \gg [V^-]_i \quad \Rightarrow \quad D(n) \gg D_i \tag{11.22}$$

This results in a compound diffusion coefficient of the form

$$D(n) = D_i^0 + D_i^+ \left(\frac{n_i}{n}\right) + D_i^- \left(\frac{n}{n_i}\right) + D_i^= \left(\frac{n}{n_i}\right)^2 \tag{11.23}$$

For an acceptor impurity such as boron (column III in the periodic chart) $\frac{n_i}{n} \gg 1$, which implies that for n-type material doped with boron only the terms involving D_i^0 and D_i^+ are significant in (11.23). For p-type silicon doped with phosphorus, arsenic, or antimony (column V elements) $\frac{n}{n_i} \ll 1$, so the term involving D_i^+ may be neglected.

The diffusion equation must now be rewritten with a concentration-dependent diffusivity $D(C)$

$$\frac{\partial C}{\partial t} = \nabla \cdot (D(C) \nabla C) \tag{11.24}$$

where D is given by (11.23) with the electron concentration determined from (11.21). Note that $D(C)$ has the same form whether one assumes a vacancy or interstitialcy mechanism or a combination of both, as long as there is no interaction between the two types of defects.

Figure 11.5 shows simulations [394] of boron diffusion under both intrinsic and extrinsic conditions and compares the results with experimental secondary ion mass spectroscopy (SIMS) data. Boron is implanted (5×10^{13} cm^{-2} at 25 keV) and diffused in silicon at 900°C for 24 hours under three different conditions: (1) intrinsic, or low concentration conditions; (2) extrinsic conditions, in the presence of a heavy arsenic background concentration of 10^{20} cm^{-3}; and (3) extrinsic conditions, in the presence of a heavy boron-11 background concentration of 2×10^{19} cm^{-3}. For this latter case, the isotope B^{11} was used for the background and the isotope B^{10} for the lower concentration implanted boron [555], since the two isotopes can be readily distinguished in a SIMS measurement. The experimental data shows a significant tail for the implant profile due to channeling along crystal axes. The diffused profiles at 900°C agree fairly well with the experimental data; similar profiles can be obtained for 950°C and 1100°C diffusions [394]. For case (2), the boron diffusion is strongly retarded due to the large number of

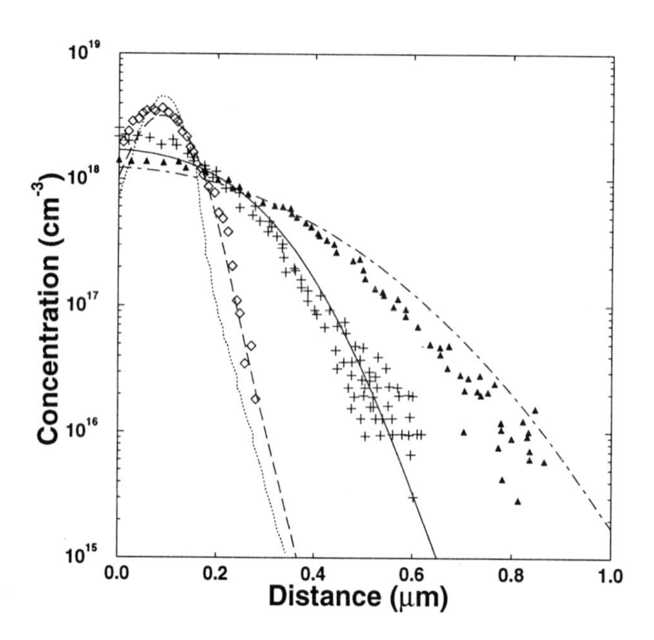

Figure 11.5: Intrinsic/extrinsic boron diffusion: the solid line (+) is simple diffusion, the short-dashed line (◇) is the profile in an arsenic background, and the dot-dashed line (△) is the boron profile in the presence of a boron-11 background.

electrons donated by the high concentration arsenic:

$$D_{boron} = D_i^0 + D_i^+ \left(\frac{n_i}{n}\right) \approx D_i^0 \tag{11.25}$$

Conversely, the boron diffusion in the presence of a very high boron-11 background is enhanced, since the number of holes increases. Although overall agreement between experiment and theory is good, quantitative discrepancies are apparent. These discrepancies cannot be explained [555] by simply adjusting the diffusion coefficients in (11.25).

11.6 Transport and Segregation Coefficients

So far only the simplest possible boundary conditions for the diffusion equation have been considered. During predeposition the surface concentration is not exactly equal to the concentration in the ambient gas. This situation can be accurately modeled by introducing an interface condition relating the flux to the concentration difference on either side of the wafer surface. At a material interface there is also a tendency for the impurity to segregate preferentially in one of the two substances. As an example, consider the upper graphs of Figure 14.2 in which the impurity prefers to reside in the silicon rather than in the oxide.

In the one-dimensional case, interface transport may be modeled [9] using first-order kinetics

$$-D\frac{\partial C}{\partial x} = h\,(C_L - C_R/m) \tag{11.26}$$

where m is the thermal equilibrium segregation coefficient, h is a transport coefficient, and C_L, C_R are the concentrations on either side of the interface. The closed-form solution to (11.8) with this boundary condition, i.e. the ratio $C(x,t)/C_s$, is

$$\operatorname{erfc}\left(\frac{x}{2\sqrt{Dt}}\right) - \operatorname{erfc}\left(\frac{ht}{\sqrt{Dt}} + \frac{x}{2\sqrt{Dt}}\right) e^{\left(ht/\sqrt{Dt}\right)^2} e^{\left(ht/\sqrt{Dt}\right)\left(x/2\sqrt{Dt}\right)} \tag{11.27}$$

which corresponds to the complementary error function solution in (11.15) with a correction term. Note that as $t \to \infty$ the subtracted term goes to zero. In higher dimensions (11.26) is replaced by $-D\nabla C \cdot \boldsymbol{n} = h\,(C_L - C_R/m)$ where it is assumed the boundary of the region Ω is orientable and has a continuous outward normal, $\boldsymbol{n}$. The solution is discontinuous across the interface and when the flux, $-D\nabla C \cdot \boldsymbol{n}$, is zero, the ratio of the concentrations across the interface is $m = C_R/C_L$, hence the name segregation coefficient.

Numerically, a flux boundary condition is easy to implement, particularly when using a finite element spatial discretization. Flux conditions are called natural boundary conditions because they enter the formulation naturally when the divergence theorem is used, as in (11.11). Dirichlet boundary conditions, specifying the value of the unknown at the boundary, are termed essential because they must be handled separately after the stiffness matrix is formed. An alternative is to use a penalty method: if $C_L = C^*$ is to be the concentration of the impurity at the surface, and $m = 1$, then taking (11.26) with an artificially large value of h, say $h = 10^{30}$, will force $C = C^*$ at the boundary. This is often the easiest way to handle Dirichlet or mixed boundary conditions in finite element programs.

11.7 Impurity Clustering

It is fortunate for VLSI that most common dopants have a high solid solubility in silicon, of the order of 10^{21} atoms cm^{-3}. Figure 11.4(b) gives the solubilities of phosphorus, arsenic, antimony, and boron. When impurities are present in silicon at very high concentrations, however, clusters of several dopant atoms can form that lower the effective electrically active concentration [227], i.e. the impurity concentration which appears in (11.21). Chemically, this is represented by the reaction

$$e^- + mS \rightleftharpoons Cl \tag{11.28}$$

in which m impurity species S combine with an electron to form a cluster Cl. Typical values for m are 2 or 3. If C and C_{cl} denote the total and clustered impurity concentrations, respectively, then the electrically active part of the impurity concentration is $N = C - mC_{cl}$. The time rate-of-change of the clustered concentration is just the difference between the rate at which clusters are formed and the rate at which they dissociate

$$\frac{\partial C_{cl}}{\partial t} = - \; k_d C_{cl} + k_c n N^m \tag{11.29}$$

where k_c and k_d are the clustering and declustering rates, and n is the electron concentration. When it is appropriate to assume that equilibrium has been reached in the above reaction, $\partial C_{cl}/\partial t = 0$. Supposing $n \approx N$, the electrically active concentration can be obtained merely by solving the algebraic equation

$$(\beta N)^{m+1} + N - C = 0 \tag{11.30}$$

with $m + 1$ the clustering exponent and $\beta = (\frac{k_c m}{k_d})^{\frac{1}{m+1}}$ the equilibrium clustering coefficient. For $m \leq 3$ this can be solved analytically, but in practice it is simpler just to use Newton's method, which works very well.

Otherwise one must solve dynamically for N in the coupled system

$$\frac{\partial N}{\partial t} = k_d(C - N) - k_c m n N^m + \frac{\partial C}{\partial t} \tag{11.31}$$

$$\frac{\partial C}{\partial t} = \nabla \cdot (D(N)\,(\nabla N \pm N \nabla \ln(n))) \tag{11.32}$$

Note that the total concentration C is $N + mC_{cl}$ and that the clusters are presumed immobile so there is no diffusion term in the first equation. For diffusion times less than a few minutes and temperatures less than 850°C dynamic clustering can be important — otherwise the approximation (11.30) is quite good.

Figure 11.6 shows simulations [394] of the clustering of arsenic using (11.31)–(11.32) and compares the results to experimental data [528]. Arsenic is implanted (2×10^{16} cm^{-2} at 140 keV) and diffused at 1000°C for 20 minutes; then it is diffused at 800°C. Chemical concentration profiles were measured using SIMS, and electrically active concentration profiles were measured using spreading resistance. After the second diffusion step, the chemical concentration of arsenic has not changed, but the electrically active concentration has decreased about 50% due to the increased clustering at the lower temperature. The sheet resistance of the diffused layer is 10.3 ohms/square before the 800°C diffusion and 15.6 ohms/square afterwards. The experimental data has a depth resolution of about 0.035μm.

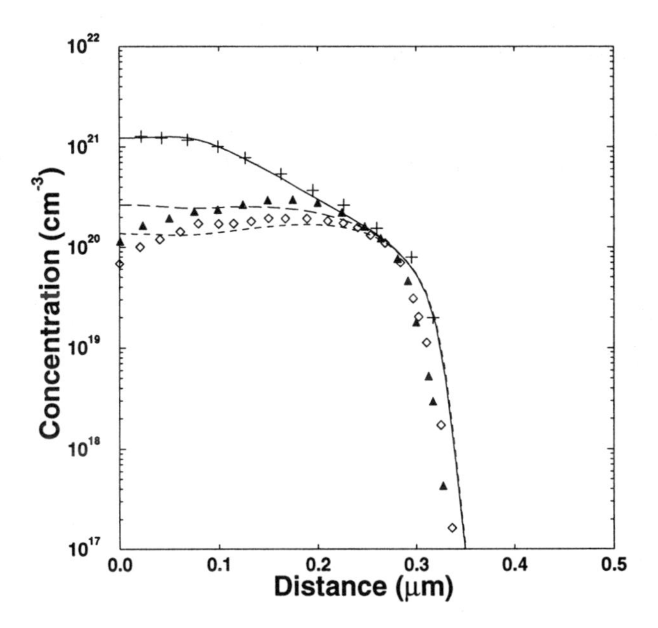

Figure 11.6: Dynamically solving for arsenic clusters. The solid line is the chemical concentration (+ SIMS); the long-dashed line is the electrically active arsenic, N_{As} ($\triangle$ spreading resistance). The short-dashed line is N_{As} after an additional diffusion of 800°C for 60 minutes ($\diamond$ spreading resistance).

11.8 Field-aided Diffusion

As the impurity atoms enter the silicon lattice, they become ionized, setting up an internal electric field. Thus, in addition to the concentration gradient term in the diffusive flux, there is an electric field term of the form $\pm\mu N_{net}\boldsymbol{E}$ that contributes to the overall impurity flux. Here μ is a mobility and $\boldsymbol{E}$ is the electric field given by $-\nabla\psi$, where ψ is the electrostatic potential; a more detailed treatment would use the chemical potential [537, 538, 539].

The continuity equation for impurity concentration becomes

$$\frac{\partial C}{\partial t} = \nabla \cdot (D(C)\nabla N + Z\mu N_{net}\nabla\psi) \tag{11.33}$$

where ψ satisfies Poisson's equation

$$\Delta\psi = \frac{q}{\varepsilon}(2n_i \sinh(\psi/U_T) + ZN_{net}) \tag{11.34}$$

This coupled system is now to be solved for (C, ψ). In (11.33) Z is the charge state for the impurity (1 for acceptors and -1 for donors), N is the electrically active impurity concentration determined from (11.30), and the net electrically active concentration is $N_{net} = \sum_A N_j - \sum_D N_j$. In (11.34) q is the electron charge, ε is the permittivity, and U_T is the thermal voltage.

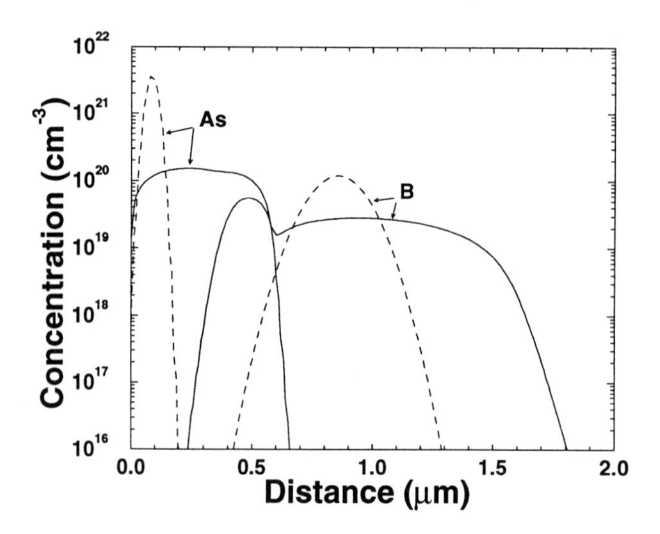

Figure 11.7: The electric field effect in a high concentration coupled arsenic-boron diffusion produces a kink in the boron profile. Dashed lines show the implanted profiles (As, 3×10^{15} at 400 keV; B, 2×10^{16} at 140 keV); the solid lines are the profiles after 20 minutes at 1000°C.

The extra work of solving for ψ is seldom justified, since this tends to be a second-order effect under most diffusion conditions. In sharp contrast to the device equations of Chapter 7, the drift term here is usually very small. It is sufficient to use space-charge neutrality to obtain n and the Einstein relation which states that $D = \mu kT/q$, so that (11.33) becomes

$$\frac{\partial N}{\partial t} = \nabla \cdot (D(\nabla N + ZN_{net}\nabla \ln(n))) \tag{11.35}$$

Figure 11.7 shows a coupled arsenic-boron diffusion in which a kink develops in the boron profile due to the presence of the n-type dopant. Note that it is possible to use (11.21) and solve directly for $\nabla \ln(n)$ in terms of N, giving

$$\frac{\partial N}{\partial t} = \nabla \cdot \left(D \left(1 + \frac{1}{\sqrt{1 + 4(n_i/N)^2}} \right) \nabla C \right) \tag{11.36}$$

where the coefficient multiplying ∇C is referred to as the effective diffusivity, D_{eff} in the literature. The quantity in brackets is sometimes termed the field enhancement factor.

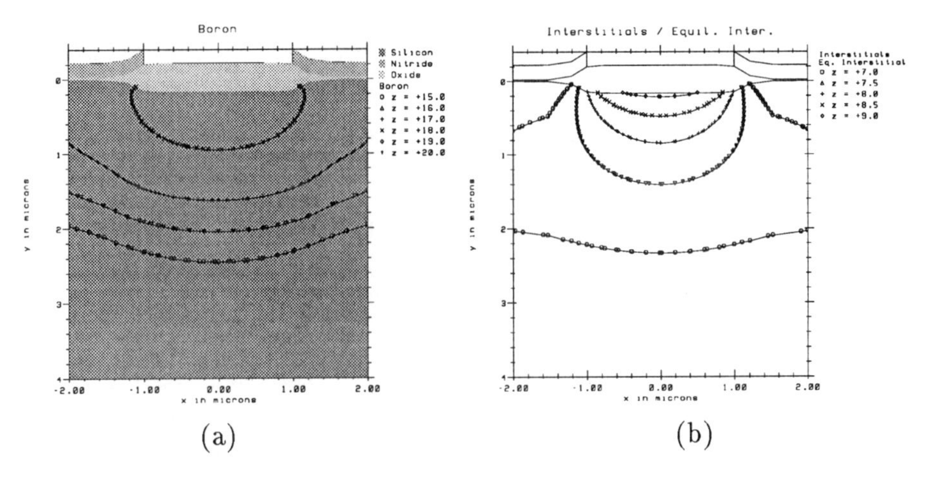

Figure 11.8: (a) A two-dimensional SSUPREM4 simulation of the lateral spread of implanted boron (3×10^{14} cm^{-2} at 50 keV; 20 min at 1100°C in H_2O). (b) Calculated interstitial profiles.

11.9 Lateral Diffusion and Emitter-Push

The process literature mentions the importance of lateral diffusion in the fabrication of integrated circuits. This implies that the true two-dimensional profile cannot be modeled simply by taking a series of one-dimensional simulations and patching them together. Early devices had one dimension of interest and usually one simulation was enough to accurately locate the junctions. As feature size decreased, diffusion under the corners of the mask became important in determining the operating characteristics of a transistor. Figure 11.8 shows the importance of this effect and stresses the need for 2D and 3D simulations for submicron devices.

A phenomenon which cannot be explained by the single-species model is the "emitter-push" effect [168] which can occur during fabrication of an *npn* bipolar transistor. In the standard buried collector process [559] a heavily doped n^+ subcollector is formed using ion implantation. Then a base is formed using ion implantation of boron (10^{12} cm^{-2}) followed by a drive-in. Finally, a phosphorus emitter is fabricated, which is heavily doped for high current gain. It was found that the boron base was "pushed" down toward the bulk in the region directly beneath the emitter. As discussed in the next chapter, this enhanced diffusion results from a supersaturation of point defects associated with the emitter diffusion. It is even possible to have the reverse situation, emitter-pull, which has been reported for gallium bases with arsenic emitters. It is important that the process engineer be aware of these deviations from results predicted by the simple model, as they make it difficult to obtain the desired base widths.

11.10 The Boltzmann–Matano Technique

The concentration-dependent diffusion equation for semiconductor processes

$$\frac{\partial C}{\partial t} = \nabla \cdot (D(C)\nabla C) \tag{11.37}$$

was discussed in Section 11.5. This nonlinear form also arises for certain heat transfer, porous flow, and boundary layer problems. In very simple cases, (11.37) can be solved exactly in one dimension. Of more importance here than the solution itself is the technique used to derive it, which furnishes a method of experimentally determining the diffusivity $D(C)$ as a function of C [4, 356].

In 1894 Boltzmann [50] used the similarity variable $\eta(x,t) = x/\sqrt{t}$ to transform (11.37) for the 1D case into an ordinary differential equation in η

$$\frac{d}{d\eta}\left(D(C)\frac{dC}{d\eta}\right) + \frac{1}{2}\eta\frac{dC}{d\eta} = 0 \tag{11.38}$$

This is similar to the Cole-Hopf transformation of Burger's equation used in fluid dynamics. Imposing boundary conditions $C(0) = 1$ and $C(+\infty) = 0$, and integrating from ∞ to η_1,

$$\int_{\infty}^{\eta_1} \frac{d}{d\eta}\left(D(C)\frac{dC}{d\eta}\right) d\eta = -\frac{1}{2}\int_{\infty}^{\eta_1} \eta\frac{dC}{d\eta}\, d\eta \tag{11.39}$$

which implies

$$D(C)\frac{dC}{d\eta}\bigg|_{\infty}^{\eta_1} = -\frac{1}{2}\int_0^{C_1} \eta\, dC \tag{11.40}$$

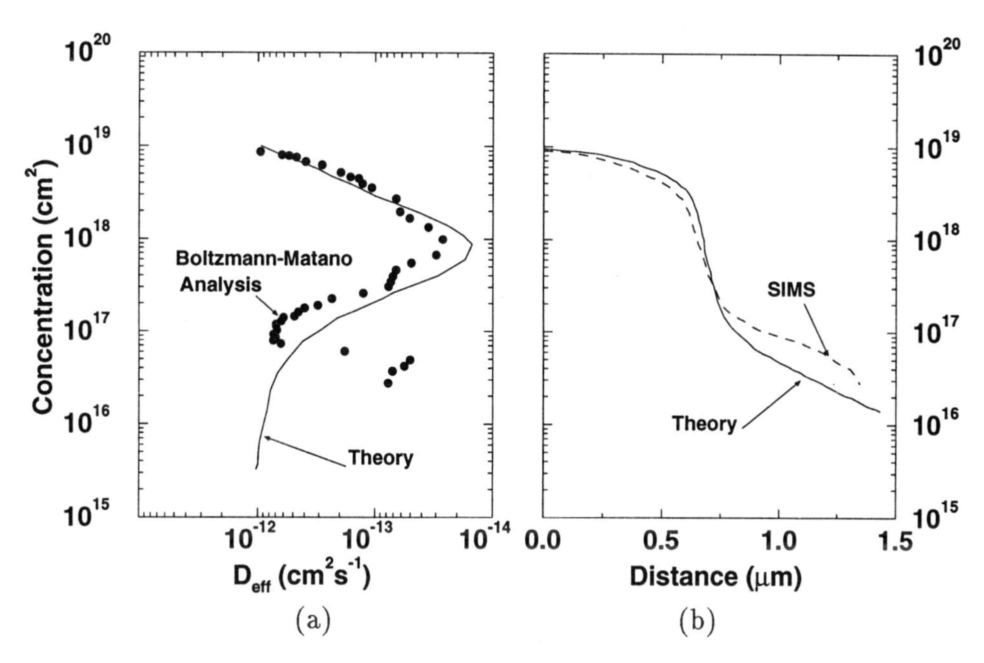

Figure 11.9: Zinc diffusion in GaAs. (a) Concentration-dependence of D_{eff} obtained by a Boltzmann–Matano analysis. (b) SIMS profile and the simulated diffusion using D_{eff} (from [283]).

so

$$D(C_1) = -\frac{1}{2}\frac{d\eta}{dC}\bigg|_{C_1} \int_0^{C_1} \eta(C)\, dC \tag{11.41}$$

where $\eta(C)$ can be any function on $0 \le C \le 1$ such that $\eta(1) = 0$ and both $\frac{d\eta}{dC}$ and $\int_0^{C_1} \eta(C)dC$ exist. For a positive flux $Q = -D\frac{\partial C}{\partial x} = -t^{-1/2}D\frac{dC}{d\eta}$, so we require $\frac{d\eta}{dC} \le 0$. Experimentally, the concentration C is tabulated as a function of the variable η and (11.41) is used to determine $D(C)$. This technique is known as the Boltzmann–Matano method. The method is applicable only when it has been established experimentally that D is a function only of C and C is a function of $x/\sqrt{t}$; it is not applicable if the diffusivity is a function of x. Figure 11.9 shows the diffusion of zinc in gallium arsenide, both measured and simulated; note the pronounced nonlinearities (similar to those of a high concentration phosphorus diffusion in silicon as discussed in the next chapter). Also shown is a Boltzmann–Matano analysis to determine an effective diffusivity.

11.11 Numerical Solution

Closed-form solutions of (11.37) for arbitrary domains are not generally possible and recourse must be made to numerical techniques. Before proceeding to describe advanced physical models for diffusion, numerical methods for solving the single-species, concentration-dependent diffusion equation are reviewed. The simplest numerical approach to solving (11.37) in one dimension with initial-boundary conditions

$$C(x,0) = \gamma(x) \quad x \in \Omega, \qquad C(x,t) = \Lambda(x) \quad x \in \partial\Omega,\ t > 0 \tag{11.42}$$

is to use the forward or explicit Euler method in time and finite differences in space. The initial condition $\gamma(x)$ gives the dopant concentration in the wafer at time zero, which can be determined experimentally using SIMS measurements or the output of a previous simulation step. Dirichlet boundary conditions hold at the surfaces where the dopant concentration in the ambient gas is fixed. The thickness of the wafer is so great compared to the width of its active region (3000 μm:1μm) that the condition $C = 0$ at ∞ may be effectively applied on the remote back surface of the wafer.

For the finite difference method in one space dimension, we first lay down a grid with timestep Δt in time and a nonuniform mesh size $\{\Delta x_i\}$ in space. Next, define $C_i^n = C(\Delta x_i, n\Delta t)$, $1 \le i \le M$, $n \ge 0$, and replace the derivatives by a forward difference in time and central difference in space to obtain

$$\frac{C_i^{n+1} - C_i^n}{\Delta t} = \frac{D(C_{i+\frac{1}{2}}^n)\left(\dfrac{C_{i+1}^n - C_i^n}{\Delta x_i}\right) - D(C_{i-\frac{1}{2}}^n)\left(\dfrac{C_i^n - C_{i-1}^n}{\Delta x_{i-1}}\right)}{\Delta x_i + \Delta x_{i-1}} \tag{11.43}$$

which is $O((\Delta x)^2, \Delta t)$ accurate as shown below. This can be written compactly in operator form as the explicit recursion

$$\boldsymbol{C}^{n+1} = \boldsymbol{C}^n + \Delta t \boldsymbol{F}(\boldsymbol{C}^n) \tag{11.44}$$

where $\boldsymbol{F}$ denotes the discretized spatial operator. This explicit Euler method is easy to implement, but by analogy with the discussion of Chapter 4, requires a small timestep for accurate and stable results. If $\boldsymbol{\mathcal{F}}$ also denotes the spatial derivative operator on the right-hand side of (11.37), then the partial differential equation may also be viewed as the ordinary differential equation $d\boldsymbol{C}/dt = \boldsymbol{\mathcal{F}}(\boldsymbol{C})$ in an appropriate function space, $H^2(\Omega)$. (For a introduction to Sobolev spaces see [412].) From this viewpoint, we seek not a function $u(\boldsymbol{x}, t) : \mathbf{R}^n \times \mathbf{R} \longrightarrow \mathbf{R}$, but a function $u(t) : \mathbf{R} \longrightarrow H^2(\Omega)$. The advantage of this approach is that conceptually many of the methods for solving ordinary differential equations discussed in Chapter 4 apply to the PDE setting as well.

To simplify the analysis, consider the linear problem with constant diffusivity and a uniform grid in space. The linear diffusion equation is the canonical example of a parabolic PDE. These are initial-boundary value problems in which time is a preferred variable. Parabolic operators smooth initial data γ; regardless of the regularity of γ one can show that for all positive time t, the function $C(\cdot, t)$ is real analytic, i.e. it has derivatives of all orders. Solutions also satisfy a maximum principle which states that if Ω is the solution domain with boundary $\partial\Omega$, then the maximum value of the solution will occur on the boundary of the "cylinder" $\Omega \times [0, t)$ and not in its interior [188]. In this case (11.43) simplifies to

$$\frac{C_i^{n+1} - C_i^n}{\Delta t} = D\,\frac{C_{i+1}^n - 2C_i^n + C_{i-1}^n}{(\Delta x)^2} \tag{11.45}$$

so that

$$C_i^{n+1} = C_i^n + \frac{D\Delta t}{(\Delta x)^2}\left(C_{i+1}^n - 2C_i^n + C_{i-1}^n\right) \tag{11.46}$$

which can be written in vector form as

$$\boldsymbol{C}^{n+1} = (\boldsymbol{I} + \Delta t\alpha \boldsymbol{A})\boldsymbol{C}^n \tag{11.47}$$

where $\boldsymbol{A}$ is the $n \times n$ tridiagonal matrix

$$\begin{bmatrix} 2 & -1 & 0 & 0 & \cdots \\ -1 & 2 & -1 & 0 & \cdots \\ 0 & -1 & 2 & -1 & \cdots \\ \cdots & 0 & -1 & 2 & -1 \\ \cdots & \cdots & \cdots & -1 & 2 \end{bmatrix} \tag{11.48}$$

and $\alpha = -D/(\Delta x)^2$. Applying (11.47) recursively, the concentration at time $n\Delta t$ is $\boldsymbol{C}^n = (\boldsymbol{I} + \Delta t\alpha\boldsymbol{A})^n\boldsymbol{C}^0$, which implies that $(\boldsymbol{I} + \Delta t\alpha\boldsymbol{A})$ is a scaling matrix applied to the initial data. An initial error or disturbance will be similarly scaled.

As discussed earlier in Chapter 4, the concept of stability for a numerical method describes whether errors in the initial data grow with time. The difference between the numerical solutions with initial data $\boldsymbol{y}^0$ and $\boldsymbol{y}^0 + \boldsymbol{\epsilon}^0$ is given at the nth step by $\boldsymbol{\epsilon}^n = (\boldsymbol{I} + \Delta t\alpha\boldsymbol{A})^n\boldsymbol{\epsilon}^0$. A sufficient condition for

$||\epsilon^n|| \leq ||\epsilon^0||$ is that $||\boldsymbol{I} + \Delta t \alpha \boldsymbol{A}|| \leq 1$, in any consistent matrix norm. For the linear diffusion equation this gives the restriction

$$\frac{D\Delta t}{(\Delta x)^2} \leq \frac{1}{2} \tag{11.49}$$

known as the Courant–Friedrich–Lewy condition. Note that (11.49) implies that the characteristic diffusion length $\sqrt{2D\Delta t}$ must be less than the grid spacing Δx for the scheme to be stable. Although the forward Euler algo-

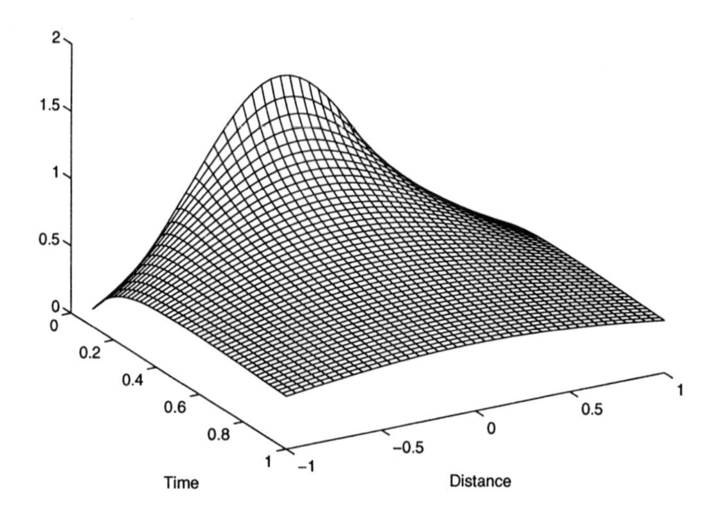

Figure 11.10: A space–time plot of the solution to the diffusion equation with a Dirac delta as the initial condition

rithm is simple to code and very cheap to calculate at each timestep, the restriction (11.49) forces so many timesteps to be taken that the scheme is impractical on scalar serial computers for most process problems. For this reason, the backward Euler and trapezoidal methods were used in the earliest process modeling programs.

For backward Euler applied to $d\boldsymbol{C}/dt = \mathcal{F}(\boldsymbol{C})$, the right-hand side $\boldsymbol{F}(\boldsymbol{C})$ in the discretized problem is evaluated at the new time, giving rise to a large system of nonlinear algebraic equations

$$\boldsymbol{C}^{n+1} = \boldsymbol{C}^n + \Delta t \boldsymbol{F}(\boldsymbol{C}^{n+1}) \tag{11.50}$$

which must be solved at each step, typically using successive approximation or Newton's method. In the linear case this reduces to solving

$$(\boldsymbol{I} - \Delta t \alpha \boldsymbol{A})\boldsymbol{C}^{n+1} = \boldsymbol{C}^n \tag{11.51}$$

which can still be a formidable task for a three-dimensional simulation. The trapezoidal (Crank–Nicolson) method represents an average of forward and backward Euler via

$$\boldsymbol{C}^{n+1} = \boldsymbol{C}^n + \frac{\Delta t}{2}\Big(\boldsymbol{F}(\boldsymbol{C}^n) + \boldsymbol{F}(\boldsymbol{C}^{n+1})\Big) \tag{11.52}$$

Both (11.50) and (11.52) are unconditionally stable: for the linear problem there is no stability restriction on the size of the timestep.

11.12 Summary

Diffusion of a substance from regions of high concentration to low, is a pervasive physical process. It is used in the processing of semiconductor wafers both to introduce impurities and to anneal the damage caused during ion implantation. Mathematically, a random walk provides an elegant and effective model of diffusion. In the limit as the stepsize goes to zero and the number of steps increases without bound, this approach yields the Fokker–Planck equation. The latter can also be derived on the basis of conservation of mass using the divergence theorem. Intrinsic diffusion is accurately modeled by the linear diffusion equation, and in 1D there are analytic solutions for many semiconductor problems. In higher dimensions and with more complicated geometries, numerical methods are required. Extrinsic diffusion is inherently nonlinear in nature with a concentration-dependent diffusivity of the form $D(C) \propto C^n$ with $n = 1$ or $n = 2$. In fact, this functional dependence may also be determined experimentally using the Boltzmann–Matano technique.

Several refinements to the extrinsic model are necessary for realistic processing conditions. The internal electric field generated during diffusion by the distribution of ionized impurity atoms adds a drift term to the basic equations. Although relatively small in magnitude, this term can be important when two or more impurities are present. Numerically, large drift terms can be treated by upwinding as discussed in Chapter 7. Simple Dirichlet boundary conditions must be replaced by flux boundary conditions which account for the tendency of an impurity to segregate across material boundaries and escape from the wafer surface. Finally, at concentrations near their solid solubility, impurities combine to form clusters consisting of several dopant atoms which are then effectively immobile. Under most processing conditions it is safe to assume equilibrium has been reached in this reaction, but it is also possible to solve dynamically for the concentration of clusters.

The simplest numerical methods for solving the parabolic diffusion equation are generalizations of the integration techniques learned earlier for the circuit equations. Forward Euler is inexpensive and easy to code, but suffers from a severe restriction on stepsize for a given grid spacing, the CFL condition. Both backward Euler and trapezoidal methods are unconditionally stable for this problem, giving excellent results at the price of solving a very large linear (more generally a nonlinear algebraic) system at each timestep. The following two chapters will discuss physical and numerical refinements to these models.

11.13 Exercises

1. Consider the classical gambler's ruin problem in which q_z and p_z are the probabilities that a particle starting at z will be absorbed at 0 and N, respectively. Derive a difference equation for q_z with appropriate boundary conditions and show how it is related to the steady-state form of (11.8), i.e. Laplace's equation $\partial^2 v/\partial x^2 = 0$. Solve the difference equation by analogy with methods for the corresponding differential equation.

2. Solve the initial value problem $u_t = u_{xx}$, $u(x,0) = \delta(x)$, $x \in \mathbf{R}$, $t \in [0,1]$ using a random walk (δ is the Dirac delta). This will require access to a routine which returns a uniformly distributed random number in $[0,1]$; for example, in the C language there is the function *drand48*.

3. Rederive the Fokker–Planck equation (11.7) by using a random walk which assumes that the random variables $X_1, X_2, ..., X_n$ are correlated. In this case, if the particle moves to the right at the nth step, there will be a greater tendency for it to continue to move to the right at step $(n+1)$.

4. Use Laplace transforms to derive the closed-form solutions given in this chapter to the one-dimensional diffusion equation with various boundary conditions.

5. Show that if u and v are solutions to the diffusion equation so are $u+v$, αu, u_x and $\int_{-\infty}^{x} u(y,t)dy$. Prove that the equation remains invariant under the change of coordinates $x \to ax$, $t \to a^2 t$ where $a > 0$. Verify that the general solution to the diffusion equation on the line is given by $u(x,t) = \int_{-\infty}^{\infty} G(x-y,t)\, u_0(y)\, dy$, where G is a Gaussian $G(x,t) = \exp(-x^2/4Dt)/\sqrt{4\pi Dt}$ with variance $2Dt$.

6. If $G(x,t)$ is the Gaussian function defined in the previous exercise, show that $S(x,y,z,t) = G(x,t)G(y,t)G(z,t)$ satisfies $u_t = D\Delta u$. Show that as $t \to 0$, S converges to the three-dimensional Dirac delta.

7. Solve $u_t = u_{xx}$ with initial condition $u(x,0) = \sin(\pi x)$, $x \in [0,1]$ and boundary conditions $u(0,t) = u(1,t) = 0$, $t \in [0,1]$, using finite differences in space and the forward Euler method in time. Take $\Delta x = 1/20$ and $\Delta t = 10^{-4}$, 10^{-3}, and 10^{-2}. Next, solve the PDE using the backward Euler scheme in time; now a tridiagonal linear system must be solved at each timestep. Contrast your results with those of the forward Euler method. Why are there no difficulties with taking large values of Δt, e.g. $\Delta t >> (\Delta x)^2/2$?

8. The Fokker–Planck equation with zero diffusivity, $u_t = -cu_x$, is called the first-order wave equation. By considering directional derivatives show that the solution is obtained by propagating the initial function $u(x,0) = f(x)$ along lines parallel to $(c,1)$.

9. Use the SIMS measurements in Figure 11.9(b) to obtain $D(C)$ from a Boltzmann–Matano analysis.

Chapter 12

Multiple Species Diffusion

12.1 Introduction

The complementary error function and Gaussian profiles of the previous chapter describe the overall behavior of the impurity during simple processing steps such as predeposition or drive-in at low concentrations. As technology has matured, the need for ever shallower junctions has required higher impurity concentrations. Under these conditions deviations from the simple linear theory were observed, which required the adoption of the vacancy-assisted diffusion mechanism to explain concentration-dependent diffusivities. Even that model could not explain all the unusual profiles that were observed experimentally.

Figure 12.1 shows a plot of concentration versus reduced distance $(x/\sqrt{t})$ for a phosphorus predeposition with a surface concentration of 4×10^{20} atoms cm^{-3} at 900°C for 10 minutes [570]. Note the pronounced plateau near the surface, the kink at 50 $\text{Å}\,s^{-1/2}$, and the deep tail that extends into the bulk. How does one account for what might even appear as two separate profiles reminiscent of the channeling that often occurs during implantation? Is a two-stream diffusion mechanism at work? The vacancy-assisted model was evoked to give a qualitative explanation: E-centers migrated into the bulk where they dissociated, resulting in enhanced diffusion in the bulk. But quantitatively this model was found to be lacking and *ad hoc* assumptions were required in order to represent accurately observed phosphorus profiles. Moreover, it was observed that if a processing step such as oxidation occurred simultaneously with diffusion, the latter could be enhanced or retarded, depending upon which impurity was present. Other impurities also exhibit this behavior, but to a lesser extent. Figure 12.2 shows the anomalous diffusion of boron [374] during a furnace anneal at 800°C. Observe the shoulders after 35 min, the absence of diffusion under the peak, and the decay of the initial transient with little diffusion occurring at longer times.

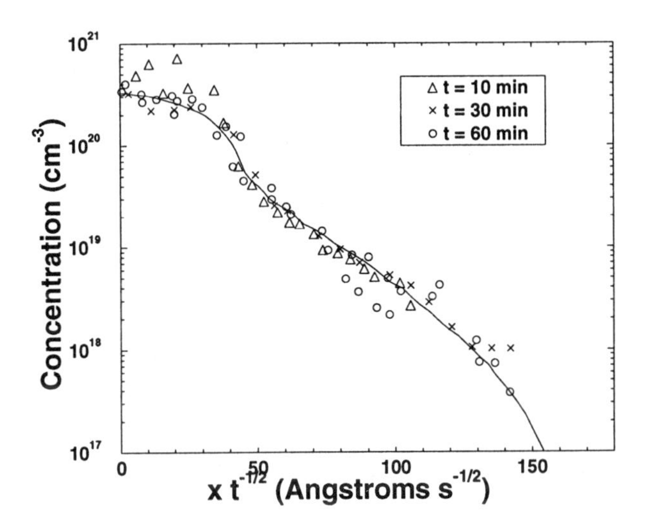

Figure 12.1: A profile of phosphorus concentration versus reduced distance for a 900°C predeposition (from [570]).

It was clear that to explain this bewildering collection of special effects in a comprehensive manner would require a deeper understanding of the diffusion mechanisms. Chapter 10 discussed how the creation of point defects, both vacancies and interstitials, during ion implantation can be modeled using the Monte Carlo method. Oxidation is known to provide a rich supply of interstitials at the silicon surface as it proceeds. Even during predeposition, both types of defects are present at equilibrium levels, as the crystal lattice trades off minimum energy against maximum entropy. Which type of defect is most important in aiding the diffusion of impurities? This question raised considerable controversy in the literature twenty years ago. It was difficult to answer, in large part because it is very difficult to measure point defects *in situ*. Experiments to measure their equilibrium concentrations are indirect in nature as compared with SIMS measurements for impurity concentrations. Indeed, it was the availability of numerical simulations using the proposed advanced models that finally resolved the issue by providing accurate profiles for these difficult-to-measure species. In fact, both vacancies and interstitials play a critical role in the diffusion process, and it is the interaction between them and impurity–defect pairs which is ultimately responsible for the observed anomalies.

12.2 Equilibrium Models

Over the past twenty years several models have been proposed to explain the anomalous diffusion of phosphorus in silicon at concentrations greater than 10^{20} atoms cm^{-3}. A few representative models are characterized here; the reader is referred to the bibliography for more details and references to other models. (In some instances the original notation used in the literature to describe a model has been modified to agree with other, more recent, models.)

One of the earliest models was developed by Schwettmann and Kendall who attributed the kink formation to diffusion of neutral and negatively charged E-centers, or phosphorus-vacancy pairs. Specifically, they proposed that: (*i*) diffusion occurred via E-centers in two charge states; (*ii*) discharge rather than dissociation of the negative E-centers produced the kink; and (*iii*) donors were electrically compensated by negative E-centers. Thus, for phosphorus to diffuse, a vacancy must occupy the nearest-neighbor site of a phosphorus atom to form an E-center.

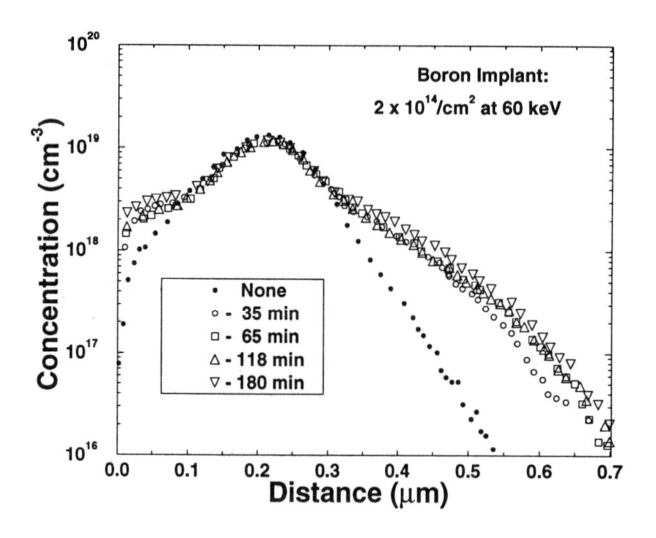

Figure 12.2: Anomalous diffusion of boron during a furnace anneal at 800°C (from [374]).

Later, Yoshida *et al.* [387, 572, 565, 566, 570, 567, 568, 569, 571] noted that vacancies are generated or annihilated at dislocations, and characterized the process via the chemical reactions

$$\begin{array}{llll} E\text{−}center & \underset{k_2}{\overset{k_1}{\rightleftharpoons}} \ phosphorus\ atom & + & vacancy \\ & & & \downarrow\uparrow k_V \\ & & & sink\ and\ source \end{array} \tag{12.1}$$

where k_1, k_2, and k_V are rate constants. The associated reaction-diffusion system has the form

$$\begin{aligned} \frac{\partial C_P}{\partial t} &= k_1\, C_E - k_2\, C_P C_V \\ \frac{\partial C_E}{\partial t} &= D_E\, \Delta C_E - k_1\, C_E + k_2\, C_P C_V \\ \frac{\partial C_V}{\partial t} &= D_V\, \Delta C_V + k_1\, C_E - k_2\, C_P C_V - k_V\, (C_V - C_V^*) \end{aligned} \tag{12.2}$$

where C_P, C_E, and C_V are the concentrations of phosphorus atoms, E-centers, and vacancies, respectively, $*$ denotes a thermal equilibrium value, and D_E and D_V are the respective diffusivities. These equations are not solved numerically in [570], but analyzed assuming that local equilibrium has been achieved in the reactions. This coupled system results in a new mechanism for phosphorus diffusion characterized by the properties: (*i*) phosphorus diffuses only via E-centers, (*ii*) when phosphorus atoms enter from the surface into the bulk, they do so as E-centers, (*iii*) E-centers flow into the bulk, and (*iv*) E-centers dissociate, generating excess vacancies, which in turn further enhance tail diffusion. Later, numerical modeling showed that diffusion via E-centers alone could not explain the plateau region near the surface or the kink. In order to obtain profiles with a marked plateau, a concentration-dependent vacancy and E-center formation energies were added to this model, charges on the various species taken into consideration. This results in the system

$$\begin{aligned} \frac{\partial}{\partial t}(C_P + C_{E^0} + C_{E^-}) &= -\nabla \cdot (\boldsymbol{J}_{E^0} + \boldsymbol{J}_{E^-}) \\ \frac{\partial}{\partial t}(C_{V^0} + C_{V^-} + C_{V^=}) &= -\nabla \cdot (\boldsymbol{J}_{V^0} + \boldsymbol{J}_{V^-} + \boldsymbol{J}_{E^0} + \boldsymbol{J}_{E^-}) \end{aligned} \tag{12.3}$$

corresponding to Fick's law for the total phosphorus and total vacancy concentrations, respectively.

Fair and Tsai [170] postulated that dissociation of $P^+V^=$ pairs created an excess of acceptor vacancies, V^-, which in turn caused the enhanced tail diffusion. The Fair–Tsai model represents a synthesis of the earlier two models with the addition of some *ad hoc* expressions for concentrations of excess vacancies and E-centers. In the article [254], the several models of phosphorus diffusion that had been advanced prior to 1983 were examined and a new model proposed, summarized by the properties: (*i*) at extremely high concentrations phosphorus diffuses into silicon in supersaturation; (*ii*) most importantly, phosphorus diffuses via a dual mechanism to explain oxidation enhanced diffusion; and (*iii*) self-interstitials become supersaturated as a result of precipitation of interstitialcy phosphorus, concurrent thermal oxidation, displacement of silicon atoms when phosphorus atoms enter the lattice at the surface, and conversion of an interstitialcy phosphorus into a substitutional phosphorus. The latter process can occur through the mechanisms

$$S + P_i \rightleftharpoons P_s + I \,, \qquad V + P_i \rightleftharpoons P_s + S \tag{12.4}$$

where S, P_i, P_s, I, and V represent a silicon atom on a lattice site, an interstitialcy phosphorus atom, a substitutional phosphorus atom, a self-interstitial and a vacancy, respectively. The following system was proposed to model these reactions mathematically

$$\begin{aligned} \frac{\partial P_s}{\partial t} &= \nabla \cdot (D_s \nabla P_s) - PPT_s + (k_1 + k_2 V) P_i - (k_1' I + k_2') P_s \\ \frac{\partial P_i}{\partial t} &= \nabla \cdot (D_i \nabla P_i) - PPT_i + (k_1' I + k_2') P_s - (k_1 + k_2 V) P_i \\ \frac{\partial I}{\partial t} &= D_I \,\nabla I + k_1 P_i - k_1' P_s - k_3 (IV - I^* V^*) + PPT_i \\ \frac{\partial V}{\partial t} &= D_V \,\nabla V + k_2' P_s - k_1 P_i V - k_3 (IV - I^* V^*) + PPT_s \end{aligned} \tag{12.5}$$

where PPT_s and PPT_i denote local rates of precipitation by agglomeration.

It was accepted that interstitials must play a significant role in the diffusion process, and Mathiot and Pfister [357, 358, 360, 359, 361, 362, 363] subsequently a percolation model. In this model, the concentrations of point defects, pairs, and substitutional phosphorus are expressed algebraically using the law of mass action and charge neutrality, to give three continuity equations:

$$\begin{aligned} \frac{\partial C_{imp}^t}{\partial t} &= -\nabla \cdot \boldsymbol{J}_{imp} \\ \frac{\partial C_V^t}{\partial t} &= -\nabla \cdot \boldsymbol{J}_V^t - \nabla \cdot \boldsymbol{J}_E^t - \frac{\partial C_E^t}{\partial t} + (G-R)_{BM} \\ \frac{\partial C_I^t}{\partial t} &= -\nabla \cdot \boldsymbol{J}_I^t - \nabla \cdot \boldsymbol{J}_{D_I}^m - \frac{\partial C_{D_I}^m}{\partial t} + (G-R)_{BM} \end{aligned} \tag{12.6}$$

where C^t_{imp}, C^t_V, C^t_I denote the total concentrations of impurity, vacancies, and interstitials, respectively; J denotes a current density or flux; E and D_I refer to E-centers and interstitial pairs; and the last term in the second and third equations accounts for bimolecular generation–recombination. Even the inclusion of interstitials into the diffusion models could account for the plateau only if the impurity diffused primarily via vacancies since the percolation effect would not occur with an interstitial mechanism. Charged species and E-centers were omitted, as was the reaction $P_s + I \rightleftharpoons P_i$. A simpler reaction-diffusion system was proposed [334] that established the importance of recombination.

Among the first work to compute explicitly the concentrations of both types of defect was that of Yeager and Dutton. The resulting system can be expressed as [563]

$$\begin{aligned} \frac{\partial C_P}{\partial t} &= \nabla\cdot\left((D^*_{PI}\frac{C_I}{C^*_I} + D^*_{PV}\frac{C_V}{C^*_V})\nabla C_P\right) \\ \frac{\partial C_I}{\partial t} &= \nabla\cdot(D_I \nabla C_I) + k(C_I\, C_V - C^*_I\, C^*_V) \\ \frac{\partial C_V}{\partial t} &= \nabla\cdot(D_V \nabla C_V) + k(C_I\, C_V - C^*_I\, C^*_V) \end{aligned} \tag{12.7}$$

Morehead and Lever [381, 382, 383, 384] proposed an interstitialcy-based model to explain the enhanced tail diffusion of phosphorus and boron. They considered the reactions

$$\begin{array}{llll} I + h^+ \rightleftharpoons I^+ & C_{I^+} &=& k_1 p C_I \\ I + B^- \rightleftharpoons (BI)^- & C_{(BI)^-} &=& k_2 C_I C_B \\ I + B^- + h^+ \rightleftharpoons (BI) & C_{BI} &=& k_3 p C_I C_B \end{array} \tag{12.8}$$

where h represents a hole and p is the concentration of holes. Rather than solve explicitly for the interstitials, they expressed C_I in terms of C_P by assuming that at any point in the crystal the flux of interstitial-impurity pairs into the bulk is equal to the flux of interstitials toward the surface. This gives a supersaturation of interstitials in the bulk, which in turn enhances the impurity diffusivity in a manner reminiscent of the Fair–Tsai vacancy model.

This model was extended [389] to include dynamic effects and relax the assumption of equality of fluxes, using the mathematical model

$$\frac{\partial C_P}{\partial t} = \nabla\cdot\left(\frac{f_I D_P}{C^*_I}\nabla(C_I C_P) + \frac{f_I D_P}{C^*_I} C_I C_P \nabla(\ln(n))\right) +$$

$$\nabla \cdot \left(\frac{(1-f_I)\, D_P}{C_V^*} \nabla (C_V C_P) + \frac{(1-f_I)\, D_P}{C_V^*} C_V C_P \nabla (\ln(n)) \right)$$

$$\begin{aligned} \frac{\partial C_I}{\partial t} &= \nabla \cdot (D_I \nabla C_I) - K_R (C_I\, C_V - C_I^* C_V^*) + \\ & \nabla \cdot \left(\frac{f_I D_P}{C_I^*} \nabla (C_I C_P) + \frac{f_I D_P}{C_I^*} C_I C_P \nabla (\ln(n)) \right) \\ \frac{\partial C_V}{\partial t} &= \nabla \cdot (D_V \nabla C_V) - K_R (C_I C_V - C_I^* C_V^*) + \\ & \nabla \cdot \left(\frac{(1-f_I)\, D_P}{C_V^*} \nabla (C_V C_P) + \frac{(1-f_I)\, D_P}{C_V^*} C_V C_P \nabla (\ln(n)) \right) \end{aligned} \tag{12.9}$$

Here, C_P, C_I, and C_V represent the concentrations of phosphorus, interstitials, and vacancies, respectively. Although phosphorus is used here for illustration, the model applies to any impurity, with an appropriate sign change on the electric field term in the case of donors. The equations are coupled through terms of the form $C_I C_P$ and the reaction terms in the two equations for defects. Later in the numerical studies we refer to (12.9) as the "three-species" model.

In the limit as the point defect concentrations approach their equilibrium values, (12.9) reduces to the standard model. If C_I and C_V are constants different from C_I^* and C_V^*, then (12.9) becomes essentially the model used at that time in SUPREM IV, i.e. the standard model of Chapter 11, but with an effective diffusivity of

$$D_{eff} = D_i \left(f_I \frac{C_I}{C_I^*} + (1-f_I) \frac{C_V}{C_V^*} \right) \tag{12.10}$$

Mulvaney draws the following conclusions concerning this model: (*i*) the strength of the nonlinear coupling is governed primarily by the quantities

$$\frac{f_I}{D_I C_I^*} D_i C_0 \,, \qquad \frac{1-f_I}{D_V C_V^*} D_i C_0 \tag{12.11}$$

where C_0 is the maximum impurity concentration; (*ii*) for smaller values of C_0 the coupling is smaller (boron exhibits less nonlinear diffusion than phosphorus, because its solid solubility is somewhat smaller); (*iii*) at lower temperatures $D_I C_I^*$ and $D_V C_V^*$ become smaller, resulting in stronger coupling (this is consistent with the fact that the phosphorus kink and tail and

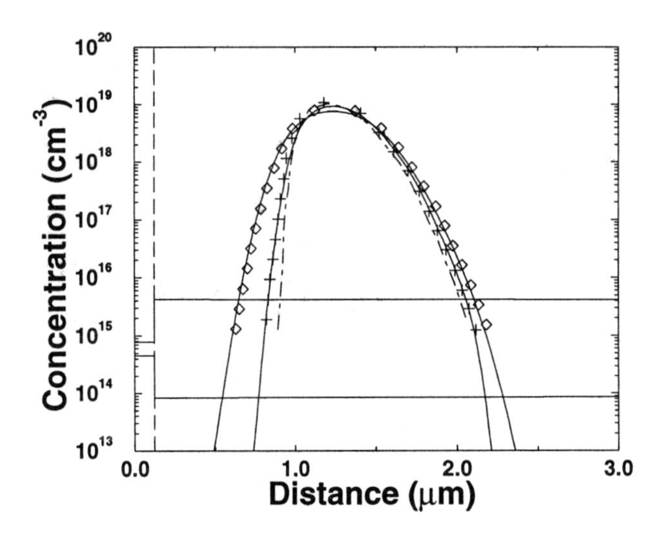

Figure 12.3: Simulation of oxidation retarded diffusion of antimony; symbols are SIMS data (from [393]).

anomalous boron tail are observed at relatively low temperatures); and (*iv*) arsenic, with an order of magnitude lower intrinsic diffusivity than phosphorus or boron, should display almost no anomalous tail, as indeed is the case. Such a model gives good results when excess defects are present due to oxidation or ion implantation. It assumes that equilibrium in defect-impurity pair formation has been attained, which is not the case during the early stages of a diffusion or during rapid thermal annealing.

Precise measurements of the relative interstitialcy factor f_I are clearly crucial to use (12.9). There is considerable debate in the literature concerning the value of f_I for the various impurities. One source of oxidation retarded diffusion (ORD) data [226] suggests that antimony diffuses primarily by a vacancy mechanism with $f_I = 0.01$. Oxidation enhanced diffusion (OED) data [10, 11, 261, 323] were used [389] to fit to Arrhenius functions for three common impurities as

$$\begin{aligned} f_I^B &= 0.0834\, e^{(0.21eV/kT)} \\ f_I^P &= 0.0147\, e^{(0.40eV/kT)} \\ f_I^{As} &= 0.35 \end{aligned} \tag{12.12}$$

These parameters should be used with care as they are based on limited experimental data. Better experiments for measuring f_I are needed if models such as (12.9) are to be used under a variety of processing conditions.

Figure 12.3 shows the result of a PEPPER simulation of one of the antimony ORD experiments reported in the literature [226]. A 500 minute 1000°C diffusion is carried out in an inert ambient and then in a dry oxidation ambient. The three-species model (12.9) is used for simulation in the latter case, while the standard model is used for the former. During oxidation excess interstitials are injected into the substrate, where they annihilate some of the vacancies, according to the reaction term $K_R(C_I C_V - C_I^* C_V^*)$. For antimony, the fractional value of the interstitial contribution to diffusion is $f_I^{Sb} = 0.01$; therefore, the smaller vacancy contribution causes a smaller diffusivity of antimony, via the first equation in (12.9). Figure 12.4 represents the other extreme of enhanced diffusion of boron due to a supersaturation of interstitials caused by a high concentration phosphorus diffusion at the surface. Boron and phosphorus are diffused for ten minutes at 900°C using the three-species point defect model (12.9). The buried boron layer shows significantly enhanced diffusion due to the excess of interstitials in the bulk.

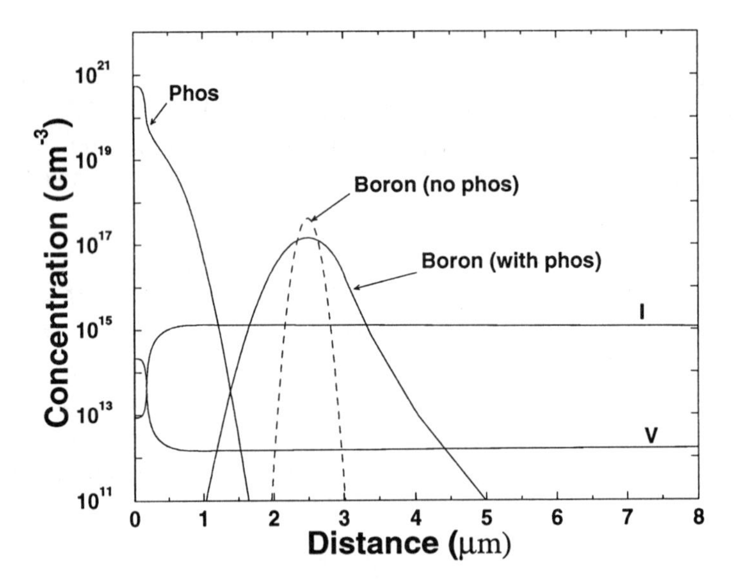

Figure 12.4: Enhanced diffusion of a buried boron layer due to a high concentration of phosphorus near the surface (from [393]).

12.3 Nonequilibrium Models

The three-species models assume the law of mass action for all reactions except bimolecular generation–recombination. $V + I \rightleftharpoons \langle 0 \rangle$. Here the forward reaction occurs when a vacancy and silicon interstitial combine to annihilate each other; while the reverse reaction takes place as the lattice creates point defects to achieve maximum entropy with lowest energy. In contrast to these models, a completely nonequilibrium model [451, 452, 391] for phosphorus diffusion in silicon may be developed. In this model, charged defects, both vacancies and interstitials, react with immobile substitutional phosphorus to form pairs, which then diffuse into the bulk. This dynamic chemical kinetic formulation results in individual equations of a simple drift-diffusion-reaction type with constant diffusivities. Most importantly, by solving for the concentrations of various species explicitly, insight is gained into the mechanisms responsible for the anomalies observed in the final impurity profile. The plateau and kink of high concentration phosphorus profiles result from recombination of the interstitial and vacancy related species. A kinetic-based approach to diffusion is applicable to other impurities in silicon and may prove very important for modeling diffusion in gallium arsenide, as discussed in Chapter 14.

Consider the following first-order reactions:

$$\begin{array}{lcll} V^0 + I^0 & \rightleftharpoons & \langle 0 \rangle & (R1) \\ P^+ + V^- & \rightleftharpoons & P^+V^- & (R2) \\ P^+ + I^- & \rightleftharpoons & P^+I^- & (R3) \\ V^0 + e^- & \rightleftharpoons & V^- & (R4) \\ I^0 + e^- & \rightleftharpoons & I^- & (R5) \end{array} \qquad (12.13)$$

Extensions to include other reaction paths or complexes are also possible. A corresponding system of partial differential equations may be derived by considering diffusion of the mobile species, drift of the charged species, and rates of production and consumption in each reaction. Consider, for example, the neutral vacancies V^0 with concentration $[V^0]$. (Thus far we have used the same symbol to denote a chemical species and its concentration; for clarity in the development of the nonequilibrium model, brackets will be used for concentration.) Since this is a mobile species, the equation for the time-rate-of-change of $[V^0]$ will contain a diffusion term $\nabla \cdot (D_{V^0} \nabla [V^0])$ together with various reaction terms. Species V^0 appears in reaction $(R4)$ where this species is produced at a rate proportional to $[V^-]$ by the reverse reaction and consumed at a rate proportional to $[V^0][e^-]$ in the forward reaction. The bimolecular generation–recombination reaction $(R1)$ is represented by the term $-k_{bi}([I^0][V^0] - [I^0]^{eq}[V^0]^{eq})$.

Continuity equations for electrons, the two vacancy species, the P^+V^- pair or E-center, and substitutional phosphorus are given below. Equations for the three interstitial-related species are obtained from these by interchanging the role of V and I. Forward and reverse rate constants for the nth reaction are denoted by k_n^f and k_n^r, respectively, and the single constant characterizing reaction $(R1)$ is k_{bi}. The last term in the first equation dynamically enforces charge neutrality, with k_{neu} taken as 10^{10} s^{-1}. Although the P^+V^0 species is mentioned in older models, only the neutral and negative E-centers, have been observed experimentally [360]. For simplicity, species associated with $V^=$ are not included. The resulting eight-species mathematical model becomes

$$
\begin{aligned}
\frac{\partial[e^-]}{\partial t} &= -k_4^f\,[e^-][V^0] + k_4^r[V^-] - k_5^f\,[e^-][I^0] + k_5^r[I^-] \\
&\quad +k_{neu}\left\{\,[P^+] + \frac{n_i^2}{n} - n - [V^-] - [I^-]\right\} \\
\frac{\partial[V^0]}{\partial t} &= \nabla \cdot \left(D_{V^0}\nabla[V^0]\right) - k_4^f[e^-][V^0] + k_4^r[V^-] \\
&\quad -k_{bi}([I^0][V^0] - [I^0]^{eq}[V^0]^{eq}) \\
\frac{\partial[V^-]}{\partial t} &= \nabla \cdot (D_{V^-}\left(\nabla[V^-] - [V^-]\nabla \ln(n)\right)) \;+\; k_4^f\,[e^-][V^0] \\
&\quad -k_4^r\,[V^-] - k_2^f\,[P^+][V^-] + k_2^r[P^+V^-] \qquad (12.14) \\
\frac{\partial[P^+V^-]}{\partial t} &= \nabla\left\{D_{P^+V^-}\nabla[P^+V^-]\right\} \\
&\quad +k_2^f[P^+][V^-] - k_2^r[P^+V^-] \\
\frac{\partial[P^+]}{\partial t} &= -k_2^f\,[P^+][V^-] + k_2^r[P^+V^-] \\
&\quad -\,k_3^f[P^+][I^-] + k_3^r[P^+I^-]
\end{aligned}
$$

12.4 Boundary and Initial Conditions

In order for a parabolic PDE system to be well-posed mathematically one must specify the initial concentrations of all species and supply appropriate boundary conditions. It is generally difficult to measure the concentrations of point defects directly. SIMS measurements are available only for the impurity concentrations. Several experiments have been devised to indirectly measure the equilibrium concentrations of defects, but values reported in the literature for the same process conditions range over several orders of magnitude. Fortunately, in the case of predeposition or a long anneal in ambient conditions, a constant defect profile is a reasonable initial condition. After a high energy implant, however, there may be several orders of magnitude more defects than impurity atoms. In order to obtain the defect profiles, a Monte Carlo simulation with damage calculations is made, tracking not only the incoming ions, but also secondary silicon atoms that are likely to become interstitials and the vacancies that are created when a silicon atom is knocked from the lattice [117]. In this way, a record of point defects is obtained which can provide the initial conditions for advanced diffusion models, as shown in Figure 12.5. Boron diffuses at 950°C for 5 and 15 seconds, following a 60 KeV, 2×10^{14} cm^{-2} implant. A 3D Monte Carlo simulation was used together with a five-species model (essentially the system (12.14) with charged species neglected). Anomalous transient diffusion saturates after 15 seconds; the 30 second curve (not shown) nearly coincides with that of 15 seconds.

It is difficult to determine the extent to which vacancies and interstitials recombine during the implantation, between the implant and subsequent furnace or RTA anneals, and during the ramping of temperature during diffusion. One approach is to consider that only a fraction of the damage produces stable defects. To avoid the difficult problem of determining point defect profiles after implantation, the simulations which follow are restricted to predeposition into undoped silicon. Research in this area is ongoing [116, 294, 255] and crucial to accurate modeling of rapid thermal annealing.

The electron concentration is initially set to the intrinsic value, $n_i = 3.8 \times 10^{18}$, and that of neutral vacancies is the thermal equilibrium value given by [536]

$$[V^0]^{eq} = n_h \, e^{\Delta S_f / k} \, e^{-\Delta H_f / (kT)} \tag{12.15}$$

with ΔS_f the entropy of formation and ΔH_f the enthalpy of formation. Assuming Fermi–Dirac statistics, the concentrations of acceptor vacancies

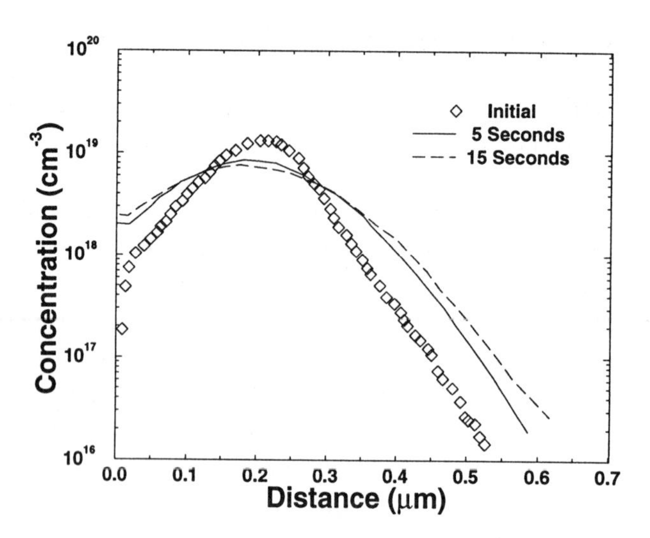

Figure 12.5: Anomalous transient diffusion using the five-species model with initial conditions for defects provided by a Monte Carlo simulation (from [117]).

is

$$[V^-]^{eq} = g_A^r \; e^{(E_F - E_{V^-})/(kT)} [V^0]^{eq} \tag{12.16}$$

with $g_A^r = 2$ a degeneracy factor, E_{V^-} the energy level of the acceptor state, and E_F the Fermi level. Using values of 1.1 k for ΔS_f, 2.66 eV for ΔH_f, and 0.56 eV for E_{V^-} yields $[V^0]^{eq} = 5.0 \times 10^{10}$ and $[V^-]^{eq} = 3.4 \times 10^{11}$, at 900°C. Given electrical equilibrium in the silicon, this amounts to assuming that the law of mass action holds, $[V^-]^{eq} \propto n\,[V^0]^{eq}$. Similar equations hold for the interstitials, with $[I^0]^{eq} = 5.0 \times 10^{11}$ and $[I^-]^{eq} = 1.0 \times 10^{11}$. Although mass action determines the initial profiles, local equilibrium is not assumed during the diffusion itself. All species involving phosphorus have an initial concentration of zero.

A zero Neumann boundary condition (no flux) is enforced at the back of the wafer for each species. At the surface the boundary conditions are

$$\boldsymbol{n} \cdot \boldsymbol{\nabla} n = \boldsymbol{n} \cdot \boldsymbol{\nabla}[P^+V^-] = \boldsymbol{n} \cdot \boldsymbol{\nabla}[P^+I^-] = 0 \tag{12.17}$$

$$[V^0] = [V^0]^{eq}, [V^-] = (\frac{n}{n_i})[V^-]^{eq} \tag{12.18}$$

$$[I^0] = [I^0]^{eq}, [I^-] = (\frac{n}{n_i})[I^-]^{eq}, [P^+] = C^* \tag{12.19}$$

where $\boldsymbol{n}$ is the outward unit normal. In the Dirichlet condition for P^+, C^* represents the concentration of phosphorus in the ambient gas; note that both $[P^+V^-]$ and $[P^+I^-]$ are much less than $[P^+]$.

12.5 Diffusivities and Reaction Rate Constants

The most difficult task in implementing a model such as (12.14) is to obtain accurate estimates of the diffusivities and rate constants. For vacancies, thermodynamic data are available to take a first-principles approach to calculating diffusion coefficients. Recalling equation (11.7), the microscopic definition of diffusivity is $\mathcal{D} = n\alpha^2 = 2D$; rewriting in terms of Δx and Δt, the macroscopic diffusivity of vacancies is given by

$$D_V = \frac{1}{2}\frac{(\Delta x)^2}{\Delta t} = \frac{1}{8}\,\alpha^2 \nu_V \tag{12.20}$$

where α is the lattice parameter and ν_V is the jump frequency of an atom adjacent to a vacancy into that vacancy. The latter can be approximated by

$$\nu_V = \sqrt{\frac{8\Delta H_m}{3\,m\,\alpha^2}}\; e^{\Delta S_m/k}\; e^{-\Delta H_m/(kT)} \tag{12.21}$$

where m is the mass of a silicon atom, ΔS_m is the entropy of migration, and ΔH_m is the enthalpy of migration. Using 4.1 k for ΔS_m and 1.2 eV for ΔH_m, results in a diffusivity of 9.5×10^{-7} cm^2 s^{-1} at 900°C.

D_V	D_I	$D_{P^+V^-}$	$D_{P^+I^-}$
4.0×10^{-8}	5.8×10^{-8}	1.4×10^{-12}	6.2×10^{-12}

Table 12.1: Diffusivities (cm^2 s^{-1}) for the eight-species model at 900°C.

Unfortunately, the existing thermodynamic data for interstitials and pairs are at present of doubtful accuracy. Hence, the diffusivities used in the simulations which follow are computed using expressions for the product of the diffusivity times the equilibrium concentration [362],

$$\begin{aligned} D_V[V^0]^{eq} &= 1.04 \times 10^{21}\, e^{-3.89/(kT)}\ \text{cm}^{-1}\text{s}^{-1} \\ D_I[I^0]^{eq} &= 2.35 \times 10^{23}\, e^{-4.40/(kT)}\ \text{cm}^{-1}\text{s}^{-1} \end{aligned} \tag{12.22}$$

Values for the equilibrium concentrations of point defects are determined as described in Section 12.4. Diffusivities for the pairs are given by [362]

$$D_{P^+V^-} = 0.5\, e^{-2.69/(kT)}\ \text{cm}^{-1}\text{s}^{-1}$$

Reaction	R2	R3	R4	R5
Forward (cm^3s^{-1})	3.0×10^{-14}	4.4×10^{-14}	1.0×10^{-8}	1.0×10^{-8}
Reverse (s^{-1})	1.4×10^2	8.8×10^1	5.6×10^{10}	1.7×10^{11}

Table 12.2: Rate constants for the eight-species model at 900°C.

$$D_{P+I^-} \quad = \quad 5.2 \times 10^{-2}\ e^{-2.31/(kT)}\ \mathrm{cm^{-1}s^{-1}} \tag{12.23}$$

Table 12.1 lists the diffusivities at 900°C; note the good agreement between D_V and the thermodynamically-based diffusivity. An acceptor defect is considered to have the same diffusion coefficient as the corresponding neutral species. The fact that four orders of magnitude separate the diffusivities of the defects from that of the pairs has several ramifications. Physically, a supersaturation of one type of defect near the surface quickly propagates to the back of the wafer, long before the pairs themselves have diffused significantly. Mathematically this gives rise to partial differential equations with radically different time constants, and hence numerically to very stiff ODE systems as discussed in Chapter 13.

Since it is difficult to measure the rate constants directly, simple kinetic approximations are used for k^f and k^r. The constants for ($R4$) are those from Shockley–Read–Hall generation–recombination theory, $k^f = v_{th}\sigma_n$ and

$$k^r = v_{th}\sigma_n n_i\ e^{-(E_i - E_{V^-})} \tag{12.24}$$

where $v_{th} \approx 10^7$ cm s^{-1} is the thermal velocity of an electron, $\sigma_n \approx 10^{-15}$ cm^2 is the capture cross-section, n_i is the intrinsic electron concentration, and E_{V^-} is the energy level of the acceptor vacancy. From the theory of diffusive kinetics [434], an estimate for the forward reaction rate constant for the reaction $A + B \rightleftharpoons C$ is

$$k^f = 4\pi R(D_A + D_B) \tag{12.25}$$

where D_A and D_B are the diffusivities of the reactants and R is the encounter distance for an interaction between the two species. Applying these results to reaction ($R2$), for example, yields

$$k^f = 4\pi R D_{(V^-)}\ , \quad k^r = \frac{k_f}{4} n_h\ e^{-E_b/(kT)} \tag{12.26}$$

where R $\approx 6 \times 10^{-8}$ cm is the encounter distance for phosphorus-vacancy interaction, $n_h \approx 5 \times 10^{22}$ cm^{-3} is the concentration of lattice sites, and E_b is the binding energy of the phosphorus-acceptor vacancy pair. Rate constants for the interstitial reactions ($R3$) and ($R5$) are handled similarly; values of the rate constants at 900°C are shown in Table 12.2.

12.6 Simulations

In order to obtain the marked plateau observed in phosphorus diffusion at high concentrations, the bimolecular generation–recombination rate, k_{bi}, was initially taken to be 1.7×10^{-7} cm^{3}s^{-1}, which is several orders of magnitude greater than the one determined from (12.25) with $R \approx 10^{-7}$cm. The values of D_V and D_I from Table 12.1 give an estimate of 7×10^{-14} cm^{3}s^{-1} for k_{bi}. It has been shown, however, that recombination is strongly dependent on impurity concentration and this dependence might well result in a much larger effective recombination rate. To understand why, note that there are two important alternative paths for defect recombination

$$\begin{array}{rclc} P^+V^- + I^0 & \rightleftharpoons & P^+ + e^- & (R6) \\ P^+I^- + V^0 & \rightleftharpoons & P^+ + e^- & (R7) \end{array} \qquad (12.27)$$

Inclusion of $(R6)$ into the model results in an additional term of the form $-k_6^f\,[P^+V^-]\,[I^0]$ in the equation for I^0, for instance. At equilibrium in reaction $(R2)$

$$[P^+V^-] = \frac{k_2^f}{k_2^r}\,[P^+][V^-] = 2 \times 10^{-16}\,[P^+]\,(\frac{n}{n_i})\,[V^0]^{eq} \qquad (12.28)$$

Assuming a surface concentration of 3×10^{20}, $[P^+V^-] \approx 10^6\,[V^0]$ for the region close to the surface and the additional term

$$k_6^f[P^+V^-]\,[I^0] \approx 10^6 k_6^f\,[V^0]\,[I^0] \qquad (12.29)$$

For the reaction of P^+V^- and I^0, orientation effects might be important as compared with the direct recombination mechanism. Still, supposing that $k_6^f \approx k_{bi}$, the effective bimolecular generation-recombination rate, k_{bi}^{eff}, would be 10^6 times the value predicted by applying (12.25) to $(R2)$. Later work [391] has shown conclusively that the addition of reactions $(R6)$ and $(R7)$ to the model do allow k_{bi} to be taken at the predicted value of $\approx 10^{-14}$ cm^3 s^{-1} and still obtain a marked plateau.

A consistent physical explanation of the phosphorus nonlinearity is therefore provided by a strong recombination of defects both directly and via reactions $(R6)$ and $(R7)$. The effect of charge seems to be secondary. This is distinct from a two-stream mechanism with fast and slow diffusing species. Here the pairs PV and PI diffuse with comparable speeds, but the profile of the former becomes kinked due to recombination. Although the simpler models give adequate results for many situations, there are cases such as rapid thermal annealing where the assumption of local equilibrium is definitely not justified. The modeling techniques used here, moreover, are applicable to any diffusion problem in which chemical reactions such as clustering or precipitation are occurring simultaneously with diffusion.

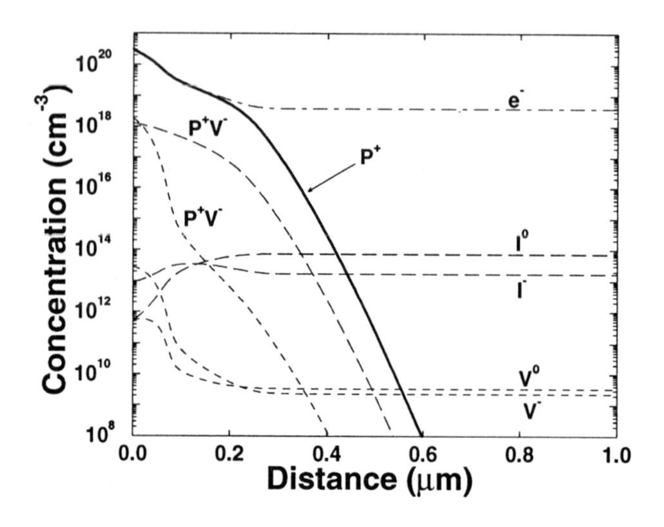

Figure 12.6: Predeposition at 900°C for 10 min using the eight-species model (from [452]).

Figure 12.6 shows the result of a 10 minute predeposition at 900°C using the eight-species model. The boundary conditions cause a brief surge of vacancies into the bulk, but the kinetics quickly force the interstitials to dominate. After a few milliseconds the defect profiles in the bulk are flat with the large recombination rate resulting in a supersaturation of interstitials and depletion of vacancies. The profile of phosphorus-interstitial pairs in Figure 12.6 is similar to a complementary error function, but the E-center profile has developed a pronounced plateau that is reflected in the substitutional phosphorus profile. This results from strong recombination; electrical effects due to the drift terms are secondary. The effect of varying the surface concentration of substitutional phosphorus, C^*, using a five species model (one which neglects charges in (12.14)) is shown in Figure 12.7(a). As C^* is increased the profile changes from a complementary error function to one in which a plateau and kink are well established. In Figure 12.7(b) profiles for several different values of k_{bi} are given. Note that for $k_{bi} < 10^{-7}$, the profile in the first 0.1 μm is concave up, as in the early simulations of Yoshida that did not include a concentration-dependent vacancy formation energy.

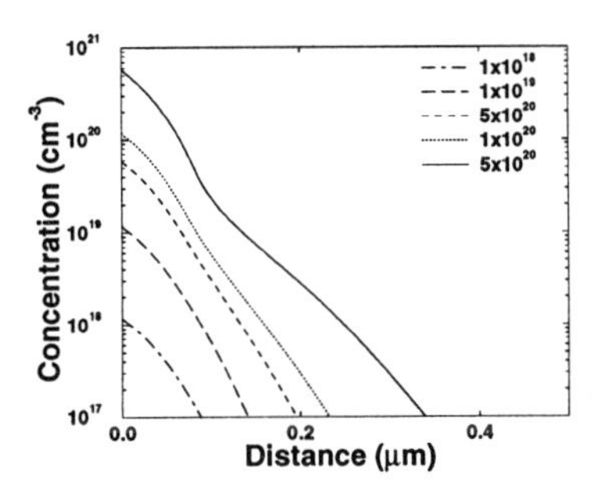

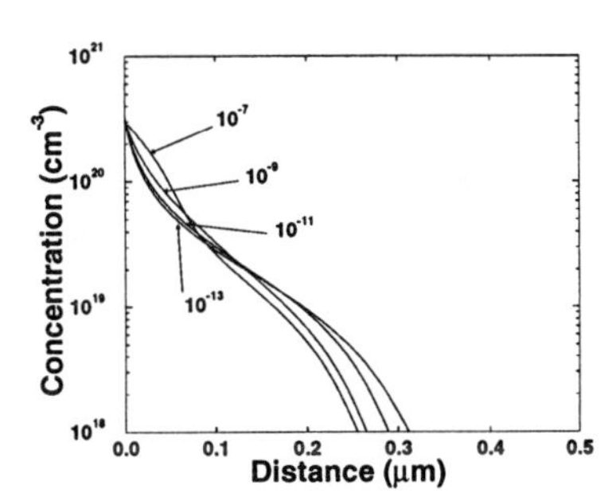

Figure 12.7: The effect of varying (a) the surface concentration of substitutional phosphorus and (b) the generation–recombination rate k_{bi} in the five-species model (from [451]).

12.7 Hierarchy of Models

The diffusion models discussed above form a hierarchy of physical and numerical models that incorporate increasing amounts of physical knowledge about the diffusion process at the expense of greater complexity and longer computation times. For convenience, in the following paragraphs we denote the eight equations of the nonequilibrium model by RM1–RM8, equations (12.6) by MP1–MP3 and equations (12.7) by YD1–YD3,

Theorem 12.1 *If* $\{e^-, V^0, V^-, P^+V^-, P^+, I^0, I^-, P^+I^-\}$ *solves (12.14), contributions of* $V^=$ *are negligible compared to those of* V^-*, and the law of mass action holds, then*

$$\begin{aligned} C^t_{imp} &= [P^+] + [P^+V^-] + [P^+I^-] \\ C^t_V &= [V^0] + [V^-] \\ C^t_I &= [I^0] + [I^-] \end{aligned} \qquad (12.30)$$

is a solution of MP1–MP3.

Proof: Rewrite the flux in the third equation of (12.14) assuming that mass action holds, so that $[V^-] = \alpha n[V^0]$, as

$$D_{V^-}\ \alpha\ n\nabla[V^0]\ +\ D_{V^-}\ \alpha\ [V^0]\ \nabla n\ -\ D_{V^-}\ [V^-]\ \frac{1}{n}\ \nabla n \qquad (12.31)$$

and note that the last two terms cancel. This allows a drift-diffusion equation with constant diffusivity to be rewritten as a concentration-dependent diffusion equation. Adding equations RM2 and RM3, one has

$$\begin{aligned} \frac{\partial([V^0]+[V^-])}{\partial t} &= \nabla\cdot\left([D_{V^0} + D_{V^-}\alpha\ n\ ]\nabla[V^0]\right) - k_2^f\ [P^+][V^-] + \\ & k_2^r[P^+V^-] - k_{bi}\ ([I^0][V^0] - [I^0]^{eq}[V^0]^{eq}) \end{aligned} \qquad (12.32)$$

Adding the fourth equation of (12.14) to this yields the second equation of (12.6). Combining this with RM4 yields MP2. MP3 is derived analogously. Finally, RM4, RM8, and RM5 imply MP1. Note that the Mathiot–Pfister model (12.6) assumes equilibrium except for bimolecular generation–recombination. Arguments that invoke mass action to simplify systems have been used before in the theory of dissociative diffusion.

Clearly YD2 and YD3 are simplified versions of RM2 and RM6. To derive the equation YD1, assume that mass action holds in the reaction $P + V \rightleftharpoons PV$ so that $[PV] = \gamma[P][V]$; then the flux of vacancy-impurity pairs, $D_{PV}\nabla[PV]$, is

$$D_{PV}\gamma\ (\ [V]\nabla[P] + [P]\nabla[V]) \approx D_{PV}\gamma[V]\nabla[P] = D_{PV}^*\frac{[V]}{[V]^{eq}}\nabla[P] \quad (12.33)$$

where it is assumed that the phosphorus gradient is much larger than that of the vacancies. This approximation to the product rule has been used extensively to give an equation in $[P]$ only. A result similar to (12.33) for interstitials can be combined to give YD2. Notice that in (12.7), D_{PI}^* and D_{PV}^* are constant, and charged species are not considered explicitly.

12.8 Theoretical Analysis

One of the first analyses of a kinetic model from a theoretical viewpoint is given in [563]. Consider the n-particle system with one reaction process $R(P_1, \ldots, P_n)$ described by the system

$$\frac{\partial P_i}{\partial t} = \nabla\cdot(D_i\nabla P_i) \pm R(P_1, \ldots, P_n) \qquad (12.34)$$

where the sign is chosen according to whether or not P_i is a reactant $(-)$ or product $(+)$ of R. They define R to be *reactive-definite* if there is a point S in $\mathbf{R}^n$ such that $R(S) = 0$ and if $\pm\frac{\partial R}{\partial P_i} < 0$ for all i. The first condition guarantees an equilibrium point for the corresponding ODE system and the second ensures stability. First examine the reaction system neglecting the diffusion operators. The linearized system about S is given by

$$\frac{d\boldsymbol{P}}{dt} = -\boldsymbol{E}(\boldsymbol{\alpha}, \mathbf{1}; 1)\boldsymbol{P}(t) \tag{12.35}$$

where $\boldsymbol{\alpha} = (\frac{\partial R}{\partial P_1}, ..., \frac{\partial R}{\partial P_n})$, $\mathbf{1} = (1, ..., 1)^T$, and $\boldsymbol{E}(\boldsymbol{u}, \boldsymbol{v}; \delta) = \boldsymbol{I} - \delta \boldsymbol{u}\boldsymbol{v}^T$ is the elementary matrix determined by the vectors $\boldsymbol{u}, \boldsymbol{v}$, and the scalar δ. Since the eigenvalues of $\boldsymbol{E}$ are known to be $-\sum_{i=1}^{n} \alpha_i$ and 0, the latter having a multiplicity of $n-1$, the system (12.35) is stable. The diffusion system, i.e. (12.34) without the reaction terms, is of course stable with global existence and uniqueness following from monotone operator theory. The same can be said for the reaction-diffusion system if the initial condition is close enough to S. Newton's method for the discrete system corresponding to (12.34) converges if the time step is small enough. For a detailed discussion of the theory behind the general reaction-diffusion PDE, $\partial \boldsymbol{u}/\partial t = D\Delta \boldsymbol{u} + \boldsymbol{F}(\boldsymbol{u})$, see [505].

It is possible to analyze the concentration-dependent diffusion equation by using similarity methods together with matched asymptotic expansions of the solution in the various regions of interest, e.g. the steep front that occurs in some high-concentration diffusions. For asymptotic analyses of several of the models presented in Chapters 11 and 12 see [5, 544, 268, 288, 291] As the exercises at the end of this chapter indicate, even relatively simple nonlinear differential equations can exhibit complex phenomena such as bifurcations, period doubling, and chaos. Solutions to reaction-diffusion PDEs often display complicated patterns and waves, and this is an area of active research [223].

12.9 Summary

A number of physical models have been proposed over the past twenty years to explain observed departures of impurity profiles from the simple complementary error function or Gaussian distributions predicted by linear theory. These so-called anomalies are actually more the rule than the exception, occurring often in high concentration diffusion steps, but also during oxidation, nitridation, and low thermal budget processing. As IC technology enters the deep submicron range, it becomes critical to understand the physics behind impurity diffusion. This, in turn, requires consideration of a host of defects in the crystal lattice including vacancies, divacancies, self-interstitials, and extended defects. It is through interactions between these defects and the impurity atoms, that the latter actually move throughout the semiconductor during the diffusion process.

This chapter has explored the historical evolution of these advanced models, which form a hierarchy with increasing levels of physical sophistication. They all involve multiple species and often postulate chemical equilibrium in some form, in order to give a more tractable theory. The diffusion models share several common features: (1) they attempt to explain the many anomalies that can occur under modern processing conditions on physical as opposed to phenomenological grounds, (2) physically they are based upon chemical reactions that occur between impurity atoms and the various defect species, (3) mathematically they result in a large, tightly coupled system of reaction-diffusion partial differential equations, and (4) they require for solution of the corresponding discretized equations highly accurate and efficient numerical methods, as discussed in Chapter 13.

12.10 Exercises

1. Compare the models of Schwettmann, Yoshida, Hu, Morehead–Lever, Mathiot–Pfister, Mulvaney–Richardson, and others cited in this chapter, providing necessary details from their papers, for the development of the multiple-species models. Most of these models employ reaction-diffusion PDEs whose solutions can be obtained by the methods discussed in Chapter 13. Could one generalize the random walk techniques of the last chapter to solve these new equations? How would the situation be complicated by the presence of two sublattices, as occur in III-V compounds?

2. Solve the diffusion equation on the line with constant dissipation, $u_t - Du_{xx} + bu = 0$ for x in $\mathbf{R}$ with $u(x,0) = f(x)$ where $b > 0$ (make the change of variable $u(x,t) = e^{-bt}v(x,t)$.)

3. The general behavior of the solution to a reaction-diffusion PDE $u_t = \Delta u + F(u)$ can often be understood by first considering the various "pieces" of the picture and then combining them into an entire canvas. These are the associated (i) steady-state problem $0 = \Delta u + F(u)$, (ii) the diffusion equation $u_t = \Delta u$, and (iii) in the absence of diffusion,

the ODE system $u_t = F(u)$. Each helps to determine the properties of the solutions to the original equation. Review the PDEs of this chapter and identify the various "pieces of the puzzle."

4. Consider the linear system $d\boldsymbol{x}/dt = \boldsymbol{A}\boldsymbol{x}$ where $\boldsymbol{A}$ is a 2×2 matrix. Show that if $\boldsymbol{A}$ has a complete set of eigenvectors $\boldsymbol{\xi}_1, \boldsymbol{\xi}_2$ with eigenvalues λ_1, λ_2, then the eigenvectors form a new coordinate system in which the differential equation has the form $d\boldsymbol{y}/dt = \boldsymbol{D}\boldsymbol{y}$ with $\boldsymbol{D} = diag\{\lambda_1, \lambda_2\}$. Draw a phase plane analysis, that is plot several trajectories $\{x_1(t), x_2(t)\}$ for the following matrices $\boldsymbol{A}$.

$$\begin{pmatrix} 3 & -2 \\ 2 & -2 \end{pmatrix}, \begin{pmatrix} 5 & -1 \\ 3 & 1 \end{pmatrix}, \begin{pmatrix} 1 & 2 \\ 0 & 2 \end{pmatrix}$$

Decide whether the origin is a sink, source, saddle, or none of these.

5. Generalizing the previous problem to nonlinear systems, investigate perhaps the simplest, yet very important, example of a predator–prey sytem

$$\begin{aligned} dH/dt &= aH - \alpha HP \\ dP/dt &= -cP + \gamma HP \end{aligned}$$

where the constants a, c, α and γ are positive. Find the rest points, i.e. points (H, P) for which the right-hand sides of (12.36) are zero. Integrate this system using forward Euler, plotting several trajectories in the neighborhoods of these rest points and showing that in one case cyclical variations in the predator–prey population can occur.

6. Write a program to integrate Duffing's system

$$\begin{aligned} x' &= y \\ y' &= x - x^3 - \epsilon y + \gamma cos(\omega t) \end{aligned}$$

on the interval $[0, 400]$ using the parameter values $\gamma = 0.3$, $\omega = 1$, and $\epsilon = 0.15,\ 0.22,\ 0.25$ successively, with $(x(0), y(0)) = (1, 1)$. Explain the behavior of this trajectory in phase space. By writing this system as a second order equation, show it is related to the oscillator discussed in Chapter 4.

7. Consider the R–D equation $u_t = u_{xx} + u(1 - u)$ on $[0, L]$ with zero Dirichlet boundary conditions. Using a phase plane analysis on the

steady-state equation (multiply by u_x, integrate, and plot solution curves in the (u, u_x)–plane) show that as L is increased, nontrivial steady-state solutions branch or bifurcate at $L^* = \pi$.

8. The equation $u_t = u_{xx}$ was studied extensively in Chapter 11. Consider the operator $\mathcal{A}u = u_{xx}$ on $L^2[0,1]$. Show it is dissipative in the sense that $\langle \mathcal{A}u, u\rangle \leq 0$ for all u. Show that $(I - \lambda\mathcal{A})^{-1}$ exists for every $\lambda > 0$. Viewing the partial differential equation as the ODE $du/dt = \mathcal{A}u$, $u(x,0) = u_0(x)$ in the function space $L^2[0,1]$, show that the backward Euler method gives the solution $u(T) = \lim_{n\to\infty} (I - \frac{T}{n}\mathcal{A})^{-n} u_0$.

9. Analyze the associated ordinary differential equation $du/dt = u(1-u)$, theoretically and numerically. Find the equilibrium points and determine their stability. For the PDE there would be the physical constraint that if u represents concentration, $u(x,t) \geq 0$. What constraints would ensure that the solution remains bounded for all time?

Chapter 13

Integrating Reaction-Diffusion Systems

13.1 Introduction

Although it is possible to integrate the discrete system that arises from a multiple-species model using the methods of Chapter 11, the larger number of PDEs and the great differences in diffusivities between the impurities and the point defects suggest that more efficient techniques are warranted. Recall that the trapezoidal (Crank–Nicolson) method is equivalent to taking one explicit Euler step of length $\Delta t/2$, followed by an implicit Euler of length $\Delta t/2$. It shares the stability of implicit Euler, but enjoys a better local truncation error (LTE). The single biggest improvement possible for the basic trapezoidal scheme is adaptive timestep control. This allows the integrator to take the largest permissible timestep while still maintaining the local truncation error below a user-specified tolerance. A crude way to test the reliability of the solution obtained using a timestep Δt is to repeat the integration with stepsize $\Delta t/2$. If the two solutions are close, then the initial Δt was probably sufficiently small, and further refinement is not likely to increase the accuracy substantially. Modern system integrators adaptively select both Δt and the order of integration for maximum efficiency. Shown in Figure 13.1 is the stepsize versus elapsed time for a typical stiff system of three partial differential equations, with diffusivities differing by nine orders of magnitude. The order of the integration method, q in equation (13.10), is also plotted [393].

More precisely, the local truncation error can be estimated numerically even though the exact solution is not known in closed-form. Consider the model initial value problem $dy/dt = f(t, y(t))$, and suppose that the system has been successfully integrated from time t_0 to t_{n-1}. Now another step must be taken and the LTE estimated. A crude trapezoidal step from t_{n-1} to t_n yields an approximate solution y_1; next, two steps are taken with $\Delta t = \frac{1}{2}(t_n - t_{n-1})$ to give a second solution y_2 at t_n. Define $y_{\text{true}} = y(t_n)$; a Taylor series expansion gives

$$\frac{y_1 - y_{\text{true}}}{y_{\text{true}}} = K_t(2\Delta t)^3, \qquad \frac{y_2 - y_{true}}{y_{true}} = K_t(\Delta t)^3 \tag{13.1}$$

where it is assumed that y'' is approximately constant on $[t_{n-1}, t_n]$. An estimate for the relative local truncation error follows as $\frac{y_1 - y_2}{y_2} = 7K_t(\Delta t)^3$. Using this we can calculate K_t and predict the actual timestep needed

$$\Delta t_{\text{next}} = \left(\frac{Tol}{K_t}\right)^{1/3} = \left(\frac{Tol}{\frac{1}{7}\left|\left|\frac{y_1 - y_2}{y_2}\right|\right|}\right)^{1/3} \Delta t \tag{13.2}$$

where Tol is the desired local truncation error. It is important to emphasize how valuable this timestep prediction can be in achieving convergence in the Newton loop and saving CPU resources. The added cost of three solves per step is usually justified, and this cost can be reduced by using more sophisticated predictor-corrector methods, as discussed in Section 13.4.

Just as the use of a variable grid in time has proven indispensable for integrating large systems of ODEs, so for partial differential equations it is desirable to work with a variable grid in space. The geometries produced by modern processing steps such as Localized Oxidation (LOCOS), Lightly Doped Drain (LDD), and trench isolation, are highly nonrectangular and require nonuniform spatial grids. The irregular geometries that are inherent in these problems suggest that using the finite element method for spatial discretization would be particularly effective. The oxidation problem in higher dimensions is essentially a problem in fluid dynamics where finite element methods have been used successfully for some years. Similarly, during diffusion sharp layers develop as point defects move in from the surface and recombine, requiring that the grid evolve with the solution. The principles of grid refinement developed for the device equations in Chapters 7 and 9 apply equally well to process modeling.

13.2 The Method of Lines

An extremely effective method of solving large parabolic systems of reaction-diffusion equations is the numerical method of lines (NMOL). Consider the

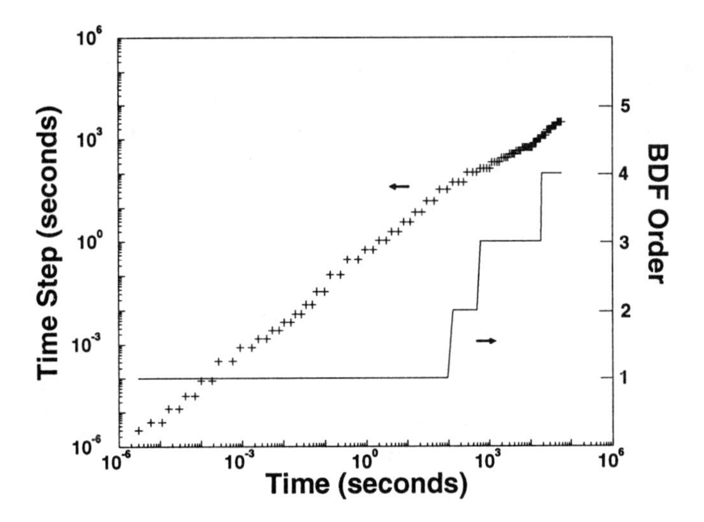

Figure 13.1: Stepsize and order versus time for a stiff system of three partial differential equations, as selected by the integrator LSODE.

system of partial differential equations

$$\frac{\partial \boldsymbol{u}}{\partial t}(x,t) = \boldsymbol{G}\left(x, t, \boldsymbol{u}(x,t), \boldsymbol{D}^1\boldsymbol{u}(x,t), \boldsymbol{D}^2\boldsymbol{u}(x,t)\right) \tag{13.3}$$

which describes the evolution of the vector-valued function $\boldsymbol{u}(x,t)$, $x \in \mathbf{R}^n$, given appropriate initial and boundary conditions. The symbols $\boldsymbol{D}^1$ and $\boldsymbol{D}^2$ denote first- and second-order spatial differential operators which are discretized using finite differences, finite elements, or spectral methods. Then the time derivative alone remains, so that the PDE system becomes the semidiscrete ODE system $d\boldsymbol{u}/dt = \boldsymbol{f}(t, \boldsymbol{u}(t))$, where $\boldsymbol{u}$ is the vector of discretized field variables. As an example, consider the continuous scalar field equation for interstitial concentration

$$\frac{\partial I}{\partial t} = D\,\frac{\partial^2 I}{\partial x^2} + IV \tag{13.4}$$

where V is the vacancy concentration. This is converted by using finite differences into the ODE system

$$\frac{dI_k(t)}{dt} = D\,\frac{I_{k+1}(t) - 2I_k(t) + I_{k-1}(t)}{(\Delta x)^2} + I_k(t)V_k(t) \tag{13.5}$$

for each grid point x_k, $k = 1, ..., n - 1$. Hence the grid point values $I_k(t)$ evolve along the vertical lines $x = x_k$ in the (x, t)–plane. The NMOL involves integrating (13.5) along these lines. The numerical method of lines is very effective on large reaction-diffusion problems, but less so when there are strong convective terms present, i.e. first derivative terms $\boldsymbol{D}^1\boldsymbol{u}$. The large coupled system of nonlinear ordinary differential equations that results from NMOL can be integrated using a system integrator package. These codes use generalizations of the integration techniques already discussed and adaptively select the stepsize and order to control error below user-specified levels. System integrators that are widely available include the LSODE family, DASSL, DVERK, and RKF45, as well as routines from the IMSL and NAG libraries. For more information on specific packages see the survey articles [76, 201, 243, 432].

From the standpoint of modeling diffusion processes, higher-order methods are preferable because of greater efficiency on difficult (but smooth) problems. Single-step methods are better for grid-refinement purposes and to accommodate a moving boundary, as occurs during oxidation. Yeager and Dutton [563] compared the standard trapezoidal rule with a two-part trapezoidal TR2 and trapezoidal–backward differentiation formula TR–BDF(2), and concluded that the latter was best for their 2D model. The PEPPER simulator uses the system integrator LSODI with up to fifth order BDF. The challenge in the area of integration is to develop an efficient one-step method which can handle very stiff reaction-diffusion problems; here implicit Runge–Kutta techniques are very promising. Another possibility is to employ an explicit–implicit scheme: where unknowns are changing rapidly and stiffness is pronounced, an implicit method is used, while deep in the bulk an explicit scheme can be employed.

13.3 Multistep Methods

The Euler and trapezoidal methods use only the value of the solution at the last timestep. In contrast, multistep methods use values at several previous points in time and have found extensive use in system integrators. To solve the system of differential equations $\boldsymbol{y}'(t) = \boldsymbol{f}(t, \boldsymbol{y}(t))$, or more generally the differential algebraic system $\boldsymbol{F}(t, \boldsymbol{y}(t), \boldsymbol{y}'(t)) = 0$, where $\boldsymbol{y}$ is a vector-valued function of time, an approximation to $\boldsymbol{y}'(t)$ must be made. Multistep methods accomplish this via the formula

$$\boldsymbol{y}_n = \sum_{j=1}^{k_1} \alpha_j \boldsymbol{y}_{n-j} + \Delta t \sum_{j=0}^{k_2} \beta_j \boldsymbol{y}'_{n-j} \tag{13.6}$$

The coefficients α_j and β_j are chosen so that (13.6) is exact for polynomials up to a given degree.

Adams formula methods result from taking $k_1 = 1$ and $\alpha_1 = 1$ in (13.6). The Adams–Bashforth method is based on numerical integration and performs well on nonstiff problems. From the fundamental theorem of calculus,

$$\boldsymbol{y}_n = \boldsymbol{y}_{n-1} + \int_{t_{n-1}}^{t_n} \boldsymbol{f}(s, \boldsymbol{y}(s))\, ds \tag{13.7}$$

To evaluate the integral, approximate the function $\boldsymbol{f}(t, \boldsymbol{y}(t))$ by the polynomial $\boldsymbol{r}(t)$ of degree $q-1$ which interpolates it at the q points $t_{n-1}, \ldots, t_{n-q}$. Then $\boldsymbol{p}(t) = \boldsymbol{y}_{n-1} + \int_{t_{n-1}}^{t} \boldsymbol{r}(s)\, ds$ is the unique polynomial of degree q such that $\boldsymbol{p}(t_{n-1}) = \boldsymbol{y}_{n-1}$ and $\boldsymbol{p}'(t) = \boldsymbol{r}(t)$, which defines the coefficients β_j in

$$\boldsymbol{y}_n = \boldsymbol{y}_{n-1} + \Delta t \sum_{j=1}^{q} \beta_j \boldsymbol{y}'_{n-j} \tag{13.8}$$

Note that this is an explicit method of global order q which only uses information at past timesteps; if $q = 1$ explicit Euler is recovered.

If instead the polynomial of degree $q-1$ which interpolates $\boldsymbol{f}(t, \boldsymbol{y}(t))$ at the times $t_n, \ldots, t_{n-q+1}$ is used, one arrives at the implicit Adams–Moulton method of order q,

$$\boldsymbol{y}_n = \boldsymbol{y}_{n-1} + \Delta t \sum_{j=0}^{q-1} \beta_j \boldsymbol{y}'_{n-j} \tag{13.9}$$

The resulting nonlinear system is usually solved for $\boldsymbol{y}_n$ by a fixed point iteration: denoting the mth iterate by $\boldsymbol{y}_n(m)$, an initial guess $\boldsymbol{y}_n(0)$ is chosen; the right-hand side of (13.9) is then evaluated to give $\boldsymbol{y}_n(1)$, and the process iterated until $\boldsymbol{y}_n(m) \approx \boldsymbol{y}_n(m-1)$. For $q = 1$ the formula (13.9) is just backward Euler; $q = 2$ gives the trapezoidal rule.

13.4 Backward Differentiation Formula Methods

From a numerical standpoint, the chief feature of the discretized systems for the multiple-species diffusion models is the stiffness of the associated system of ordinary differential equations. Recall from Chapter 4 that stiffness refers to the fact that physically some components have short characteristic time scales while others have very long ones. In this instance, stiffness is caused by (*i*) the great differences in the diffusivities of the various species, (*ii*) the strong reactive terms, and (*iii*) the large variation in the grid spacing that is

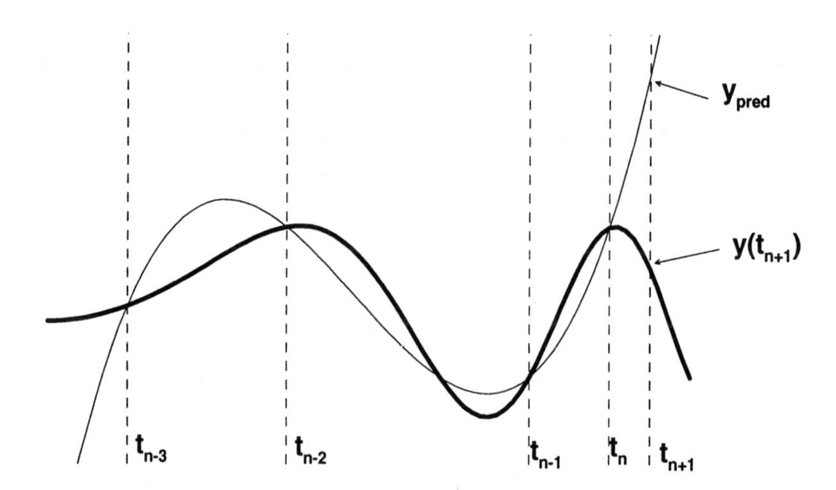

Figure 13.2: The formulas for multistep methods are all based on polynomial approximation.

required for accurate simulations. Because backward differentiation formula (BDF) methods have been very successful in handling stiff problems, they are discussed here in some detail.

BDF methods follow from (13.6) by taking $k_1 = q$ and $k_2 = 0$, to give

$$\boldsymbol{y}_n = \sum_{j=1}^{q} \alpha_j \boldsymbol{y}_{n-j} + \Delta t \beta_0 \boldsymbol{y}'_n \qquad 1 \leq q \leq 5 \tag{13.10}$$

where $\boldsymbol{y}'_n = \boldsymbol{f}(t_n, \boldsymbol{y}_n)$. In essence, this approximates the derivative $\boldsymbol{y}'$ by a linear combination of $\boldsymbol{y}_n, ..., \boldsymbol{y}_{n-q}$, chosen to be exact for a polynomial of degree q. The resulting nonlinear algebraic system may be solved by a Newton iteration, called the inner loop, at each step of which a linear system is solved using direct or iterative methods. This increases the work per timestep as compared to explicit methods, but for stiff problems is offset by the faster convergence characteristics of Newton. The basic algorithm procedes as follows: suppose that we have integrated successfully to time t_{n-1}; choose a trial stepsize $\Delta t \equiv t_n - t_{n-1}$.

Algorithm 13.1 (BDF)

(1) Set $t_n = t_{n-1} + \Delta t$

(2) Predict the solution at t_n *using* $\boldsymbol{y}_n^P = \sum_{i=1}^{q+1} \gamma_i \, \boldsymbol{y}_{n-i}$

(3) *Correct by solving* $\boldsymbol{f}(t_n, \boldsymbol{y}_n, \boldsymbol{y}'_n(\boldsymbol{y}_n)) = \boldsymbol{0}$ *for* $\boldsymbol{y}_n$ *using Newton's method or successive approximation with* $\boldsymbol{y}_n^P$ *as the initial guess.*

(4) *Estimate the local truncation error using*

$$E_{\Delta t} = \frac{\Delta t}{t_n - t_{n-k-1}}(\boldsymbol{y}_n - \boldsymbol{y}_n^P) + O((\Delta t)^{q+2})$$

then compare with user-specified tolerance; select a new timestep Δt *and order* q*; if* $E_{\Delta t}$ *is too large, reject this step.*

(5) *Set* $n - 1 \Leftarrow n$*, go to* (1).

Note the use of a predictor in step 2 above. Because multistep methods are based on polynomial approximation, they afford a better initial guess for the inner loop (or corrector iteration) than merely using the value of the solution at the last timestep. These predictor–corrector methods also estimate the local truncation error by using the difference between the predicted and corrected values at each timestep. For further details on multistep methods, and in particular BDF schemes, consult the texts [203, 241, 176].

The reader has already encountered the notion of stability in connection with dynamical systems. Recall that the forward Euler scheme is conditionally stable, while the backward Euler and trapezoidal methods both enjoy unconditional stability. There are other definitions of stability that refine the basic concept. Consider the scalar test problem $y'(t) = \lambda y(t)$ with $Re(\lambda) < 0$, and the one-step method $y_{n+1} = p(h\lambda)y_n$ where p is a rational function and $h = \Delta t$. The method is said to be A-stable if $|p(h\lambda)| < 1$. It is L-stable if it is both A-stable and $|p(h\lambda)| \to 0$ as $|h\lambda| \to 0$. In the case of the forward Euler algorithm, $p(z) = 1 + z$, and the requirement for A-stability is $|1 + z| < 1$. The region of stability consists of all z in the complex plane whose distance from -1 is less than 1. The corresponding rational functions for the backward Euler and trapezoidal methods are $p(z) = 1/(1 - z)$ and $p(z) = (1 + z)/(1 - z)$, respectively. Regions of stability are shown in Figure 13.3; note that using this more discriminating definition, the backward Euler method is more stable than the trapezoidal scheme.

13.5 Solving Nonlinear Algebraic Systems

Many of the processes of interest in the semiconductor industry involve nonlinear interactions, and lead to large sparse nonlinear semidiscrete ODE systems. Implicit methods for discretizing such systems give rise to corresponding nonlinear systems of algebraic equations as discussed in Chapter

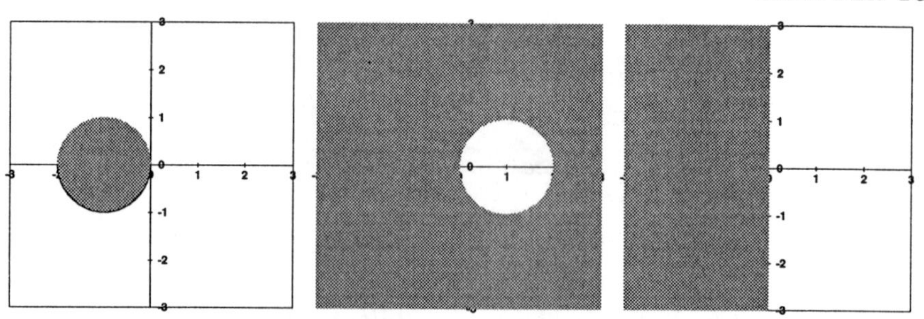

Figure 13.3: Regions of stability (hatched) for forward Euler, backward Euler, and the trapezoidal methods.

5. We obtain a nonlinear function $\boldsymbol{f} : \mathbf{R}^n \rightarrow \mathbf{R}^n$ and seek a vector $\boldsymbol{\xi}$ in $\mathbf{R}^n$ such that

$$\boldsymbol{f}(\boldsymbol{\xi}) = \mathbf{0} \tag{13.11}$$

In general such nonlinear equations have several characteristics not shared by their linear counterparts:

1. Unlike the linear system $\boldsymbol{Ax} = \boldsymbol{b}$, with $\boldsymbol{A}$ nonsingular, equation (13.11) may have no solution, one solution, or many solutions. The set of all solutions is no longer a subspace, but can have a quite complicated geometry. Existence and uniqueness for the nonlinear case are often far more difficult to establish.

2. Gaussian elimination for solving (13.11) with $\boldsymbol{f}(\boldsymbol{x}) = \boldsymbol{Ax} - \boldsymbol{b}$ was mentioned in the chapter on circuit simulation. This is a direct method that produces the exact solution in finitely many steps, assuming infinite precision arithmetic. This is not possible for a nonlinear system, which must instead be solved iteratively; at each step of the iteration a (hopefully) better approximation to the solution is produced. From a practical standpoint, one is satisfied with an ϵ-zero of $\boldsymbol{f}$; that is, problem (13.11) is more accurately stated as, given an error tolerance $\epsilon > 0$, find a vector $\boldsymbol{\xi}$ such that $\|\boldsymbol{f}(\boldsymbol{\xi})\| < \epsilon$.

3. The theory of linear transformations on a vector space has been refined over the last hundred years; a corresponding calculus for nonlinear transformations with analogues of the spectral theorem, etc. is not as yet mature. Recent work in dynamical systems has shown that even a simple nonlinear equation can exhibit a variety of phenomena including strange attractors, period doubling, and bifurcation, that have no linear counterparts.

Of the many methods for solving nonlinear systems (e.g., see [420]), successive approximation and Newton's method are the most widely used. As with integration techniques, there is an inevitable tradeoff between simplicity and ease of implementation versus accuracy and convergence rate. For most TCAD problems, the quadratic convergence enjoyed by a Newton iteration more than offsets its higher complexity and greater overhead. Successive approximation still has certain attributes, as seen in Gummel's method for device simulation. Conceptually it provides a unifying framework for a large class of methods for solving linear and nonlinear equations, as well as differential and integral equations. In fact, Newton's method can be viewed as a special case of successive approximation.

To solve $\boldsymbol{f}(\boldsymbol{\xi}) = \mathbf{0}$, let us pose the equivalent problem, find $\boldsymbol{\xi} \in \mathbf{R}^n$ such that $\boldsymbol{\xi} = \boldsymbol{\xi} - \boldsymbol{f}(\boldsymbol{\xi})$. Defining $\boldsymbol{g}(\boldsymbol{x}) = \boldsymbol{x} - \boldsymbol{f}(\boldsymbol{x})$, a solution $\boldsymbol{\xi}$ to the algebraic system is seen to be a vector which is fixed under the action of $\boldsymbol{g}$

$$\boldsymbol{g}(\boldsymbol{\xi}) = \boldsymbol{\xi} \tag{13.12}$$

and hence is called a fixed-point of $\boldsymbol{g}$. Given a nonlinear equation $\boldsymbol{f}(\boldsymbol{\xi}) = \mathbf{0}$, the fixed-point iteration function is not unique; $\boldsymbol{g}(\boldsymbol{x}) = \boldsymbol{x} + \boldsymbol{f}(\boldsymbol{x})$ or $\boldsymbol{g}(\boldsymbol{x}) = \boldsymbol{x} - \boldsymbol{A}\boldsymbol{f}(\boldsymbol{x})$ where $\boldsymbol{A}$ is a nonsingular matrix would have worked equally well. For an important class of iteration functions, a theorem of Banach assures a unique fixed-point.

Theorem 13.1 (Contraction Mapping Principle) *If $\boldsymbol{g}$ is a function from a (closed) subset S of $\mathbf{R}^n$ into S and there exists a number K, $0 < K < 1$, such that for $\boldsymbol{u}$ and $\boldsymbol{v}$ in S,*

$$\|\boldsymbol{g}(\boldsymbol{u}) - \boldsymbol{g}(\boldsymbol{v})\| \leq K \|\boldsymbol{u} - \boldsymbol{v}\| \tag{13.13}$$

then there is a unique point $\boldsymbol{\xi}$ in S such that $\boldsymbol{g}(\boldsymbol{\xi}) = \boldsymbol{\xi}$, and if $\boldsymbol{x}_0$ is in S, then $\boldsymbol{\xi} = \lim_{n\to\infty} \boldsymbol{g}^n(\boldsymbol{x}_0)$, where $\boldsymbol{g}^n$ is the n-fold composition of $\boldsymbol{g}$.

Because $\boldsymbol{g}$ shrinks distances between points, starting with any initial guess $\boldsymbol{x}_0$ in S and applying the iteration $\boldsymbol{x}_{n+1} = \boldsymbol{g}(\boldsymbol{x}_n)$, the iterates should converge to a unique point $\boldsymbol{\xi}$ such that $\boldsymbol{\xi} = \boldsymbol{g}(\boldsymbol{\xi})$. The iteration function $\boldsymbol{g}$ may fail to be a contraction in one norm, but in another norm condition (13.13) will be satisfied. If $\boldsymbol{g}$ is a contraction in any norm, then $\{\boldsymbol{g}^n(\boldsymbol{x}_0)\}$ will converge in that norm to the fixed-point $\boldsymbol{\xi}$.

To apply successive approximation to backward Euler for the problem $d\boldsymbol{y}/dt = \boldsymbol{f}(\boldsymbol{y})$, recall that the solution $\boldsymbol{y}_n$ is known at time t_n, and one must determine at t_{n+1} the solution $\boldsymbol{y}_{n+1}$ such that $\boldsymbol{y}_{n+1} = \boldsymbol{y}_n + \Delta t \boldsymbol{f}(\boldsymbol{y}_{n+1})$.

Pick $\boldsymbol{y}_{n+1}(0)$ and iterate using $\boldsymbol{y}_{n+1}(m+1) = \boldsymbol{y}_n + \Delta t \boldsymbol{f}(\boldsymbol{y}_{n+1}(m))$; in other words, use the iteration function $\boldsymbol{g}(\boldsymbol{x}) = \boldsymbol{y}_n + \Delta t \boldsymbol{f}(\boldsymbol{x})$. For vectors $\boldsymbol{u}$ and $\boldsymbol{v}$

$$\begin{aligned} \|\boldsymbol{g}(\boldsymbol{u}) - \boldsymbol{g}(\boldsymbol{v})\| &= \|\boldsymbol{y}_n + \Delta t \boldsymbol{f}(\boldsymbol{u}) - \boldsymbol{y}_n - \Delta t \boldsymbol{f}(\boldsymbol{v})\| \\ &= \Delta t \|\boldsymbol{f}(\boldsymbol{u}) - \boldsymbol{f}(\boldsymbol{v})\| \qquad (13.14) \\ &= (\Delta t) \|\boldsymbol{u} - \boldsymbol{v}\| \max \{ \|\boldsymbol{f}'(\alpha)\| : t_n \leq \alpha \leq t_{n+1} \} \end{aligned}$$

If the timestep Δt is small enough relative to the size of $\|\boldsymbol{f}'\|$, then $\boldsymbol{g}$ will be a contraction with a unique fixed-point, the value of $\boldsymbol{y}_{n+1}$ at t_{n+1}. The cost of each iteration is essentially that of evaluating the function $\boldsymbol{f}$, which is relatively cheap. Unfortunately, many iterations are usually required and if Δt is not small enough, $\boldsymbol{g}$ may fail to be a contraction.

The rate of convergence of the standard fixed point iteration is linear, that is, $\|\boldsymbol{\xi} - \boldsymbol{x}_{n+1}\| \leq C \|\boldsymbol{\xi} - \boldsymbol{x}_n\|^p$, with $p = 1$, for some constant C and sufficiently large n. Newton's method for solving $f(\boldsymbol{\xi}) = 0$ has the attractive feature that the error decreases quadratically at each step,

$$\|\boldsymbol{\xi} - \boldsymbol{x}_{n+1}\| \leq C \|\boldsymbol{\xi} - \boldsymbol{x}_n\|^2 \qquad (13.15)$$

For a scalar function $f : \mathbf{R} \to \mathbf{R}$, Newton's algorithm has a simple geometric derivation as noted previously in the discussion of nonlinear circuit solution: to find a zero of f, pick an initial guess x_0 and follow the tangent line at $(x_0, f(x_0))$ until it intersects the x-axis at the point $x_1 = x_0 - f(x_0)/f'(x_0)$; continue this process recursively, converging to a zero ξ.

A Taylor series expansion provides a simple derivation for a vector-valued function $\boldsymbol{f} : \mathbf{R}^n \to \mathbf{R}^n$. Starting at $\boldsymbol{x}_0$, determine an approximate solution $\boldsymbol{x}_1 = \boldsymbol{x}_0 + \boldsymbol{h}$ by using

$$\boldsymbol{f}(\boldsymbol{x}_0 + \boldsymbol{h}) = \boldsymbol{f}(\boldsymbol{x}_0) + \boldsymbol{f}'(\boldsymbol{x}_0)\boldsymbol{h} + O(\|\boldsymbol{h}\|^2) \qquad (13.16)$$

and solving the "linearized" equation $\mathbf{0} = \boldsymbol{f}(\boldsymbol{x}_0) + \boldsymbol{f}'(\boldsymbol{x}_0)\boldsymbol{h}$. The iterates

$$\boldsymbol{x}_{n+1} = \boldsymbol{x}_n - \boldsymbol{f}'(\boldsymbol{x}_n)^{-1} \boldsymbol{f}(\boldsymbol{x}_n) \qquad n = 1, 2, 3, ... \qquad (13.17)$$

converge to a zero $\boldsymbol{\xi}$ provided $\boldsymbol{x}_0$ is close enough to $\boldsymbol{\xi}$ (in the domain of attraction of $\boldsymbol{\xi}$), $\boldsymbol{f}'$ is continuous, and $\boldsymbol{f}'(\boldsymbol{\xi})$ is invertible. If $\boldsymbol{f}''$ is continuous, there is a $C > 0$ so that (13.15) holds and the number of significant digits in the approximate solution doubles at each iteration.

Chapter 5 introduced Newton's method for the circuit simulation problem and described some of the issues related to convergence. Several difficulties also arise when using a Newton iteration for process or device simulation. For time-dependent problems a good initial guess is available, but this may not be true for steady-state or elliptic PDEs. In this case one has to employ continuation techniques which solve the continuous one-parameter family of equations $\boldsymbol{g}(\boldsymbol{x}, \lambda) = \boldsymbol{0}$, where the equation $\boldsymbol{g}(\boldsymbol{x}, 0) = \boldsymbol{0}$ is very easy to solve compared to $\boldsymbol{g}(\boldsymbol{x}, 1) \equiv \boldsymbol{f}(\boldsymbol{x}) = 0$. Damping, which has been mentioned in connection with circuit simulation, can also be employed. If using $\boldsymbol{x}_{n+1} = \boldsymbol{x}_n + \boldsymbol{g}$, where $\boldsymbol{g} = -\boldsymbol{f}'(\boldsymbol{x}_n)^{-1}\boldsymbol{f}(\boldsymbol{x}_n)$, leads to an increase in the residual error, that is, $\|\boldsymbol{f}(\boldsymbol{x}_{n+1})\| > \|\boldsymbol{f}(\boldsymbol{x}_n)\|$, then examine the vectors $\boldsymbol{x}_n + \frac{1}{2^i}\boldsymbol{g}, i = 1, 2, 3, ...$, and choose $\boldsymbol{x}_{n+1}$ to be the first which decreases the residual error.

A second difficulty with Newton's method is correctly constructing the Jacobian matrix $\boldsymbol{f}'(\boldsymbol{x})$ and then solving the associated linear system. For some classes of problems the Jacobian can be computed analytically, although the programming can be time consuming and error prone. Excellent results can often be obtained with a difference quotient approximation

$$\frac{\partial f_i}{\partial x_j}(\boldsymbol{x}) = \frac{f_i(\boldsymbol{x} + \varepsilon \boldsymbol{e}_j) - f_i(\boldsymbol{x})}{\varepsilon} \tag{13.18}$$

which only uses calls to the routine that evaluates $\boldsymbol{f}$ at a vector $\boldsymbol{x}$. When using a direct linear solver, if the Jacobian changes little over several Newton iterations or timesteps, it makes sense to reuse the earlier factorization of $\boldsymbol{f}'(\boldsymbol{x})$: this results in the modified or quasi-Newton procedure. Finally, since the Newton algorithm is itself an inexact, iterative method, it is reasonable to use iterative solvers for each "innermost" linear system solved, particularly on very large problems.

13.6 Solving Linear Systems

The price of the improved stability of implicit methods and the quadratic convergence of Newton's algorithm is that of solving many sparse linear systems. This is by far the most computationally intensive part of most process simulations, often consuming 60 to 70% of the CPU time. Matrices arising from finite difference or finite element spatial discretizations and from circuit simulation are always sparse, usually with less than 5% nonzero elements. This fortuitous circumstance occurs because the nonzeros in a row represent connections between a node and the rest of the circuit or between a gridpoint and its neighbors. In the case of the PDE, it reflects the local character of the discrete approximations to a differential, as opposed to an integral, operator. In nature, as in integrated circuits, one pays a heavy price for excessively many interconnects to remote nodes.

In some cases, the matrices have other special properties that can be used to advantage. Finite difference methods on rectangular domains give rise to banded matrices with a regular sparsity pattern. The spatial derivatives in the linear diffusion problem $C_t = D\Delta C$ and Poisson's equation $\Delta\psi = g$ both generate a symmetric ($A_{ij} = A_{ji}$), positive definite (strictly positive eigenvalues) matrix $\boldsymbol{A}$. Some techniques for solving $\boldsymbol{Ax} = \boldsymbol{b}$ make special use of this symmetric, positive definite (SPD) property, but break down in the case of a general matrix.

Many excellent texts are devoted solely to solving linear systems; the purpose of the following sections is to provide the process-device engineer with an overview of this area, so as to better understand the workings of current simulators. For instance, to what do the options *[min. fill][symmetric]* or *(|lu|sip|SOR|iccg|plucg|auto)* on the SUPREM IV command *symbolic* refer? Answer: the system to be solved is symmetric, minimum fill is desired in the factorization, and the solution technique can be LU decomposition, Stone's implicit method, successive over-relaxation, or conjugate gradient acceleration with incomplete Cholesky or LU preconditioning.

Although several classifications are possible, techniques for solving linear systems are best divided into three categories. *Direct methods* are those which produce the exact answer, neglecting roundoff error, after finitely many operations. *Iterative methods* produce increasingly accurate approximations to the exact solution, often giving reasonable results after a few iterations. Note that if the linear solver is used as part of a Newton algorithm, a solution to machine accuracy is probably overkill, since at each step Newton's method only gives an approximate solution to the nonlinear system. Finally, *projection methods* build an orthogonal sequence of vectors in an appropriate inner product space, and so must terminate with the exact solution in at most n steps if $\boldsymbol{A}$ is an $n \times n$ matrix. Of great importance in this third case is the fact that, as the algorithm proceeds, an increasingly accurate approximate solution is constructed and, unlike Gaussian elimination, a projection method can be halted early, if sufficient accuracy has been obtained.

13.7 Direct and Iterative Methods

The canonical direct method, Gaussian elimination, is based on the fact that any nonsingular $n \times n$ matrix $\boldsymbol{A}$ can be factored uniquely as $\boldsymbol{PLU}$ where $\boldsymbol{P}$ is a permutation matrix, $\boldsymbol{L}$ is a lower triangular matrix, and $\boldsymbol{U}$ is a unit upper triangular matrix, in $O(n^3)$ operations. The system $\boldsymbol{PLUx} = \boldsymbol{b}$ can then be solved in $O(n^2)$ operations using forward and backward substitution. For details on the $\boldsymbol{LU}$ factorization algorithm see [110, 510, 541]; for the related Cholesky and Doolittle–Crout procedures see [251]. Recall that a quasi-Newton method only updates the Jacobian matrix periodically and hence many linear systems must be solved with the same matrix $\boldsymbol{A}$, but different right-hand sides $\boldsymbol{b}$. Sparseness and symmetry, if present, are exploited in implementations of direct solvers. Structural sparseness occurs when the nonzero entries lie in bands or a narrow envelope — usually the result of finite differences or finite elements applied on a regular grid. For a system with bandwidth β the above operation estimates are reduced to $O(\beta^2 n)$ and $O(\beta n)$, respectively. Finite difference and finite element schemes with random node numbering give rise to general sparseness which must be handled by an appropriate data structure and system of pointers. It is possible to renumber the equations and unknowns so as to minimize bandwidth or fill using Cuthill–McKee, minimum degree, or nested dissection algorithms [89]. Finally, although direct methods do not lend themselves as readily to parallel implementations as iterative methods, research in this area is now being actively pursued.

As technology pushes simulation into three dimensions, iterative methods have become increasingly attractive. They are based on successive approximation and often have straightforward nonlinear counterparts. If $\boldsymbol{A}$ is an $n \times n$ matrix, an approximate inverse for $\boldsymbol{A}$ is any matrix $\boldsymbol{C}$ such that $\|\boldsymbol{I} - \boldsymbol{C}\boldsymbol{A}\| < 1$ in some matrix norm. If $\boldsymbol{C}$ is an approximate inverse for $\boldsymbol{A}$, then the function $\boldsymbol{G}(\boldsymbol{x}) = \boldsymbol{x} + \boldsymbol{C}(\boldsymbol{b} - \boldsymbol{A}\boldsymbol{x})$ is a contraction and for any initial guess $\boldsymbol{x}_0$, $\boldsymbol{G}^n(\boldsymbol{x}_0) \to \boldsymbol{\xi}$, where $\boldsymbol{A}\boldsymbol{\xi} = \boldsymbol{b}$. The matrix $\boldsymbol{C}$ should be "close" to $\boldsymbol{A}^{-1}$, but have the property that $\boldsymbol{C}\boldsymbol{x} = \boldsymbol{d}$ is much easier to solve than $\boldsymbol{A}\boldsymbol{x} = \boldsymbol{b}$. For example, decompose $\boldsymbol{A}$ as $\boldsymbol{L} + \boldsymbol{D} + \boldsymbol{U}$ where $\boldsymbol{L}$ is strictly lower triangular, $\boldsymbol{D}$ diagonal, and $\boldsymbol{U}$ strictly upper triangular. If none of the diagonal entries of $\boldsymbol{A}$ are zero, the Jacobi iteration or method of simultaneous displacements uses $\boldsymbol{C} = \boldsymbol{D}^{-1}$. It can be interpreted as choosing an initial guess for the solution and computing a better approximation by looping over the equations: in the ith equation setting all unknowns except the ith to their old values, and then solving for a new x_i. Although Jacobi is easy to program and vectorize or parallelize, convergence is slow and deteriorates very rapidly as the size of the matrix increases. This can be ameliorated to some extent by preconditioning, as discussed below. The nonlinear variant of Jacobi which solves

$$f_k\left(x_1^{(m)}, \ldots, x_{k-1}^{(m)}, x_k^{(m+1)}, x_{k+1}^{(m)}, \ldots, x_n^{(m)}\right) = 0, \qquad k = 1, \ldots, n \tag{13.19}$$

for the new estimate $\boldsymbol{x}^{(m+1)}$, using a one-dimensional Newton's method and then sets $\boldsymbol{x}^{(m)} \leftarrow \boldsymbol{x}^{(m+1)}$. It has received increasing attention as attempts are made to parallelize existing codes.

The obvious improvement to Jacobi is to use the most current values available to compute the new estimate. The Gauss–Seidel iteration or method of successive displacements uses $\boldsymbol{C} = (\boldsymbol{L} + \boldsymbol{D})^{-1}$ and converges much faster than Jacobi when $\boldsymbol{A}$ is strictly row diagonally dominant. Now, however, the algorithm no longer lends itself to parallelizing. A nonlinear version of Gauss–Seidel is

$$f_k\left(x_1^{(m+1)}, \ldots, x_{k-1}^{(m+1)}, x_k^{(m+1)}, x_{k+1}^{(m)}, \ldots x_n^{(m)}\right) = 0, \qquad k = 1, \ldots, n \quad (13.20)$$

and has been used extensively in circuit simulation [404, 405, 554].

For certain matrices $\boldsymbol{A}$, it is possible to accelerate Gauss–Seidel by optimally choosing $\omega > 1$ so that

$$\boldsymbol{x}^{m+1} = \boldsymbol{x}^{m} + \omega(\boldsymbol{L} + \boldsymbol{D})^{-1}(\boldsymbol{b} - \boldsymbol{A}\boldsymbol{x}^{m}) \quad (13.21)$$

converges faster than the standard algorithm. This successive over-relaxation or SOR scheme typically uses $1.2 \leq \omega \leq 1.6$; it is possible to adaptively choose a sequence $\{\omega_m\}$ in order to obtain even faster convergence. Local relaxation schemes have also been proposed [55] in which the relaxation factor is allowed to vary from equation to equation. Contrast over-relaxation with the under-relaxation discussed earlier under the name of damping for Newton's method.

Many variants of these two methods have been used in process and device codes. Line Jacobi has its origins in 2D finite differences on a regular grid. The unknowns on one line, either in the $x-$ or $y-$direction, are calculated concurrently before moving to the next line. This generalizes to a line Gauss–Seidel and the alternating-direction-implicit (ADI) scheme, in which sweeps are made first in the $x-$ and then in the $y-$direction. For a full matrix this would hardly be worth the extra complexity, but in the PDE system an unknown on a line is only coupled to its nearest neighbors, assuming off-line elements are known. The resulting tridiagonal systems can be solved very quickly in $O(n)$ operations. These line techniques have been successfully applied to 2D point defect models [563].

In the general case one may similarly define block iterative methods. At the PDE level this forms the basis of the Gummel iteration, discussed in Chapter 7, and various "decoupling" strategies in process modeling. Here, the idea is applied to the algebraic system. Consider a large linear system

$\boldsymbol{Ax} = \boldsymbol{b}$ and partition the matrix $\boldsymbol{A}$ into k^2 blocks

$$\begin{bmatrix} A_{11} & A_{12} & \cdots & A_{1k} \\ A_{21} & A_{22} & \cdots & A_{2k} \\ \vdots & \vdots & & \vdots \\ A_{k1} & A_{k2} & \cdots & A_{kk} \end{bmatrix} \tag{13.22}$$

in such a way that the diagonal blocks are square matrices and inexpensive to factor. Extensions of scalar Jacobi and Gauss–Seidel iterations to the block system are straightforward. In some applications, for example the multiple-species diffusion models, there are several unknowns per node, and these form a natural nodal block structure in (13.22). For nonlinear systems, the iterations (13.19) and (13.20) can be performed similarly, where x_i represents not a scalar but a block vector quantity.

To summarize, iterative methods can be quite effective when it is known that $\boldsymbol{A}$ possesses certain properties, such as symmetry or diagonal dominance. Unfortunately, for TCAD problems this is sometimes not the case and even when it is, projection methods generally give faster convergence. For futher details on iterative methods see [217, 230, 575, 574]. The article [509] describes Stone's implicit method which has found some popularity in solving the linear systems which arise in device modeling.

13.8 Preconditioners

Regardless of the method used to solve $\boldsymbol{Ax} = \boldsymbol{b}$, a certain error will be introduced by the algorithm and we need to estimate the quantity $\boldsymbol{e} = \boldsymbol{\xi} - \hat{\boldsymbol{x}}$ where $\boldsymbol{\xi}$ is the true solution and $\hat{\boldsymbol{x}}$ is the approximation. Since $\boldsymbol{\xi}$ is not known exactly, use the fact that the residual $\boldsymbol{r} = \boldsymbol{b} - \boldsymbol{A}\hat{\boldsymbol{x}}$ satisfies $\boldsymbol{r} = \boldsymbol{b} - \boldsymbol{A}\hat{\boldsymbol{x}} - (\boldsymbol{b} - \boldsymbol{A}\boldsymbol{\xi}) = \boldsymbol{Ae}$, so that $\boldsymbol{e} = \boldsymbol{A}^{-1}\boldsymbol{r}$. The following inequalities give precise bounds on the relative error in terms of the relative residual $||\boldsymbol{r}|| \, / \, ||\boldsymbol{b}||$,

$$\frac{1}{||\boldsymbol{A}|| \, ||\boldsymbol{A}^{-1}||} \frac{||\boldsymbol{r}||}{||\boldsymbol{b}||} \leq \frac{||\boldsymbol{e}||}{||\boldsymbol{x}||} \leq ||\boldsymbol{A}|| \, ||\boldsymbol{A}^{-1}|| \frac{||\boldsymbol{r}||}{||\boldsymbol{b}||} \tag{13.23}$$

The number $cond(\boldsymbol{A}) = ||\boldsymbol{A}|| \, ||\boldsymbol{A}^{-1}||$ is called the condition number of $\boldsymbol{A}$; it is always ≥ 1 and a large value indicates an ill-conditioned problem. For a poorly-conditioned system, small changes in the input data $\boldsymbol{b}$ produce large changes in the solution $\boldsymbol{\xi}$, in analogy with the concept of instability for a dynamical system. The definition of condition number given above depends on the choice of a matrix norm and for this reason $cond(\boldsymbol{A})$ is often taken to be the related quantity $\rho(\boldsymbol{A})\rho(\boldsymbol{A}^{-1})$, where $\rho(\boldsymbol{A})$ is the spectral radius,

$$\rho(\boldsymbol{A}) = \max\,\{|\lambda| : \lambda \text{ is an eigenvalue of } \boldsymbol{A}\} \tag{13.24}$$

If $\boldsymbol{A}$ is a symmetric positive definite matrix, $cond(\boldsymbol{A})$ is then the ratio of the largest to the smallest eigenvalue of $\boldsymbol{A}$.

Current research on iterative methods focuses on finding a "splitting" matrix $\boldsymbol{Q}$ such that the preconditioned system $\boldsymbol{Q}^{-1}\boldsymbol{A}\boldsymbol{x} = \boldsymbol{Q}^{-1}\boldsymbol{b}$ is better-conditioned than the original system or enjoys faster convergence because of a better distribution of eigenvalues (called the spectrum of $\boldsymbol{A}$) in the complex plane. The requirement for $\boldsymbol{Q}$ is that it be easily invertible and "close" to $\boldsymbol{A}$ in the sense that $\rho(\boldsymbol{I} - \boldsymbol{Q}^{-1}\boldsymbol{A})$ is small. Although less common than left-preconditioning, two-sided preconditioning is also possible

$$(\boldsymbol{Q}_L^{-1}\boldsymbol{A}\boldsymbol{Q}_R^{-1})\boldsymbol{Q}_R\boldsymbol{u} \;=\; \boldsymbol{Q}_L^{-1}\boldsymbol{b} \tag{13.25}$$

The error bound (13.31) given below for conjugate gradient indicates that the iterates can converge quite slowly, even when the condition number is moderate, and that it would be advantageous to precondition. Often used preconditioners include Jacobi ($\boldsymbol{Q} = \boldsymbol{D}$), SOR ($\boldsymbol{Q} = \frac{1}{\omega}(\boldsymbol{D} + \boldsymbol{L})$), and incomplete Cholesky or LU decomposition with various levels of fill-in allowed. The multigrid and multilevel schemes [65, 370] discussed in Chapter 7 may also be viewed as a type of preconditioning. See [29] for an application of preconditioning techniques to 3D process simulation.

13.9 Conjugate Gradient and Least-squares

If $\boldsymbol{A}$ is symmetric positive definite (SPD), one of the most effective methods for solving $\boldsymbol{A}\boldsymbol{x} = \boldsymbol{b}$ is the conjugate gradient (CG) algorithm, developed in the 1950s by Hestenes and Stiefel [242]. A detailed derivation for CG is given below before reviewing its generalizations to non-SPD systems. First, reformulate the problem in variational form by considering the quadratic functional $f : \mathbf{R}^n \to \mathbf{R}$ given by

$$f(\mathrm{x}) = \frac{1}{2}\langle \boldsymbol{A}\boldsymbol{x}, \boldsymbol{x}\rangle - \langle \boldsymbol{b}, \boldsymbol{x}\rangle + C \tag{13.26}$$

where C is an arbitrary constant and $\langle , \rangle$ denotes the standard Euclidean inner product. Minimizing f is equivalent to solving the linear system $\boldsymbol{A}\boldsymbol{x} = \boldsymbol{b}$. To see this, note that the gradient of f at $\boldsymbol{x}$ is $\boldsymbol{A}\boldsymbol{x} - \boldsymbol{b}$ and, from calculus, a necessary condition for $\boldsymbol{\xi}$ to be a minimum of f is that $\nabla f(\boldsymbol{\xi}) = \mathbf{0}$. This is also sufficient if $\boldsymbol{A}$ is symmetric positive definite, for then $\boldsymbol{A}$ is nonsingular and $Q(\boldsymbol{x}, \boldsymbol{y}) \equiv \langle \boldsymbol{A}\boldsymbol{x}, \boldsymbol{y}\rangle = \langle \boldsymbol{x}, \boldsymbol{A}\boldsymbol{y}\rangle$ defines a new inner product with norm $N(\boldsymbol{x}) = Q(\boldsymbol{x}, \boldsymbol{x})^{\frac{1}{2}}$ in which $f(\boldsymbol{x}) = \frac{1}{2}N(\boldsymbol{x} - \boldsymbol{A}^{-1}\boldsymbol{b})^2$. This follows from

$$
\begin{aligned}
N(\boldsymbol{x} - \boldsymbol{A}^{-1}\boldsymbol{b})^2 &= \langle \boldsymbol{A}(\boldsymbol{x} - \boldsymbol{A}^{-1}\boldsymbol{b}), (\boldsymbol{x} - \boldsymbol{A}^{-1}\boldsymbol{b})\rangle \\
&= \langle \boldsymbol{A}\boldsymbol{x} - \boldsymbol{b}, \boldsymbol{x} - \boldsymbol{A}^{-1}\boldsymbol{b})\rangle \\
&= \langle \boldsymbol{A}\boldsymbol{x}, \boldsymbol{x}\rangle - \langle \boldsymbol{A}\boldsymbol{x}, \boldsymbol{A}^{-1}\boldsymbol{b}\rangle - \langle \boldsymbol{b}, \boldsymbol{x}\rangle + \langle \boldsymbol{b}, \boldsymbol{A}^{-1}\boldsymbol{b}\rangle \\
&= \langle \boldsymbol{A}\boldsymbol{x}, \boldsymbol{x}\rangle - 2\langle \boldsymbol{b}, \boldsymbol{x}\rangle + \langle \boldsymbol{b}, \boldsymbol{A}^{-1}\boldsymbol{b}\rangle \\
&= 2f(\boldsymbol{x})
\end{aligned}
$$

where C in (13.26) is taken as $\frac{1}{2}\langle \boldsymbol{b}, \boldsymbol{A}^{-1}\boldsymbol{b}\rangle$. Clearly $\boldsymbol{\xi} = \boldsymbol{A}^{-1}\boldsymbol{b}$ is the unique minimizer of f.

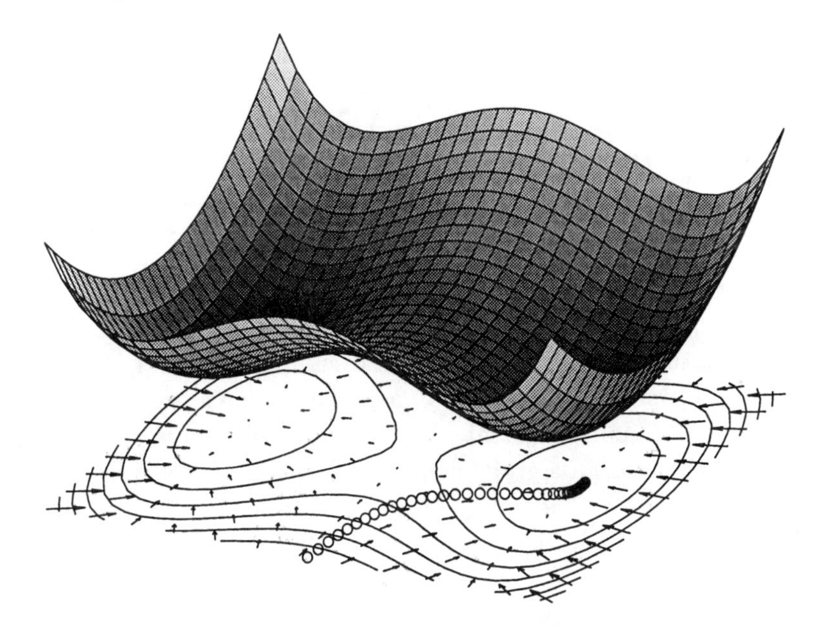

Figure 13.4: Minimization using steepest descent.

To minimize a nonlinear functional such as (13.26), one could follow Cauchy's lead and try the method of steepest descent [100]. Pick a starting point $\boldsymbol{x}_0$ and a small $\alpha > 0$; then move a distance α in the direction $\boldsymbol{g}_0$ in which f decreases most rapidly at $\boldsymbol{x}_0$. From the calculus of several variables, this direction is given by the negative of the gradient of f at $\boldsymbol{x}_0$, $\boldsymbol{g}_0 = -\boldsymbol{\nabla} f(\boldsymbol{x}_0)$. The current location is now $\boldsymbol{x}_1 = \boldsymbol{x}_0 + \alpha \boldsymbol{g}_0$, and the process is continued recursively. Figure 13.4 shows the results of this strategy appled to the functional $f(x, y) = (x^2 - 4)^2 + 4y^2$, which has two global minima, $(-2,0)$,$(2,0)$, and a saddle point. In addition to a surface representation, the isolines and selected gradient vectors are shown in the $z = 0$ plane. The circles enclose points given by the steepest descent algorithm with stepsize 0.01 starting at $(0.1, -3)$. Note that the method converges, but rather slowly.

Because the $\boldsymbol{g}$ defines an "instantaneous" descent direction, there is the question: how large should α be ? The answer is to move a distance α_n chosen to minimize the scalar functional $\phi(t) = f(\boldsymbol{x}_n + t\boldsymbol{g}_n)$. This is termed a "line search" in the gradient direction; it is a definite improvement over using a fixed α but convergence can still be agonizingly slow. There are various ways to ameliorate this situation, including, for instance, the Levenburg–Marquadt algorithm for parameter extraction described in Chapter 6. In the quadratic case (13.26), there is a closed-form expression for the α_n which minimizes f during the line search. Nevertheless, difficulties can arise. Consider the simple quadratic functional on $\mathbf{R}^2$

$$f\begin{pmatrix} x \\ y \end{pmatrix} = \left\langle \begin{bmatrix} 2 & 1 \\ 1 & 0.5 \end{bmatrix} \begin{pmatrix} x \\ y \end{pmatrix}, \begin{pmatrix} x \\ y \end{pmatrix} \right\rangle - \left\langle \begin{pmatrix} 2 \\ -1 \end{pmatrix}, \begin{pmatrix} x \\ y \end{pmatrix} \right\rangle \tag{13.27}$$

which upon expanding can be written as $f(x, y) = 2x^2 + xy + 0.5y^2 - 2x + y$. Figure 13.5(a) shows the results of applying optimal steepest descent to this f. The algorithm requires 14 steps (some too small to be seen without magnification) in order to locate the global minimum $(1, 2)$ to a tolerance of 10^{-6}. The path to the global minimum oscillates back and forth between the walls of the "valley" defined by f, a process known as "hemstitching". In this case the optimal local strategy for minimizing f is not a good global strategy and the descent algorithm must be modified to another scheme such as the conjugate gradient algorithm. Figure 13.5(b) shows the results of applying the latter method to the same problem: it converges in exactly two steps.

This section has discussed the use of steepest descent to minimize nonlinear functionals on $\mathbf{R}^n$. Note that the problem of solving many ordinary and partial differential equations can be reformulated as solving an equation $\boldsymbol{F}(\boldsymbol{u}) = \boldsymbol{0}$, where $\boldsymbol{F}$ is a nonlinear operator on a function space H. A least-squares formulation then becomes $\min_{\boldsymbol{x}} \phi(\boldsymbol{x}) = 1/2\|\boldsymbol{F}(\boldsymbol{x})\|^2$. Steepest descent has a natural generalization to this setting, but now the gradient of ϕ is defined in terms of the inner product on H. It is essential to use an inner product that is appropriate to the particular problem at hand. For solving the linear system $\boldsymbol{A}\boldsymbol{x} = \boldsymbol{b}$, it is the $\boldsymbol{A}$ inner product defined below. In the case of a partial differential equation it is the Sobolev inner product of H^1 or H^2 rather than that of L^2 (for details see [403, 448]).

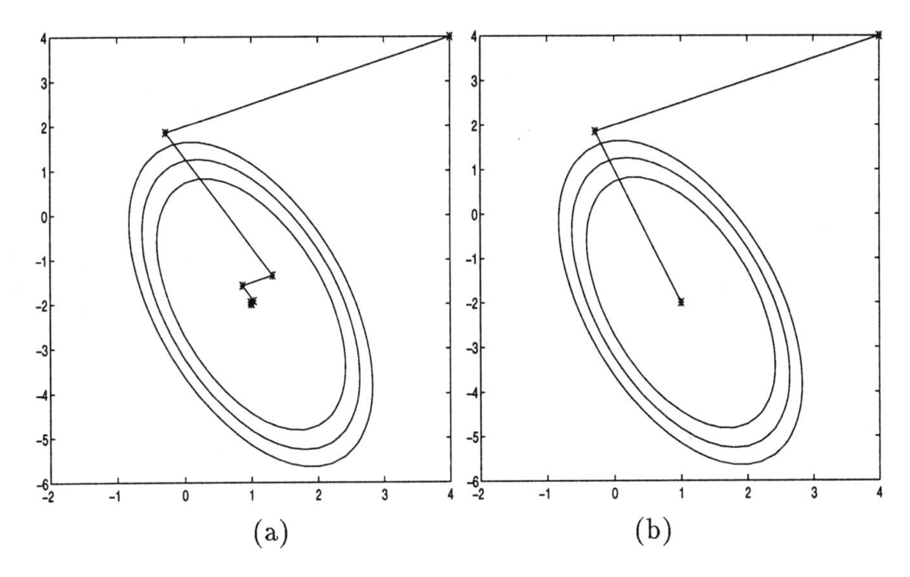

Figure 13.5: (a) Optimal steepest descent versus (b) conjugate gradient for minimizing $f(x, y) = 2x^2 + xy + 0.5y^2 - 2x + y$.

There are several excellent texts on the conjugate gradient method, including the extensive treatment by Hestenes [242]. The derivation given below is based upon orthogonal projections in a Hilbert space and the reader familiar with finite elements will doubtless recognize connections with the Galerkin method for solving PDEs. The following theorem provides the definition of a projection operator onto a subspace.

Theorem 13.2 (Projection) *If L is a (closed) subspace of a Hilbert space $\{H, \langle , \rangle\}$ and $\boldsymbol{y} \in H$, then each two of the following statements are equivalent for a point $\boldsymbol{w}$ in L:*
(i) for every $\boldsymbol{x}$ in L, $||\boldsymbol{y} - \boldsymbol{w}|| \leq ||\boldsymbol{y} - \boldsymbol{x}||$
(ii) for every $\boldsymbol{x}$ in L, $\langle \boldsymbol{y} - \boldsymbol{w}, \boldsymbol{x} \rangle = 0$
(iii) if $\{\boldsymbol{u}_i\}$ is an orthonormal basis of L, then $\boldsymbol{w} = \sum_i \langle \boldsymbol{y}, \boldsymbol{u}_i \rangle \boldsymbol{u}_i$

Proof: $(i) \Rightarrow (ii)$ Let $\boldsymbol{w}$ be a point of L which satisfies (i). Let $\boldsymbol{x} \in L$ and $\boldsymbol{z} = \boldsymbol{x} + \boldsymbol{w}$. For t in $\mathcal{R}$, define $h(t) = \frac{1}{2}||\boldsymbol{y} - \{(1-t)\boldsymbol{z} + t\boldsymbol{w}\}||^2$, then $h'(t) = \langle \boldsymbol{y} - \{(1-t)\boldsymbol{z} + t\boldsymbol{w}\}, \boldsymbol{z} - \boldsymbol{w} \rangle$ From (i), the minimum of h occurs when $t = 1$, so that $0 = h'(1) = \langle \boldsymbol{y} - \boldsymbol{w}, \boldsymbol{z} - \boldsymbol{w} \rangle = \langle \boldsymbol{y} - \boldsymbol{w}, \boldsymbol{x} \rangle$.
$(ii) \Rightarrow (iii)$ Suppose that $\langle \boldsymbol{w} - \boldsymbol{y}, \boldsymbol{x} \rangle = 0$ for $\boldsymbol{x}$ in L.

$$\boldsymbol{w} - \sum_i \langle \boldsymbol{y}, \boldsymbol{u}_i \rangle \boldsymbol{u}_i = \sum_i \langle \boldsymbol{w}, \boldsymbol{u}_i \rangle \boldsymbol{u}_i - \sum_i \langle \boldsymbol{y}, \boldsymbol{u}_i \rangle \boldsymbol{u}_i = \sum_i \langle \boldsymbol{w} - \boldsymbol{y}, \boldsymbol{u}_i \rangle \boldsymbol{u}_i = \boldsymbol{0}$$

$(iii) \Rightarrow (i)$ Let $\{\boldsymbol{u}_i\}$ be an orthonormal basis for L. Let $\boldsymbol{w} = \sum_i \langle \boldsymbol{y}, \boldsymbol{u}_i \rangle \boldsymbol{u}_i$ and $\boldsymbol{x}$ be in L. There exist scalars $\{\alpha_i\}$ such that $\boldsymbol{x} = \sum_i \alpha_i \boldsymbol{u}_i$. Now,

$$\begin{aligned} ||\boldsymbol{y} - \boldsymbol{x}||^2 &= \langle \boldsymbol{y} - \sum_i \alpha_i \boldsymbol{u}_i, \boldsymbol{y} - \sum_i \alpha_j \boldsymbol{u}_j \rangle \\ &= \langle \boldsymbol{y}, \boldsymbol{y} \rangle - 2 \sum_i \alpha_i \langle \boldsymbol{y}, \boldsymbol{u}_i \rangle + \sum_i \alpha_i^2 \\ ||\boldsymbol{y} - \boldsymbol{w}||^2 &= \langle \boldsymbol{y}, \boldsymbol{y} \rangle - \sum_i \langle \boldsymbol{y}, \boldsymbol{u}_i \rangle \end{aligned}$$

So that the difference $||\boldsymbol{y} - \boldsymbol{x}||^2 - ||\boldsymbol{y} - \boldsymbol{w}||^2$ is

$$\sum_i \alpha_i^2 - 2 \sum_i \alpha_i \langle \boldsymbol{y}, \boldsymbol{u}_i \rangle + \sum_i \langle \boldsymbol{y}, \boldsymbol{u}_i \rangle^2 = \sum_i \{\alpha_i - \langle \boldsymbol{y}, \boldsymbol{u}_i \rangle\}^2 \geq 0$$

The sum will be zero only if $\alpha_i = \langle \boldsymbol{y}, \boldsymbol{u}_i \rangle$, that is, when $\boldsymbol{x} = \boldsymbol{w}$.

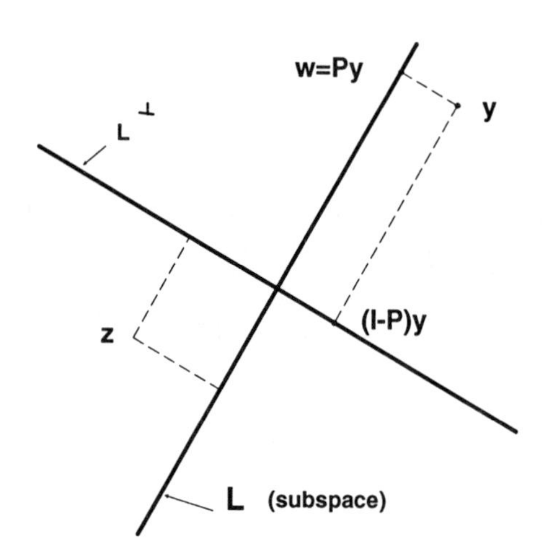

Figure 13.6: The projection theorem: the closest point to y in L is $\mathcal{P}\boldsymbol{y} = \sum_i \langle \boldsymbol{y}, \boldsymbol{u}_i \rangle \boldsymbol{u}_i$, where $\{u_i\}$ is an orthornomal basis of L.

This theorem characterizes the closest point $\boldsymbol{w}$ in L to $\boldsymbol{y}$. The existence of $\boldsymbol{w}$ is guaranteed by the following theorem: if S is a closed, convex subset of H and $\boldsymbol{y}$ is a point of H, then there is a unique point $\boldsymbol{w}$ of S which is closest to $\boldsymbol{y}$. Note that subspaces are necessarily convex. The point $\boldsymbol{w}$ that satisfies any one of the three conditions of the theorem is the closest point in L to $\boldsymbol{y}$

and denoted by $\mathcal{P}\boldsymbol{y}$. This defines a bounded linear transformation $\mathcal{P}$ from H onto L, called the orthogonal projection onto L, which is idempotent, $\mathcal{P}^2 = \mathcal{P}$, and symmetric, $\mathcal{P}^* = \mathcal{P}$. In fact, these two properties are sufficient to define an orthogonal projection. The set of all vectors orthogonal to L is called the orthogonal complement of L in H and is denoted by $L^\perp$; H is the internal direct sum of L and $L^\perp$, $H = L \oplus L^\perp$. Most importantly, if $\boldsymbol{I}$ denotes the identity transformation $\boldsymbol{I}\boldsymbol{x} = \boldsymbol{x}$, then $\boldsymbol{I} - \mathcal{P}$ is the orthogonal projection onto $L^\perp$. These relationships are summarized geometrically in Figure 13.6.

Examples:

1. Take $H = L^2[0,1]$, the set of square integrable functions defined on the interval $[0,1]$ with the inner product given by $\langle f, g\rangle = \int_0^1 f(x)g(x)\,dx$. Pick a positive integer n and for $1 \le j \le n$, define ϕ_j to be the function which is $n^{-\frac{1}{2}}$ on $I_j = [\frac{j-1}{n}, \frac{j}{n}]$ and zero elsewhere. If L is the set of all linear combinations of the basis functions, denoted by $\mathrm{span}(\{\phi_j\})$, then $\{\phi_j\}$ is an orthonormal basis for L and $\mathcal{P}f$ is the piecewise constant function whose value on I_j is just the average value of f on I_j.

2. Take $H = L^2[-1,1]$ with the same inner product as before. Then the set $S = \{\frac{1}{\sqrt{2}}, \frac{1}{\sqrt{2}}\cos(j\pi x), \frac{1}{\sqrt{2}}\sin(j\pi x)\}_{j\ge 1}$ of "trigonometric polynomials" is an orthonormal basis for $L = H$ (L need not be a proper subspace of H) and part *(iii)* yields Fourier's theorem

$$(\mathcal{P}f)(x) = f(x) = \frac{1}{2}a_0 + \sum_{j=1}^{\infty} a_j \cos(j\pi x) + \sum_{j=1}^{\infty} b_j \sin(j\pi x) \qquad (13.28)$$

where the Fourier coefficients are given in terms of inner products.

3. Take $H = H_0^1[0,1]$ the set of continuous functions on $[0,1]$ which are zero at 0 and 1, and have derivatives in $L^2[0,1]$. Now, $\langle f, g\rangle = \int_0^1 f(x)g(x) + f'(x)g'(x)\,dx$, which involves derivatives as well as the functions themselves. For n a positive integer, let $h = \frac{1}{n}$ and for $1 \le j \le n+1$, let ϕ_j be the piecewise linear function which is 1 at jh and 0 at kh for $k \ne j$. This is the standard piecewise linear finite element basis, which is not orthonormal. L is $\mathrm{span}(\{\phi_j\})$, and the finite element approximate solution to the differential equation $-y'' = g, y(0) = 0 = y(1)$ is just $y_h = \mathcal{P}y$, where y is the exact solution.

Now, suppose that H is finite dimensional, and let us solve the minimization problem for (13.26) by minimizing f on successively larger hyperplanes.

Theorem 13.3 *If $\boldsymbol{x}_0 \in \mathbf{R}^n$ and $\{\boldsymbol{p}_0, ..., \boldsymbol{p}_{k-1}\}$, $\boldsymbol{p}_i \neq \mathbf{0}$, is a conjugate orthogonal set, in the sense that $\langle \boldsymbol{p}_i, \boldsymbol{A}\boldsymbol{p}_j \rangle = 0$ for $i \neq j$, then the minimizer of f on the k-dimensional hyperplane*

$$\pi_k = \boldsymbol{x}_0 + t_0 \boldsymbol{p}_0 + \cdots + t_{m-1}\boldsymbol{p}_{k-1}$$

occurs at the point $\boldsymbol{z}$, where ∇f is orthogonal to π_k; $\boldsymbol{z}$ is given explicitly by $\boldsymbol{x}_0 + a_0\boldsymbol{p}_0 + \cdots + a_{k-1}\boldsymbol{p}_{k-1}$ with $a_i = \dfrac{\langle \boldsymbol{y}, \boldsymbol{A}\boldsymbol{p}_i \rangle}{\langle \boldsymbol{p}_i, \boldsymbol{A}\boldsymbol{p}_i \rangle}$ where $\boldsymbol{y} = \boldsymbol{A}^{-1}\boldsymbol{b} - \boldsymbol{x}_0$ $= \boldsymbol{\xi} - \boldsymbol{x}_0$.

Proof: Minimizing f over π_k is equivalent to minimizing $f(\boldsymbol{x}+\boldsymbol{x}_0)$ over $L = \mathrm{span}(\{\boldsymbol{p}_0, ..., \boldsymbol{p}_{k-1}\})$, that is, to finding the closest point to $\boldsymbol{y} = \boldsymbol{A}^{-1}\boldsymbol{b} - \boldsymbol{x}_0$ in the $\boldsymbol{A}$ inner product. The minimization problem can be successively rewritten as $\min_{\boldsymbol{x}\in\pi_k} \frac{1}{2}N(\boldsymbol{x} - \boldsymbol{A}^{-1}\boldsymbol{b})^2 \equiv \min_{\boldsymbol{x}\in L} N(\boldsymbol{x} + \boldsymbol{x}_0 - \boldsymbol{A}^{-1}\boldsymbol{b}) \equiv \min_{\boldsymbol{x}\in L} N(\boldsymbol{x} - \boldsymbol{y})$
The set $\{\boldsymbol{p}_i\}$ is orthogonal in the $\boldsymbol{A}$inner product, and to form an orthonormal basis, define $\tilde{\boldsymbol{p}}_i = \dfrac{1}{N(\boldsymbol{p}_i)}\boldsymbol{p}_i$. From Theorem 13.2(*iii*),

$$\boldsymbol{w} = \sum_i Q(\boldsymbol{y}, \tilde{\boldsymbol{p}}_i)\tilde{\boldsymbol{p}}_i = \sum_i \langle \boldsymbol{y}, \boldsymbol{A}\tilde{\boldsymbol{p}}_i \rangle \tilde{\boldsymbol{p}}_i = \sum_i \frac{\langle \boldsymbol{y}, \boldsymbol{A}\boldsymbol{p}_i \rangle}{\langle \boldsymbol{p}_i, \boldsymbol{A}\boldsymbol{p}_i \rangle}\boldsymbol{p}_i$$

Once again translating by $\boldsymbol{x}_0$ yields $\boldsymbol{z} = \boldsymbol{x}_0 + \boldsymbol{w}$, the closest point to $\boldsymbol{A}^{-1}\boldsymbol{b}$ in π_k using the Q norm. The orthogonality condition stated in Theorem 13.3 requires that $\nabla f(\boldsymbol{z})$ be orthogonal to vectors which lie *in the hyperplane*, those of the form $\boldsymbol{v} = t_0\boldsymbol{p}_0 + \cdots + t_{k-1}\boldsymbol{p}_{k-1}$, which means

$$\langle \nabla f(\boldsymbol{z}), \boldsymbol{v} \rangle = \langle \boldsymbol{A}\boldsymbol{z} - \boldsymbol{b}, \boldsymbol{v} \rangle = \langle \boldsymbol{A}((\boldsymbol{x}_0 + \boldsymbol{w}) - \boldsymbol{A}^{-1}\boldsymbol{b}), \boldsymbol{v} \rangle = Q(\boldsymbol{w} - \boldsymbol{y}, \boldsymbol{v}) = 0$$

This must be zero by Theorem 13.2(*ii*).

This theorem states that the global minimizer of f must lie in the conjugate orthogonal hyperplane to π_k containing $\boldsymbol{z}$. Of course, the vector $\boldsymbol{y}$ is not explicitly computed; if $\boldsymbol{A}^{-1}\boldsymbol{b}$ were known, the problem would already be solved. It suggests, however, that we should constantly project onto the $\perp_A$ hyperplane. This idea of projection also forms the basis of Gram–Schmidt orthogonalization: given a linearly independent set $\{\boldsymbol{x}_0, \boldsymbol{x}_1, ..., \boldsymbol{x}_j\}$, an orthonormal set with the same span is constructed by *(i)* normalizing, $\boldsymbol{w}_0 = \boldsymbol{x}_0$, $\boldsymbol{v}_0 = \boldsymbol{w}_0/||\boldsymbol{w}_0||$; *(ii)* projecting the next vector in the sequence onto the orthogonal complement of the span of the orthonormal vectors currently constructed, $\boldsymbol{w}_k = \boldsymbol{x}_k - \sum_i \langle \boldsymbol{x}_k, \boldsymbol{v}_i \rangle \boldsymbol{v}_i$, $\boldsymbol{v}_k = \boldsymbol{w}_k / ||\boldsymbol{w}_k||$; *(iii)* noting that span($\{\boldsymbol{x}_i\}$) = span($\{\boldsymbol{v}_i\}$) for the $\boldsymbol{v}_i$ that have been generated; and *(iv)* continuing this process of projecting and normalizing. Projection also underlies the conjugate gradient algorithm.

Algorithm 13.2 (Conjugate Gradient Method) *To solve* $\boldsymbol{A}\boldsymbol{x} = \boldsymbol{b}$ *where* $\boldsymbol{A}$ *is an* $n \times n$ *symmetric positive definite matrix, recursively perform the following steps. Pick an initial vector* $\boldsymbol{x}_0$*; define* $\boldsymbol{r}_0 = \boldsymbol{b} - \boldsymbol{A}\boldsymbol{x}_0$ *and* $\boldsymbol{p}_0 = \boldsymbol{r}_0$*; for* $k \geq 0$ *compute* $\boldsymbol{x}_{k+1}, \boldsymbol{r}_{k+1}, \boldsymbol{p}_{k+1}$ *by*

$$
\begin{array}{lll}
(1) & \alpha_k & = \langle \boldsymbol{r}_k, \boldsymbol{p}_k \rangle / \langle \boldsymbol{p}_k, \boldsymbol{A}\boldsymbol{p}_k \rangle \\
(2) & \boldsymbol{x}_{k+1} & = \boldsymbol{x}_k + \alpha_k \boldsymbol{p}_k \\
(3) & \boldsymbol{r}_{k+1} & = \boldsymbol{r}_k - \alpha_k \boldsymbol{A}\boldsymbol{p}_k \\
(4) & \beta_k & = -\langle \boldsymbol{r}_{k+1}, \boldsymbol{A}\boldsymbol{p}_k \rangle / \langle \boldsymbol{p}_k, \boldsymbol{A}\boldsymbol{p}_k \rangle \\
(5) & \boldsymbol{p}_{k+1} & = \boldsymbol{r}_{k+1} + \beta_k \boldsymbol{p}_k
\end{array}
$$

Notice that in step 1, α_k is chosen so that $\boldsymbol{x}_{k+1}$ is the unique minimizer of f on the line $\boldsymbol{\ell}(t) = \boldsymbol{x}_k + t\boldsymbol{p}_k$. The vector $\boldsymbol{r}_{k+1}$ is the negative of the gradient of f at $\boldsymbol{x}_{k+1}$, and will be linearly independent of $\boldsymbol{p}_0, ..., \boldsymbol{p}_k$, unless $\boldsymbol{r}_{k+1} = \boldsymbol{0}$, in which case the solution has been found. Supposing that $\boldsymbol{p}_0, ..., \boldsymbol{p}_k$ are conjugate orthogonal, $\boldsymbol{r}_{k+1}$ must be orthogonal to the hyperplane $\pi_k = \boldsymbol{x}_0 + t_0\boldsymbol{p}_0 + \cdots + t_k\boldsymbol{p}_k$. Steps 4 and 5 are perhaps the most important. Since the global minimizer of f must have a nonzero component in the $\perp_A$ complement of π_k, one should move not in the direction $\boldsymbol{r}_{k+1}$, but rather project in the $\boldsymbol{A}$ inner product, Q, onto this complement.

$$
\boldsymbol{p}_{k+1} = (\boldsymbol{I} - \mathcal{P})\boldsymbol{r}_{k+1} = \boldsymbol{r}_{k+1} - \sum_{j=0}^{k} \frac{\langle \boldsymbol{r}_{k+1}, \boldsymbol{A}\boldsymbol{p}_j \rangle}{\langle \boldsymbol{p}_j, \boldsymbol{A}\boldsymbol{p}_j \rangle} \boldsymbol{p}_j \tag{13.29}
$$

so $\boldsymbol{p}_{k+1} = \boldsymbol{r}_{k+1} - \sum_{j=0}^{k} \langle \boldsymbol{r}_{k+1}, \boldsymbol{A}\tilde{\boldsymbol{p}}_j \rangle \tilde{\boldsymbol{p}}_j$ where $\tilde{\boldsymbol{p}}_j$ is the normalized (with respect to Q) version of $\boldsymbol{p}_j$.

The rather surprising fact is that using this recursion only the last term of the sum is nonzero, leaving $\boldsymbol{p}_{k+1} = \boldsymbol{r}_{k+1} + \beta_k \boldsymbol{p}_k$. To see how this simplification occurs, note that L_k = span($\{\boldsymbol{p}_0, ..., \boldsymbol{p}_k\}$) is just the Krylov subspace span($\{\boldsymbol{p}_0, \boldsymbol{A}\boldsymbol{p}_0, \boldsymbol{A}^2\boldsymbol{p}_0, ..., \boldsymbol{A}^k\boldsymbol{p}_0\}$). Since $\boldsymbol{r}_{k+1}$ is orthogonal to L_k, it is also orthogonal to span($\{\boldsymbol{A}\boldsymbol{p}_0, ..., \boldsymbol{A}\boldsymbol{p}_{k-1}\}$) $\subset L_k$ and $\langle \boldsymbol{r}_{k+1}, \boldsymbol{A}\boldsymbol{p}_j \rangle = 0$ for $j = 0, ..., k-1$. An exercise in induction shows that the sequences of vectors $\{\boldsymbol{r}_k\}$ and $\{\boldsymbol{p}_k\}$ satisfy the relations

$$
\begin{array}{llcll}
\langle \boldsymbol{r}_j, \boldsymbol{r}_k \rangle = 0 & j \neq k & , & \langle \boldsymbol{p}_j, \boldsymbol{A}\boldsymbol{p}_k \rangle = 0 & j \neq k \\
\langle \boldsymbol{p}_k, \boldsymbol{r}_j \rangle = \langle \boldsymbol{p}_k, \boldsymbol{r}_k \rangle & j \leq k & , & \langle \boldsymbol{p}_j, \boldsymbol{r}_k \rangle = 0 & j < k
\end{array} \tag{13.30}
$$

and for the sequence $\{x_k\}$, one obtains the error bound [242]

$$N(y - x_k) = \min_{deg(q) \le k} N(y - q(A)r_0) \le 2 \left(\frac{1 - \sqrt{C}}{1 + \sqrt{C}} \right)^k N(y) \qquad (13.31)$$

with C being the reciprocal of the condition number of the matrix A. This follows because the vector x_k is the minimizer of $N(y - x)$ over L_k which can be rewritten as $\{q(A)r_0 : q \text{ is a polynomial of degree} \le k\}$.

13.10 Krylov Projection Methods

When A is nonsingular, but not necessarily SPD, solving $Ax = b$ can be accomplished in theory by using least-squares. Define the functional f by

$$f(x) = \frac{1}{2}\|Ax - b\|^2 = \frac{1}{2}\langle A^*Ax, x\rangle - \langle A^*b, x\rangle + \frac{1}{2}\langle b, b\rangle \qquad (13.32)$$

where $*$ denotes the adjoint or transpose matrix. This f has the same general form as (13.26). A^*A is indeed symmetric and positive definite, but if A is poorly conditioned, then A^*A will be much more ill-conditioned since $cond(A^*A) = cond(A)^2$. The estimate (13.31) shows that convergence for such a least-squares problem can be quite slow.

Conjugate gradient and biconjugate gradient are special cases of a family of projection methods [72, 73] which are conceptually similar to the Galerkin finite element method for solving $\mathcal{A}u = g$ with $\mathcal{A}$ a differential operator defined on a functional Hilbert space H. A projection method for solving $Ax = b$ takes a proper, m-dimensional subspace K_m of H and seeks an approximate solution x_m which belongs to K_m and has the property that the residual at x_m, $b - Ax_m$, is orthogonal to K_m. That is, the problem becomes: find $x_m \in K_m$ such that

$$\langle v_i, b - Ax_m\rangle = 0 , \qquad i = 0, 1, ..., m - 1 \qquad (13.33)$$

where $\{v_0, v_1, ..., v_{m-1}\}$ is a basis for K_m. Let V_m be the matrix consisting of the column vectors $v_0, ..., v_{m-1}$. If $\xi = A^{-1}b$ denotes the solution in the full space and $\mathcal{P}_m$ the orthogonal projection onto K_m, then (13.33) can be written in terms of $\mathcal{P}_m$ as

$$\mathcal{P}_m x_m = x_m , \quad \mathcal{P}_m(b - Ax_m) = 0 \qquad (13.34)$$

If A is symmetric and positive definite, $x_m = \mathcal{P}_m\xi$, but for the nonsymmetric case this is not necessarily true. Since x_m belongs to K_m there is a linear combination of the $\{v_i\}_{i=0}^{m-1}$ which is x_m, i.e. a vector y_m in $\mathbf{R}^m$ such that $x_m = V_m y_m$. Then $(AV_m y_m - b)^* V_m = \mathbf{0}$ or

$$V_m^* A V_m y_m = V_m^* b \tag{13.35}$$

Keep in mind that H is a high-dimensional space, infinite-dimensional for the differential operator and of dimension perhaps 10^5 when A arises from a 3D discretized diffusion problem. The dimension of K_m is small and in practice m is in the range $5 - 20$; however, if K_m is properly chosen, then $b - Ax_m$ will be close to the zero vector in the space H. Notice that the exact solution ξ in H satisfies $\langle v, b - A\xi \rangle = 0$ for all v in H, not just those in K_m. Also, note that if A is symmetric positive definite, the orthogonality condition (13.33) is equivalent to minimizing a quadratic functional such as f in (13.26) over K_m. Let $A_m = \mathcal{P}_m A|_{K_m}$ be the restriction of A to K_m followed by $\mathcal{P}_m$. An estimate on the approximate residual is

$$\|b - A_m \mathcal{P}_m \xi\| \leq \|\mathcal{P}_m A(I - \mathcal{P}_m)\| \; \|(I - \mathcal{P}_m)\xi\| \tag{13.36}$$

while the solution satisfies the inequality [467]

$$\|\xi - x_m\|^2 \leq \left(1 + \|A_m^{-1}\|^2 \; \|\mathcal{P}_m A(I - \mathcal{P}_m)\|^2\right) \; \|(I - \mathcal{P}_m)\xi\|^2 \tag{13.37}$$

so that if ξ is close to K_m, then x_m will be a good approximation to ξ.

One of the best known choices for K_m was proposed by Arnoldi and it gives rise to the incomplete orthogonalization method. Choose an initial guess x_0; take K_m to be the Krylov subspace generated by applying powers of A to an appropriate initial residual vector, $r_0 = b - Ax_0$. Introduce the variable $z = x - x_0$, so that solving $Ax = b$ is equivalent to solving $Az = r_0$. K_m is then span($\{ r_0, Ar_0, ..., A^{m-1} r_0\}$), a subspace that should contain the largest component of the error vector. Next, build an orthonormal basis $\{v_k\}_{k=0}^{m-1}$, setting $w_0 = r_0, v_0 = w_0/\|w_0\|$ and recursively defining

$$w_k = Av_{k-1} - \sum_{i=0}^{k-1} \langle Av_{k-1}, v_i \rangle v_i, \quad v_k w_k / \|w_k\| \tag{13.38}$$

The $\{v_i\}$s are orthogonal and conjugate in the sense that $\langle Av_i, v_j \rangle = 0$ for $i > j+1$. Define the $m \times m$ upper Hessenberg matrix H by $H_{ij} = \langle Av_i, v_j \rangle$, then $H = V_m^* A V_m$ so that the solution $x_m \equiv z_m + x_0$ is given explicitly by $\|r_0\| V_m^* H^{-1} \, e_1 + x_0$ with $e_1 = (1, 0, ..., 0)$. An approximate solution of the linear system can be computed at the cost of building the orthonormal set $\{v_i\}_{i=0}^{m-1}$: namely, $(m-1)$ calls to compute a matrix-vector product Av

and then $m(m+1)/2$ inner products, and finally computing $\boldsymbol{H}^{-1}$, which can be done cheaply via $\boldsymbol{LU}$ decomposition.

The storage requirements of these Krylov methods are very modest and since only the matrix-vector product $\boldsymbol{Av}$ is needed, vectorization and parallelization are easy. Much of the work in this algorithm results from the orthogonalization of $\boldsymbol{v}_m$ against all the previous vectors $\{\boldsymbol{v}_i\}_{i=0}^{m-1}$. If instead, $\boldsymbol{v}_m$ is only orthogonalized against $\{\boldsymbol{v}_i\}_{i=m-k}^{m-1}$, for a fixed $1 < k < m-1$, one has the incomplete orthogonalization method or IOM. There are many variants of IOM which use slightly different subspaces K_m, including ORTHOMIN, ORTHORES, ORTHODIR, GMRES, and BCGS. They are often termed accelerators in the literature on iterative methods. When the problem is symmetric, IOM reduces to conjugate gradient and in general is equivalent to ORTHORES and ORTHOMIN [468, 575].

The following guidelines have been suggested [415] for selecting an accelerator. If $\boldsymbol{A}$ is symmetric positive definite, then any of the projection methods can be used with conjugate gradient being the simplest. When $\boldsymbol{A}$ is symmetric indefinite, that is some eigenvalue is nonpositive, either IOM or GMRES is appropriate. If $(\boldsymbol{A} + \boldsymbol{A}^*)/2$, the symmetric part of $\boldsymbol{A}$, or its negative is SPD, then both ORTHOMIN and GMRES are guaranteed to converge. The remaining matrices comprise a fourth class, which is the most difficult to handle and represents an area of active research. This group also includes many matrices arising in TCAD simulation, e.g. reaction-diffusion systems, convection-dominated device equations, as well as the nonlinear ODEs solved by SPICE, can all give rise to highly asymmetric matrices. The following methods, listed in order of increasing reliability and decreasing overall efficiency, should be used for these matrices: (1) Lanczos methods (biconjugate gradient) use $\boldsymbol{A}^*$, have a short recurrence, and will converge in n iterations if the process does not "break down"; (2) Biconjugate gradient squared (BCGS). circumvents the need for the transpose matrix-vector product, however, it often exhibits oscillation in the residual norm which decreases, but not necessarily monotonically as desired; (3) ORTHOMIN, ORTHORES, ORTHODIR, and IOM; and (4) normal equation methods which are based on solving $\boldsymbol{A}^*\boldsymbol{Ax} = \boldsymbol{A}^*\boldsymbol{b}$ and are guaranteed to converge in exact arithmetic if $\boldsymbol{A}$ is nonsingular – in practice, convergence is slow and roundoff error may preclude convergence altogether.

13.11 Numerical Experiments in 3D

The first experiments described below are computed using the LSODP package [72]. This variant of LSODE uses basically the same integration and quasi-Newton methods as its predecessor, but in place of the direct solver, a scaled incomplete orthogonalization method (SIOM) is used. A group of three test problems is used to compare the iterative methods. In each case phosphorus diffuses from the surface of a $2 \times 2 \times 5$ μm silicon brick, modeling a predeposition for 10 min at 900°C with a surface concentration of 3.2×10^{20} cm^{-3}. In problem (A) diffusion occurs from the entire upper surface; in (B) from two narrow strips along the edges of the brick with the area between the strips capped by nitride, which prevents any diffusion; and finally, in problem (C) there is only an "L" shaped region through which the dopant can enter (See Figure 13.7). The level of difficulty of the problems increases: results from problem (A) can be compared with 1D simulations, while (C) must be solved fully in 3D for accurate profiles because of the lack of geometric symmetry. A digit 1 or 5 after the problem letter refers to the single-species model or five-species model, respectively.

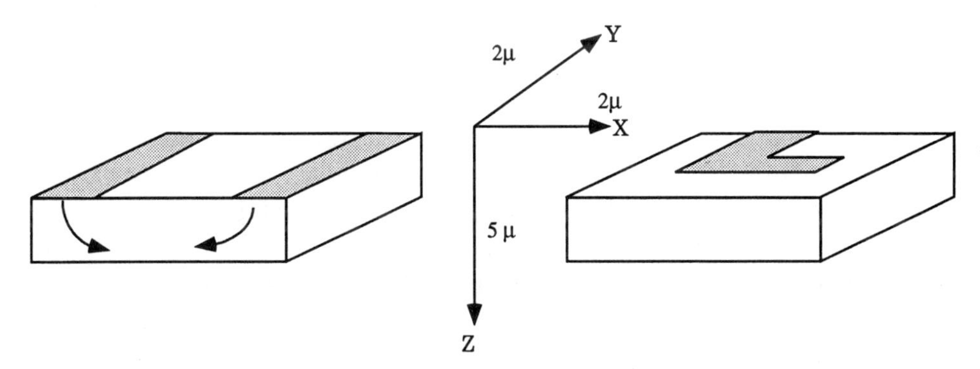

Figure 13.7: Schematic of the geometry for problems B and C. Phosphorus diffuses through the shaded areas into silicon below (from [450]).

Table 13.1 from [450] shows the result of using Adams–Moulton and BDF integrators on problem A1. As expected, BDF performs better as the problem size, and hence the system stiffness, increases. For both methods there is a point of diminishing return in increasing the maximum order. Not only does the overhead increase with order, but the region of stability (see Section 13.4) shrinks as well. Occasionally the user can actually improve performance by requiring the integrator to avoid costly attempts to increase order beyond three. All remaining data was collected using only the BDF

Grid	10^3	20^3	30^3
Adams(1)	53	927	5973
Adams(2)	25	577	3935
Adams(3)	20	523	3795
Adams(4)	22	524	4543
BDF(1)	116	1503	5843
BDF(2)	76	992	4235
BDF(3)	56	683	2970
BDF(4)	52	685	3140

Table 13.1: Problem A1 showing Adams–Moulton versus backward differentiation formula integration methods; values are user CPU seconds on a SPARC station1 (from [450]).

integrator with the maximum order parameter of LSODP set to five. Step size versus elapsed time for problems (A1), (A5), and (C5) is graphed in Figure 13.8(a). For (A1), aggressive increases in the timestep Δt are made throughout the integration; also shown is the order of integration. In the latter two problems, the ill-conditioning of the Jacobian is apparent in the final stages of the diffusion, producing "ringing" in the stepsize. This illustrates that the task of adaptively choosing the timestep and order in a system integrator may properly be regarded as a problem in control theory, and indeed there has been research along these lines.

There are two user-adjustable parameters in the LSODP package [72] which determine the efficiency of the algorithm. The first, m, is the maximum number of vectors in the SIOM algorithm, i.e. the dimension of the Krylov subspace. The second, i_0, is the number of vectors against which $\boldsymbol{v}_m$ is orthogonalized. The default values are $m = 5$ and $i_0 = m$, which have been used for all test problems unless otherwise noted. Increasing either m or i_0 results in more work and storage, but also in greater accuracy. Variations obtained by letting m range from 1 to 7 with $i_0 = m$ and i_0 range from 1 to 4 with $m = 5$, using the single-species model, are given in Table 13.2. Funcs is the number of functional evaluations performed and $Fail_{linear}$ is the number of convergence failures in the innermost loop that iteratively solves the linear system Choosing $m = 1$ results in faster execution on problem (A1), but only at the expense of more failures in the linear solver; as m increases run times increase somewhat with no convergence difficulties. Note that it is usually better to take $i_0 = m$ rather than orthogonalize against just one or two previous vectors. The default value of $m = 5$ proved to be inadequate for the five-species model.

The reaction-diffusion PDEs yield a very stiff system of ODEs, making it harder to solve the linear system at each timestep and effectively requiring a larger Krylov subspace. For the SIOM algorithm in LSODP, the quantity $K_{avg} = \boldsymbol{Avs}/\text{Calls}$ is the average Krylov dimension and represents a measure of the efficiency of the method. Figure 13.8(b) shows K_{avg} for both models on several problems, each with a grid of 10^3 points [450].

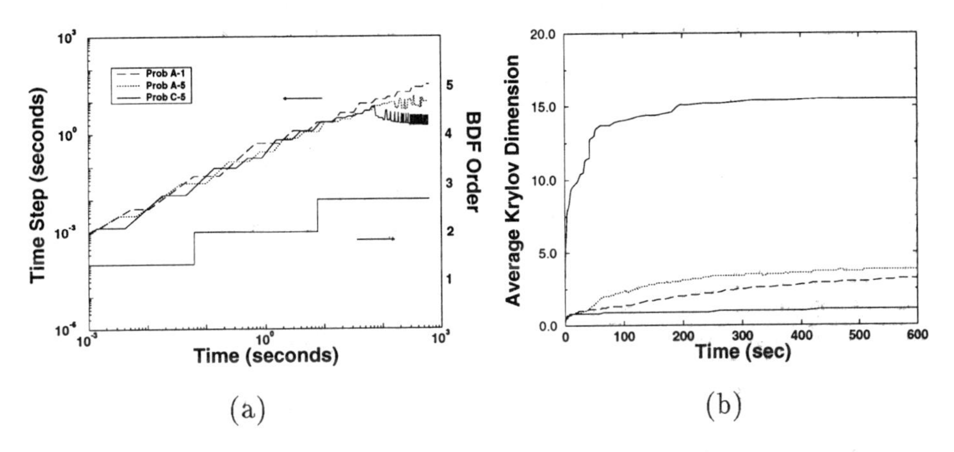

Figure 13.8: (a) Step size versus elapsed time for problems (A1), (A5), (C5) – note the "ringing". (b) Average Krylov dimension m_{avg} as a function of time.

Figure 13.9(a) shows [450] the effects on vacancy and interstitial profiles of varying m from 10 to 20 on problem (A5) with 20 gridpoints in the $z-$direction. Notice that the solutions with $m = 20$ agree well with earlier PEPPER simulations, but as m is decreased the results become poorer, and when $m = 8$ the method fails to converge on a grid of 20^3 points unless the maximum stepsize is severely reduced. Refining the grid does not improve the situation: as shown in Figure 13.9(b) a grid of 80 points is used for the $z-$direction and now m must be 30 before relatively flat defect profiles are achieved. This illustrates the danger that if the Krylov dimension is large enough for convergence, it may still be insufficient to produce a physically correct solution. It is important to solve the linear system accurately at each step and this cannot be done simply by reducing the tolerance in the SIOM algorithm; one must either enlarge the subspace or cut back on the stepsize in the outer loop.

As the geometry becomes more complex with sharp corners and discontinuities, all models have more difficulty with integrating the ODE system. This is to be expected, since multistep methods become more efficient as the

m	i_0	Time	Steps	Funcs	$Fail_{linear}$
1	1	529	75	255	86
2	2	589	72	286	25
3	3	564	71	276	6
4	4	562	71	275	0
6	6	562	71	275	0
7	7	562	71	275	0
5	1	683	75	332	1
5	2	597	72	290	1
5	3	577	72	281	0
5	4	562	71	275	0

Table 13.2: Comparisons of various choices for the dimension, m, of the Krylov subspace and the number of vectors, i_0, against which $\boldsymbol{v}_m$ is orthogonalized for problem (A1) with a $20 \times 20 \times 20$ grid.

solution smoothness (regularity) increases. The error estimates, on which stepsize selection and error control are based, assume a corresponding level of differentiability. For "rougher" problems, it is necessary to refine the grid in space. In the test cases, a nonuniform rectangular mesh is used for convenience, but to handle very irregular geometries, as well as sharp interior layers and steep fronts that develop with time, use of an unstructured, graded finite element discretization is warranted. This would allow for an adaptive grid, which evolves with the solution and permits refinement and deletion of nodes.

The test problems also illustrate that there is, however, a point of diminishing return in mesh refinement. As the mesh increases from 10^4 to 27×10^3 points, runtime increases exponentially and in some cases convergence is not achieved unless the Krylov dimension is increased to more than 20. This can be explained by noting that when the Laplacian operator is discretized, the resulting matrix $\boldsymbol{A}$ has a spectrum that depends on the mesh spacing, and $\boldsymbol{A}$ becomes more ill-conditioned as the mesh is refined. This, of course, carries over to the operators

$$\nabla \cdot (D(C)\nabla C), \; Diag\,(D_V, D_I, D_E, D_F, 0)(\Delta V, \Delta I, \Delta E, \Delta F, 0)^T \quad (13.39)$$

and gives rise to an increasingly stiff system of ODEs. (Recall that the system is stiff if the eigenvalues are widely separated; see the exercises at the end of this chapter.)

There are several preconditioning approaches to alleviate the perfor-

Prob	Method	Time	F(u)'s	Av's	Calls
A	BCGS	132.7	314	244	69
	GMRES	109.3	253	183	69
	ME	135.2	314	244	69
	ORTHOMIN	107.7	253	183	69
B	BCGS	1452.0	3489	2784	704
	GMRES	1087.4	2589	1935	653
	ME	1448.3	3489	2784	704
	ORTHOMIN	1167.3	2793	2088	704
C	BCGS	1898.4	4544	3628	915
	GMRES	1555.7	3661	2739	921
	ME	1899.9	4544	3628	915
	ORTHOMIN	1537.6	3637	2721	915

Table 13.3: Comparison of NSPCG routines on problems (A–C) for the single-species model.

mance degradation that comes with mesh refinement. The first is to use a multigrid strategy, as discussed in Chapter 7. With multigrid preconditioning, it is known that, for the linear diffusion problem the amount of work grows only as a constant times n, the number of nodes. Domain decomposition may be similarly used to develop block preconditioners. Finally, other strategies may be devised to transform the spectrum of $\boldsymbol{A}$ so that SIOM performs better. Preconditioning the linear system has more of an effect on runtime than various improvements to the quasi-Newton method itself. The modular form of the LSODE family of integrators allows other linear solvers to be substituted with little change. A comparison of ORTHOMIN, GMRES, BCGS, and ME on a 10^3 grid for problems (A–C) with the single-species model [449] is given in Table 13.3. Time in the table is in seconds on a SPARC station, "$\boldsymbol{F}(\boldsymbol{u})$'s" is the number of evaluations of the right-hand side, "$\boldsymbol{Av}$'s" is the number of times the matrix-vector product is formed, and "Calls" is the number of calls made to the iterative solver. All four methods performed well, with roughly equivalent runtimes, although GMRES and ORTHOMIN were slightly faster.

Several conclusions can be drawn from these numerical experiments. The resources required to model a diffusion step in 3D clearly depend upon: (*i*) the complexity of the mathematical model used; (*ii*) the complexity of the physical geometry; and (*iii*) the difficulty of solving the linear system in the inner loop. The different models discussed in Chapter 12 incorporate

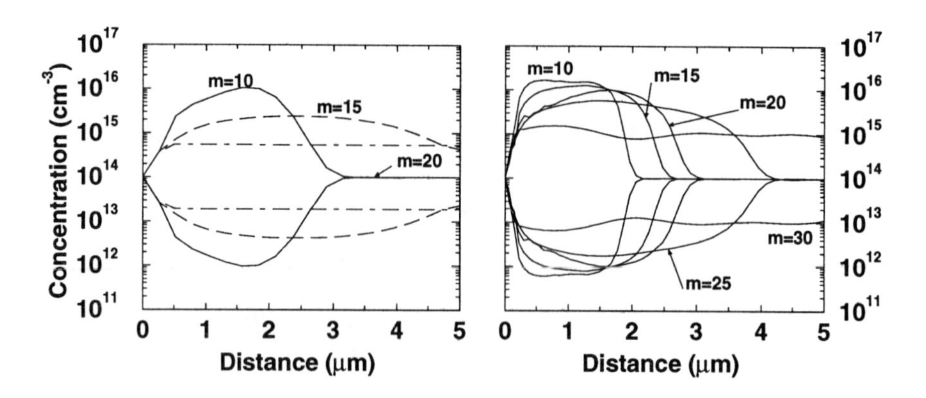

Figure 13.9: The effect on vacancy and interstitial profiles of increasing the Krylov subspace dimension.

varying degrees of what is known about the physics of impurity diffusion. The process engineer must choose a model based on the inevitable accuracy versus expense tradeoff.

13.12 Summary

This chapter has covered those numerical methods most often used to integrate reaction-diffusion partial differential equations, such as those which arise from the multiple-species diffusion models of Chapter 11. A powerful technique for solving these numerically is the numerical method of lines. Continuous-in-time and discrete in space, NMOL converts the PDE into a large coupled system of ODEs by replacing the spatial derivatives using finite difference or finite element discretizations. The resulting ODEs can be integrated using the methods of Chapter 4, but the linear multistep (LMS) methods of Section 13.3 provide a useful generalization that is particularly effective when the user requests a very low error tolerance. A subclass of LMS, backward differentiation formula methods have proven to be very successful for stiff systems.

The most desirable integration techniques are implicit in nature and require the repeated solution of a coupled system of nonlinear algebraic equations. Two approaches have historically been used in this inner loop: successive approximation and Newton's method. The former is a unifying concept, subsuming linear and nonlinear variants of Jacobi and Gauss-Seidel, concept, subsuming linear and nonlinear variants of Jacobi and Gauss-Seidel, waveform relaxation, and even Newton itself. Based on the contraction mapping principle, it generally gives only a linear convergence rate. Because of this fact, it is desirable in most TCAD applications to implement a special fixed-point iteration which enjoys quadratic convergence: Newton's method.

Newton's method requires the solution of the linear system $\boldsymbol{J}\boldsymbol{\delta x} = \boldsymbol{f}(\boldsymbol{x})$ where J is the Jacobian; hence access to very efficient algorithms for solving large ($> 10^5$ unknowns) systems of linear equations. There are three classes of such algorithms: (*i*) direct, i.e. LU decomposition, (*ii*) iterative (based on a fixed-point iteration), and (iii) projective. The best known example of the third class is the conjugate gradient method, applicable when the matrix is symmetric positive definite. Its generalizations, Krylov subspace techniques, are finding increasing use in many areas of TCAD simulation.

13.13 Exercises

1. Write a program which calls a system integrator such as LSODE or GEARB (available from netlib) to solve $u_t = u_{xx} + u(1-u)$ on $[0, L]$ with zero Dirichlet boundary conditions. First use $L = \pi/2$ and show that as $t \to \infty$, the solution decays to zero. Then take $L = 2\pi$ and use your program to locate one of the nontrivial steady-state solutions to the equation.

2. Show that if one starts close enough to a solution ξ of $f(x) = 0$, a modification of Newton's method given by $x_{n+1} = x_n - f'(\xi)^{-1} f(x_n)$ is a contraction.

3. Show that the Newton direction $\boldsymbol{f}'(\boldsymbol{x})^{-1}\boldsymbol{f}(\boldsymbol{x})$ for the least-squares formulation $\min_{\boldsymbol{x}} \phi(\boldsymbol{x}) = 1/2\ \|\boldsymbol{f}(\boldsymbol{x})\|^2$ is a descent direction in the sense that it has a positive component in the direction $-\boldsymbol{\nabla}\phi(\boldsymbol{x})$.

4. Show that if $\{v_j\}$ is a basis (not necessarily orthonormal) of a subspace L, then the closest point to y in span($\{v_j\}$) is given in terms of the Gramian matrix, $M_{ij} = \langle v_i, v_j \rangle$. How is this related to the stiffness matrix used in finite element methods?

5. Derive the inequalities (13.23) which characterize the condition number of a matrix $\boldsymbol{A}$.

6. Consider Poisson's equation on the unit square Ω, $\Delta u = g$ with Dirichlet boundary conditions $u = \phi$ on $\partial\Omega$. Choose g and ϕ so that the exact

solution is $u(x,y) = x^2y^2$. Using the standard 5-point finite difference discretization solve this equation using the Jacobi iteraton. At each step print the quantities $\|u - u_h^{(n)}\|_\infty$ and $\|u_h^{(n+1)} - u_h^{(n)}\|_\infty$. Estimate the rate of convergence. Repeat with the Gauss–Seidel method.

7. Let $\boldsymbol{A}$ be a symmetric matrix of order n, with eigenvalues $\lambda_1 \leq \lambda_2 \leq \cdots \leq \lambda_n$ and define the Rayleigh quotient $R(\boldsymbol{x}) = \langle \boldsymbol{A}\boldsymbol{x}, \boldsymbol{x}\rangle / \langle \boldsymbol{x}, \boldsymbol{x}\rangle$, $\boldsymbol{x} \neq \boldsymbol{0}$. Show that $\max_{\boldsymbol{x}} R(\boldsymbol{x}) = \lambda_n$ and $\min_{\boldsymbol{x}} R(\boldsymbol{x}) = \lambda_1$.

8. Theorem 13.1 asserts the existence of a fixed-point $\boldsymbol{\xi}$. Assuming that in fact $\boldsymbol{\xi}$ does exist, show that starting at any $\boldsymbol{x}_0$ in the set S, $\|\boldsymbol{\xi} - \boldsymbol{g}^n(\boldsymbol{x}_0)\| \to 0$ as $n \to \infty$.

9. Let $\mathcal{A}$ be the operator defined by $\mathcal{A}u = -u''$ acting on the space $C_0^2[0,1]$ of functions satisfying zero Dirichlet boundary conditions. If K is a positive integer show that $\lambda_k = k^2\pi^2$ is an eigenvalue with associated eigenfunction $\sin(k\pi x)$. Fix n and consider the corresponding discrete problem on a finite difference grid of $n-1$ interior points. Show that the eigenvalues of the matrix $n^2\boldsymbol{A}$, where $\boldsymbol{A}$ is defined in (11.48), are given by $\lambda_k = 2(1 - \cos(\frac{\pi k}{n}))$ for $k = 1, ..., n-1$ with associated eigenvector $(u_k)_j = \sin(\pi k \frac{j}{n})$ for $j = 1, ..., n-1$. What happens in the limit as $n \longrightarrow \infty$?

Chapter 14

Specialized Diffusion Topics

14.1 Introduction

This chapter discusses several specialized topics all of which involve diffusion in some form. Low thermal budget technology seeks to anneal damage while minimizing redistribution of impurities. It is particularly attractive for fabricating very shallow junctions and minimizing the redistribution of dopants already incorporated into the lattice during previous processing steps. Polysilicon is used extensively in modern integrated circuits, and either intentionally or as an unwanted side effect it becomes doped with the impurities used to impregnate the crystalline silicon. Poly is composed of many fine silicon grains and the impurity atoms will diffuse not only through these grains, but along grain boundaries as well. To a first approximation, the linear diffusion equation is appropriate for modeling such impurity movement, but a more detailed approach is required to explain the observed pile-up that occurs at the polysilicon–silicon boundary.

There are several process steps during which some diffusion occurs, although this may not be the primary effect. Epitaxy, the growing of crystalline silicon on top of existing structures, is accompanied by diffusion because of the elevated temperatures involved. The theory of diffusion near a moving interface has a long history that includes investigations on the melting of polar ice [508], metal solidification, and many other problems. Diffusion during oxidation is even more delicate, because impurity concentration will have an effect on the rate at which oxidation proceeds, hence on the location of the silicon–oxide boundary. In both instances, the numerical methods used for simulation must allow for the evolution of the grid with insertion/deletion of nodes in the vicinity of the interface.

At one time gallium arsenide was predicted to supplant silicon as the semiconductor of choice for many applications. The development of novel device structures for silicon has enabled designers to reduce feature sizes to the half micron range. If history is a guide, extensions to this technology will work into the deep submicron region. Still, there are some applications for which GaAs will have definite advantages over silicon. These include microwave devices and those which require radiation hardening. Diffusion in III-IV compounds is more complicated than silicon: now there are two sublattices and a much wider assortment of point defects. However, the multiple-species models developed in Chapter 12 readily extend to this situation.

14.2 Rapid Thermal Processing

To achieve the shallowest possible junctions, the process engineer needs to anneal the damage done to the silicon lattice during implantation, while at the same time minimizing diffusion. This is successfully accomplished using rapid thermal annealing (RTA) techniques developed during the past twenty years. With this technology the implanted wafer is subjected to an intense burst of incoherent light of several seconds duration, which elevates the wafer temperature to values found during a furnace anneal. If the aim of RTA, annealing damage rather than redistributing the impurity, were perfectly achieved, there would be no diffusion to model. Invariably some diffusion does occur during RTA and often results in anomalous profiles.

To a first approximation the single-species model of Chapter 11 will handle rapid thermal annealing, if care is taken to ramp the temperature and re-evaluate the diffusivities at each timestep. The underlying assumption, that chemical equilibrium has been reached in many competing reactions, is however, not justified on the time scales involved. For this reason, advanced models that allow for dynamic solutions are more appropriate.

To accurately model an RTA step following an implantation, initial data for the defects and the temperature profile of the wafer during the step are required. Large numbers of point defects, approximately 10^{22} atoms cm^{-3}, are created during implantation and although they diffuse so rapidly that the defect profiles are essentially flat after five or ten seconds, the initial concentration data for the defects can dominate the solution of the reaction-diffusion equations for a 15 second anneal. Initial data for defects [116] can be obtained from Monte Carlo calculations as discussed in Chapter 10, and wafer temperature is measured experimentally or simulated [394] as shown in Figure 14.1. Three different simulations are run for different time regimes of rapid thermal processing. The adiabatic anneal regime, on the order of 100 nanoseconds, is appropriate for pulsed laser annealing. The flux anneal regime, on the order of milliseconds, is appropriate for scanned laser or electron beam annealing. Finally, the isothermal regime, on the order of seconds or more, is appropriate for rapid isothermal annealing as well as conventional furnace annealing. The initial condition is $T(x,0) = 300°\text{K}$, and boundary conditions are specified as $T(0,t) = 1273°\text{K}$ and $\frac{\partial T}{\partial t}(\infty,t) = 0$. After one second, the entire 0.1 cm thick wafer has reached 1273°K

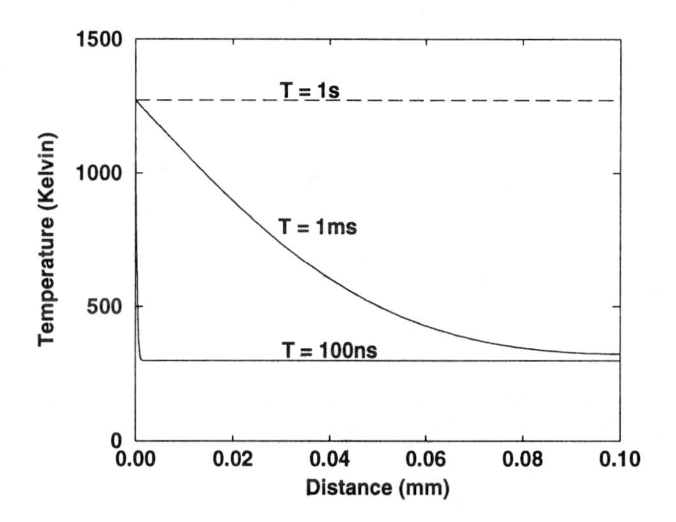

Figure 14.1: Temperature profiles for different rapid thermal processing regimes.

In the case of very short anneals, and particularly laser anneals, it clearly may become necessary to solve explicitly for the temperature of the wafer as a function of space as well as time. Moreover the Dirichlet boundary condition at the surface should be replaced by an ODE specifying dT/dt. The temperature T of the silicon near the surface of the wafer during power absorption (neglecting thermal gradients) obeys [478]

$$C \frac{dT}{dt} = P_a - P_i \tag{14.1}$$

where C is the heat capacity per unit area, P_a is the instantaneous power absorbed, and P_i is the power lost by radiation or convection. P_a is approximated by

$$P_a = \int_0^\infty I_\lambda \, (1 - R_\lambda)(1 - e^{-\alpha_\lambda x}) \, d\lambda \qquad (14.2)$$

where λ is the wavelength, I_λ is the incident emission spectrum from the lamp, R_λ is the silicon surface reflectivity, α_λ is the absorption coefficient of silicon, and x denotes the wafer thickness. To date most work modeling low thermal budget processes has been experimental in nature, rather than numerical. See however [1, 134, 255, 294, 493].

14.3 Diffusion in Polysilicon

Polysilicon is composed of many fine grains of silicon ranging in size from 0.001 μm to 0.5 μm and having random orientation. The region separating two adjacent grains is the grain boundary. An impurity diffuses through the interior of a grain, but also tends to move very rapidly along grain boundaries, giving an effective diffusivity that is much larger than that of crystalline silicon. There is also a strong tendency for the dopant to segregate, preferring the grain boundary over the interior. A first-order approximation treats the diffusivity as constant in the poly, and yields the linear diffusion equation in that region. Although this generally works well, it does not explain the observed "piling-up" of dopant at the silicon-poly interface.

The model [393] used in PEPPER solves for C^g the local concentration of the impurity within the grain, and C^{gb} the local concentration at the grain boundary, via the coupled PDEs

$$\frac{\partial C^g}{\partial t} = \nabla \cdot \left(D \left(\nabla C^g \pm N^g \nabla \ln(n) \right) \right) + \frac{1}{\tau} \left(\frac{C^g}{P_{seg}} - C^{gb} \right) \qquad (14.3)$$

$$\frac{\partial C^{gb}}{\partial t} = D^{gb} \Delta C^{gb} - \frac{1}{\tau} \left(\frac{C^g}{P_{seg}} - C^{gb} \right) \qquad (14.4)$$

where D^{gb} is the diffusivity along the grain boundary, P_{seg} is the segregation coefficient, and τ is a "relaxation rate" constant that represents the rate of segregation. Note that the last term in each equation serves to couple the two species and has the general form of the transport boundary condition discussed in Section 11.6. Since the average poly grain size $L(t)$ generally grows as $O(\sqrt{t})$ during the diffusion and $P_{seg} \propto \frac{1}{L(t)}$, the segregation terms in (14.3) – (14.4) are time-dependent and P_{seg} must be recalculated at each new timestep. Mathematically the system is similar to the multi-material diffusion that occurs in polymers and also diffusion through a homogeneous medium with cracks or fissures. A more rigorous analysis would include homogenization techniques. Boundary conditions for C^{gb} and C^{g}, are zero flux and the interface transport condition of (11.26), respectively. The silicon bulk is treated as a very large grain of polysilicon and a thin interface material is added between the poly and bulk. At the interface material–poly boundary the model enforces the conditions

$$C^{g} = C^{gb} \ , \quad \nabla C^{gb} \cdot \boldsymbol{n} = 0 \tag{14.5}$$

where $\boldsymbol{n}$ is the outward unit normal to the boundary, while at the interface material–silicon boundary $C = C^{g}$. The thin layer is then removed at the end of the diffusion.

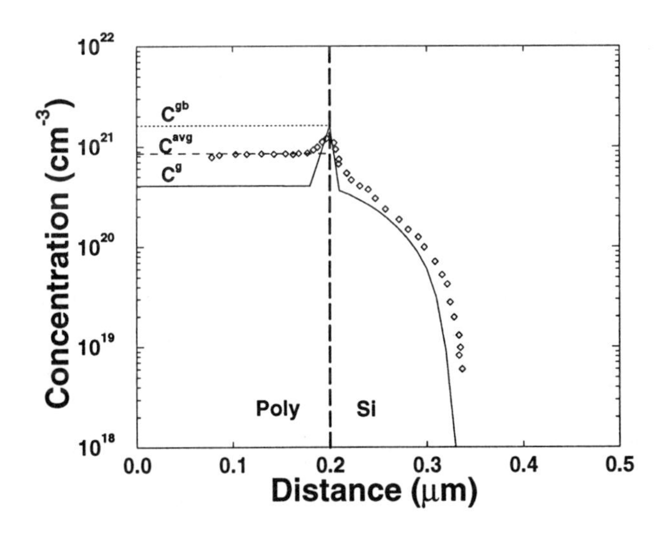

Figure 14.2: "Piling-up" of arsenic at a polysilicon–silicon interface.

Figure 14.2 shows a PEPPER simulation [393] of the diffusion of implanted arsenic from a polysilicon layer into the silicon bulk, a process used to form a high-efficiency emitter in a bipolar technology. The arsenic is diffused at 950°C for 30 minutes, following an implantation (2×10^{16} cm^{-2} at 100 keV) into a 0.2 μm poly layer over silicon. Plotted is C^{gb}, C^{g}, and the weighted average of these two, along with SIMS data from [438]. C^{gb} diffuses rapidly across the poly and a peak is produced at the interface, because C^{gb}

is larger than C^g due to segregation. This provides a high concentration diffusion source for the silicon layer and results in a deeper junction than otherwise expected. For other approaches to modeling impurity diffusion in poly see [272, 344, 345].

14.4 Impurity Diffusion During Oxidation

As a first-order approximation for a drive-in step with an oxide layer capping the silicon, it is assumed that the flux of impurity into the oxide is zero. This is reasonable since most impurities diffuse much slower in SiO_2 than in silicon and the oxide layer will effectively seal the impurities into the silicon. The actual situation is more complex and the resulting profile is governed by several factors, including:

(1) the difference in diffusivities of the impurity in each of the materials;

(2) the tendency for most impurities to segregate preferentially in one of the two materials;

(3) movement of the SiO_2 interface into the silicon during the oxidation;

(4) volume expansion of the silicon as it is oxidized.

Figure 14.3 shows SSUPREM4 simulations of four different cases of impurity redistribution during thermal oxidation in the 1D case. In the top two plots the segregation coefficient m is greater than 1 and the oxide rejects the impurity. In the lower plots boron ($m < 1$) is taken up into the oxide. In each case a discontinuity at the interface results, but the exact shape and magnitude depends on the rate at which the oxidation proceeds. Such discontinuities occur naturally in many physical processes and are easily handled by a finite element numerical formulation.

To model diffusion properly during oxidation the diffusion equation, or system of equations in the case of the advanced models, is solved simultaneously with the creeping-flow Navier–Stokes equation discussed in Chapter 15. This is a difficult problem, sometimes requiring an order of magnitude increase in computer resources. In some cases the processes can be decoupled by performing the oxidation and then the diffusion, with acceptable results. This is not generally possible because the oxidation rate, through the stress and pressure dependent reaction rates, is dependent upon impurity concentration in the silicon. Diffusion, in turn, is strongly influenced by thermal oxidation because copious amounts of interstitials are injected into

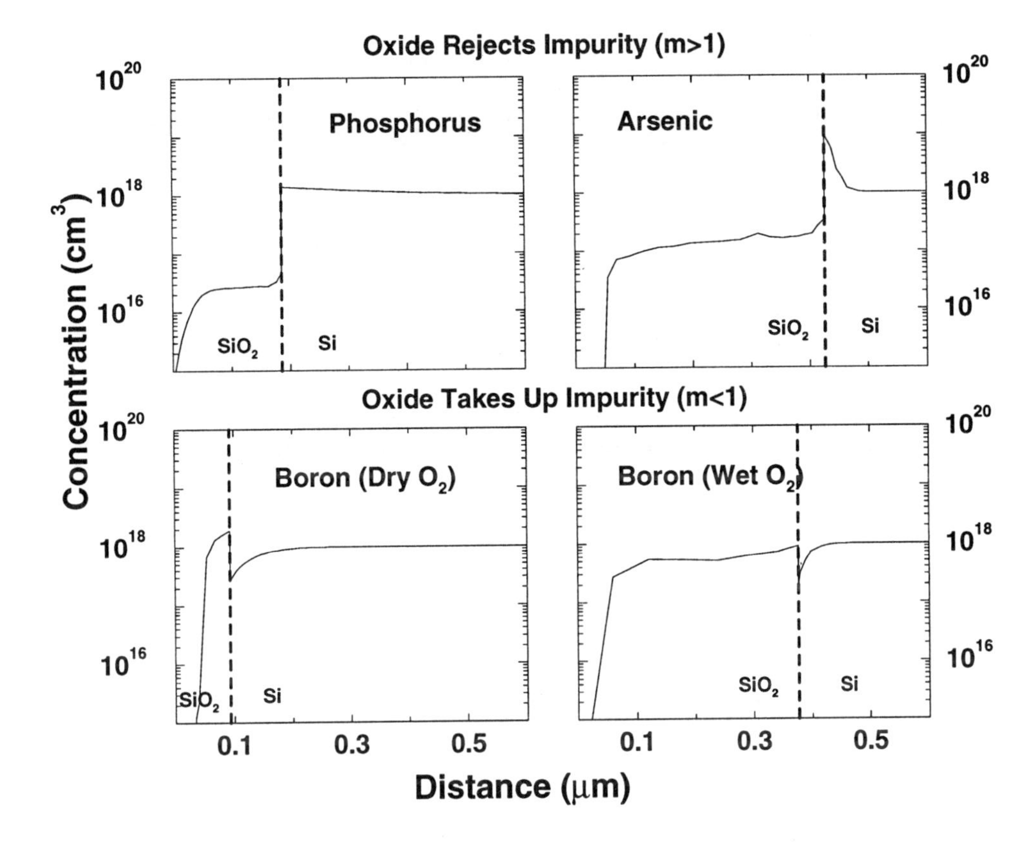

Figure 14.3: Simulated impurity redistribution in Si during thermal oxidation.

the silicon, resulting in oxidation enhanced or oxidation retarded diffusion (OED/ORD). The standard flux boundary condition for interstitials must be replaced by

$$D_I \nabla C_I = K_{si}(C_I \; - \; C_I^* f_0) \tag{14.6}$$

where K_{si} is the interface recombination velocity for interstitials and f_0 is an interstitial injection factor given by [146]

$$f_0 = 1 + \frac{k_1 dx_o/dt}{(\rho - 1)^{1/2}(\rho - 1 + k_3)} \tag{14.7}$$

with $\rho = \sqrt{1 + k_2 dx_o/dt}$, where dx_o/dt is the rate of growth of the oxide layer and the k's are temperature-dependent constants. For more details on the kinetics of diffusion during oxidation see [10, 11, 353, 352].

14.5 Diffusion During Epitaxy

Epitaxy consists of growing a thin layer of silicon on top of the existing structure at some point during the processing of the wafer. Silane gas, either in pure form or mixed with one of the common dopants, is introduced into the furnace and silicon leaves the gas to bond to the surface in crystalline form. Although epitaxy is an older technology than ion implantation, it still is extremely useful for producing silicon layers with a prescribed distribution of impurities. The profile resulting from an epitaxy step will depend on processing times and temperatures, diffusivities of the various impurities, evaporation and segregation rates, and the rate at which the layer is grown. A special phenomenon, autodoping, occurs when lightly doped films are grown on top of a heavily doped substrate.

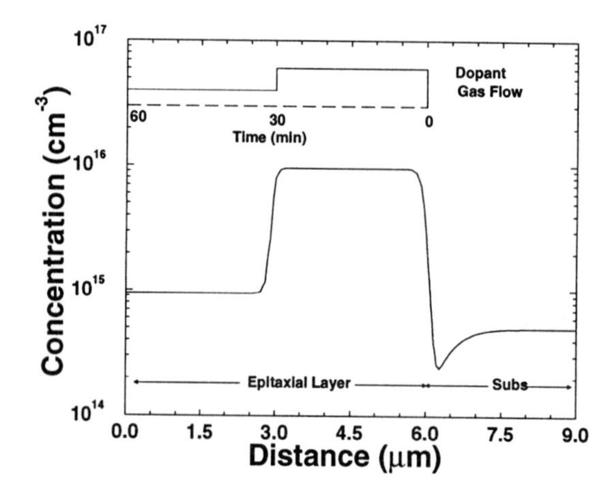

Figure 14.4: Simulated boron profile resulting from a step change in gas flow during epitaxy.

Mathematically, epitaxy is particularly interesting because of the moving boundary. Such problems have a long history, beginning with work by Stefan [508] on the temperature distribution near melting ploar ice. Modeling diffusion during epitaxy is not as difficult as during oxidation, in as much as the location of the boundary is known and does not have to be solved for as the simulation proceeds. Silicon is assumed to be deposited at a constant rate so that growth occurs uniformly in a direction normal to the wafer surface. A more accurate numerical model of epitaxy requires consideration of surface chemistry, gas dynamics, and boundary layer flow. At that level of complexity, it can be considered as a special case of chemical vapor deposition [267].

Figure 14.4 shows results from a SSUPREM4 simulation of a decreasing step change in gas flow during epitaxy. A wafer doped uniformly with boron (5×10^{14} cm^{-3}) was diffused at 1200°C for 30 minutes. The temperature was reduced to 1000°C and an epitaxial layer grown at a rate of 0.1 μm per min with impurity concentration of 10^{16} cm^{-3} for 30 min, followed by another 30 minutes at 10^{15} cm^{-3}. Note the outdiffusion at the original interface that occurs before the heavily doped layer is established.

As with oxidation, the problem is greatly simplified in one dimension and there is an analytic solution [99, 224]. The diffusion equation

$$\frac{\partial C}{\partial t} = \frac{\partial}{\partial x}\left(D(C)\frac{\partial C}{\partial x}\right) , \qquad -\infty < x < x_f , \qquad t > 0 \tag{14.8}$$

is solved with $x = 0$ being the metallurgical interface between the substrate and the epitaxial film ($x < 0$ corresponds to the substrate and $x > 0$ to the film). The interface location is given by $x_f(t) = Vt$ where V is the film growth rate. The substrate has a uniform initial doping, $C(x,0) = C_s$ for $x < 0$, and boundary conditions are taken to be

$$C(x,t) \to C_s \text{ as } x \to -\infty , \qquad -D\frac{\partial C}{\partial x}\bigg|_{x_f} = (h+V)C \tag{14.9}$$

where h is a transport coefficient as defined in Section 11.6. The second boundary condition in (14.9) implies that the material diffusing to the surface either escapes into the ambient gas or is incorporated into the growing film. This can be derived rigorously by considering the total amount of impurity in the solid at time t, given by $Q(t) = \int_{-\infty}^{x_f} C(x,t)dx$. The rate of escape of impurities from the solid into the ambient gas, $-\frac{dQ}{dt}$, is found to be

$$hC(x_f,t) = -\int_{-\infty}^{x_f} \frac{\partial C}{\partial t}(x,t)dx - VC(x_f,t) \tag{14.10}$$

$$= -\int_{-\infty}^{x_f} \frac{\partial}{\partial x}\left(D(C)\frac{\partial C}{\partial x}\right) dx - VC(x_f,t) \tag{14.11}$$

where Leibniz's rule has been applied.

If $D(C)$ is a constant, (14.8) and (14.9) can be solved analytically using the coordinate transformation $(x, t) \to (Vt - x, t)$ and one obtains

$$\frac{C(x,t)}{C_s} = \frac{1}{2}\text{erfc}\left(\frac{x}{2\sqrt{Dt}}\right) - \frac{h+v}{2h}\exp\left(\frac{V}{D}(Vt-x)\right)\text{erfc}\left(\frac{2Vt-x}{2\sqrt{Dt)}}\right)$$

$$+\frac{V+2h}{2h}\exp\left(\frac{V+h}{D}[(V+h)t-x]\right)\text{erfc}\left(\frac{2(V+h)t-x}{2\sqrt{Dt}}\right) \quad (14.12)$$

For two or more dimensions with complicated surface geometries, the moving interface may be tracked in time and gridpoints added as needed. An alternative approach is to work in the transformed coordinate system with equations that now have drift terms, and, if the growth rates are high enough, that require upwinding to prevent numerical oscillations.

14.6 Gallium-Arsenide Diffusion Models

In the quest for faster devices, there is considerable interest in semiconductors made of III-V materials, such as gallium arsenide. Certainly the transistors of choice for most of the microwave band, GaAs devices have the additional advantages of high temperature operation and tolerance for high levels of radiation. The processing of a GaAs transistor is similar to that of Si, with several notable exceptions. The theory of ion implantation carries over without change. It is very difficult to form oxides on GaAs, but the compound's good insulating qualities obviate the need for oxides and provide still another advantage over silicon. It is in modeling diffusion using the single-species model that the greatest differences between the two semiconductors are observed. The common dopants for GaAs — Zn, Mn, S, Si, and Al — exhibit a bewildering array of complicated profiles including steep fronts, kinks, tails, and concave spikes near the surface, that preclude a simple complementary error function fit. Figure 14.5(a) shows experimental diffusion profiles [578] of sulfur in GaAs at 1120°C and an arsenic pressure of 0.25 atm for one hour. Curves 1–3 correspond to surface concentrations of 6.0×10^{19}, 6.2×10^{18}, 1.1×10^{18}, respectively. A simulation of silicon diffusion in gallium arsenide (1050°C for minutes) using the Greiner–Gibbons model [221] is given in Figure 14.5(b); triangles represent SIMS data.

The multiple-species models, however, extend conceptually to GaAs and are complicated only by the presence of two sublattices and the large number of possible species. To illustrate this, the early work of Jordan and Nikolakopoulou [274] on manganese diffusion in GaAs is reviewed here and several proposed mechanisms for Si diffusion in GaAs are discussed, giving a flavor of the extensive work in this field. For further information see the articles [6, 16, 122, 219, 220, 221, 273, 532, 533, 534, 573]. The edited texts [157, 395] provide summaries of GaAs IC technology with applications to high-speed, military, and radiation-hardened devices.

Currently, ion implantation and molecular beam epitaxy (MBE) techniques have been used to fabricate GaAs Field Effect Transistors (FETs) and monolithic ICs with very thin, precisely doped layers on a semi-insulating GaAs substrate. Among impurities there is considerable interest in manganese, which is usually present as an undesirable contaminant in GaAs or as a p-type dopant. The first continuous wave GaAs–$Al_xGa_{1-x}As$ double heterostructure lasers grown by molecular beam epitaxy (MBE) and operating at room temperature contained a p-type layer that was Mn-doped. Because Mn exhibits a marked tendency to redistribute during high-temperature processing, it is important to be able to accurately simulate its diffusion.

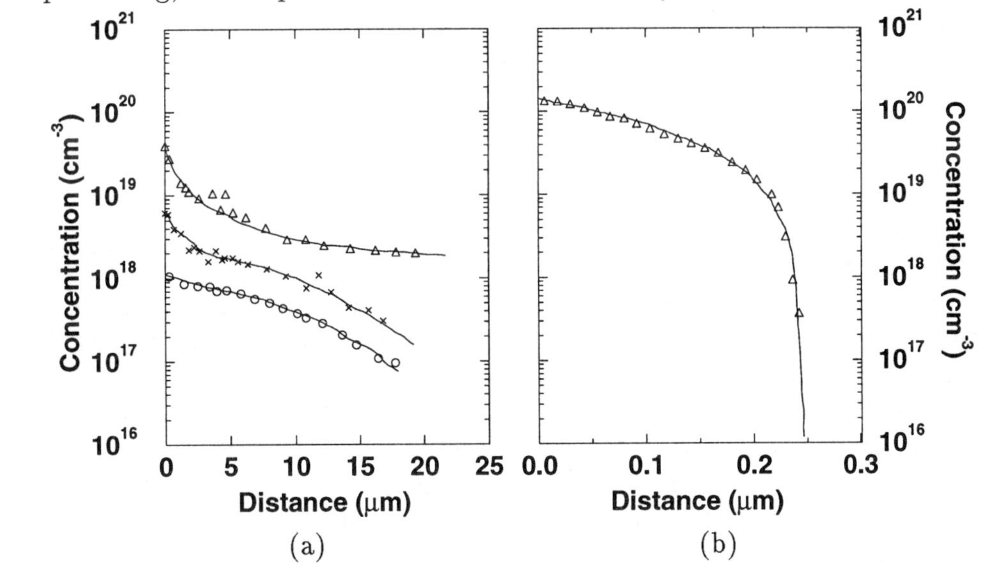

Figure 14.5: Diffusion of sulphur and silicon in gallium arsenide.

It is generally thought that direct diffusion of substitutional Mn is not thermodynamically favorable. Jordan and Nikolakopoulou [274] invoke the Frank–Turnbull interstitial mechanism that was originally proposed to explain the diffusion of Cu in Ge. When a Mn acceptor occupying a Ga site moves, it leaves behind a gallium vacancy and enters an interstitial site as Mn_i^+ via the reaction

$$Mn_{Ga}^- + 2e^+ \rightleftharpoons V_{Ga} + Mn_i^+ \tag{14.13}$$

Recent work suggests that equilibrium in (14.13) is achieved during the diffusion, which is therefore kinetically controlled. This associative–dissociative diffusion has been discussed previously in the case of phosphorus in silicon, and gives rise to three equations (charges have been neglected and [ ] denotes concentration)

$$\begin{aligned}
\frac{\partial[Mn_{Ga}]}{\partial t} &= k_r[V_{Ga}][Mn_i] - k_f[Mn_{Ga}] \\
\frac{\partial[Mn_i]}{\partial t} &= D_{Mn_i}\Delta[Mn_i] - k_r[V_{Ga}][Mn_i] - k_f[Mn_{Ga}] \qquad (14.14) \\
\frac{\partial[V_{Ga}]}{\partial t} &= D_{V_{Ga}}\Delta[V_{Ga}] - k_r[V_{Ga}][Mn_i] - k_f[Mn_{Ga}]
\end{aligned}$$

where k_f and k_r are forward and reverse rate constants, respectively. Jordan used numerical methods to solve this system and obtained good agreement with experimental results. In addition to (14.13), an alternative path by which vacancies are immobilized was considered:

$$V_{Ga} + D_{As} \rightleftharpoons V_{Ga} : D_{As} \qquad (14.15)$$

and this results in a fourth PDE being coupled to the system (14.14). Here D_{As} represents a donor such as S or Se and $V_{Ga} : D_{As}$ is the complex species. Figure 14.6 shows output [520] from PEPPER, which was used to reproduce Jordan's results, as described below.

Column IV dopants such as silicon are amphoteric in a III-V semiconductor. That is, they are able to reside on either sublattice. Silicon forms a donor when occupying a gallium site and an acceptor when on an arsenic site. Two silicon atoms on nearest-neighbor donor–acceptor sites form a neutral pair. In the Greiner–Gibbons model [220, 221] it was assumed that the pairs are the mobile diffusing species, in much the same way that the E- and F-centers are mobile during phosphorus diffusion in silicon. Consider the reaction

$$Si_{Ga}^{+} + Si_{As}^{-} \rightleftharpoons Si_{Ga}^{+} : Si_{As}^{-} \qquad (14.16)$$

where the pairs are assumed to be in equilibrium with donors and acceptors, i.e. $[Si_{Ga} : Si_{As}] = k_p[Si_{Ga}^{+}][Si_{As}^{-}]$, and the dopant concentration is so high that the semiconductor is heavily compensated, $[Si_{Ga}^{+}] = [Si_{As}^{-}]$. The flux of total silicon concentration can then be expressed in the form

$$-D_{eff}\nabla C_T = -D_p\left(1 - \left(1 + \frac{2C_T}{k_p}\right)^{-1/2}\right)\nabla C_T \qquad (14.17)$$

where D_{eff} is a concentration-dependent diffusion coefficient and C_T is the total silicon concentration. Note that the neutral pair can move substitutionally by pairing with either a gallium or arsenic vacancy via the reaction

$$Si_{Ga}^{+} : Si_{As}^{-} + V_{As}^{-} \rightleftharpoons V_{As}^{-} + Si_{As}^{-} : Si_{Ga}^{+} \tag{14.18}$$

with a similar reaction for gallium vacancies. Although exchanges with both types of vacancies are possible, experiments have suggested that the arsenic vacancy concentration exceeds that of gallium vacancies by an order of magnitude and therefore (14.18) should be the rate-limiting step. This model is able to explain the steep diffusion front that is experimentally observed, but requires that silicon be present at high concentrations so that the semiconductor is heavily compensated.

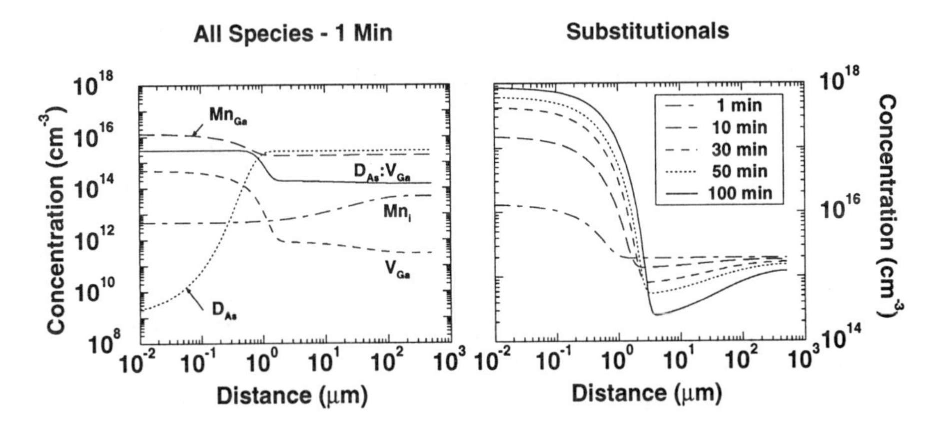

Figure 14.6: PEPPER simulation of substitutional concentration as a function of depth from the surface for a manganese diffusion in gallium arsenide, using the model of Jordan and Nikolokopoulou (from [520]).

Deppe *et al.* [136, 135] performed studies of silicon diffusion in GaAs containing varying amounts of zinc. As the background zinc level is increased, they found that silicon diffusion increases and proposed a second mechanism based on the neutral $Si_{Ga}^{+}:V_{Ga}^{-}$ pair. Later studies showed that as the background doping in GaAs is changed from p- to n-type, the silicon diffuses deeper into the crystal for otherwise identical annealing conditions. This is consistent with the $Si_{Ga}^{+}:V_{Ga}^{-}$ mechanism and the known fact that at low silicon concentrations, silicon atoms preferentially occupy the gallium sites.

At higher concentrations, silicon atoms also occupy arsenic sites and can form mobile pairs with arsenic vacancies. Tatti and Mulvaney [520] proposed a model in which all three mechanisms play an important role:

$$\begin{aligned} Si_{Ga}^{+} + Si_{As}^{-} &\rightleftharpoons Si_{Ga}^{+} : Si_{As}^{-} \\ Si_{Ga}^{+} + V_{Ga}^{-} &\rightleftharpoons Si_{Ga}^{+} : V_{Ga}^{-} \\ Si_{As}^{-} + V_{As}^{+} &\rightleftharpoons Si_{As}^{-} : V_{As}^{+} \end{aligned} \tag{14.19}$$

If conditions of equilibrium exist, a concentration-dependent diffusivity can be derived that will give the observed silicon profile. In many cases the assumption of equilibrium is not justified and then the chemical reactions should be used to derive a system of reaction-diffusion PDEs, as in the kinetic model for phosphorus diffusion in silicon.

14.7 Rapid Prototyping of Diffusion Models

As the reader has observed in this overview of GaAs diffusion, there are many competing theories which attempt to explain diffusion in III-IV compounds. Because of this fact it is important that the process engineer be able to change not only physical parameters for existing models, but also enter entirely new systems of PDEs in order to test proposed diffusion mechanisms. The PDE solver in the proprietary simulator PEPPER [394] was built around the dial-an-operator concept [206]. Since this feature of the simulator foreshadowed current work on object-oriented codes, it is described in some detail. The PDE solver in PEPPER contains a library of C functions representing common spatial operators including

$$\frac{\partial^2 f}{\partial x^2}, \frac{\partial f}{\partial x}, \frac{\partial}{\partial x}\left(H(f)\frac{\partial f}{\partial x}\right), k \cdot H(f) \cdot f, k \cdot H(f) \cdot f \cdot g$$

Here, f and g are dependent variables, and k is a constant. The functions $H(f)$ are defined for the particular problem; for process simulation these include nonlinear diffusion coefficients, the electric field , surface coefficients, and so forth. Using this library of operators, the process engineer constructs the system of PDEs to be solved by specifying the operators on the diffusion command line as the input to the program.

Consider the five reaction-diffusion PDEs modeling manganese diffusion in gallium arsenide using the two reactions (14.13) and (14.15). They are represented on the PEPPER input file as shown in Table 14.1. The operator `dfdt` refers to the time derivative, `d2fdx2` the second spatial derivative,

```
title 5 SPECIES KINETIC MODEL FOR Mn REDISTRIBUTION
init silicon th=500 dx=1e-5 xdx=1e-4 nodes=480
+ u0=1 u1=2e15 u2=1e14 u3=3e15 u4=1.
diff time=100 temp=750 print special atol=1e6 rtol=0.05
+ pde=dfdt(0,0,1.)=d2fdx2(0,0,6e3,afun15)
+ pde=k1fg_k2h(0,0,2,1,-6e-12,-6e-2)
+ pde=dfdt(1,1,1.)=d2fdx2(1,1,6e-10,afun15)
+ pde=k1fg_k2h(1,0,2,1,6e-12,6e-2)
+ pde=dfdt(2,2,1.)=d2fdx2(2,2,4.2,afun15)
+ pde=k1fg_k2h(2,0,2,1,-6e-12,-6e-2)
+ pde=k1fg_k2h(2,2,3,4,-3.6e-14,0.)
+ pde=dfdt(3,3,1.)=d2fdx2(3,3,6e-4,afun15)
+ pde=k1fg_k2h(3,2,3,4,-3.6e-14,0.)
+ pde=dfdt(4,4,1.)=d2fdx2(4,4,6e-6,afun15)
+ pde=k1fg_k2h(4,2,3,4,3.6e-14,0.)
+ pde=dirich(r,0,2,10000,5e14)
```

Table 14.1: Rapid prototyping using the dial-an-operator concept.

and `k1fg_k2h` the forward and reverse terms which arise from a first-order chemical reaction. The arguments of these operators identify the partial differential equation $(0, 1, ..., N-1)$ and unknowns $(0, 1, ..., N-1)$ to which the operator refers, and give a constant by which the term is multiplied. The nonlinear coefficients such as diffusivities are represented by the auxiliary functions `afun0,...` Thus the system of PDEs can be readily altered without extensive reprogramming (except perhaps coding a new auxiliary function).

As described in these chapters, this approach has allowed rapid prototyping of several new models. The system integrator LSODI is used with the BDF option which automatically selects both the stepsize and order, as discussed in Chapter 13. A difference quotient approximation to the Jacobian and LU decomposition routines from LINPACK yield a robust quasi-Newton method, with the only intervention on the part of the user being the specification of the operators as shown above. Once a new model has been evaluated and tested using this dial-an-operator approach, it is hardwired into PEPPER as a default model with a simple keyword representing the whole set of equations. This approach proved to be more convenient than other general-purpose process codes that required the user to supply routines for computing the residual and an analytic Jacobian.

14.8 Summary

This chapter has covered IC fabrication steps in which diffusion occurs simultaneously with other processes. One important example is diffusion during movement of a material boundary, as occurs during epitaxy or oxidation. The former is easily handled in 1D by a coordinate transformation to a fixed domain, and this approach has even been used with some success for the oxidation problem in one or two dimensions. In general, however, detailed modeling of impurity diffusion during these processes requires the solution of a system of PDEs which describe the transport, reaction, and diffusion of multiple species, and can be solved numerically using the methods of Chapters 7 and 12–16.

Rapid thermal processing seeks to anneal lattice damage with the application of a burst of incoherent light of a few seconds duration, thereby minimizing diffusion. Inevitably, some impurity redistribution does occur and is best modeled using the advanced multiple-species models discussed in Chapter 12. These models do not assume that equilibrium has been reached in various competing reactions and hence are appropriate for the very short time scales of RTA (typically tens of seconds). However, particular care must be taken to accurately determine initial conditions for the various species, in particular for point defects.

The multiple-species models are also essential for detailed modeling of diffusion in gallium arsenide. In this case there are two sublattices, and therefore a much richer variety of chemical species than are usually present in silicon. The numerical methods discussed in Chapter 13 are extremely effective for implementing the reaction-diffusion systems required for these models. Rapid prototyping and testing of new diffusion models is enhanced if some form of object-oriented programming is incorporated into the software, thereby allowing the user to modify these PDEs in a relatively painless fashion.

14.9 Exercises

1. Show that in 1D the change of coordinates $y = x - Vt$ transforms the linear diffusion equation with a moving interface into an equation with a fixed boundary, but with an added drift term. (See Chapters 15 and 16 for higher dimensional analogues in which the determination of boundary position is part of the solution process.)

2. Use a moving frame $y = x - ct$ to solve the heat equation with con-

vection, $u_t + cu_x - Du_{xx} = 0$ for x in **R** with $u(x,0) = f(x)$ where $b > 0$.

3. Solving $u_t = -cu_x$ with $c > 0$ numerically using finite differences, presents three possibilities for discretizing the spatial derivative, the forward, central, and backward difference,

$$(i)\ \frac{u_{i+1}^n - u_i^n}{\Delta x}, \qquad (ii)\ \frac{u_{i+1}^n - u_{i-1}^n}{2\Delta x}, \qquad (iii)\ \frac{u_i^n - u_{i-1}^n}{\Delta x}$$

   Show that the central difference is more accurate, $O((\Delta x)^2)$.

4. Let E denote the forward shift operator on a vector $\{v_j\}$, $(Ev)_j = v_{j+1}$. Using a forward difference in time, show that when using (i) from the previous exercise the approximate solution on $[0, T]$ after n timesteps is $((1 + \lambda c) - \lambda cE)^n\, u(x,0)$ where $\lambda = \Delta t/\Delta x$ and $\Delta t = T/n$. Use the binomial theorem on this expression and the corresponding formulas for (ii) and (iii) to show that instability can result unless the backward difference is used.

5. Using the three reactions of (14.19), derive a PDE system which describes the diffusion of Si in gallium arsenide. Reasoning on physical grounds, what are the relative diffusivities of the various species? Is the mathematical system likely to be stiff? Solve the PDE system in 1D using NMOL with a system integrator such as LSODE.

6. The term *homogenization* describes various physical and mathematical methods for modeling macroscopic diffusion and other processes in heterogeneous materials containing numerous fissures or voids. Investigate the applicability of these ideas to modeling diffusion in polysilicon?

Chapter 15

Silicon Oxidation

15.1 Introduction

Since silicon dioxide is an excellent insulator, the process of making a transistor involves the formation of an oxide layer on a silicon substrate. As an example, trench oxidation has received considerable attention in recent years because of its usefulness for storing charge in high-density memory devices and for isolating a device electrically from its neighbors. A trench is first formed in silicon by chemically etching the substrate as shown in Figure 15.1(a). The silicon surface is then exposed to an oxidizing gas such as steam or dry oxygen. The film that results after oxidation is the dielectric component of a capacitor that can be charged or discharged, thereby forming a solid-state memory device. The geometry of the trench permits a large film surface area with small trench width. Therefore, many capacitors can be fabricated on a small silicon chip, leading to a high-density device. Typical dimensions of trench capacitors are from one two μms wide and 10 μms deep. There are simpler geometric structures besides the trench that are also of interest. A well-known example is the "birds beak" configuration in Figure 15.1(b). A mask covers part of the silicon and the exposed surface is in contact with the oxidizing gas. As oxidation occurs the layer grows and deforms to the shape shown.

The shape and structure of the silicon dioxide layer significantly influences device processing and the resulting electrical characteristics. For instance, when silicon is doped to form transistor junctions, the presence of an oxide layer influences the thermal diffusion of p- and n-type impurities, which tend to segregate at the silicon oxide interface. These surface effects become more important as devices are fabricated on smaller scales and more

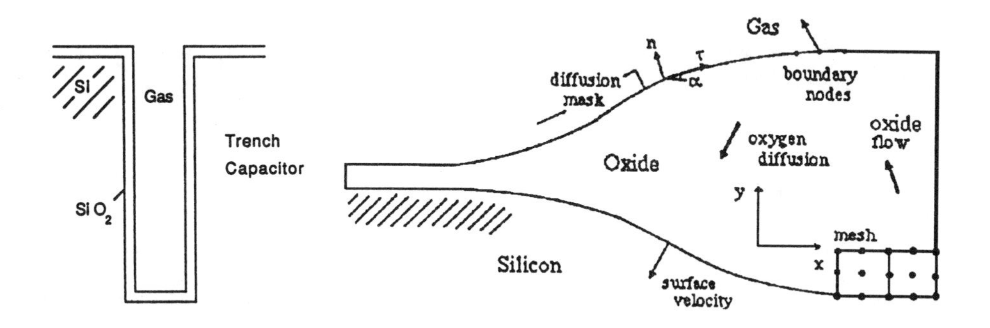

Figure 15.1: (a) Geometry of the trench capacitor. (b) Silicon oxidation and the "bird's beak" configuration.

of the processing occurs at or near interfaces. As a consequence, it is critical to model the oxidation process accurately for a wide range of conditions such as temperature, pressure and oxidation time. In particular, oxide film stress can produce a thinning effect at the trench corner, which in turn can lead to charge leakage in the capacitor.

Consider the basic oxidation process indicated in Figure 15.1(b). Initially, a thin pad of oxide a few hundred angstroms thick is formed on a uniform silicon substrate. Then a layer of silicon nitride is deposited on the oxide pad to act as a diffusion mask. A portion of the silicon nitride is etched to expose a region in which further oxidation will occur. An oxidizing gas at high temperature is brought into contact with the etched region and reaction occurs, producing a thin oxide film. Oxygen continues to diffuse through the oxide layer to react at the underlying silicon surface. The oxide produced by the reaction occupies a greater volume than the original silicon, so the oxide layer deforms during the process. As the reaction proceeds, lateral oxidation occurs under the mask and the oxide tapers off to a negligible thickness sufficiently far from the mask opening. The silicon nitride is deformable and lifts as a result of oxide expansion, giving local oxidation its characteristic, two-dimensional "bird's beak" geometry.

The reaction rate at the silicon surface is obtained from chemical kinetics, and the surface velocities are obtained from a mass balance which relates the reaction rate to the conversion of silicon to oxide. Approximately 44% of the oxide formed is accounted for by motion of the retreating silicon surface as reaction occurs. The remaining 56% of the oxide volume is due to expansion that occurs during the reaction, since the ratio of Si to SiO_2 by volume is

4:9. This net volume increase from silicon to silicon dioxide causes the bulk oxide to flow towards the gas surface leading to the geometrical shape shown in Figure 15.1(b).

The basic oxidation problem involves several complex, coupled processes: diffusion of oxidant through the silicon dioxide; reaction at the silicon interface; volumetric expansion due to the reaction; flow of the resulting oxide layer; and movement of the free and reacting surfaces. The behavior of the oxide layer depends strongly on the processing temperatures, which are commonly in the range 700° to 1200°C. Experiments on the bending of silicon wafers during oxidation demonstrate that elastic stress in the oxide layer is relieved very quickly during oxidation in steam at temperatures above 1000°C. The oxide flow can therefore be characterized using the Navier–Stokes equations. At lower temperatures, a Navier–Stokes model is no longer appropriate and a visco-elastic representation that more correctly characterizes the stress relaxation rates should be used [120, 259].

The molecular structure of vitreous silica lacks long-range chrystallographic order and groups of molecules can move relative to each other as in a fluid. Strong forces are required for this relative movement to occur, so the viscosity of oxide is large. At lower temperatures, however, these local forces are so strong that the time scale for flow is much greater and it is more appropriate to treat oxide as an elastic material. Numerous measurements have been performed to determine the material properties of oxide for a wide range of temperatures. During oxidation in steam, some water molecules dissociate to produce hydroxyl groups bonded to silicon atoms. This reaction has the effect of decreasing the viscosity of "wet oxide" as opposed to "dry oxide," e.g. the viscosities of wet oxide containing 0.12% hydroxyl by weight at temperatures $T = 800°C, 900°C, \ldots, 1200°C$ are 1.3×10^{18}, 10^{16}, 1.7×10^{14}, 5.0×10^{12} and 2.4×10^{4} poise, respectively.

At lower temperatures, visco-elastic behavior is significant and the effects of stresses are more pronounced. Stress in the oxide can influence both the shape of the oxide layer and thinning at corners, as previously indicated for trench oxidation. Experimental evidence suggests, moreover, that lower oxidation rates occur at sharp corners in the low temperature range, and that this may be attributable to viscous stresses. In fact, the diffusivity of the oxidant in oxide, the reaction rate at the silicon surface, the solubility of the oxidant in oxide, and the viscosity of the oxide may all be significantly influenced by stress.

In the present development we focus on the high temperature oxidation problem. As indicated above, viscous flow in the high temperature regime can be adequately modeled with the Navier–Stokes equations, which describe the nonlinear transport of momentum and conservation of mass in a viscous fluid. Moreover, since the viscosity of the oxide is very large and the flow velocity is relatively small, the inertial terms in the equations of motion may be neglected, and the problem simplifies to the linear case of Stokes flow. Measurements of film stress have been obtained for oxide films on planar surfaces after cooling to room temperature. Changes in temperature cause a thermal stress that can be estimated using the difference in thermal expansivity between silicon and oxide. Subtracting the thermal stress from the total stress yields an estimate of the intrinsic stress during oxidation. Average film stresses have been measured in oxide samples at a thickness as small as 100 angstroms. Extrapolating these experimental results to zero oxide thickness, a limit value of film stress of approximately 4×10^9 dynes/cm^2 is obtained irrespective of temperature.

A viable model of physical oxidation involves characterization of the viscous flow of the oxide layer, coupled with the diffusion of oxide through the layer, and the reaction and expansion at the moving boundary. A corresponding mathematical model will then permit analysis, simulation and testing of processing techniques of interest to the process engineer.

The driving process in oxidation is transport of oxidant in the oxide layer and reaction at the interface. Transport of oxidant occurs by diffusion and convection in the oxide layer. In the processes of interest here, diffusion of oxidant is very rapid in comparison with transport due to convection. Furthermore, using film thickness s as the length scale together with the kinematic viscosity $\frac{\mu}{\rho}$ and velocity V, the time scale for transport of momentum and the time scale for reaction are $\tau_f = s^2/(\mu/\rho)$ and $\tau_r = s/V$, respectively, where $\tau_f \ll \tau_r$. This implies that in a numerical simulation, the time period of oxidation can be divided into a number of timesteps, and within each timestep the fluid flow and diffusion rapidly adjust to a quasi-steady state. Following each timestep, the reaction and surface boundaries may be incrementally advanced based on the present solution. In the next section we confine our attention to the diffusion problem.

15.2 Diffusion of Oxidant

It is important to recognize that oxidation is still a time-dependent problem even though the model suggested above for fluid flow and oxidant transport uses quasi-steady assumptions. This is because both the boundaries of the domain and the boundary conditions themselves are functions of time. Consider the case of two-dimensional modeling of the transport of oxidant. Since the oxide flow is very slow, convection is negligible compared with diffusion and the governing transport equation reduces to the diffusion equation considered previously in Chapters 11–14 for species diffusion. Let us consider the two-dimensional case, where the concentration c of oxidant then satisfies the classical diffusion equation

$$\frac{\partial c}{\partial t} - \frac{\partial}{\partial x}\left(D\frac{\partial c}{\partial x}\right) - \frac{\partial}{\partial y}\left(D\frac{\partial c}{\partial y}\right) = 0 \tag{15.1}$$

with boundary conditions

$$c = c_0 \tag{15.2}$$

at the gas surface and a reaction boundary condition

$$-D\frac{\partial c}{\partial n} = k\,r(c,\sigma) \tag{15.3}$$

at the silicon surface. Here $D = D(c,\sigma)$ is the diffusivity, c_0 is the solubility of oxidant in oxide, $r(c,\sigma)$ is the reaction rate at the silicon surface, k is the reaction rate coefficient, σ is the silicon surface stress, and n is the surface coordinate in the normal direction. This problem statement can be compared with the diffusion reaction systems considered in Chapter 11 where the reaction is distributed through the domain rather than localised at a surface.

For convenience, it is assumed that initially there is a uniform, thin oxide layer with zero oxidant concentration. The reaction boundary condition (15.3) equates the species mass flux to the consumption of the species due to reaction at the interface. As a result of the reaction, silicon is converted into silicon oxide and the reaction interface moves. The normal velocity V of the moving, reacting boundary depends upon the rate at which silicon is being converted into oxide, and therefore on the flux of oxidant at the silicon surface. That is, the oxide growth rate V is determined by an additional ordinary differential equation for conversion of silicon at the silicon–oxide interface

$$-D\frac{\partial c}{\partial n} = NV \tag{15.4}$$

where N is the molar constant.

The volume of a silicon atom is 20 cubic angstroms and the volume of a SiO_2 molecule is 45 cubic angstroms giving a volume ratio of 4:9. Hence, oxide occupies a larger volume than the reacting silicon. This implies that only a fraction γ of the velocity V in (15.4) may be associated with the moving reaction surface, which then has velocity components

$$v_n = \gamma V\,, \qquad v_s = 0 \tag{15.5}$$

where s, n are the tangential-normal coordinates. The remaining part $u_n = (1-\gamma)V$ can be interpreted as a velocity boundary condition for the viscous flow problem.

The reaction boundary condition in (15.3) merits further comment. The simplest first-order reaction is given by $r(c) = kc$, with the rate coefficient k defined by the Arrenhius formula

$$k = k_0 = ce^{-A/(\kappa T)} \tag{15.6}$$

where κ is the Boltzmann constant, T is temperature in degrees Kelvin, e is the exponential, and A is the activation energy. More sophisticated models can be devised that include a dependence of the reaction rate on the oxide stress. It can be argued, for instance, that additional energy is required to break silicon bonds which are under compression. Hence, the effect of surface stress could be included by adding an energy of expansion pV_s where $V_s = 25$ cubic angstroms is the volume difference between an oxide molecule and a silicon atom and p is the normal surface pressure. In planar oxidation, the oxide is free to expand without resistance and therefore, the reaction rate constant is given by (15.6). If the silicon surface is curved and concave, compressive pressure lowers the reaction rate and models have been suggested, of the form [396]

$$k = ce^{(-A+pV_s)/(\kappa T)} = k_0 e^{pV_s/(\kappa T)} \tag{15.7}$$

The governing diffusion equation together with boundary conditions, initial condition and the stated relationships for k constitute a mathematical model for this class of reaction-diffusion processes. Before further considering this problem and its quasi-steady form, it is useful to briefly review some relevant fundamentals on diffusion from classical PDE theory.

15.3 Mathematical Analysis of Diffusion

Consider the general second-order PDE written in the form

$$au_{xx} + 2bu_{xy} + cu_{yy} + du_x + eu_y + gu = f \tag{15.8}$$

where the coefficients a, b, c, d, e, f and g are given functions of x and y, $u_x = \partial u/\partial x$ $u_y = \partial u/\partial y$ $u_{xx} = \partial^2 u/\partial x^2$, etc. Under a change of variables $x = x(\xi, \eta)$, $y = y(\xi, \eta)$ with use of the chain rule, the partial derivatives in (15.8) transform as

$$u_x = u_\xi \xi_x + u_\eta \eta_x \,,$$

$$u_{xx} = \xi_x^2 u_{\xi\xi} + 2\xi_x \eta_x u_{\xi\eta} + \eta_x^2 u_{\eta\eta} + \xi_{xx} u_\xi + \eta_{xx} u_\eta$$

and so forth. Substituting into (15.8), the transformed partial differential equation is

$$Au_{\xi\xi} + 2Bu_{\xi\eta} + Cu_{\eta\eta} + Du_\xi + Eu_\eta + Gu = F \tag{15.9}$$

with $A = A(\xi_x, \xi_y)$, etc.

Classification theory is based on the observation that the transformed equation simplifies when the coefficients of $u_{\xi\xi}$ and $u_{\eta\eta}$ vanish. Setting $A = 0$,

$$a\xi_x^2 + 2b\xi_x\xi_y + c\xi_y^2 = 0 \tag{15.10}$$

yields the characteristic ordinary differential equation (since $dy/dx = -\frac{\xi_x}{\xi_y}$),

$$\frac{dy}{dx} = \frac{\left[b \pm (b^2 - ac)^{1/2}\right]}{a} \tag{15.11}$$

The same result follows on setting $C = 0$ with $dy/dx = -\eta_x/\eta_y$.

A family of real solutions to (15.11) exists provided the discriminant $b^2 - ac \geq 0$. Otherwise, there is no real family $\xi(x, y) =$ constant, $\eta(x, y) =$ constant, satisfying the characteristic differential equation. When $b^2 - ac = 0$ the characteristic families degenerate to a single family and equation (15.8) is termed parabolic. If $b^2 - ac > 0$, two real characteristic families exist and the partial differential equation is hyperbolic. Finally, $b^2 - ac < 0$ implies complex characteristics and the equation is called elliptic. These classical definitions are based on the corresponding properties of the coefficients for the parabola, hyperbola and ellipse in the theory of conic sections.

For planar oxidation at constant diffusivity $D = 1$, (15.1) simplifies to the model problem

$$c_t - c_{xx} = 0 \tag{15.12}$$

which corresponds to (15.8) with $a = -1$, $b = 0$, $c = 0$, $d = 0$, $e = 1$ and $f = 0$. Then $b^2 - ac = 0$ and this problem is of parabolic type. Substituting in (15.11) and integrating, the single characteristic family is the set of horizontal lines $t =$ constant. Since initial data propagate along characteristics with velocity equal to the reciprocal of the slope of the characteristics, it follows that, for the parabolic problem, a disturbance in the initial concentration will propagate in all directions with infinite velocity. That is, the mathematical model implies that the effect of a small change in the initial conditions will be experienced instantaneously throughout the solution domain. This behavior can be contrasted with wave propagation for hyperbolic problems in which information propagates at finite speed.

To gain further insight into the nature of a solution to the diffusion equation, consider the case of a fixed layer of depth L with boundary conditions $c(0,t) = c(L,t) = 0$, and an initial concentration profile

$$c(x,0) = \sin(n\pi x/L) \tag{15.13}$$

Assuming a separated form for the solution, $c(x,t) = \alpha(t)\sin(n\pi x/L)$, and substituting into the homogeneous parabolic equation $c_t = c_{xx}$, an ordinary differential equation is obtained that describes the behavior of the coefficient function α with time. On solving this ordinary differential equation,

$$\alpha(t) = e^{-\left(\frac{n\pi}{L}\right)^2 t} \tag{15.14}$$

so that the PDE solution has the form

$$c(x,t) = e^{-\left(\frac{n\pi}{L}\right)^2 t}\sin(n\pi x/L) \tag{15.15}$$

and the initial profile decays exponentially as a function of time. Hence, the qualitative behavior for the parabolic problem is that the effect of a disturbance propagates instantaneously throughout the domain, but decays exponentially as a function of time. The maximum height of the initial profile in the previous example has fallen from its initial value of unity to a value e^{-1} in time $t = (L/n\pi)^2$, which defines the time scale for the diffusion process. This can be compared with the other relevant time scales for reaction and for viscous flow in the oxidation problem.

For more general initial data, Fourier series methods or transform techniques can also be conveniently applied. Consider the same problem (15.12) on the interval $-L < x < L$ with $c(-L,t) = c(L,t) = 0$ and initial data $c(x,0) = 1$. Laplace transforming (multiplying $c_t = c_{xx}$ by e^{-st}, integrating with respect to time over $0 \le t \le \infty$ and simplifying),

$$-1 + s\hat{c}(x,s) = \hat{c}_{xx}(x,s) \tag{15.16}$$

where s is the transform variable and, by definition,

$$\hat{c}(x,s) = \int_0^\infty e^{-st}c(x,t)dt$$

Solving the second order linear ordinary differential equation (15.16) gives the transform function

$$\hat{c}(x,s) = \frac{1}{s} + A\cosh(\sqrt{s}x) + B\sinh(\sqrt{s}x) \tag{15.17}$$

Using the transformed end conditions, the final solution in the transform plane has the form

$$\hat{c}(x,s) = \frac{1}{s} - \frac{\cosh(\sqrt{s}x)}{s\cosh(\sqrt{s}L)} \tag{15.18}$$

This transform can be inverted to obtain the final solution. Alternatively, the hyperbolic cosines in the transform expression can be expanded in series form and then inverted term-by-term to obtain (see [98]),

$$c(x,t) = \sum_{j=0}^{\infty} \frac{4(-1)^j}{(2j+1)\pi} e^{\frac{-(2j+1)^2\pi^2}{4L^2}t} \cos\left(\frac{(2j+1)\pi x}{2L}\right) \tag{15.19}$$

Material diffuses from regions of high concentration to regions of low concentration. In the absence of sources or sinks in the interior of the domain, the maximum and minimum concentration should be achieved on the boundary Γ in the space-time region defined by segments $0 < x < L$ for $t = 0$, and $x = 0$ and $x = L$ for $t > 0$. The proof of this result follows using argument by contradiction. Assume there exists a point $(\bar{x}, \bar{t})$ in the interior or on the "final time" boundary $0 < x < L$ at $t = T$, where the maximum is achieved. Next define the auxiliary function

$$v(x,t) = c(x,t) - \varepsilon(t - \bar{t}) \tag{15.20}$$

where $\varepsilon > 0$. Since this auxiliary function takes on the maximum value at the point $(\bar{x}, \bar{t})$ and exceeds the greatest value of $c(x,t)$ on the boundary Γ, ε can be chosen sufficiently small that v is greater than the maximum value obtained by c on Γ. This implies that v attains its maximum at some point not on the boundary, and at this point

$$v_{xx} \leq 0\,, \quad v_t > 0$$

which implies

$$c_{xx} \leq 0\,, \quad c_t > 0 \tag{15.21}$$

This contradicts the governing equation $c_t = c_{xx}$ at this point.

In some of the approximate models considered later, oxidant diffusion takes place on a rapid time scale so that the system quickly reaches a steady state relative to the other transient processes. It follows that the steady state form of the oxidant diffusion problem can then be used for modeling. Returning to the two-dimensional diffusion problem and applying the steady state assumption, the governing equation becomes

$$-\nabla \cdot (D\nabla c) = f \tag{15.22}$$

where the diffusivity D is a positive function. In the case where the diffusivity is constant, (15.22) simplifies further to Poisson's equation

$$-\Delta c = f/D \tag{15.23}$$

which is the prototype for elliptic partial differential equations under the classification scheme described earlier. Ellipticity can be easily verified by setting $a = c = -1$, $b = d = e = f = 0$ in (15.8), so that equation (15.11) reduces to $\frac{dy}{dx} = \sqrt{-1}$ and the characteristics are complex. The absence of real characteristics for elliptic problems implies that the solution at any interior point depends continuously on the data prescribed on the entire boudary of the domain. In the case where $f = 0$, (15.23) is Laplace's equation.

Such elliptic problems arise in a variety of applications and have been the subject of extensive study using techniques from complex variable theory. The relevance of analytic function theory can be simply demonstrated from the fundamental property that the directional derivative, at a point for an analytic function $w = u + iv$ of a complex variable $z = x + iy$, is independent of the path of approach in the complex plane. Using this property, consider the special case where the directional derivative is taken along the x and y directions respectively. Equating real and imaginary parts of the directional derivatives, we obtain the Cauchy–Riemann equations

$$u_x = v_y\,, \qquad u_y = -v_x \tag{15.24}$$

Differentiating these equations with respect to x and y and combining the results,

$$\Delta u = 0\,, \quad \Delta v = 0 \tag{15.25}$$

hold in the domain of analyticity of w. Therefore, the real and imaginary parts of an analytic function satisfy Laplace's equation and are termed harmonic functions [97].

The steady state solution c satisfying Laplace's equation also satisfies a local mean-value theorem. On integrating the potential equation $\Delta c = 0$ over any sub-region with boundary Γ and applying the divergence theorem

$$\oint_\Gamma \frac{\partial c}{\partial n} ds = 0 \tag{15.26}$$

The fact that the integral of the normal flux around any closed contour is zero is consistent with the idea that the solution is at a steady state and that mass is conserved. To demonstrate the mean-value property, consider a point in the interior and construct a circle of radius r, centered at this point. The average value of c on this circle is

$$\bar{c}(r) = \frac{1}{2\pi} \int_0^{2\pi} c(r,\theta) d\theta \tag{15.27}$$

Differentiating with respect to r and setting $c_r = \frac{\partial c}{\partial n}$ at $r = R$

$$\bar{c}_r(R) = \frac{1}{2\pi} \int_0^{2\pi} c_r(R,\theta) d\theta = \frac{1}{2\pi R} \int_0^{2\pi} \frac{\partial c}{\partial n} ds \tag{15.28}$$

Applying the previous result, (15.26), $\bar{c}(R)$ is independent of radius R. It follows that the value of the harmonic function at a point is equal to the average value on any circle centered at that point. By a similar argument, the value at the point is equal to the average value over a disk surrounding that point.

Using these mean-value results, it is also possible to show that the maximum principle holds. That is, the maximum values of the concentration must occur on the boundary of the region. Intuitively, if a maximum were achieved at some point in the interior, then mass would diffuse from this high concentration point and therefore a steady state could not be maintained. More formally, if c has a maximum at interior point P, then the average value on a circle centered at P is less than the value of c at P, and this contradicts the mean-value property. By the same reasoning, the minimum must also occur on the boundary.

These principles can be used to show that the solution to the corresponding Dirichlet problem, in which the concentration is specified on the boundary, must be unique. Let c_1 and c_2 be solutions to the Dirichlet problem. Then the difference $d(x,y) = c_1(x,y) - c_2(x,y)$ between these two functions is also harmonic and it vanishes on the boundary. Since d attains its maximum and minimum on the boundary, it follows that d must be identically zero in the interior and uniqueness follows. For the Neumann problem, the flux, rather than the concentration, is specified everywhere on the boundary, and conservation of mass implies a compatibility condition on the flux boundary data of the form $\int_{\partial\Omega} \frac{\partial c}{\partial n} ds = \int_{\Omega} f dx dy$. Moreover, the solution is unique only within an arbitrary constant.

15.4 Approximate Models

The most commonly used numerical techniques for solving partial differential equations are the finite difference and finite element methods so the present treatment is restricted to these strategies. Related discussion is provided in the earlier chapters on modeling dopant diffusion and device transport.

Finite Differences

We begin with the basic finite difference strategy applied to the one-dimensional diffusion problem, $c_t = Dc_{xx}$ on the interval $[0, L]$ with specified Dirichlet data $c(0,t) = a(t)$, $c(L,t) = b(t)$ and initial condition $c(x,0) = g(x)$. By introducing a uniform finite difference grid $x_i = ih$, $i = 0, 1, \ldots N$ with $h = L/N$, Taylor series expansions for $c(x_{i+1}, t)$ and $c(x_{i-1}, t)$ about $x = x_i$ yield

$$\begin{aligned} c(x_{i+1}, t) &= c(x_i, t) + hc_x(x_i, t) + \frac{h^2}{2!}c_{xx}(x_i, t) \\ &\quad + \frac{h^3}{3!}c_{xxx}(x_i, t) + \frac{h^4}{4!}c_{xxxx}(x_i, t) + \ldots \end{aligned} \tag{15.29}$$

$$\begin{aligned} c(x_{i-1}, t) &= c(x_i, t) - hc_x(x_i, t) + \frac{h^2}{2!}c_{xx}(x_i, t) \\ &\quad - \frac{h^3}{3!}c_{xxx}(x_i, t) + \frac{h^4}{4!}c_{xxxx}(x_i, t) + \ldots \end{aligned} \tag{15.30}$$

Adding these equations, a representation for the second derivative in terms of discrete grid point values is

$$c_{xx}(x_i, t) = \left[c(x_{i+1}, t) - 2c(x_i, t) + c(x_{i-1}, t)\right]/h^2 + \tau(x_i, t) \tag{15.31}$$

The remainder or truncation error term

$$\tau(x_i, t) = \frac{h^2}{12}c_{xxxx}(x_i, t) + O(h^4) \tag{15.32}$$

is the error at grid point x_i when the difference expression is taken as an approximation for c_{xx}.

For the parabolic diffusion equation, at grid point x_i

$$(c_t - Dc_{xx})_{(x_i,t)} = 0 \tag{15.33}$$

so that the exact solution satisfies

$$c_t(x_i, t) - D\left(c(x_{i+1}, t) - 2c(x_i, t) + c(x_{i-1}, t)\right)/h^2 = D\tau(x_i, t) \tag{15.34}$$

at the grid points. The approximate problem is obtained by setting the right-hand side of (15.34) to zero. That is, the approximate solution $c_i(t)$ for $c(x_i, t)$ satisfies

$$(c_i)_t - D\left(c_{i+1} - 2c_i + c_{i-1}\right)/h^2 = 0 \tag{15.35}$$

for each interior grid point $i = 1, 2, 3, \ldots, N-1$ where the end values $c_0(t)$ and $c_N(t)$ are given by $a(t)$ and $b(t)$ respectively.

The previous representations yield an expression for the error $e_i = c(x_i, t) - c_i(t)$ at grid point x_i

$$\frac{\partial e_i}{\partial t} - \frac{D}{h^2}(e_{i-1} - 2e_i + e_{i+1}) = D\tau \tag{15.36}$$

If good bounds for the derivatives appearing in the truncation error are known, then a useful estimate of the error at the grid points can be determined.

System (15.35) is referred to as a semidiscrete system of ordinary differential equations for the grid point values. This ODE system is to be integrated in (x, t) space along grid lines $x_i =$ constant from prescribed initial data $c_i(0) = g(x_i)$, and, therefore, the approach is sometimes termed the method of lines (MOL). In matrix form, the resulting semidiscrete system may be written

$$\frac{d\boldsymbol{c}}{dt} + \boldsymbol{A}\boldsymbol{c} = \boldsymbol{0} \tag{15.37}$$

with initial vector $\boldsymbol{c}(0) = \boldsymbol{c}_0$, given.

There are a variety of techniques for integrating such semidiscrete systems of ordinary differential equations. In recent years, several software packages have been developed to implement these methods. The software is widely available in computer libraries and can be incorporated naturally into the solution procedure with appropriate calls to the library routines. From a practical standpoint, this is generally recommended because it frees the model developer to concentrate on spatial discretizations. Further details on system integrators are given in the chapter on multiple species diffusion models.

It is instructive to examine the stability and oscillatory properties of the explicit forward Euler method and the midstep Crank–Nicholson method for the oxidation process. Consider a finite difference approximation with respect to time for the system (15.37). A forward difference approximation yields

$$\frac{\boldsymbol{c}(t + \Delta t) - \boldsymbol{c}(t)}{\Delta t} + \boldsymbol{A}\boldsymbol{c}(t) = \boldsymbol{0} \tag{15.38}$$

so that at grid point x_i the representative equation becomes

$$c_i(t + \Delta t) = \left(\frac{\Delta t}{h^2} D c_{i-1}(t) + (1 - 2\frac{\Delta t}{h^2} D) c_i(t) + \frac{\Delta t}{h^2} D c_{i+1}(t)\right) \tag{15.39}$$

Hence, the value of c_i at grid point x_i and the new time can be determined explicitly from adjacent grid point values at the previous time. The resulting system vectorizes and parallelizes immediately, but there is a stability restriction on the size of the timestep.

To analyze the stability of the method, consider an oscillatory disturbance on the domain $[0, L]$ at time $t = t_j = j\Delta t$ of the form

$$c(x, t_j) = \varepsilon_j \sin(k\pi x/L) \tag{15.40}$$

where the coefficient represents the amplitude at time level j and a single representative Fourier mode has been considered. Substituting into the difference equation (15.39),

$$\begin{aligned}\varepsilon_{j+1} \sin\left(\frac{k\pi x_i}{L}\right) &= \varepsilon_j \left(\frac{\Delta t D}{h^2} \sin(\frac{k\pi x_{i-1}}{L}) + (1 - 2\frac{\Delta t D}{h^2}) \sin(\frac{k\pi x_i}{L})\right. \\ &\quad \left. + \frac{\Delta t D}{h^2} \sin(\frac{k\pi x_{i+1}}{L})\right)\end{aligned} \tag{15.41}$$

Simplifying, this implies

$$\varepsilon_{j+1} = \left(1 - \frac{4\Delta t D}{h^2} \sin^2\left(\frac{k\pi h}{L}\right)\right) \varepsilon_j \tag{15.42}$$

The magnitude of the amplification factor in the above equation should be less than unity to prevent errors from growing. This condition implies the stability restriction

$$\Delta t \leq \frac{1}{2} h^2/D \tag{15.43}$$

Therefore, for stability, the timestep varies with the square of the grid size and inversely with the diffusivity. In the oxidation problem, the value of diffusivity is large, which implies that the time step for an explicit method is quite restrictive. Instead, an implicit scheme that is unconditionally stable can also be used to integrate the system (15.37). Introducing a central difference scheme with respect to time,

$$\frac{\mathbf{c}(t_{j+1}) - \mathbf{c}(t_j)}{\Delta t} + \frac{\boldsymbol{A}(\mathbf{c}(t_{j+1}) + \mathbf{c}(t_j))}{2} + O((\Delta t)^2) = \mathbf{0} \tag{15.44}$$

Hence at each time timestep, the sparse algebraic system

$$\left(\boldsymbol{I} + \frac{\Delta t}{2}\boldsymbol{A}\right) \mathbf{c}_{j+1} = \left(\boldsymbol{I} - \frac{\Delta t}{2}\boldsymbol{A}\right) \mathbf{c}_j \tag{15.45}$$

is solved for $\mathbf{c}_{j+1}$, where $\mathbf{c}_j$ is known from the previous timestep. Repeating the previous Fourier stability analysis demonstrates that the scheme (15.45) is unconditionally stable; that is, there is no restriction on the timestep Δt for this method. As with the explicit system integrators, the size of the timestep can be varied adaptively during the integration process to meet a specified accuracy requirement with high efficiency.

For the steady state diffusion problem in two dimensions with $D = 1$, the governing equation (15.23) reduces to the potential problem

$$\Delta c = f \tag{15.46}$$

For convenience, consider a rectangular domain with constant diffusivity. Introducing a uniform grid with mesh size h and using Taylor series expansions for the grid points adjacent to the interior grid point (x_i, y_j),

$$\begin{aligned}(-\Delta c + f)_{ij} &= h^{-2}\left(4c\left(x_i, y_j\right) - c\left(x_{i-1}, y_j\right) - c\left(x_i, y_{j-1}\right)\right.\\ &\qquad \left. - \, c\left(x_{i+1}, y_j\right) - c\left(x_i, y_{j+1}\right)\right) + f_{ij} + \tau\end{aligned} \tag{15.47}$$

where $\tau = h^2\left[c_{xxxx}\left(x_i, y_j\right) + c_{yyyy}\left(x_i, y_j\right)\right] + O(h^4)$ is the truncation error. The corresponding finite difference approximation at interior grid point (i, j) is

$$4c_{ij} - (c_{i-1j} + c_{ij-1} + c_{i+1j} + c_{ij+1}) = -h^2 f_{ij} \tag{15.48}$$

where c_{ij} denotes the finite difference approximation to $c(x_i, y_j)$. Note that (15.48) can be rewritten as

$$c_{ij} = \frac{1}{4}(c_{i-1j} + c_{ij-1} + c_{i+1j} + c_{ij+1}) - \frac{h^2}{4} f_{ij} \tag{15.49}$$

When $f = 0$ this implies that the central value is the average of its four neighbors and can be interpreted as a discrete form of the mean-value property.

The scheme in (15.48) involves the familiar five-point stencil for the Laplacian and the scheme is second-order accurate; i.e. the difference approximation converges at an asymptotic rate that is $O(h^2)$. Other stencils and higher-order difference schemes can also be developed. For example, if the difference stencil is rotated 45 degrees, a new grid is produced that yields a similar five-point formula for the Laplacian but with grid spacing now $h\sqrt{2}$. We also remark that the Laplacian is special since it is invariant under rotation. The truncation errors for these two five-point formulas are different. By taking an appropriate linear combination of the two representations and averaging, various nine-point schemes can be constructed.

In particular, for the case $f = 0$, optimizing the weights in the averaging formula to cancel the leading truncation error leads to the scheme

$$\begin{aligned} 20c_{ij} - 4\left(c_{i-1,j-1} + c_{i+1j-1} + c_{i+1j+1} + c_{i-1j+1}\right) & \\ - \left(c_{i-1j} + c_{ij-1} + c_{i+1j} + c_{ij+1}\right) &= 0 \end{aligned} \qquad (15.50)$$

that is $O(h^6)$ accurate for Laplace's equation. Of course, the matrix for the nine-point operator is less sparse but the additional computation is small and the nodal solution is dramatically improved.

After incorporating the boundary conditions in (15.48) or (15.50), the resulting system of linear equations can be written in matrix form as

$$\boldsymbol{A}\boldsymbol{c} = \boldsymbol{b} \qquad (15.51)$$

where $\boldsymbol{A}$ is sparse, symmetric and positive definite. The system can be solved by sparse elimination using the Choleski algorithm to exploit symmetry and positive definiteness. Since $\boldsymbol{A}$ is symmetric only the symmetric upper or lower triangular part of $\boldsymbol{A}$ is stored. The basic algorithm can be subdivided to two main steps: (1) factorization of the coefficient matrix $\boldsymbol{A}$ as

$$\boldsymbol{A} = \boldsymbol{L}\boldsymbol{L}^T \qquad (15.52)$$

where $\boldsymbol{L}$ is lower triangular and can be stored over $\boldsymbol{A}$; (2) then inexpensive forward and backward substitution sweeps

$$\boldsymbol{L}\boldsymbol{v} = \boldsymbol{b}, \quad \boldsymbol{L}^T\boldsymbol{c} = \boldsymbol{v} \qquad (15.53)$$

to determine $\boldsymbol{c}$. The grid points (equations) can be reordered using the Cuthill–McKee algorithm or similar strategies to minimize the system bandwidth. Other sparse elimination methods such as nested dissection are also effective.

The system is also amenable to solution by a variety of iterative methods including the familiar Jacobi, Gauss–Seidel and successive over-relaxation (SOR) point iteration schemes, as well as the conjugate gradient method discussed earlier in Chapter 13. Let us consider the five-point scheme (15.49) to illustrate some basic iterative methods. The Jacobi point iterative method is based on the averaging formula (15.49) with

$$c_{ij}^{\text{new}} = \frac{1}{4}\left(c_{i-1j}^{\text{old}} + c_{ij-1}^{\text{old}} + c_{i+1j}^{\text{old}} + c_{ij+1}^{\text{old}}\right) - \frac{h^2}{4} f_{ij} \qquad (15.54)$$

where the neighbor values are obtained from the previous iterate and used to construct the new central point value. Assuming, for convenience, that

Dirichlet data are specified on the boundary, the scheme loops over the interior points in the grid in a single sweep to construct a new approximate solution and then the iteration is repeated. In the Gauss–Seidel iteration, the most recently computed values at the neighboring grid points are used so that

$$c_{ij}^{\text{new}} = \frac{1}{4}\left(c_{i-1j}^{\text{new}} + c_{ij-1}^{\text{new}} + c_{i+1j}^{\text{old}} + c_{ij+1}^{\text{old}}\right) - \frac{h^2}{4} f_{ij} \tag{15.55}$$

Successive iterates can also be averaged by introducing a relaxation parameter ω so that

$$c_{ij} = \omega c_{ij}^{\text{old}} + (1 - \omega)\, c_{ij}^{\text{new}} \tag{15.56}$$

By carefully picking the optimization parameter ω, the Gauss–Seidel scheme can be accelerated. This combination of (15.55) and (15.56) defines the successive over-relaxation method. For the particular example of the Laplacian in a square with Dirichlet boundary data, the optimum choice of relaxation parameter ω is

$$\omega^* = 2 - 2\sin(\pi h) \tag{15.57}$$

If the system is solved utilizing Jacobi iteration, the number of iterations required to reduce the error by a factor of 100 is $n = O(h^{-2})$, while for the SOR scheme with optimal relaxation $n = O(h^{-1})$. A variety of preconditioners and acceleration schemes are available for system solution (see [574]).

It is also important to note that for vector and parallel supercomputing some of the advantages of the accelerated schemes are either lost or more difficult to realize. Clearly both vector and parallel algorithms can be constructed very easily for the Jacobi iteration, since it uses only the old solution vector. On the other hand, the Gauss–Seidel method uses the most recently computed values at the adjacent grid points during the iterative sweep. This will destroy vectorization if a natural ordering of grid points from left to right in rows is used. If the grid points are grouped into two categories with alternate grid points in the mesh colored red and black, the adjacent grid points in the five-point operator are then decoupled, resulting in a red–black SOR scheme that will vectorize easily. Alternatively, if the grid points in the Gauss–Seidel scheme are numbered by diagonals instead of the natural numbering, then vectorization can again be exploited.

Returning to the time-dependent diffusion problem $c_t - D\Delta c = 0$ in higher dimensions, procedures analogous to those developed for the one-dimensional diffusion equation can again be employed. Now the spatial differencing is in two dimensions, and explicit or implicit schemes for time integration can also be developed as before. Alternating-direction line schemes can also be introduced to construct stable, efficent vector and vector-parallel algorithms. For example, the spatial operator in the respective cartesian directions can be "lagged" to yield

$$\frac{c_{ij}^{n+1/2} - c_{ij}^{n}}{\Delta t/2} - D\left(\frac{c_{i+1j}^{n+1/2} - 2c_{ij}^{n+1/2} + c_{i-1j}^{n+1/2}}{h^2} + \frac{c_{ij+1}^{n} - 2c_{ij}^{n} + c_{ij-1}^{n}}{h^2}\right) = 0 \tag{15.58}$$

for the first sweep using horizontal lines from bottom to top and

$$\frac{c_{ij}^{n+1} - c_{ij}^{n+1/2}}{\Delta t/2} - D\left(\frac{c_{i+1j}^{n+1/2} - 2c_{ij}^{n+1/2} + c_{i-1j}^{n+1/2}}{h^2} + \frac{c_{ij+1}^{n+1} - 2c_{ij}^{n+1} + c_{ij-1}^{n+1}}{h^2}\right) = 0 \tag{15.59}$$

for the second sweep using vertical lines from left to right. The pair of tridiagonal systems (15.58) and (15.59) constitutes one sweep of this alternating direction-scheme. This operator splitting can also be employed to construct alternating direction, line iterative schemes for the steady-state problem.

Finite Elements

The finite element method was introduced in Chapter 7 for approximating carrier transport in the device problem. The same ideas can be applied for diffusion in the oxidation problem using unstructured grids on irregular evolving geometries. For simplicity of exposition, consider the steady diffusion problem $-\Delta c = f$ in domain Ω with $c = g$ on boundary $\partial\Omega$ as a representative case. The basic idea is to develop an integral formulation of the problem and then approximate this on a partition of the domain into finite elements. Let c be an admissible trial function and define the residual $r = -\Delta c - f$. Clearly $r = 0$ for the solution itself and, in general, the residual may be interpreted as a measure of the extent to which an admissible function satisfies the differential equation. An integral formulation may be introduced by constructing the weighted residual statement

$$\int_\Omega (-\Delta c - f)\, w\, dx dy = 0 \tag{15.60}$$

for admissible test functions w. Integrating (15.60) by parts

$$\int_\Omega \nabla c \cdot \nabla w\, dx dy - \int_{\partial\Omega} \frac{\partial c}{\partial n} w\, ds = \int_\Omega f w\, dx dy \tag{15.61}$$

where n is the outward normal direction to the boundary $\partial\Omega$.

From the form of (15.61), the admissible functions belong to the Hilbert space $H^1(\Omega)$ of functions with square integrable first derivatives. Furthermore, $-\Delta c = f$ can be interpreted as the Euler–Lagrange equation from the corresponding variational problem where w corrresponds to the variation in solution c. Since c is given on $\partial\Omega$, this implies that the variation $w = 0$ on $\partial\Omega$. The weak (variational) problem becomes: find $c \in H^1(\Omega)$ with $c = g$ on $\partial\Omega$ and such that

$$\int_\Omega \nabla c \cdot \nabla w \, dxdy = \int_\Omega fw \, dxdy \tag{15.62}$$

holds for all $w \in H^1(\Omega)$ with $w = 0$ on $\partial\Omega$. The boundary term in (15.61) is zero since the test functions w for the Dirichlet problem are required to be zero on the boundary.

A specified flux $\partial c/\partial n = \gamma$ on $\partial\Omega_2$ can be included directly in the integral statement by substituting in (15.61). The weak or variational statement then becomes: find $c \in H^1(\Omega)$ such that $c = g$ on $\partial\Omega_1$ and satisfying

$$\int_\Omega \nabla c \cdot \nabla w \, dxdy = \int_\Omega fw \, dxdy + \int_{\partial\Omega_2} \gamma w \, ds \tag{15.63}$$

for all admissible test functions w with $w = 0$ on $\partial\Omega_1$. Note that the variation w is now arbitrary on $\partial\Omega_2$.

A finite element formulation for (15.62) or (15.63) can be constructed by discretizing the domain Ω into a union of E finite elements and introducing an appropriate basis on each of the elements. In practice, simple low-degree polynomials are commonly used on triangular or quadrilateral elements for two-dimensional applications or tetrahedral and hexahedral elements in three dimensions. For example, Ω can be triangulated and the vertex values used to define linear approximations within each triangle. This results in a global piecewise-linear continuous approximation.

For example, the integral statement (15.62) can then be approximated by accumulating element contributions as

$$\sum_{e=1}^{E} \int_{\Omega_e} \nabla c^e \cdot \nabla w^e \, dxdy = \sum_{e=1}^{E} \int_{\Omega_e} f^e w^e \, dxdy \tag{15.64}$$

Let $\{\psi_i^e(x,y)\}$, $i = 1, 2, \ldots, N_e$, denote the set of basis functions for the approximation on element e. Expanding the restriction of the approximate solution on element e in terms of this basis

$$c^e(x,y) = \sum_{j=1}^{N_e} c_j^e \psi_j^e(x,y) \tag{15.65}$$

For the Galerkin finite element method the test functions are chosen from the same subspace. That is,

$$w^e(x,y) = \psi_i^e(x,y) \tag{15.66}$$

By substituting (15.65) and (15.66) in the element integrals (15.64), the corresponding element matrix and vector contributions are

$$\begin{aligned} A_{ij}^e &= \int_{\Omega_e} \nabla\psi_i^e \cdot \nabla\psi_j^e dxdy \\ b_i^e &= \int_{\Omega_e} f^e \psi_i^e dxdy \end{aligned} \tag{15.67}$$

Adding element contributions according to (15.64) and enforcing the Dirichlet boundary conditions for c at nodes on the boundary yields the sparse symmetric algebraic system $\boldsymbol{Ac} = \boldsymbol{b}$ to be solved for nodal concentration vector $\boldsymbol{c}$.

To summarize, the basic structure of the finite element method for this Dirichlet problem is:

1. Discretize the domain.
2. Compute element matrix and vector contributions.
3. Assemble element contributions into a sparse global matrix and global right-hand side vector.
4. Apply the Dirichlet boundary conditions.
5. Solve the resulting sparse linear system.

It is easy to verify that a uniform grid of right isosceles triangles with a linear basis yields precisely the five-point difference operator encountered previously for Laplace's equation. On the other hand, the finite element method also allows discretizations of irregular domains as well as higher-degree element bases. These features enhance the usefulness of the method for practical applications. For example, in the oxidation problem the domain shape changes as a result of the moving reaction and free surfaces. The diffusion and oxide flow problems are solved successively on the evolving domain.

15.5 Viscous Oxide Flow

Solution of the viscous flow problem defined by the stationary Stokes equations is now considered. As noted in the introductory remarks, the relaxation time scale for oxidant diffusion is short relative to that associated with the reaction at the moving boundary. Hence, a quasi-steady analysis is possible. It was also established that the motion of the oxide is very slow, since it is produced by the flux of oxide associated with volumetric expansion due to the reaction at the silicon surface. This implies that the nonlinear terms in the Navier–Stokes equations can be neglected and the problem reduces to the simpler Stokes flow.

In primitive variable form, steady, viscous, incompressible Stokes flow is governed by

$$\begin{aligned} -\nu\Delta \boldsymbol{u} + \nabla p = \boldsymbol{f} \qquad &\text{in } \Omega \\ \nabla \cdot \boldsymbol{u} = 0 \qquad &\text{in } \Omega \end{aligned} \tag{15.68}$$

where $\boldsymbol{u}$ is the velocity, p is the pressure, ν is the kinematic viscosity and $\boldsymbol{f}$ is a forcing term. In component form

$$-\nu\Delta u_1 + \frac{\partial p}{\partial x} = f_1, \quad -\nu\Delta u_2 + \frac{\partial p}{\partial y} = f_2, \quad \frac{\partial u_1}{\partial x} + \frac{\partial u_2}{\partial y} = 0 \tag{15.69}$$

Alternative forms of the Stokes flow equations may be derived. For example, if the first equation in (15.69) is differentiated with respect to y and the second with respect to x and the resulting equations subtracted

$$-\nu\Delta\zeta = F \tag{15.70}$$

is obtained, where the vorticity component ζ in (15.70) is

$$\zeta = \frac{\partial u_2}{\partial x} - \frac{\partial u_1}{\partial y} \tag{15.71}$$

and F is similarly defined by replacing u_1, u_2, by f_1, f_2 in (15.71). Introducing the stream function ψ, the velocity components satisfy

$$u_1 = \frac{\partial \psi}{\partial y}, \quad u_2 = -\frac{\partial \psi}{\partial x} \tag{15.72}$$

so that, from the definition of vorticity above,

$$-\Delta\psi = \zeta \tag{15.73}$$

With the stream function as a variable, conservation of mass is automatically guaranteed. The governing equations become the stream function - vorticity equations (15.70), (15.73). Furthermore, substituting the expression for vorticity into the vorticity transport equation yields a fourth-order equation for the stream function

$$\nu \Delta^2 \psi = F \tag{15.74}$$

where $\Delta^2 = \Delta(\Delta)$ is the biharmonic operator. Appropriate boundary conditions for the stream function (e.g. the stream function is constant on a solid boundary) and for the flow velocities (no flow through a solid boundary and tangential velocity specified there) complete the mathematical statement of the problem.

In the decoupled stream function-vorticity iteration, the stream function is first computed from (15.73) with the previous vorticity iterate specified as the forcing function. Next the vorticity is determined from (15.70) with vorticity boundary data computed from the stream function solution at the previous stage. This process is repeated until convergence within a specified tolerance or an iteration limit is achieved. Alternatively, one can solve either the fully-coupled stream function-vorticity system directly, or the fourth-order stream function equation, or the mixed velocity-pressure primitive variable form of the equations (15.68). Here we focus on the decoupled scheme since it can be conveniently developed as an extension of the previous potential problem. Moreover, the decoupled algorithm is in some respects similar to the decoupled Gummel–Scharfetter scheme for device transport considered in the chapter on drift-diffusion models for semiconductor devices. The previous finite difference or finite element formulations for the associated Poisson problems (15.70), (15.73), then yield a corresponding pair of algebraic systems for iterate pair $(\psi^{(n)}, \zeta^{(n)})$ as

$$\boldsymbol{A}\boldsymbol{\psi}^{(n)} = \boldsymbol{\zeta}^{(n-1)} \quad \text{and} \quad \nu \boldsymbol{A}\boldsymbol{\zeta}^{(n)} = \boldsymbol{F}^{(n-1)} \tag{15.75}$$

respectively. Here the coefficient matrix $\boldsymbol{A}$ again corresponds to the discretized Laplacian and the superscript n indicates the iteration step in the decoupled scheme.

An implicit decoupled scheme for the corresponding time-dependent formulation leads to a very similar algorithm: the vorticity transport equation (15.70) is replaced by the corresponding evolution equation which is discretized in time. Within each timestep, the vorticity field is advanced with iterative adjustments based on "vorticity boundary data" computed from the current stream function iterate. The basic structure of the algorithm is the same.

Under the quasi-steady-state scenario described previously for the oxidation problem, the location of the reaction boundary and the free surface are adjusted incrementally during the period of oxidation processing. At each timestep the decoupled stream function-vorticity equations are solved on the associated grid. This implies that either the grid is moved or the domain is remeshed as the boundaries change. The algebraic systems (15.75) in the decoupled discrete stream function-vorticity formulation are recomputed at each timestep.

The structure of the discrete systems can be easily exploited by the solution algorithms. For example, if Dirichlet data is specified for the stream function, then the coefficient matrices of the two systems are identical. The matrix is also symmetric, sparse and positive definite. Hence, the Choleski algorithm can again be used to factor the matrix $\boldsymbol{A} = \boldsymbol{L}\boldsymbol{L}^T$ and the solution for the stream function iterate requires only forward and backward substitution sweeps on the corresponding right-hand side vector:

$$\boldsymbol{L}\boldsymbol{v} = \boldsymbol{\zeta}\,, \qquad \boldsymbol{L}^T\boldsymbol{\psi} = \boldsymbol{v} \tag{15.76}$$

The same triangular factors can be used in the solution for the vorticity iterate

$$\boldsymbol{L}\boldsymbol{v} = \boldsymbol{F}/\nu\,, \qquad \boldsymbol{L}^T\boldsymbol{\zeta} = \boldsymbol{v} \tag{15.77}$$

Thus, within a given timestep, a single Choleski factorization is required followed by repeated forward and backward substitutions and right-hand side evaluations. If a band-solver is used, the Choleski factorization requires $O(\frac{1}{3}w^2N)$ operations where w is the half-bandwith and N is the size of the system. Each pair of substitution sweeps requires only $O(wN)$ arithmetic operations, so the computational complexity for convergence in k iterations is $O((w+k)wN)$ per timestep.

Furthermore, since the change in the domain will be small between successive timesteps in the moving boundary problem, the stream function and vorticity solutions will not change markedly. The stream function and vorticity fields computed on the previous domain can be extrapolated to provide starting iterates for the new problem after the boundary has been moved. This implies that the number of decoupled iterations should not be large and hence the scheme will be efficient for this class of problems. Iterative sparse solution of the linear algebraic systems in (15.75) clearly is also possible and may be quite effective since the solution from the previous timestep or decoupled iteration provides a good starting iterate for the sparse system solver.

In some situations, it may be preferable to solve the coupled system directly without block iteration. Consider again the pair of governing partial differential equations (15.70) and (15.73) for stream function and vorticity with the associated boundary conditions. Let u_s denote the tangential velocity component on the boundary ($u_s = 0$ for a stationary solid boundary). Then the weak variational statement for the coupled problem is: find the pair (ψ, ζ) satisfying the boundary conditions for ψ and such that

$$\int_\Omega \nabla\psi \cdot \nabla v \, dx dy - \int_\Omega \zeta v \, dx dy = \int_{\partial\Omega} u_s v ds \tag{15.78}$$

$$\nu \int_\Omega \nabla\zeta \cdot \nabla w \, dx dy = \int_\Omega F w \, dx dy \tag{15.79}$$

holds for all v and w in an appropriate space of test functions.

Introducing the finite element expansions for the stream function and vorticity for a discretization with N nodes and with basis $\{\phi_i\}$

$$\psi_h = \sum_{j=1}^{N} \psi_j \phi_j(x, y), \quad \zeta_h = \sum_{j=1}^{N} \zeta_j \phi_j(x, y) \tag{15.80}$$

and the associated test functions v_h, $w_h = \phi_i(x, y)$, the resulting fully-coupled system has the form

$$\begin{bmatrix} \boldsymbol{A} & \boldsymbol{M} \\ \boldsymbol{0} & \nu\tilde{\boldsymbol{A}} \end{bmatrix} \begin{bmatrix} \boldsymbol{\psi} \\ \boldsymbol{\zeta} \end{bmatrix} = \begin{bmatrix} \boldsymbol{b} \\ \boldsymbol{F} \end{bmatrix} \tag{15.81}$$

where $\boldsymbol{A}$ and $\tilde{\boldsymbol{A}}$ differ since the terms involving the specified boundary values of ψ are included in the vector $\boldsymbol{b}$. This system can be solved directly by sparse Gaussian elimination or by iterative methods such as the preconditioned conjugate gradient method. The cost of factorization and substitution sweeps is increased with problem size, but only a single solution per viscous flow calculation is required, instead of the block decoupled iteration of (15.75).

Moving Boundary Treatment

For the oxidation problem the flow domain changes at each incremental timestep in the moving boundary problem. Moreover, the shape of the boundary will generally not have a simple cartesian geometry. If a finite difference scheme is used, then the problem domain may be mapped to a rectangular computational domain on which a uniform rectangular cartesian grid is defined. There is an extensive literature associated with the problem of grid generation and the construction of the appropriate mapping functions between the physical and computational domains [82, 522]. For example, if the domain is convex, then a transformation exists which maps the physical domain into a rectangular computational domain, and the Laplacian in the new coordinate sytem remains unchanged. Such "closed-form" conformal maps have been determined only for very special domains. For more general convex domains, the mapping function can be approximated by numerically solving Laplace's equation for the coordinate functions $x = x(\xi, \eta)$, $y = y(\xi, \eta)$ in the computational (ξ, η) domain; that is, solving

$$\Delta x = 0 \quad \text{and} \quad \Delta y = 0 \tag{15.82}$$

on a rectangular grid in the computational (ξ, η) domain to obtain the grid point coordinates in the physical domain. This defines a grid in the physical domain which is an approximation to the orthogonal curvilinear grid corresponding to the conformal map.

When the domain is not convex, grid lines from the computational domain may be mapped by the solution of (15.82) outside the physical domain boundary. That is, the analytic map $x(\xi, \eta)$, $y(\xi, \eta)$ may have a "fold". This problem can be circumvented by considering instead the harmonic problem for othogonal curvilinear grid families $\xi(x, y) = \text{constant}$ and $\eta(x, y) = \text{constant}$ in the physical domain. Then these curvilinear grid families satisfy

$$\Delta \xi = 0 \,, \quad \Delta \eta = 0 \tag{15.83}$$

in the physical domain and by the maximum principle, ξ and η attain their maximum and minimum values on the actual boundary. The approximate solution of (15.83) in the physical domain Ω still requires a grid so the basic problem is not resolved. However, this does provide an intermediate step to replace the problem in (15.82). Now, the pair of equations (15.83) may be transformed to a pair of quasi-linear partial differential equations in the computational domain of the form

$$Ax_{\xi\xi} + 2Bx_{\xi\eta} + Cx_{\eta\eta} + Dx_{\xi} + Ex_{\eta} + Fx = G \tag{15.84}$$

with coefficients determined by the metrics of the map between the physical and computational domains. A simple rectangular grid can be defined on the computational (ξ, η) domain. This pair of nonlinear partial differential equations in the computational domain can then be solved to obtain the approximate coordinates (x_i, y_j) of the curvilinear orthogonal coordinate set $\xi_i(x, y)$, $\eta_j(x, y)$ in the physical domain. Clearly, the complexity of the problem (15.84) increases significantly compared with (15.82) when such irregular domains are considered, since a pair of nonlinear partial differential equations must now be solved to determine the appropriate grid.

There are other strategies that involve generating a nonorthogonal grid in the physical domain directly, defining a map between this grid and a rectangular grid in the computational domain, and then mapping the governing equations over to the computational domain. In the oxide flow problem, the stream function-vorticity equations are then mapped to a pair of quasi-linear partial differential equations in the computational domain. The simplicity of the Laplace operator and the factorizations noted previously are lost, but the problem can still be solved on a rectangular domain using standard finite difference strategies.

Instead of using these mapping techniques, the geometric flexibility of the finite element method can be exploited. For example, the physical domain can be discretized using triangular elements to approximate the irregular oxide domain. The solution algorithm for the stream function-vorticity scheme then proceeds precisely as before. Following this, the domain boundary is updated using the reaction and free surface boundary conditions from the diffusion-reaction solution. The triangulated mesh points can then be adjusted based on the boundary motion and a new mesh defined. A trial solution iterate at the new mesh points can be interpolated from the previous grid and its solution. This provides a starting iterate for the viscous flow solution on the new triangulation. If the boundary motion is significant, as in the birds-beak problem, elements may become so slender that accuracy is degraded.

The Delaunay procedure was introduced in Chapter 9 for grid generation using point insertion. The main idea in this procedure is to test adjacent triangles using a circumcircle construction to see if a diagonal swap will improve the triangulation (recall Figure 9.2). This circumcircle test is equivalent to maximizing the minimum angle of the two triangulations possible for the convex quadrilateral. A similar test can be devised for the case where the quadrilateral defined by the two triangles is not convex. The diagonal swap test can be applied throughout the new triangulation to improve local grid quality. This simple procedure is fast and alleviates some of the problems associated with distorted elements.

Primitive Variables

The primitive variable formulation offers some advantages for viscous flow computation. A finite difference approximation can be constructed easily if the domain and grid are rectangular and uniform. Typically, staggered grids are used with pressure grid points offset from velocity grid points. Once again, a finite element formulation offers greater flexibility in handling the irregular geometries encountered in the oxidation problem. A weak variational formulation for the primitive variable Stokes problem is constructed by first forming the weighted residual statements for the momentum and continuity equation,

$$\begin{aligned}
&\int_\Omega (-\nu \Delta u_1 w + p_x w)\, dx dy = \int_\Omega f_1 w \, dx dy \\
&\int_\Omega (-\nu \Delta u_2 w + p_y w)\, dx dy = \int_\Omega f_2 w \, dx dy \\
&\int_\Omega ((u_1)_x + (u_2)_y)\, q dx dy = 0
\end{aligned} \tag{15.85}$$

Integrating by parts in the weighted residual expression for the momentum equations,

$$\begin{aligned}
&\nu \int_\Omega \nabla u_1 \cdot \nabla w \, dx dy + \int_\Omega p w_x \, dx dy = \int_\Omega f_1 w \, dx dy - \int_{\partial\Omega} \nu \frac{\partial u_1}{\partial n} w \, ds - \int_{\partial\Omega} p w \, ds \\
&\nu \int_\Omega \nabla u_2 \cdot \nabla w \, dx dy + \int_\Omega p w_y \, dx dy = \int_\Omega f_2 w \, dx dy - \int_{\partial\Omega} \nu \frac{\partial u_2}{\partial n} w \, ds - \int_{\partial\Omega} p w \, ds \\
&\int_\Omega ((u_1)_x + (u_2)_y)\, q \, dx dy = 0
\end{aligned} \tag{15.86}$$

For simplicity, consider the case where Dirichlet data for the velocity is specified on the boundary. The variational problem becomes: find u_1, $u_2 \in H$ satisfying the essential boundary conditions and pressure field $p \in G$ such that

$$\begin{aligned}
&\nu \int_\Omega \nabla u_1 \cdot \nabla w \, dx dy + \int_\Omega p w_x \, dx dy = \int_\Omega f_1 w \, dx dy \\
&\nu \int_\Omega \nabla u_2 \cdot \nabla w \, dx dy + \int_\Omega p w_y \, dx dy = \int_\Omega f_2 w \, dx dy \\
&\int_\Omega ((u_1)_x + (u_2)_y)\, q \, dx dy = 0
\end{aligned} \tag{15.87}$$

hold for all admissible test functions $w \in H$ and $q \in G$ with $w = 0$ on $\partial\Omega$. Other traction-type and free surface boundary conditions can be included through the boundary integral arising from the integration by parts procedure and by allowing w to vary on $\partial\Omega$. For example, if we rewrite (15.68) in terms of the stress tensor τ_{ij} then the momentum equations can be expressed in cartesian tensor notation as $-\tau_{ij,j} = f_i$ in Ω. Integrating the stress tensor by parts in the weighted-residual statement,

$$-\int_\Omega \tau_{ij,j} w_i \, dx dy = \int_\Omega \tau_{ij} w_{i,j} \, dx dy - \int_{\partial\Omega} \tau_{ij} n_j w_i \, ds \tag{15.88}$$

where summation is implied on a repeated index. Normal tractions $T_i = \tau_{ij} n_j$ can be directly specified in the boundary integral. In the oxidation problem, reaction at the silicon-oxide interface produces a volumetric expansion which acts as a source along the reacting interface. That is, there is a mass flux of oxide at the interface due to the reaction or, equivalently, the normal velocity resulting from the reaction is specified at the interface. Therefore, the boundary condition at the reacting interface for the viscous flow problem corresponds to no slip in the tangential direction and a specified normal velocity due to the volumetric expansion. A pressure boundary condition applies at the free surface.

Introducing the finite element expansions for the velocity components

$$u_{1h} = \sum_{j=1}^{N} u_1^j \phi_j(x,y), \qquad u_{2h} = \sum_{j=1}^{N} u_2^j \phi_j(x,y) \tag{15.89}$$

and for the pressure

$$p_h = \sum_{k=1}^{M} p_k \chi_k(x,y) \tag{15.90}$$

into the weak variational problem, the approximate formulation follows after replacing trial functions u_1, u_2, p by u_{1h}, u_{2h}, p_h and test functions w, q by $w_h = \phi_i$ and $q_h = \chi_i$ in (15.87). Evaluating the corresponding integrals, the linear algebraic system has the form

$$\begin{bmatrix} \mathbf{A} & \mathbf{0} & \mathbf{B}_1 \\ \mathbf{0} & \mathbf{A} & \mathbf{B}_2 \\ \mathbf{B}_1^T & \mathbf{B}_2^T & \mathbf{0} \end{bmatrix} \begin{bmatrix} \mathbf{u}_1 \\ \mathbf{u}_2 \\ \mathbf{p} \end{bmatrix} = \begin{bmatrix} \mathbf{F}_1 \\ \mathbf{F}_2 \\ \mathbf{0} \end{bmatrix} \tag{15.91}$$

where

$$A_{ij} = \int_\Omega \nabla\phi_i \cdot \nabla\phi_j \, dx dy, \qquad B_{ik}^1 = \int_\Omega (\phi_i)_x \chi_k \, dx dy \tag{15.92}$$

and so on. This coupled system for the velocity components and the pressure at the node points can then be solved using sparse elimination to obtain a primitive variable approximate solution for the Stokes flow problem.

One can also derive the primitive variable formulation by first defining a Lagrangian

$$L(\boldsymbol{u}, p) = \frac{\nu}{2} \int_{\Omega} \nabla \boldsymbol{u} : \nabla \boldsymbol{u} \, dxdy - \int_{\Omega} p \nabla \cdot \boldsymbol{u} \, dxdy - \int_{\Omega} \boldsymbol{F} \cdot \boldsymbol{u} \, dxdy \quad (15.93)$$

where ":" denotes the dyadic product $\sum_{i,j} \frac{\partial u_i}{\partial x_j} \frac{\partial u_i}{\partial x_j}$. The solution $(\boldsymbol{u}, p)$ is the stationary point of this functional. Taking variations with respect to $\boldsymbol{u}$ and p, the previous weak variational statement is obtained, where now the test functions are respective variations in the velocity and pressure. The original problem can also be posed as minimizing the classical variational integral, subject to the constraint that the velocity field be divergence-free. Using this interpretation, the mixed variational functional (15.93) again follows with the pressure p now interpreted as the Lagrange multiplier.

Since the gradient of the pressure appears in the governing equations, pressure can be determined only within an arbitrary constant. In practice this constant can be specified by requiring the pressure at an arbitrary node to have some reference value. Numerical experiments with this mixed finite element formulation demonstrate that the choice of bases is important. For some choices the solution is observed to converge; for other choices of finite element spaces the velocity converges, but the pressure exhibits spurious numerical oscillations. Analogous behavior is observed for finite difference discretizations.

The spurious pressure oscillations can be interpreted as follows: let the matrix $\boldsymbol{B} = [\boldsymbol{B}_1 \, \boldsymbol{B}_2]^T$ in the mixed formulation have an oscillatory eigenmode $\boldsymbol{p}_0$ corresponding to the zero eigenvalue. Then $\boldsymbol{p}_0$ is in the nullspace of $\boldsymbol{B}$. A scalar multiple of this oscillation mode could be added to the true pressure solution $\boldsymbol{p}$ without influencing the equations, since $\boldsymbol{B}\boldsymbol{p}_0 = \boldsymbol{0}$ for $\boldsymbol{p}_0 \neq \boldsymbol{0}$, implies $\boldsymbol{B}(\boldsymbol{p}+\boldsymbol{p}_0) = \boldsymbol{B}\boldsymbol{p}$. The entries in the matrix $\boldsymbol{B}$ depend upon not only the basis functions for the velocity expansion but also those for the pressure expansion. An incompatible choice of basis functions can lead to a matrix $\boldsymbol{B}$ with mode $\boldsymbol{p}_0 \neq \boldsymbol{0}$ and therefore to spurious oscillations. This pitfall has been analyzed mathematically and is related to the fact that the variational statement is a saddle-point problem. The balance between the momentum equations and the incompressibility constraint should be preserved in the discrete model. This implies a compatibility restriction on the respective velocity and pressure bases. A richer pressure approximation with a large number of basis functions implies a disproportionately larger number of discrete algebraic constraint equations. In some instances the solution can "lock", producing the velocity approximation $\boldsymbol{u} = \boldsymbol{0}$ which trivially satisfies incompressibility but fails to satisfy the momentum equations.

Inconsistent choices of basis functions are referrred to as "unstable."

As noted previously, the viscous flow problem with reduced quadrature rules can be formulated as a constrained optimization problem using the Lagrange multiplier method. Other techniques from constrained optimization theory are also applicable. In particular, the constraint $\nabla \cdot \boldsymbol{u} = \mathbf{0}$ can be enforced by adding a penalty functional to the variational statement. As the penalty parameter ϵ approaches zero, the constraint is more strongly enforced. The advantage of the penalty method is that only the velocity appears in the formulation of the problem, and therefore the size of the resulting discrete algebraic system is smaller. Similar problems to those noted for the mixed method may arise. Special "under-integration" techniques are needed to reduce the rank of the penalty matrix and circumvent these difficulties. Some common examples of stable and unstable elements are given in Figure 15.2. Finally, we remark that, the scaling due to the penalty coefficient also adversely affects roundoff error.

Biharmonic Stream Function

We conclude this discussion on the approximation of the viscous flow problem with a brief treatment of the fourth-order stream function formulation,

$$\nu \Delta^2 \psi = F \tag{15.94}$$

It is easy to derive a finite difference expression on a regular grid using Taylor series expansions, but now the difference stencil is larger since higher-order derivatives need to be approximated. When mapping strategies are used to transform the irregular domain to a computational domain, the form of the governing stream function equation becomes more complicated due to the additional terms generated by repeated application of the chain rule to derivatives.

As in the previous approaches, finite element formulations may be preferable due to their flexibility in handling irregular geometries with arbitrary triangular or quadrilateral grids. Beginning from a weighted residual statement for the fourth order biharmonic operator, and integrating by parts twice,

$$\int_{\Omega} \nu \Delta\psi \Delta w \, dx dy + \int_{\partial\Omega} \frac{\partial}{\partial n}(\Delta\psi) w \, ds - \int_{\partial\Omega} \Delta\psi \frac{\partial w}{\partial n} ds = \int_{\Omega} F w \, dx dy \tag{15.95}$$

In the particular case of an enclosed flow, where the stream function and normal derivative of the stream function are specified on the boundary, the

Velocity Approx.	Quadrature Rule (Pressure Approx.)	β_h	Rate of Convergence
	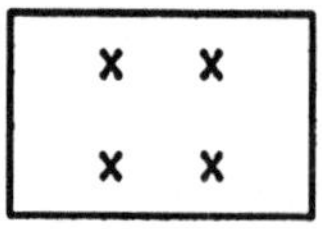	$O(1)$	Locks for small ϵ; ϵ must be taken as dependent on h
	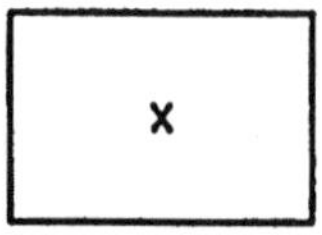	$O(h)$	Unstable pressure
	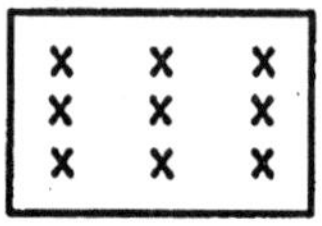	$O(1)$	Locks for small ϵ; ϵ must be taken as dependent on h
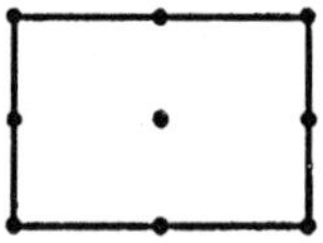	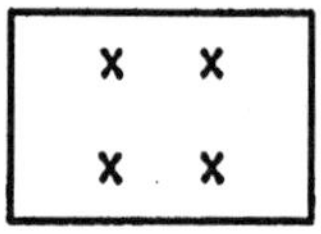	$O(h)$	Unstable pressure
	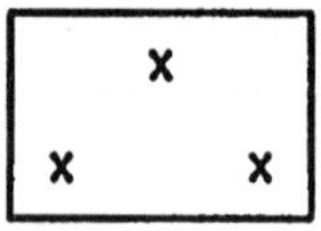	$O(h)$	Unstable pressure
	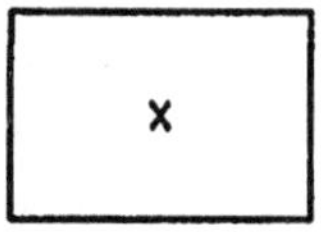	$O(1)$	Suboptimal ($O(h)$) for velocity error in energy norm
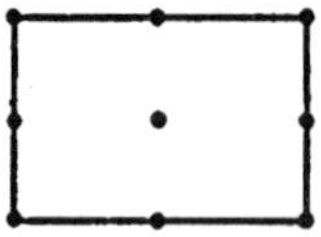	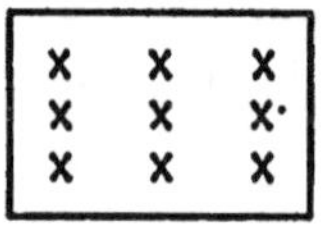	$O(1)$	Locks for small ϵ; ϵ must be taken as dependent on h

Figure 15.2: Examples of unstable and stable elements: velocity node $(\cdot)$, pressure node $(\times)$ (from [90]).

weak statement becomes: find $\psi \in H(\Omega)$ satisfying the boundary conditions and such that

$$\int_\Omega \nu \Delta\psi \Delta w \, dx dy = \int_\Omega F w \, dx dy \tag{15.96}$$

holds for all test functions $w \in H(\Omega)$, with both w and $\partial w/\partial n$ zero on the boundary $\partial\Omega$ of Ω. Here the space $H \equiv H^2(\Omega)$, the Hilbert space of functions with square-integrable second derivatives.

The integral expression (15.96) involves products of the second derivatives of the stream function. This implies a stronger restriction on the smoothness of the finite element basis functions than in the previous cases. A sufficient condition is to require that the first derivatives of the basis functions be continuous across the interelement boundaries. This, in turn, implies that high-degree, smoother polynomial basis functions are needed, complicating the formulation and solution. For these reasons the biharmonic stream function approach is rarely used, although recent activity in high-degree bases for second-order PDEs may lead to a resurgence of interest in this scheme.

15.6 Free Surface Conditions

The stream function-vorticity decoupled formulation is applied later to the oxide flow calculation. As boundary conditions, the normal component of oxide velocity at the silicon surface is specified as $u_n = \gamma V$, where V is the normal component of the velocity associated with the oxide growth rate in (15.4). The tangential and normal flow velocities can be represented in terms of the stream function as $u_s = \frac{\partial\psi}{\partial n}$, $u_n = -\frac{\partial\psi}{\partial s}$. The stream function boundary condition at the silicon surface follows on integrating $\partial\psi/\partial s = -\gamma V$ where $\gamma = 4/9$ is the volume expansion ratio for oxidation. At the gas surface, the surface stress in the normal direction satisfies

$$p - p_g - 2\mu\left(\frac{\partial u_n}{\partial n} + u_s\frac{\partial\alpha}{\partial n}\right) = -\beta\frac{\partial\alpha}{\partial s} \tag{15.97}$$

where β is the surface tension, α is the radius of curvature, $\partial\alpha/\partial s$ is the surface curvature, and p_g is the gas pressure. The surface stress condition in the tangential direction for an oxide with viscosity μ is

$$\mu\left(\frac{\partial u_n}{\partial s} + u_s\frac{\partial\alpha}{\partial s} + \frac{\partial u_s}{\partial n} - u_n\frac{\partial\alpha}{\partial n}\right) = 0 \tag{15.98}$$

Using (15.97) in conjunction with the continuity equation $\nabla \cdot \boldsymbol{u} = 0$, the gas surface pressure becomes

$$p - p_g = -\beta \frac{\partial \alpha}{\partial s} - 2\mu \left(\frac{\partial u_s}{\partial s} - u_n \frac{\partial \alpha}{\partial s} \right) \tag{15.99}$$

and from (15.98) the surface vorticity is

$$\zeta = 2 \left(\frac{\partial u_n}{\partial s} - u_s \frac{\partial \alpha}{\partial s} \right) \tag{15.100}$$

The basic solution algorithm proceeds as follows: at a given timestep the surface velocities are set equal to those obtained from the solution of the governing transport equations at the previous timestep. Using these velocities at both the silicon and gas surfaces, the surface nodal points are advanced to obtain a new oxide configuration. All interior nodes are moved proportionately, based on the motion of the silicon and gas interfaces, and the mesh is smoothed if the deformation of the elements is excessive. The positions of surface nodal points are also adjusted to maintain a well graded mesh, otherwise, nodes on the surface may coalesce in regions of high surface curvature. The viscous flow and reaction-diffusion problems are solved as quasi-steady problems within the current timestep. The flux and the reaction boundary condition then determine the silicon-oxide boundary velocity for the next step and the above procedure is repeated. For further details consult [396].

For the results in the next section, biquadratic quadrilateral elements for each of the solution variables are utilized. The number of elements in the mesh remains constant throughout the calculation and the nodal positions are adjusted in response to the surface motion as described above. At interelement nodes along the surface, normal and tangential directions are not uniquely defined, since the surface is not continuously differentiable at the interfaces between the adjacent elements. In the computations, a unique surface normal direction is chosen as a weighted-average of the values on adjacent elements. Since the solution for the oxide flow and oxidant concentration is a steady-state solution within each timestep, the interior solution information need not be retained from one timestep to the next. A moving mesh would normally introduce a convective term in the governing equations to account for the transformation associated with the moving boundary. This is not the case here, since the time scales associated with the steady state assumption imply that the effect of node motion on transport is negligible.

15.7 Silicon Oxidation Results

As an example, consider oxide growth on an inside trench corner. Oxidation in steam at one atmosphere pressure and various temperatures is simulated, resulting in an approximate oxide thickness of up to 0.3 μm. The mesh consists of 80 quadrilateral elements and the timestep varies with the temperature. For the particular case of oxidation at 1000°C, the timestep is fixed at 6 sec. The trench walls are 3 μm in length and covered with an initial thin oxide layer of 100 angstroms. This is a reasonable assumption since the process is extremely rapid for the first few hundred angstroms of oxide formation. Different crystal planes of the silicon oxidize at different rates. Near the corner, the surface orientation will not necessarily coincide with the principal plane, so an average value is taken in the numerical calculations.

Surface tension for glass has been measured, for various compositions of silicon dioxide with oxides of other metals, to be approximately 300 dynes cm^{-1}, and this value has been used in the test calculations. In fact, surface tension does not appreciably affect the flow because of the extremely high viscosity of oxide. Indeed, surface tension is a stabilizing force that resists thinning at the corners. The oxide layer at two distinct times in the process is given in Figure 15.3, which illustrates the thinning effect at the trench corner. Results at T = 1200°C do not exhibit this thinning as shown in Figure 15.3. The higher oxide viscosity at the lower temperature generates large stresses on the silicon surface near the corner, resulting in a depression of the reaction rate and more pronounced thinning. As a second example, the "bird's beak" problem was simulated, beginning with a uniform oxide layer, to produce the final configuration and mesh shown in Figure 15.5.

For oxidation at lower temperatures, stress effects in the oxide layer may be important in depressing the reaction rates and the diffusivity. This subject remains a topic of current research. Experiments in which annular silicon structures are oxidized, illustrate this point [276]. Here film thickness is measured for layers of oxide formed at various temperatures and surface curvatures. Using these results, data relating film thickness to surface curvature at various temperatures are tabulated. The difference between film thickness on a cylindrical surface as compared with a planar surface is small for oxidation temperatures above 1100°C. This difference can be substantial, however, for temperatures below 800°C and increases with increasing curvature.

Thinning in the corner of the trench is promoted by stress effects due to surface curvature. It is apparent that there is a correlation between surface curvature and oxide thickness, but from a modeling standpoint it is more

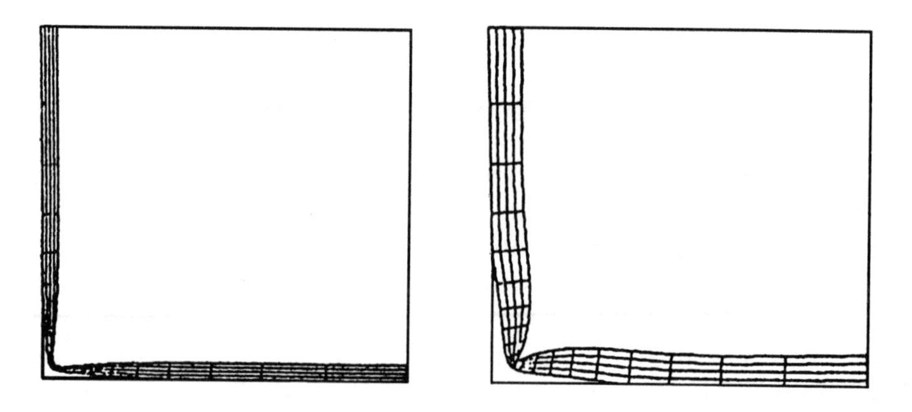

Figure 15.3: Oxide configurations at $T = 1000°C$ for trench corner at times t_1 and t_2 (from [396]).

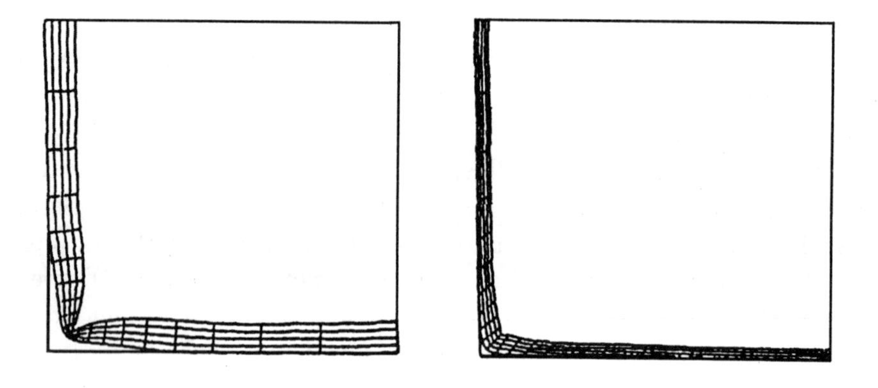

Figure 15.4: Oxide configurations for 1200°C (from [396]).

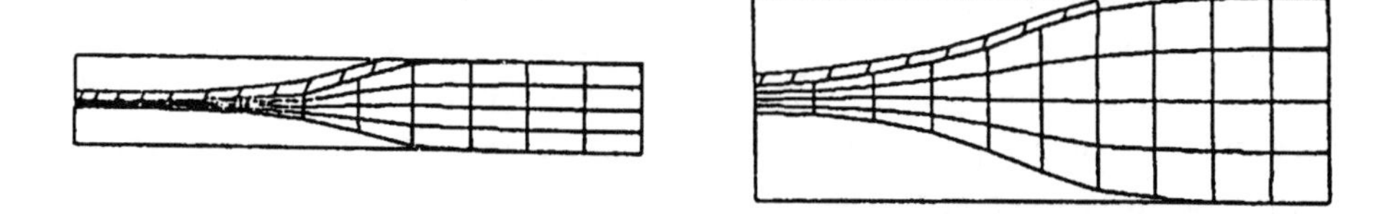

Figure 15.5: Sequence of meshes for bird's beak at 1000°C (from [396]).

appropriate to correlate oxide thickness with stress. The film stress depends on temperature and oxidation time. As temperature increases, the viscosity of the oxide decreases and the stresses relax more rapidly. Experiments

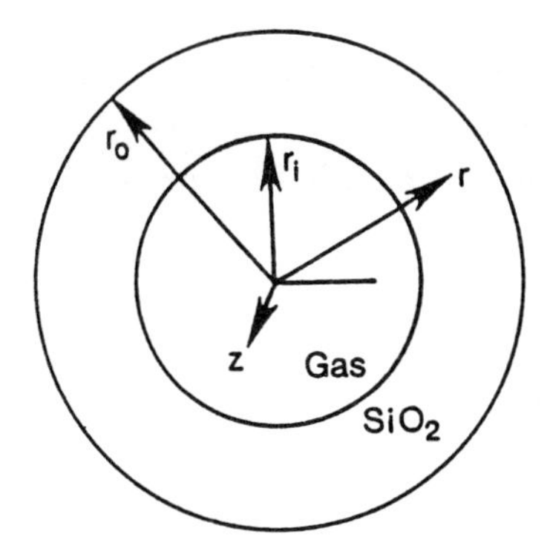

Figure 15.6: Geometry for cylindrical oxidation.

confirm that a thinner oxide is formed during oxidation at low temperatures when the stress is larger. It has been conjectured that oxide film stress enhances the surface reaction rate since, at a given temperature, a strained surface bond will react more readily than an unstrained bond. There is also evidence that the effect of stress on mass transport is to reduce the diffusivity of oxidant by straining the oxide network.

The volume change accompanying the reaction at the silicon surface results in motion of the oxide in the direction perpendicular to the silicon surface, but the component of velocity parallel to the surface is zero. As a result, a compressive normal stress exists parallel to the surface. To examine this effect, consider the planar oxidation problem for an oxide layer with gas and silicon surfaces at $x = x_g$ and $x = x_s$, respectively. An average film stress can be defined as

$$\sigma_f = \frac{1}{(x_g - x_s)} \int_{x_s}^{x_g} \sigma_y \, dx \tag{15.101}$$

where σ_y is the normal stress in the y direction. The curvature induced by this film stress can be measured by optical techniques to yield the strain in the silicon substrate. Using the material properties of silicon, the substrate stress can then be determined and the oxide film stress is obtained. This oxide film stress can be removed by subsequent annealing at high temperature.

The difference between high and low temperature behavior in the oxide may be attributed to differences in the viscosity at these temperatures. At temperatures near 1200°C, the stress relaxes rapidly and the total force in the film is not large enough to cause observable bending of the substrate. At temperatures below 700°C, the stress during oxidation persists due to the long relaxation times. Film stresses on the order of 7×10^9 dynes cm^{-2} have been observed and average film stresses in oxide samples with thicknesses as small as 100 angstroms have been measured. The value of film stress extrapolated to zero oxide thickness is approximately 4×10^9 dynes cm^{-2}.

The compressibility of the oxide can be defined as

$$\beta = -\frac{1}{v_0}\frac{\partial v}{\partial p} \tag{15.102}$$

where v is the specific volume, p is the pressure and the derivative is evaluated at constant temperature. The compressibility of oxide has been measured to be $2.7 \times 10^{-12}(\text{dynes cm})^2)^{-1}$ at room temperature, and has been observed to remain nearly constant at pressures up to 10^{10} dynes cm^{-2}. Furthermore, the increase in oxide density remains after removal of very high pressure, with an observed density increase of approximately 7%. Changes in density on the order of 3% at 600°C have been reported using measurements of the index of refraction of the oxide. These experiments imply that the density of oxide increases as a result of the increase in pressure.

There are several other relevant physical properties that must be considered. The thermal expansivity at constant pressure is defined by

$$\varepsilon = \frac{1}{v_0}\frac{\partial v}{\partial T} \tag{15.103}$$

and experiments confirm that diffusivity decreases with increasing pressure. This functional dependence of diffusivity on pressure can clearly be an important issue in modeling oxide growth. In theoretical work, the dependence of diffusivity D on pressure has been hypothesized of the form

$$D(T,p) = D_0(T)\exp\left(\frac{-(p-p_0)\Delta V}{RT}\right) \tag{15.104}$$

where D_0 is the diffusivity at standard pressure p_0, T is the absolute temperature, R is the gas constant and

$$\Delta V = -RT\frac{\partial}{\partial p}\left(\ln\frac{D}{D_0}\right) \tag{15.105}$$

is the activation volume. Such a physical model can be incorporated in a simulator to assess the effect of pressure on the diffusion process.

15.8 Planar and Cylindrical Oxidation

An analysis and simulation of compressibility effects in silicon for both planar and cylindrical surfaces is given in [396] and is summarized here. As reference velocity, the surface reaction velocity $\bar{u} = kc_g/N$ is used, where c_g is the gas solubility and N is the density of oxygen molecules in oxide. An appropriate time scale for compressibility effects is the stress relaxation time

$$\tau = \frac{\mu}{p_s} \tag{15.106}$$

where μ is the viscosity and p_s is the oxide pressure at a planar silicon surface. The length scale is $\ell = k\gamma\tau$, where $\gamma = c_g/N$ is the solubility fraction and k is the reaction rate coefficient. The associated Peclet number, $Pe = (k^2 c_g \mu)/(D_0 N^2 p_s)$, defines the ratio of the time scale for diffusion to that of surface motion. The Biot number may be expressed in terms of the Peclet number as $Bi = Pe/\gamma$, while the Schmidt number, $Sc = \mu/\rho_0 D_0$, is the ratio of the diffusion time scale to that of momentum transport.

Consider planar oxidation of a silicon surface aligned with the y axis. The silicon surface velocity is $\dot{x}_s = -\frac{4}{9}V$ where V is the velocity associated with the reaction. For this 1D case, the continuity equation simplifies to $(\rho u)' = 0$ so that $\rho(x)u(x) = \text{constant} = \rho_0 \frac{5}{9}V$. For convenience, a gas surface pressure $p(x_g) = 1$ atm is taken as the reference pressure, so that $\delta p = p - p(x_g) = 0$ at the gas surface in the following analysis. Using the equation of state, the velocity is

$$u(x) = \frac{5}{9}V(1 - \delta p(x)) \tag{15.107}$$

After considering equilibrium of viscous and pressure forces in the oxide film and using (15.107),

$$p(x) = p(x_s)e^{-(x-x_s)/a} \tag{15.108}$$

with $a = (20/27)\delta V$.

Combining (15.107) and (15.108),

$$u(x) = \frac{5}{9}V(1 - \delta e^{-(x_g - x_s)/a}) \tag{15.109}$$

The diffusivity is assumed to depend on the pressure in a form similar to (15.104), as

$$D = (Pe)^{-1} e^{-\lambda p(x)} \tag{15.110}$$

where $\lambda = p_s \Delta V/RT$ for activation volume ΔV.

Using Fick's law $q = -D(dc/dx)$ to relate concentration gradient to mass flux q, for a first order reaction

$$\frac{dc}{dx} = Bi\, e^{\lambda p(x)} c(x_s) \tag{15.111}$$

Table 15.1: Comparison of thickness ratio η for cylindrical and planar oxidation

Bi	α	t	s	η
2000	∞	0.01	0.00265	1.0
2000	0.1	0.01	0.00248	0.94
2000	0.01	0.01	0.00164	0.62
100	∞	100	1.723	1.0
100	10.0	100	1.744	1.01
100	5.0	100	1.769	1.03

with $c(x_g) = 1$ at the surface. Solving,

$$c(x) = \left(1 - \int_x^{x_s} Bi\, e^{\lambda p(x)} c(x_s) dx\right)^{-1} \tag{15.112}$$

and since $q(x_s) = -(Bi\, V)/Pe$, the velocity V is given by

$$V = \left(1 + Bi \int_{x_s}^{x_g} e^{\lambda p(x)} dx\right)^{-1} \tag{15.113}$$

Therefore, the surface position is the solution of $\dot{x}_s = \frac{4}{9}V$ with V computed implicitly by (15.113).

The treatment for cylindrical oxidation is similar, leading to

$$V = \left(1 + Bi \int_{r_s}^{r_g} \frac{r_s}{r} e^{\lambda p(r)} dr\right)^{-1} \tag{15.114}$$

where $r = r_g$ is the radial position of the gas surface and $r = r_s$ is the position of the silicon surface.

The relative film thickness $\eta = (r_g - r_s)/(x_g - x_s)$ can be used to compare planar and cylindrical oxidation. This ratio is computed in Table 15.1 for several Biot numbers with a dimensionless time t at which a thin oxide layer of thickness s is produced. The value of λ is fixed at 0.1. The effect of curvature can be examined by varying the radius of curvature α ($\alpha = \infty$ is the planar case). This model suggests that the increased pressure generated during oxidation for cylindrical films can decrease transport of oxidant and lead to thinning. Thinning occurs predominantly at high Bi because stress relaxation is inhibited by the high viscosity of oxide. (Little thinning is observed in experiments at low Bi.)

15.9 Summary

The silicon oxidation problem is one of the most complex modeling challenges in technology CAD since it involves coupled viscous flow, species transport, a moving reaction boundary and a deformable free surface. Not only are there the usual difficulties arising from computer simulation of Newtonian and non-Newtonian flows but the moving boundary problem increases the degree of difficulty significantly. Here we describe a basic model for the coupled flow and transport problem including the consistency requirements on the velocity-pressure spaces for finite element analysis. The treatment of the moving reacting boundary and related moving grid concepts are included. Some exploratory results for the bird's beak, trench and cylindrical oxidation problems conclude the treatment.

The formulation for coupled viscous flow and transport and the associated numerical techniques presented here apply to a variety of other applications of interest to the semiconductor industry. These include chemical vapor deposition, chemical–mechanical polishing of wafers, laser bonding for packaging, and crystal growth by the Bridgman process. This crystal growth problem is considered in Chapter 16.

15.10 Exercises

1. Compare the solution to a planar diffusion problem for a slab of thickness L to the corresponding solution for radial diffusion through an annulus of inner radius R and outer radius $R + L$. Consider the cases of large and small R.

2. Let $g = u + iv$, $i = \sqrt{-1}$ be an analytic function of complex variable $z = x + iy$. Derive the Cauchy–Riemann equations in (15.24) and verify that u and v are harmonic functions.

3. Carry out a Fourier stability analysis of the trapezoidal integration (Crank–Nicholson) scheme (15.45) for the diffusion equation to show that the scheme is unconditionally stable.

4. In practice it has been observed that high frequency error modes persist when the trapezoidal scheme is used. Apply the recursion in (15.45) to a high frequency eigenmode to explain this result. Comment on the nature of this transient behavior. Carry out a similar analysis for the forward and backward schemes and compare your conclusions.

5. Consider the stationary reaction diffusion equation (15.46) with reaction term $f = f(c)$ on the right. Assuming the nonlinear reaction is mild, modify the solution scheme in (15.51)–(15.53) to obtain an efficient iterative scheme.

6. Verify theoretically that ω^* in (15.57) provides the optimum choice of relaxation parameter for Gauss–Seidel iteration of Laplace's equation. Consider a low-, mid- and high-frequency error mode and determine the relative reduction in amplitude for one iteration of the Gauss–Seidel scheme.

7. Write down the Jacobi iteration with relaxation for the steady state diffusion problem and compare this with the forward-Euler explicit integration recursion for the transient problem.

8. Assume that the diffusivity in exercise 7 is no longer constant but is now a function of the solution c. Propose a local time-stepping scheme to obtain rapid "iterative" acceleration to the steady state. Comment on the utility of this local adaptive point relaxation scheme.

9. Use the weak form of the diffusion problem in (15.62) and the corresponding statement of the finite element problem to derive the orthogonality condition

$$\int_\Omega \nabla e \cdot \nabla w_h \, dx dy = 0$$

for error $e = c - \tilde{c}$ where $\tilde{c}$ denotes the papproximation and w_h is the test function in the approximation space. Introduce the finite element interpolant and use ellipticity and the orthogonality condition to show that $\|e\|_1 \leq C\|e_I\|_1$ where e_I is the interpolation error, C is a constant and $\|\cdot\|_1$ denotes the corresponding H^1 norm.

10. Write down the boundary conditions at a "no-slip" surface for the biharmonic stream function formulation in (15.74) and for the coupled stream function/vorticity equations in (15.70)–(15.73). Construct finite difference schemes for these two forms of the governing equations.

11. Develop an appropriate treatment for the "boundary effect" in the decoupled stream function/vorticity iterative algorithm (15.76)–(15.77). Explain your reasoning.

12. The "bird's beak" oxidation problem involves an oxide layer with moving free and reacting surfaces (Figure 15.5). Assume that the position

$x(s)$, $y(s)$ of these surfaces is known as a function of arc length s at each step in the simulation. Construct a mapping of the oxide layer to a rectangular computational domain. Give the corresponding form of the mapped viscous flow and diffusive transport equations.

13. Describe how the strategy in exercise 12 would be modified to use an elliptic PDE solution technique to compute the mapping (i.e. compute the grid point coordinates in the physical domain).

14. Construct the element matrix $\boldsymbol{B}$ for the Galerkin primitive formulation in (15.91) with constant pressure and bilinear velocity approximation on each element. Consider a patch of four elements and verify by substitution that the matrix $\boldsymbol{B}$ can support a spurious oscillatory "checkerboard" mode on this patch.

15. Show that the lowest-degree complete polynomial approximation on a triangle that will satisfy the inter-element C^1 continuity is quintic. Sketch the element and mark appropriate vertex, mid-side and interior degrees-of-freedom for the element. Comment on the structure and properties of the element and system matrices and the treatment of boundary conditions. How would this scheme be modified to handle the free and reacting surfaces in the silicon oxidation problem?

Chapter 16

Crystal Growth

16.1 Introduction

An important class of processing problems for the semiconductor industry involves the growth of high quality semiconductor crystals. A standard approach is to grow crystals from a liquid "melt" by solidification, using an imposed thermal gradient. This is the basic mechanism for crystal growth in the horizontal Bridgman process. This process, therefore, involves a moving phase-change boundary which separates the molten and solidified semiconductor regions. Heat transfer due to the applied thermal gradient occurs by conduction in the solid phase and by both conduction and convection in the melt. Moreover, buoyancy forces lead to circulatory flows in the melt zone and these flows may have complicated cell structures similar to those observed in Rayleigh–Benard and Benard–Marangoni flows for heated fluid layers. Hence, the process involves coupled fluid flow and heat transfer with a moving phase-change boundary. Evidently, there are several similarities to the silicon oxidation process in the previous chapter. Both processes involve: (1) a viscous flow domain; (2) a driving transport process; and (3) a moving interface. Therefore, the analysis and numerical techniques for modeling either process are fundamentally similar.

16.2 Coupled Flow and Heat Transfer

Perhaps the most significant distinctions between the crystal growth and oxidation processes from the standpoint of mathematical modeling are in the hydrodynamics. Both processes involve flow of an incompressible viscous fluid that can be mathematically represented by the Navier–Stokes

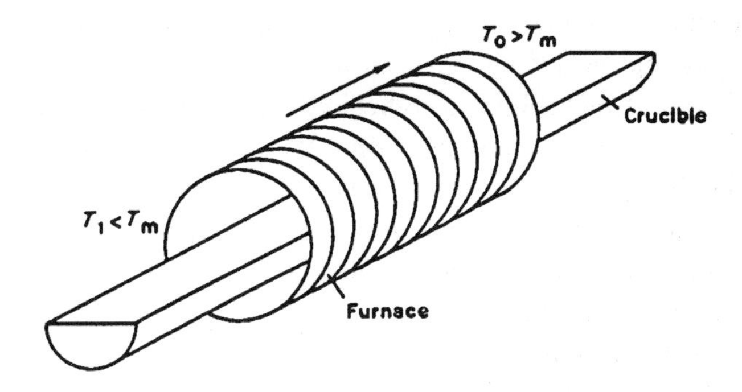

Figure 16.1: Bridgman process for crystal growth.

equations. Temperature differences cause density gradients, which produce buoyancy forces to drive the flow in the crystal melt. On the other hand, in the oxidation problem, reaction at the silicon interface produces a volumetric expansion that induces flow in the oxide. Both flows are slow relative to those encountered in other hydrodynamic problems such as chemical-mechanical polishing. Oxide growth is truly a creeping flow that can be represented by linear Stokes flow, but with complications associated with thin film stresses. Flow in the melt involves buoyancy driven cells, and the extreme (Rayleigh and Grashoff) parameters that govern the flow prohibit linear analysis. Consequently, nonlinear effects in the Navier–Stokes equations are important in determining flow and heat transfer for the crystal growth problem. In fact, for practical ranges of these parameters, the melt flow for gallium arsenide crystal growth may involve bifurcation from a steady-state solution to a transient periodic solution. This, in turn, may have a strong impact on the heat transfer process, phase change and geometry of the moving interface.

In the horizontal Bridgman process, a furnace supplies a thermal gradient to a crucible containing the melt. The crucible is fixed and shaped like a "boat," e.g. a long horizontal half-cylinder with semicircular section. The furnace moves slowly along the crucible followed by the solidification interface (see Figure 16.1). It has been observed that, for many different crystal growth configurations, the flow becomes unstable and periodic oscillations may develop when critical temperature gradients exist across the crucible. Theoretical issues related to the onset of this instability have received some consideration, but there remain many open questions. As with other hydrodynamic instabilities, there are uncertainties associated with the dimensionality of the flow; e.g. in Rayleigh–Benard flows the cell structures in two dimensions are cylindrical rolls, while in three dimensions they are arrays of hexagonal cells like those of a beehive. In some two-dimensional flows, the dominant instability is to break into a three-dimensional flow and, therefore, it is not readily analyzed using 2D models. However, experimental studies of the Bridgman process do suggest that a two-dimensional model provides some insight into the flow behavior and dynamics. Accordingly, the present treatment focuses on the coupled heat transfer and hydrodynamics in the melt zone to examine cell structure and bifurcation from the steady state to a periodic flow behavior.

The length of the "boat" in the Bridgman process is typically around 20 cm with the radius of the semi-circular section near 2.5 cm. As the crystal grows, the length of the melt decreases correspondingly. The longitudinal temperature gradient imposed by the furnace is approximately 2°C cm^{-1}, and the associated buoyancy forces generate vortices in the melt. It is instructive to first consider the classical two-dimensional Rayleigh–Benard geometry for a heated layer of fluid.

Benard [40] carried out experiments to study the stability of a horizontal fluid layer of depth d between two plates, with the lower plate at temperature T_1, and the upper plate at temperature T_0, $T_1 > T_0$. For steady flow of an incompressible fluid, the governing stationary Navier–Stokes and energy equations for velocity $\boldsymbol{u}$, pressure p and temperature T can be written as

$$-\Delta \boldsymbol{u} + \boldsymbol{u} \cdot \nabla \boldsymbol{u} + \nabla p - \frac{Ra}{Pr} T \boldsymbol{j} = \boldsymbol{0} \tag{16.1}$$

$$\nabla \cdot \boldsymbol{u} = 0 \tag{16.2}$$

$$-\frac{1}{Pr} \Delta T + \boldsymbol{u} \cdot \nabla T = 0 \tag{16.3}$$

where Ra and Pr are the dimensionless Rayleigh and Prandtl numbers

$$Ra = \frac{\beta g d^3 (T_1 - T_0)}{\nu k}, \quad Pr = \frac{\nu}{k} \tag{16.4}$$

for a fluid with kinematic viscosity ν, thermal conductivity k and thermal expansion coefficient β; g is the acceleration due to gravity and $\boldsymbol{j}$ is a unit gravitational vector. The Rayleigh number Ra reflects the importance of buoyancy relative to viscous dissipation and heat conduction. The Prandtl number Pr indicates the relative importance of heat conduction versus viscous dissipation. The ratio $Gr = Ra/Pr$ in the momentum equation (16.1) defines the Grashoff number $Gr = Ra/Pr = (\beta g d^3(T_1 - T_0))/(\nu^2)$ and measures the importance of buoyancy forces within the flow. For the Bridgman growth process, typical values are $Pr \sim 10^{-2}$ and $Gr \sim 10^6$.

In the Rayleigh–Benard problem, the fluid is stationary at low Rayleigh number, e.g. for small temperature gradients. The temperature field then corresponds to the simple one-dimensional conduction solution $T(x, y) = T_0 y/d + T_1(1 - y/d)$ of the steady 1D equation $T_{yy} = 0$ through the layer. If T_1 is gradually increased, then Ra increases and above some critical value Ra_{cr} the buoyancy forces become sufficiently strong to destabilize the viscous forces so that fluid motion occurs in the form of steady convective cells. From a mathematical standpoint Ra_{cr} is an eigenvalue of the linear problem, and there is a bifurcation from the trivial flow solution $\boldsymbol{u} = \mathbf{0}$, to a steady vortex flow structure. As the temperature T_1 is further increased, the velocities become stronger, and at higher critical values further cells are generated.

Note that the trivial solution $\boldsymbol{u} = \mathbf{0}$ is still an admissible solution branch for $Ra > Ra_{cr}$, but it is unstable relative to an arbitrary perturbation. The nature of the perturbation determines the orientation of flow within adjacent vortex pairs. Another classical flow problem that is relevant to the present work is natural convection in a cavity with heated side walls. This corresponds to rotating the Rayleigh–Benard problem through an angle $\pi/2$. In this case, flow is initiated immediately upon application of the temperature boundary conditions and cell structures evolve.

An approach to modeling the Rayleigh–Benard problem is to solve numerically the stationary equations (16.1)–(16.3). This presumes that bifurcations, if present, proceed to genuine steady solutions. Introducing a finite difference, finite element, or finite volume approximation to the PDE system (16.1)–(16.3), leads, as before, to a coupled system of sparse algebraic equations for the vectors $\boldsymbol{v}$, $\boldsymbol{p}$, $\boldsymbol{T}$ of nodal velocity, pressure and temperature on the computational grid. The resulting algebraic system is nonlinear and has the form

$$\tilde{\boldsymbol{A}}\boldsymbol{v} + \boldsymbol{c}(\boldsymbol{v}) + \boldsymbol{B}\boldsymbol{p} + Gr\boldsymbol{M}\boldsymbol{T} = \boldsymbol{b} \tag{16.5}$$

$$\boldsymbol{B}^T\boldsymbol{v} = \mathbf{0} \tag{16.6}$$

$$(Pr)^{-1}\boldsymbol{A}\boldsymbol{T} + \boldsymbol{E}(\boldsymbol{v})\boldsymbol{T} = \boldsymbol{f} \tag{16.7}$$

where $\tilde{\boldsymbol{A}}$ is a block diagonal matrix with blocks $\boldsymbol{A}$ corresponding to the discrete negative Laplacian in the respective momentum equations; $\boldsymbol{c}(\boldsymbol{v})$ and $\boldsymbol{E}(\boldsymbol{v})$ embody the nonlinear terms; $\boldsymbol{B}$ is the discrete gradient operator; $\boldsymbol{B}^T$ is the discrete divergence operator; Gr, Pr are Grashoff and Prandtl numbers; and $\boldsymbol{b}$, $\boldsymbol{f}$ are associated with prescribed boundary conditions. The matrix $\boldsymbol{M}$ is the identity matrix if the finite difference method is employed or the Gram matrix for a finite element method.

The discrete system (16.5)–(16.7) can be written compactly as the nonlinear system for vector $\boldsymbol{w} = (\boldsymbol{v}\,\boldsymbol{p}\,\boldsymbol{T})$

$$\boldsymbol{F}(\boldsymbol{w}) = \boldsymbol{0} \tag{16.8}$$

The behavior of the solution $\boldsymbol{w}$ depends on the problem parameters Ra, Pr, Gr. For example, if $Ra < Ra_{cr}$, then the trivial (no flow, conducting) solution is obtained. If $Ra > Ra_{cr}$ there are convective cells, so $\boldsymbol{v} \neq \boldsymbol{0}$ and the nonlinear system (16.8) must be solved from a specified starting iterate $\boldsymbol{w}^{(0)}$. Using Newton's method, iterates $\boldsymbol{w}^{(k)}$, $k = 1, 2, \ldots$, are obtained by solution of the linearized systems

$$\boldsymbol{J}^{(k-1)}\left(\boldsymbol{w}^{(k)} - \boldsymbol{w}^{(k-1)}\right) = -\boldsymbol{F}\left(\boldsymbol{w}^{(k-1)}\right) \tag{16.9}$$

where $\boldsymbol{J}^{(k-1)} \equiv J_{ij}(\boldsymbol{w}^{(k-1)}) = \left(\frac{\partial Fi}{\partial w_j}\right)_{\boldsymbol{w}=\boldsymbol{w}^{(k-1)}}$ is the Jacobian matrix evaluated at the current iterate.

Convergence of (16.9) depends on the strength of the nonlinearity and the quality of the initial guess, which must be in the domain of attraction of the solution. The size of this region will depend on the nature of the nonlinearity. Continuation in a parameter such as the Rayleigh number, or equivalently, the temperature T_1 of the lower plate is often necessary for such problems. Recall that a similar incremental continuation scheme in the applied voltage was successful for the semiconductor device transport problem in Chapter 7. However, since the Jacobian in (16.9) is singular at the critical points, this incremental continuation approach will fail near the bifurcation point. Let the arc-length parameter s be introduced so that v and Ra are now both functions of s and an additional equation is added to (16.8). The critical point is then transformed into a regular point with respect to the new parameter. Continuation now proceeds using the arc-length as a parameter to trace out the proper solution branch (Figure 16.2).

16.3 Transient Analysis

As indicated previously, the Bridgman process may undergo bifurcation to a periodic unsteady solution state. That is, a stable steady-state post-critical solution cannot be achieved and instead a periodic unsteady flow occurs. Transient analysis techniques can be introduced to compute the unsteady, periodic flow. This approach can then be used to provide an accurate description of the time evolution of the solution as well as the final steady state or periodic state. This also obviates the need for special

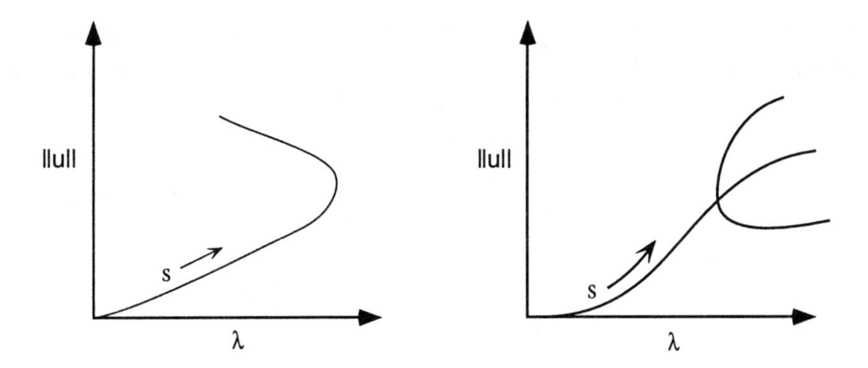

Figure 16.2: Representative diagrams for solution measure $||u||$ and parameter λ showing turning point and bifurcation behavior (s is arc length).

techniques such as parameter continuation or bifurcation detection, since the numerical solution models the gradual evolution of the flow through time. Of course, accuracy and stability requirements for the time integration procedure must be satisfied if a viable numerical solution is to be achieved.

The time-dependent form of the governing equations becomes (compare (16.1)–(16.3)):

$$\frac{\partial \boldsymbol{u}}{\partial t} - \Delta \boldsymbol{u} + \boldsymbol{u} \cdot \boldsymbol{\nabla} \boldsymbol{u} + \boldsymbol{\nabla} p - GrT\boldsymbol{j} = \mathbf{0} \tag{16.10}$$

$$\boldsymbol{\nabla} \cdot \boldsymbol{u} = 0 \tag{16.11}$$

$$\frac{\partial T}{\partial t} + \boldsymbol{u} \cdot \boldsymbol{\nabla} T - \frac{1}{Pr} \Delta T = 0 \tag{16.12}$$

Discretizing in space, the semi-discrete system of ordinary differential equations for $\boldsymbol{v}(t)$, $p(t)$, $T(t)$ has the form (compare (16.5)–(16.7)):

$$\boldsymbol{G}\dot{\boldsymbol{v}} + \tilde{\boldsymbol{A}}\boldsymbol{v} + \boldsymbol{c}(\boldsymbol{v}) + \boldsymbol{B}\boldsymbol{p} + Gr\boldsymbol{M}\boldsymbol{T} = \boldsymbol{b} \tag{16.13}$$

$$\boldsymbol{B}^T \boldsymbol{v} = \mathbf{0} \tag{16.14}$$

$$\boldsymbol{M}\dot{\boldsymbol{T}} + (Pr)^{-1}\boldsymbol{A}\boldsymbol{T} + E(\boldsymbol{v})\boldsymbol{T} = \boldsymbol{f} \tag{16.15}$$

where $\boldsymbol{G}$ is block-diagonal with block submatrices $\boldsymbol{M}$. More compactly, for $\boldsymbol{w}(t) = (\boldsymbol{v}(t)\, \boldsymbol{p}(t)\, \boldsymbol{T}(t))$,

$$\boldsymbol{H}\dot{\boldsymbol{w}} + \boldsymbol{F}(\boldsymbol{w}) = \mathbf{0} \tag{16.16}$$

where $\boldsymbol{v}$ satisfies the incompressibility constraint (16.14) at each timestep, $\boldsymbol{H}$ consists of block submatrices $\boldsymbol{M}$ and zero blocks, and the ODE system

(16.16) is to be integrated with respect to time from a prescribed initial time to the final time of interest. Explicit, predictor–corrector or fully-implicit techniques can be introduced to integrate (16.16). In [121], a predictor–corrector strategy is used for 2D transient calculations. The basic steps in the algorithm are as follows. For given solutions $\boldsymbol{w}_{n-1}, \boldsymbol{w}_n$ at the previous and current time levels t_{n-1}, t_n, and derivative $\dot{\boldsymbol{w}}_{n-1}$ at t_{n-1}:

1. Estimate $\dot{\boldsymbol{w}}_n$ using the trapezoidal rule through timestep Δt_{n-1}

$$\frac{1}{2}(\dot{\boldsymbol{w}}_{n-1} + \dot{\boldsymbol{w}}_n) = (\boldsymbol{w}_n - \boldsymbol{w}_{n-1})\Delta t_{n-1} \tag{16.17}$$

2. Predictor step: predict $\boldsymbol{w}_{n+1}$ using

$$\boldsymbol{w}_{n+1}^P = \boldsymbol{w}_n + \frac{\Delta t_n}{2}\left((2 + \frac{\Delta t_n}{\Delta t_{n-1}})\dot{\boldsymbol{w}}_n - \frac{\Delta t_n}{\Delta t_{n-1}}\dot{\boldsymbol{w}}_{n-1}\right) \tag{16.18}$$

3. Corrector step: trapezoidal integration of (16.16) yields

$$\boldsymbol{H}(\boldsymbol{w}_{n+1} - \boldsymbol{w}_n) + \frac{\Delta t_n}{2}(\boldsymbol{F}(\boldsymbol{w}_{n+1}) + \boldsymbol{F}(\boldsymbol{w}_n)) = \boldsymbol{0} \tag{16.19}$$

where the nonlinear system (16.19) is solved by Newton iteration using $\boldsymbol{w}_{n+1}^P$ as the starting iterate.

The computationally expensive parts of the algorithm are the formation of the Jacobian matrix and associated linear system solution in step 3. The efficiency of this calculation may be improved by replacing the full Newton iteration by a block-Newton decoupled scheme. Specifically, we can decouple the temperature calculation and "lag" the Boussinesq body force term $GrMT$ as a forcing vector in the momentum equations. This is a common block scheme for coupled flow and transport processes. On scaling by Pr, it is clear from (16.12) that the coupling term $(Pr)\boldsymbol{u} \cdot \nabla T$ associated with the velocity will be small for $0 < Pr \ll 1$. Therefore, the block decoupled scheme is appropriate for this problem.

This method has been applied in [121] to a class of two-dimensional test problems corresponding to flow and heat transfer in rectangular domains of length L and width w where $L/w = 2$, 4 or 8. A graded mesh of 22×9 biquadratic nine-node elements in Figure 16.3 is used for velocity and temperature, with a discontinuous linear basis for pressure. The nondimensional temperature is specified as $T = 1$ on the left wall and $T = 0$ on the right wall, so the problem models the example of natural convection in a rectangular cavity discussed earlier. The free surface is assumed insulated

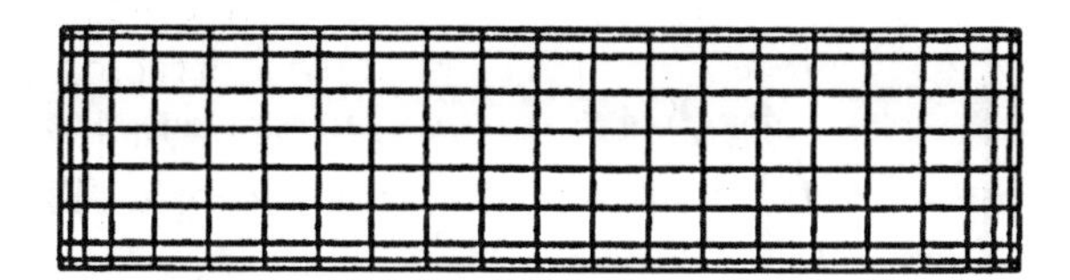

Figure 16.3: Mesh for Bridgman calculation.

and the temperature variation is linear on the lower surface. Following [121], a dimensionless form of the kinetic energy per unit thickness in the direction perpendicular to the plane of the flow may be introduced as

$$K^* = \int_{\Omega^*} \frac{1}{2}(u^{*2} + v^{*2})dx \tag{16.20}$$

where Ω^* is the non-dimensional flow domain and $\frac{1}{2}(u^{*2} + v^{*2})$ is the dimensionless kinetic energy. This functional provides a useful scalar quantity for characterizing the flow and for constructing bifurcation diagrams. Other quantities such as the L^2 or L^∞ norm of the solution could also be used.

For the test problem, the Prandtl number is taken as 0.015 and the values of Ra are 1000, 5000, 10,000, 20,000 for $w = 2.5$ cm and actual temperature differences $T_0 - T_1 = 0.57, 2.86, 5.72, 11.44°C$. The corresponding values of Gr are $\frac{1}{3}(2 \times 10^5, 10^6, 2 \times 10^6, 4 \times 10^6)$. For $T_0 - T_1$ small, Gr is low and a weak single cell is formed, but by $T_0 - T_1 \sim 0.28°C$ at $Gr = 33,333$, a weak double cell is present. Increasing the temperature difference further, at $Gr = 666,666$, the time evolution of K^* exhibits oscillations with a period of roughly 90 s, as indicated in Figure 16.4. The oscillations decay and the eventual steady state solution corresponds to the streamline and isotherm plot shown in Figure 16.5.

The oscillations become more persistent as Gr increases and sustained periodic oscillations are evident at $Gr = 316,666$ with a period of 33 s in Figure 16.4. Attempts to solve the Newton system for the steady problem at this value of Gr fail. If $L/h = 4$ and Gr is further increased to $\frac{4}{3} \times 10^6$, the sustained periodic flow has a main vortex adjacent to the cold wall on the right and several other vortices that alternate in size and strength. The streamlines and isotherms at two-second intervals during a period are shown in Figure 16.6.

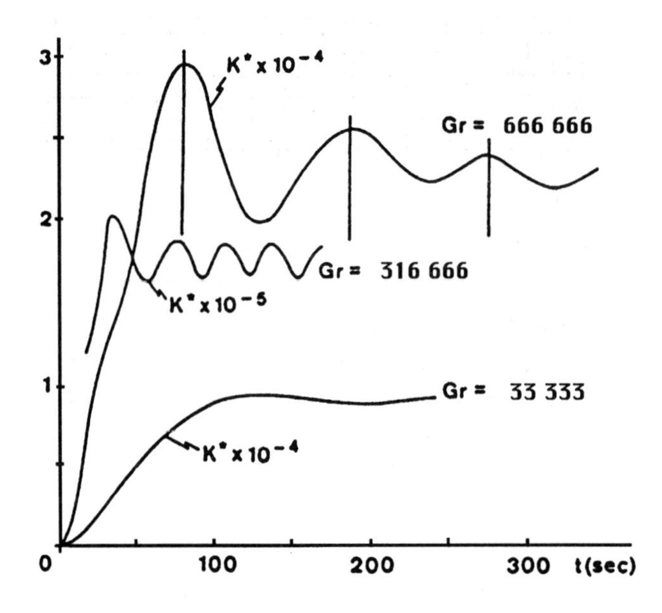

Figure 16.4: Variation of K^* with increasing t for flow at different Gr (from [121]).

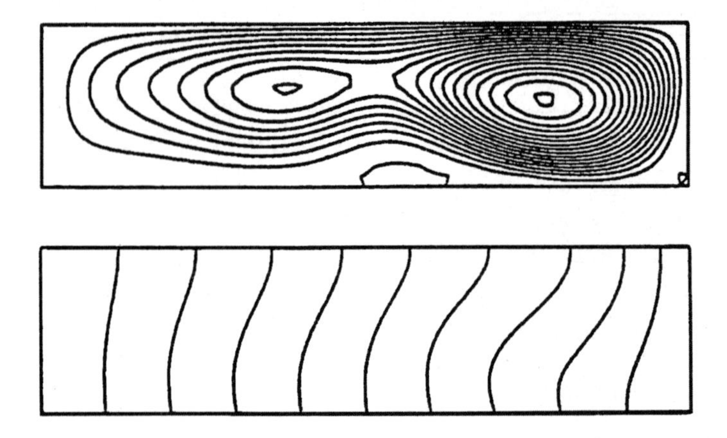

Figure 16.5: Steady state streamline and isotherm contours for process at $Gr = 666,666$ (from [121]).

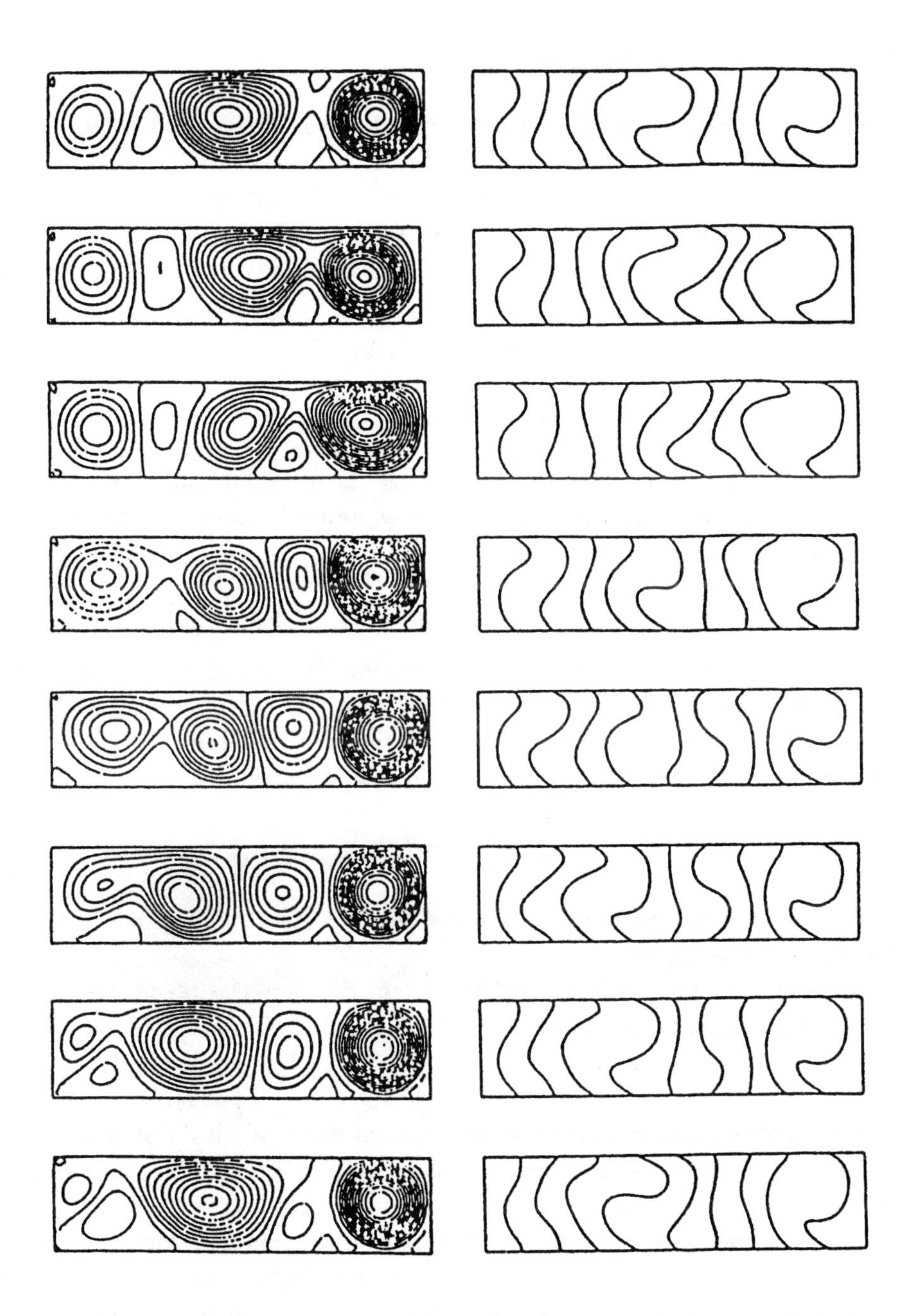

Figure 16.6: Streamline and isotherm contours at two-second intervals during a period (from [121]).

16.4 Phase Change

The preceding treatment has been directed to the flow structure in a melt, but in practical crystal growth applications there is a moving phase-change boundary separating the melt and crystal. In this two-phase problem heat transfer takes place by conduction in the solid region but by both convection and conduction in the melt. That is, the governing equations for temperature are

$$\rho_s c_s \frac{\partial T}{\partial t} = k_s \Delta T \tag{16.21}$$

$$\rho_\ell c_\ell (\frac{\partial T}{\partial t} + \boldsymbol{u} \cdot \boldsymbol{\nabla} T) = k_\ell \Delta T \tag{16.22}$$

where subscripts s and ℓ indicate coefficients for solid and liquid phases, respectively, ρ is density, c is heat capacity, and k is thermal conductivity.

At the phase-change boundary

$$T = T_{\mathrm{m}} \tag{16.23}$$

where T_{m} is the melt temperature. Moreover, the velocity $\boldsymbol{V}$ of the phase boundary is governed by the Stefan condition. This energy balance statement at the phase boundary relates the thermal flux jump in the normal direction to the velocity via the latent heat L of fusion by

$$\left[\!\left[k \frac{\partial T}{\partial n} \right]\!\right] = L \boldsymbol{V} \cdot \boldsymbol{n} \tag{16.24}$$

where $[\![\cdot]\!]$ denotes the jump across the phase boundary and $\boldsymbol{n}$ is the unit normal to the boundary.

Note that in the crystal growth problem the material properties change at the phase-change boundary. This change, together with the velocity condition (16.24), suggests that the phase-change boundary should coincide with a connected set of element boundary segments. That is, it is preferable that the phase-change boundary does not intersect the interior of elements when using a finite element or control volume method. This implies that nodes on the phase change boundary move in the direction normal to the phase boundary with velocity $\boldsymbol{V} \cdot \boldsymbol{n}$ based on (16.24) and that the meshes in the liquid and solid subdomains move accordingly or be regenerated. For example, elements in the interior of the solid region can be linearly stretched and those in the liquid region linearly compressed based on the motion $(\boldsymbol{V} \cdot \boldsymbol{n})\Delta t$ of the interface in timestep Δt.

An effective approach is to uncouple the phase boundary motion from the flow and transport equations in each timestep Δt of the integration scheme. As a simpler illustrative example, in the one-dimensional Stefan problem (16.24) becomes

$$\left[\left[k\frac{\partial T}{\partial x}\right]\right] = LV \tag{16.25}$$

so the velocity V of the phase boundary satisfies

$$V = \frac{1}{L}\left[\left[k\frac{\partial T}{\partial x}\right]\right] \tag{16.26}$$

Let $s(t)$ denote the position of the phase boundary at time t. As $V = ds/dt$, integrating (16.26) numerically with respect to time through a timestep Δt, the forward Euler scheme yields

$$s(t+\Delta t) = s(t) + \frac{1}{L}\left[\left[k\frac{\partial T}{\partial x}\right]\right]\Delta t \tag{16.27}$$

where the flux jump is approximated as constant throughout the step and is computed from the temperature field at the end of the previous step. Let the domain be normalized to $[0,1]$ with the solidification boundary moving towards the left. Grid points to the left of phase boundary $s(t)$ may be linearly compressed and grid points to the right linearly stretched by interpolating the grid velocity $V(x)$ at time t as $U(x) = Vx$ for $x \leq s(t)$ and $U(x) = V(1-x)/(1-s)$ for $x \geq s(t)$. Then the new grid is obtained by convective adjustment of the old grid following (16.27) according to $x_i(t+\Delta t) = x_i(t) + U(x_i)\Delta t$. The generalization to higher dimensions is conceptually clear, but introduces another level of complexity in terms of implementation. In the finite element model, nodes on the phase boundary move in the normal direction through distance $\delta s = \left[\left[\frac{k}{L}\frac{\partial T}{\partial n}\right]\right]_{S(t)}\Delta t$ and the grid is convectively adjusted with velocity U.

It is important to recognize that the grid adjustment can be interpreted as a transformation from a fixed reference frame or grid $(\boldsymbol{\xi},\tau)$. This implies that there will be a convective grid adjustment in the heat transfer equation. Introducing the transformation $\boldsymbol{x} = \boldsymbol{x}(\boldsymbol{\xi},\tau)$, $t = \tau$ and using the definition of the material derivative (or the chain rule and transposing)

$$\frac{\partial T}{\partial t} = \frac{\partial T}{\partial \tau} - \sum_{i=1}^{N}\frac{\partial T}{\partial x_i}\frac{\partial x_i}{\partial \tau} = \frac{\partial T}{\partial \tau} - \boldsymbol{U}\cdot\boldsymbol{\nabla} T \tag{16.28}$$

where $\boldsymbol{U}$ is the transformation velocity. Note that the transformation velocity enters as a convective adjustment $\boldsymbol{U}\cdot\boldsymbol{\nabla} T$ to the heat transfer equations.

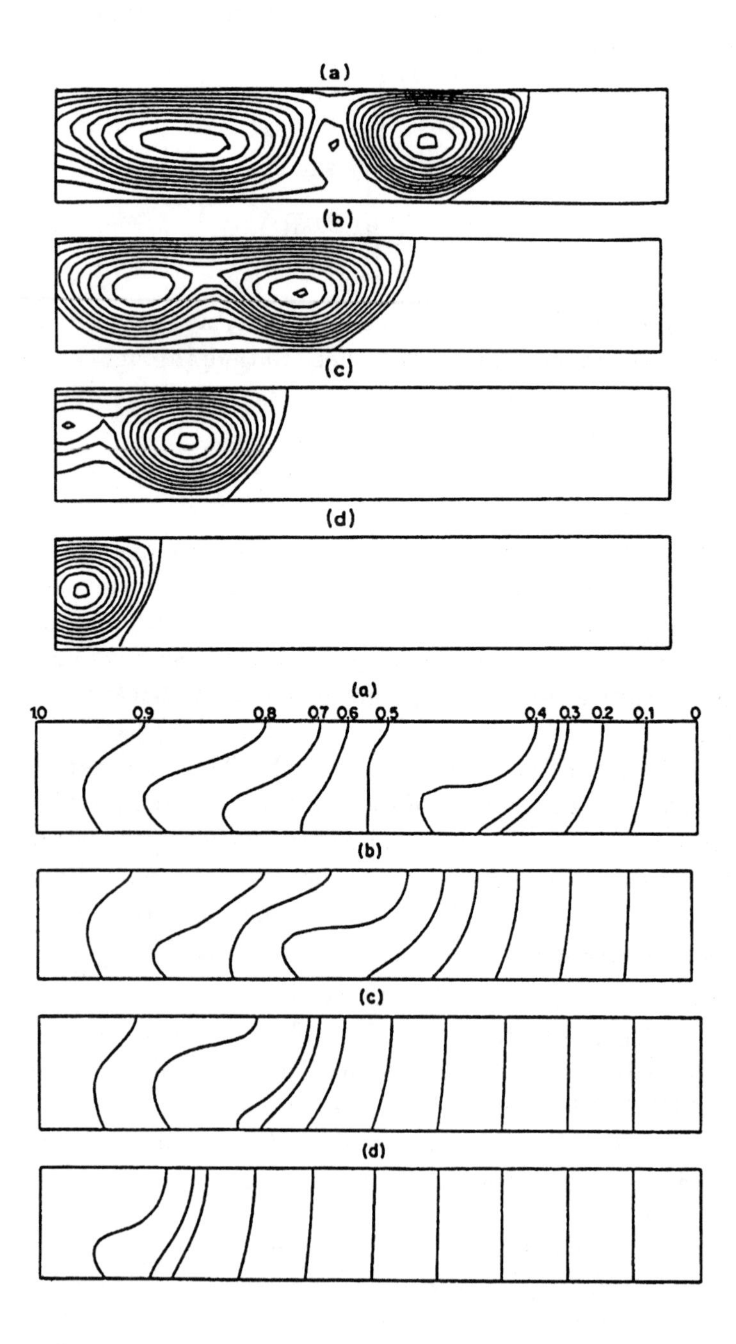

Figure 16.7: Streamline and isotherm patterns (from [119]).

Substituting into (16.21) and (16.22), the governing heat transfer equations in the moving mesh frame are

$$\rho_s c_s \left(\frac{\partial T}{\partial \tau} - \boldsymbol{U} \cdot \nabla T \right) = k_s \Delta T \tag{16.29}$$

$$\rho_\ell c_\ell \left(\frac{\partial T}{\partial t} - \boldsymbol{U} \cdot \nabla T + \boldsymbol{u} \cdot \nabla T \right) = k_\ell \Delta T \tag{16.30}$$

The time derivatives in the Navier–Stokes equations transform similarly to include a convective adjustment for the moving reference frame or grid. Introducing the finite element expansion $\boldsymbol{u}_h$, p_h, T_h into the corresponding integral statement and simplifying, leads once again to a semidiscrete system of ODEs of fixed size for the moving mesh problem. Within each timestep Δt, the momentum equations in the melt and the energy equation in both subregions are integrated. A decoupled algorithm can be constructed as follows: at timestep Δt_n

1. solve the momentum equations in the melt to determine $\boldsymbol{u}^{n+1}$, p^{n+1} with T^n and $\boldsymbol{U}^n$ specified from the previous step ("lag" T and $\boldsymbol{U}$)
2. solve for T^{n+1} in both phases using $\boldsymbol{u}^{n+1}$ from step 1
3. update $\boldsymbol{V}^{n+1}$ from (16.24) and compute the grid velocity $\boldsymbol{U}^{n+1}$
4. convect the mesh and solution values
5. proceed to the next timestep

Higher-order predictor–corrector schemes can be introduced to advance the interface in step 4. If the mesh becomes excessively distorted as a result of the convective grid adjustment, then a new grid can be generated. Values are interpolated from the old solution and the integration continued. The procedure of interpolating between grids requires some elaboration: a given grid point in the new mesh will be located in an element of the old mesh and the solution interpolated from the grid point values associated with that element. This reduces to a "search and interpolate" process.

One variant of the scheme above has been used for crystal growth computation [121]. A rectangular domain is considered with the upper boundary thermally insulated and the temperature on the remaining sides as in the previous studies. The problem parameters are $L/h = 6$, $Ra = 2 \times 10^4$, $Pr = 0.015$, and $Gr = (4/3) \times 10^6$. Initially, the interface is assumed to be vertical with a graded mesh discretizing both the thin solid region and the remaining melt. As solidification takes place, the mesh in the solid region stretches and the mesh in the melt compresses. The streamline and isotherm patterns for a representative stage during the crystal growth process are given in Figure 16.7.

16.5 Summary

The problem of growing very pure silicon crystals is the first step in what we now recognize as a long chain of manufacturing processes that eventually lead to a circuit. As in the silicon oxidation problem, simulating crystal growth from a melt involves modeling coupled viscous flow and transport with a moving boundary. Here we show how the Stefan condition at the moving solidification boundary can be used in a predictor-corrector strategy to approximate the evolving boundary shape in a step-wise manner. We also discuss techniques for computing buoyancy-driven flows in which there may be bifurcations to flows with multiple cells. The related concepts of limit and bifurcation points as well as path-following techniques and arc-length continuation are briefly introduced. We remark that crystal growth problems are also a research topic of interest in NASA's space program. Here the micro-gravity environment implies that buoyancy effects will be very small and hence will not promote strong circulatory cells. However, temperature gradients along the free surface will lead to surface-tension driven flows and this will influence the solidification process. Modeling surface-tension driven flows (the Marangoni effect) then becomes important. In closing we again emphasize that the techniques discussed in this book for viscous flow and transport are amenable to wafer polishing, chemical vapor deposition, etching and many other processes.

16.6 Exercises

1. The phase change condition (16.26) can be used to estimate the velocity of the melt interface at each timestep. This suggests a predictor or predictor–corrector strategy to decouple the temperature solution from the phase boundary calculation. Use this idea in the one-dimensional form of the heat transfer problem in (16.21)–(16.22) to develop a finite difference scheme for this moving interface problem.

2. Use a finite difference or finite element scheme in a two-dimensional domain to discretize the coupled viscous flow and thermal system in

(16.1)–(16.3) and obtain a discrete system of the form (16.5)–(16.7). Construct the corresponding Jacobian matrix contributions for the Newton solver. Show how this scheme can be extended to the time-dependent case.

3. In the crystal growth problem there is an additional feature associated with the phase change from melt to crystal. Describe how the transient formulation in exercise 2 can be extended to include this term.

Chapter 17

Technology Computer Aided Design

17.1 Introduction

The present chapter serves several purposes. The first is that of tying together the diverse material discussed in earlier chapters on circuit, device, and process simulation. Although these divisions are useful pedagogically, the working engineer must not see TCAD design as a linearly ordered sequence in which process simulation is "completed" before device simulation is begun. It is rather an intricate iterative cycle with several levels of feedback between the different areas. Figure 17.1 shows a schematic of the design sequence for an integrated circuit, together with the information output at each stage. Note that this chapter starts at the bottom of the "food chain" with process simulation and works up to the top, circuit simulation. To this end the reader is presented with a sequence of simulations: processing of a lightly doped drain (LDD) MOSFET, generation of I–V curves for this device using a device simulator, and finally combining several of these NMOS transistors to build a ring oscillator.

A second purpose of the chapter is to provide the reader with a brief history of TCAD and a list of practical considerations when using a simulator. Many of these concerns have already been addressed in previous chapters, but will be collected here for convenience. Finally, a representative survey of simulation software is presented. Many, many man-years of work have been spent over the last thirty years to write new codes and improve existing TCAD simulators. Space limitations preclude an exhaustive discussion of all these codes, and although an effort was made to include many of those an

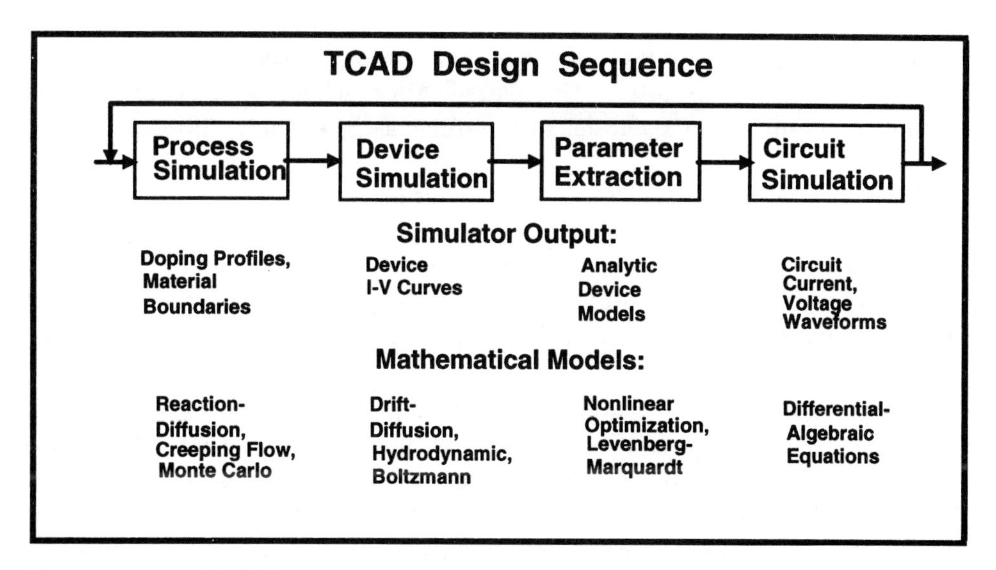

Figure 17.1: Modeling of an integrated circuit component: process, device, and circuit simulation.

engineer might encounter on the job, unintentional omissions undoubtably will occur. Readers should consult both the literature and documentation at their facility for more detailed information on a particular program.

17.2 Process Simulation

It is difficult to write topically about simulators because they are changing as rapidly as IC technology itself. Each of the major simulators is referenced in the literature by at least one major article describing its features. Realize that such an article usually appears a couple of years after one cyle of program development is complete, that proprietary considerations often result in omission of many details, and that developers start adding features shortly after the current generation of the code is released to users. Establishment of world-wide-web sites as repositories for user manuals and articles will speed the dissemination of information about simulators, both within a company, and for nonproprietary codes, to the general user community.

Process and device simulators can be categorized in several different ways. The former can be grouped by the generality of their diffusion models (single- or multiple-species, etc.), number of dimensions (1, 2 or 3D) necessary to describe the geometries and relevant physics incorporated into the simulator, simplifying assumptions used to model oxidation, total number of process steps the simulator can handle, and the ease with which new models can be incorporated. One classification scheme is perhaps as good as another, and we have chosen to use dimension, as shown in Table 17.1.

One-dimensional Process Simulators

Perhaps the most widely used process modeling program, SUPREM III [246] had its inception at Stanford's Integrated Circuits Laboratory in 1971 with SUPREM I. The third generation simulator is a general-purpose code that models multilayer structures and many process steps including diffusion, oxidation, ion implantation, epitaxy, deposition, and etch. Diffusion is modeled using the single-species model with modifications for OED/ORD and the Fair–Tsai model for phosphorus. The Deal–Grove model for oxidation is implemented with special treatment of the thin-oxide regime. Implantation is modeled using either look-up tables based on Lindhard–Scharff–Schiott (LSS) theory to construct range distributions for implanted ions or a 1D solution of the Boltzmann transport equation under simplifying assumptions.

The integration technique in SUPREM III is a trapezoidal scheme. Lagging the diffusivities (that is, evaluating $D(C)$ at the previous timestep) makes the Jacobian matrix quite easy to form and invert. When several impurities are present, the system of PDEs is solved in a decoupled fashion. Generally, this works very well, but when, say, arsenic and boron are present at high concentration, the electric field term can introduce oscillations where the profiles cross. These oscillations will not decay with time and must be eliminated by cutting back on the timestep. The technique of latency is used to good advantage: grid points deep in the bulk are neglected in the solution process until the concentration there becomes physically relevant, thus reducing the size of the linear system considerably during the early steps of the diffusion.

PEPPER is a one-dimensional process simulator for VLSI developed at the Microelectronics and Computer Technology Corporation (MCC) and which emphasizes physically-based models. The 3D Monte Carlo algorithm for implantation includes effects such as channeling and lattice damage. Because it treats implantation at the atomic level, it can handle virtually any type of ion, composition of target, or geometry. For diffusion of impurities in silicon the user can select from a new model for polysilicon and two new models for point defect assisted diffusion. The dial-an-operator concept was used in the diffusion module so the user can change the PDEs being solved by making changes in the input data. This allows for rapid prototyping

One-dimensional simulators			
SUPREM I,II,III	71	Stanford	GP, FD [246]
ZOMBIE	85	TU Vienna	GP, FD
FABRICS	85	CMU	statistical
PEPPER	86	MCC	3D MC, prototyping [393]
PREDICT	86	MCNC	phenomenological

Two-dimensional simulators			
CREEP	80	Berkeley	oxidation only
SUPRA	81	Stanford	Q-2D analytic [306]
SOAP	83	Stanford	2D oxidation
ROMANS II	83	Rockwell	diffusion–oxidation [341]
BICEPS	83	ATT	GP, coordinate trans. [431]
FEDSS	83	IBM	GP, FE [472]
SAFEPRO	83	IBM	GP, FE [409]
COMPOSITE	85	Siemens	GP, FD [332]
PROMIS	85	TU Vienna	GP, FD [433]
SUPREM IV	85	Stanford	GP, FV, Navier-Stokes
SSUPREM4	87	Silvaco	GP, proprietary
TSUPREM-4	87	TMA	GP, proprietary
TITAN	86	CNET	GP
FLOOPS	92	U Florida	GP, object-oriented
PROPHET	93	ATT	GP
PROCESS WIZ.	95	PDF Solutions	proprietary
DIOS	95	Zurich	GP

Three-dimensional simulators			
OPUS	89	OKI	limited 3D [406]
SMART-P	88	Mats.	coordinate trans. [410]

Table 17.1: Representative process simulators. Abbreviations are as follows: GP–general purpose, FE–finite element, FD–finite difference, FV–finite volume, MC–Monte Carlo. Note that dates of introduction in the second column are approximate, and that development groups are constantly adding capabilities to their simulators.

of new models without reprogramming. The backward differentiation formula integration techniques of LSODE allow efficient solution of these stiff nonlinear problems. In addition, there are modules for deposition, etching, oxidation, and epitaxy.

PREDICT is a process simulation program written at the Microelectronics Center of North Carolina. The 1D version uses a collection of phenomenological models and a decision-tree organization to choose empirically determined parameters that are appropriate for the user-specified conditions. The 2D version also incorporates several physical models for dopant diffusion.

Two-dimensional Process Simulators

SUPRA [306] was developed at Stanford in 1981 as a quasi-2D analytic and numerical process modeling code. Knowledge gained from it was later incorporated into the design of the device modeling program PISCES. SOAP (Stanford, 1983) used a Green's function boundary-integral method to solve the 2D oxidation problem.

In addition to work at universities, simulation programs were developed at most large semiconductor companies. Developed at Bell Labs in the early eighties [431], BICEPS is a general purpose process code that solves the oxidation problem in 2D by using a coordinate transformation approach. Although this sacrifices some of the physics of later models, it is extremely fast and gives good results on the birds-beak problem. Two codes were written by IBM research groups at approximately the same time as BICEPS was developed; both IBM programs used a finite element approach for discretizing the spatial operators. FEDSS [472] uses an analytic model for 2D ion implantation and a 1D Monte Carlo algorithm, the diffusion model includes clustering and Chin's model was employed for oxidation (Chorin's pressure-velocity formulation). SAFEPRO [409] (1984) has much the same functionality as FEDSS but a limited oxidation capability.

One of the most widely used 2D simulators is the SUPREM IV program developed at Stanford University. It was designed with the same philosophy as PISCES, the 2D device simulator, which in turn had its roots in the finite element code, PLTMG [30]. Spatial discretization is via a finite volume approach and integration uses a trapezoidal or TR-BDF(2) method. For the implicit methods in the time integration, the user can choose between several linear solvers including SOR and ILUCG. Figure 17.2 shows a simulation of a lightly doped drain structure using a commerical version of SUPREM IV.

The first oxide deformation model in SUPREM IV used three-node triangles for compatibility with the diffusion code. However, this does not allow the incompressibility condition to be enforced and "grid-locking" occurs, a situation well known in the finite element fluid mechanics literature [86]. As a result, triangles having three vertex nodes and three midside nodes for velocity, with three vertex nodes for pressure, were implemented with moderate success. The matrix sparsity pattern is now different for the oxidation and diffusion sections of the code and some conversion is required at each timestep. Finally, to further improve stresses, an additional node was added at the center of each triangle.

As a result of the study by Kao for oxides grown on structures of varying radii of curvature, the oxidation module in SUPREM IV was modified to include stress effects on the oxide viscosity. This increases the nonlinearity of the physical model, but gives far more accurate results. Much work needs to be done before a comprehensive theory of the effect of stress on reaction rate, viscosity and diffusivity of oxygen in SiO_2 is complete.

COMPOSITE [332] is a 2D process modeling program developed in Germany that simulates doping, oxidation, lithography, etching, and deposition. For diffusion, finite differencing with undamped Newton is used to solve the nonlinear PDEs; linear systems are solved by block-SOR. The moving boundary in diffusion during oxidation is handled via a conformal-mapping technique, with a new diffusion grid for each timestep. On this grid, the diffusion module, including segregation and the moving-boundary condition at the interface, is run to compute a new profile. In addition, optical lithography can be simulated by calling SAMPLE from the main program.

PROMIS [433] was developed by the simulation group at the Technische Universität Wien and is currently used at Siemens. It was designed to handle partial differential equations having the general form of a continuity equation with a general current relation

$$\sum_{j=1}^{N} a_{ij}\frac{\partial C_j}{\partial t} + \nabla \cdot \boldsymbol{J}_i + G_i - R_i = 0 \tag{17.1}$$

$$\boldsymbol{J}_i = \sum_{j=1}^{N}(d_{ij}\nabla C_j + \mu_{ij}C_j\nabla\psi) \tag{17.2}$$

where C_j is the concentration of the jth unknown, ψ is the electrostatic potential, and G_i,R_i represent generation and recombination terms, respectively. It allows the user to develop new models by coding FORTRAN routines to compute the quantities appearing in (17.1) and (17.2), as well as the partial derivatives, say $\partial G_i/\partial C_j$, for use in the Jacobian matrix. For complicated models, writing routines to compute an analytic Jacobian is an arduous, error prone task; difference quotient approximations to the Jacobian offer a much more user-friendly avenue for developing new process or device models. Time integration is handled by BDF of up to order six for the one-dimensional version and in the two-dimensional code by backward Euler with adaptive timestep selection. Spatial discretization is by finite differences and the novel feature of a dynamic (adaptive) grid is employed. Strict use of finite difference rather than finite element or a control volume approach makes adaptive refinement much more difficult in higher dimensions.

Three-dimensional Process Simulators

OPUS [406] was written at OKI Electric, first as a 2D process simulator; later a 3D version was developed. The simulator employs a five-point finite difference stencil in space with either forward or backward Euler in time for diffusion and oxidation — the creeping flow problem is not solved. The 3D version expands a 2D simulation by effectively "shifting," "turning" or "rotating" the 2D profile, so the problem must possess symmetry for accurate results.

SMART-P [410] is a 3D process simulator developed at Matsushita. Diffusion of impurities is via the standard model with a diffusivity that is modified under oxidizing conditions to account for the interstitial-assisted OED effect, as in equation (1.10):

$$D = D_N + f_I \; D_i \left(\frac{C_I}{C_I^*} - 1 \right) \tag{17.3}$$

and the diffusion of interstitials is modeled by the linear diffusion equation $\partial C_I / \partial t = D_I \Delta C_I$. Oxidation is simulated as a slow incompressible viscous flow [104] and implemented using the coordinate-transformation method of Penumalli [431]. Finite differences are used for spatial discretization with upwinding for the convective terms. The rational Runge–Kutta method is used for time integration; this is an explicit scheme with greater stability than forward Euler. When the interstitials must be solved for, this is carried out using a separate grid and the backward Euler method using biconjugate gradient iteration with incomplete LU factorization for preconditioning.

Statistical Simulators

Of increasing importance in recent years is the need of the process engineer to provide statistical data on how the actual yield of a process will vary

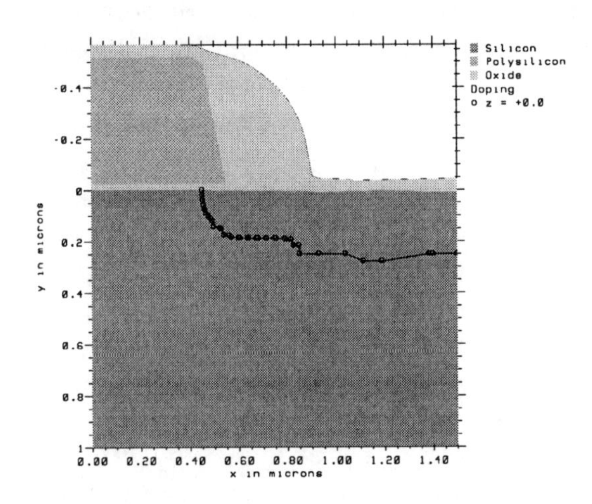

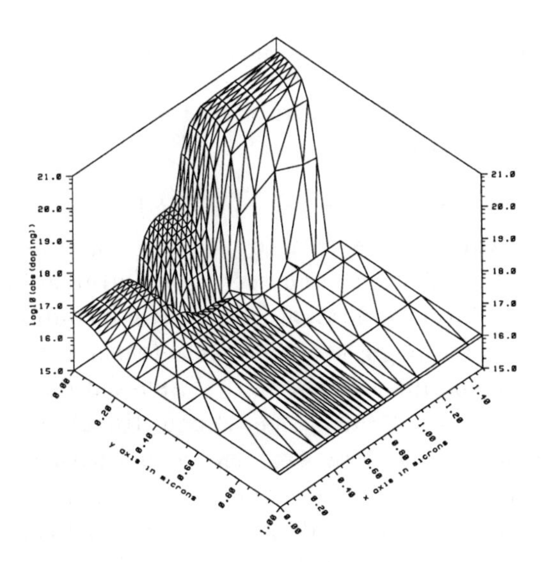

Figure 17.2: Simulating a lightly doped drain structure (LDD) using SSUPREM4. (a) Material boundaries and junction position. (b) Plot of ln | N_{net} |.

around the expected values predicted by a deterministic process simulator. The values of certain physical constants — impurity diffusivity, segregation coefficients, and initial concentration of an impurity after an implantation or predeposition — are often best modeled, not as physical constants at a given temperature, but as random variables with mean equal to the value used by a deterministic model and variance which measures the spread of measured values caused by defects in the wafer, variations in process conditions, and the "lumping" together of three-dimensional effects such as irregular geometries when using a one-dimensional model.

One method of obtaining the desired statistical distributions is to take a program such as SUPREM III and run it many times, varying each of the parameters over a range of values. Even in one space dimension this can be computationally intensive, and in higher dimensions it becomes prohibitive when using the appropriate physical models. It is here that physics must be sacrificed for speed and older analytical models can be used to advantage.

FABRICS is a statistical simulator of IC fabrication processing developed in 1984 at Carnegie-Mellon. In the fabrication step, analytical functions relate physical parameters (a dopant profile) to process parameters, (time and temperature of diffusion), with the parameters being random variables. Similarly, in circuit element models, analytical functions relate electrical parameters such as threshold voltage to physical parameters.

17.3 Device Simulation

The development of device simulators is considerably more advanced than that of process simulators, due in part to the fact that the physics of the latter is much less well understood and there are a greater variety of phenomena to model. Moreover, additional impetus for device simulation was provided early because, although is was possible to make measurements of doping profiles in 1D experimentally, one cannot even now make detailed measurements of potential and carrier concentrations within a device during its operation. One is limited to what can be measured at the various contacts, which is hardly enough to provide an understanding of such phenomena as pinchoff, punchthrough, and avalanche breakdown.

A reasonable classification scheme for device simulators can be based on the number of PDEs solved. For the subthreshold and linear regions of a MOSFET's operation solving Poisson's equation for the potential alone is sufficient, because current flow is low. At higher current levels, for example in the saturation region, one must couple the current continuity equation to

Poisson's equation. For MOS devices this is one additional PDE, since most current is carried by a majority carrier. Bipolar technology requires the full set of three equations in the drift-diffusion model; transport by both holes and electrons is important to device operation. Simulators can also be classified as DC only, small signal AC, and full time domain, depending upon assumptions made concerning the size of the $\partial n/\partial t$, $\partial p/\partial t$ terms in the continuity equations and the nature of the boundary conditions. However, these groupings are not static: once a simulator is written to solve Poisson's equation, a natural enhancement is to add one continuity equation, and finally two. It is more difficult to take a 1D simulator and "add a dimension" inasmuch as the various data structures necessary to define the topology of the grid and the device geometry become considerably more complicated as the dimension is increased. Because such a classification would always be changing, and to provide compatibility with the scheme for process simulators, a grouping by dimension will again be used. The following paragraphs describe a few representative device simulators that are in common use at many companies. No attempt is made at an exhaustive list. By now almost every semiconductor company, national laboratory, or large university research group has either written their own device simulator or modified a public domain code to solve a particular problem of interest. The reader should consult such journals as *IEEE Transactions on CAD* and *Electron Devices* or such conferences as NUPADS and NASECODE for futher details.

One- and Two-dimensional Device Simulators

SEDAN is a 1D simulator developed at Stanford for modeling steady-state and transient operation of bipolar devices. It solves the fully-coupled problem (three equations) and in addition to I–V characteristics will generate bipolar gain, cutoff frequency, Gummel number, and Early voltage.

BIPOLE, a bipolar simulator developed at Waterloo University, is a quasi-2D code which solves the full set of three equations in the vertical (emitter to collector) direction and couples the result to another 1D solution for the majority carrier in the quasi-neutral base (horizontal base current). It accounts for bandgap narrowing at high doping levels and adequately includes 2D effects without the expense of a full 2D simulation. For this reason it can even be run on a small personal computer.

GEMINI is a 2D simulator for MOS devices in the subthreshold and linear regions. It solves not only Poisson's equation, but can also simulate punchthrough, junction capacitance, and even the avalanche breakdown of reverse-biased junctions. It uses a variable mesh finite difference formulation

One-dimensional simulators			
SEDAN	71	Stanford	bipolar
BIPOLE	72	Waterloo	bipolar

Two-dimensional simulators			
CADDET	82	Hitachi	Poisson
GEMINI	80	Stanford	Poisson
MINIMOS	80	TU Vienna	GP, MOS
PISCES	81	Stanford	GP, MOS and bipolar
MEDICI	85	TMA	GP, proprietary
PADRE	85	ATT	GP
FIELDAY	81	IBM	GP
SIFCOD	81	Sandia	GP, mixed-mode
FLOODS	92	U Florida	GP, object-oriented
DESSIS	93	Zurich	GP

Three-dimensional simulators			
CADDETH	85	Hitachi	GP, FD, ICCG, BCG
FEDAS	81	NTT	GP, FD, SSIP
MINIMOS-3	87	Tech. Un. Vien	GP, FD
SIERRA	88	TI	GP, FD, General
TOPMOST	85	Toshiba	GP, FD, ICCG
TRANAL	82	NTT	GP, FD, ICCG

Table 17.2: Representative device simulators. Abbreviations are as follows: GP–general purpose, FE–finite element, FD–finite difference, FV–finite volume, MC–Monte Carlo, BCG–biconjugate gradient, ICCG–incomplete Cholesky with conjugate gradient. Note that dates of introduction in the second column are approximate, and that development groups are constantly adding capabilities to their simulators.

and can import impurity profiles from SUPREM or SUPRA. Like the 1D simulators discussed above, it is very computationally efficient in its problem class, as compared with general purpose simulators.

MINIMOS [479, 481] is a two-dimensional planar MOS transistor simulator developed at Tech. Universität Wien in the late seventies. It solves the nonlinear Poisson equation coupled with one steady-state continuity equation. Finite differences are used with the the Gummel–Scharfetter discretization for the continuity equation. A nonuniform structured mesh is allowed and Stone's implicit method was chosen to solve the linearized equations. There is now a three-dimensional version [235] which includes hot carrier effects.

PISCES-II is a general purpose two-dimensional, two-carrier simulator developed at Stanford University in the early seventies. It handles both steady-state and transient problems, calculating such characteristics as I–V curves, MOS threshold voltage, bipolar current gain, I_{Dsat}, and parasitic capacitances/resistances, in addition to the standard outputs of potential, carrier concentrations, and current densities. It may be used to model such phenomena as punchthrough, hot carrier effects, and CMOS latchup.

Based on the earlier finite element code PLTMG [30], PISCES uses a hybrid finite element-finite difference scheme, called a box or control volume method, to discretize the differential operators that occur in the drift-diffusion model. A general triangular grid conveniently handles irregular geometries such as trenches and LOCOS structures. Conceptually, each PDE is integrated against a test function over a small polygon enclosing each node while ensuring that conservation of flux is maintained. The continuity equations are discretized using Scharfetter–Gummel along triangle edges. If doping profiles, geometries, and locations of material interfaces are input from a process simulator such as SUPREM IV or SUPRA no user intervention is required to set up the initial grid. It is possible to regrid on the doping profile or on any of the unknowns during the solution process.

The system of nonlinear algebraic equations which results from discretization can be solved using a full Newton or Gummel iteration. As discussed in Chapter 7, the latter iteration decouples the three PDEs, solving sequentially for ϕ, n, and p using a fixed-point iteration. This is very efficient and works well at zero bias or in the subthreshold region of a MOS device. The added expense of using a fully-coupled Newton scheme is justified physically when there is significant current flow in either MOS or bipolar technology devices. Mathematically, this situation results in large drift terms and hence tight coupling of the unknowns. The algorithm is very stable with solution times practically independent of bias condition.

A quasi-Newton or Newton–Richardson option refactors the Jacobian only when necessary, with the decision being made on the basis of decrease in error norm at each iteration. Damping is included to reject Newton updates that would cause the error norm to increase. Convergence is determined using both the residual norm and the so-called X-norm which measures the size of the update at each step. Linear systems are solved using either sparse LU factorization or, for the Gummel iteration, conjugate gradient with incomplete Cholesky preconditioning.

Figure 17.3 shows the mesh used for device simulation of the LDD MOSFET. Note the very fine spacing in the vicinity of the junctions and in the channel, with gradual grading toward the substrate. Most of today's codes perform much of the work of setting up the initial mesh with little user intervention, automatically refining on the impurity concentration gradient. Later, further adaptation is possible during the solution process, and this is a topic of current research interest. In Figure 17.4 potential contours obtained using the commercial simulator MEDICI are shown. Here, the drift-diffusion equations have been solved to obtain ϕ, n and p. Such a simulation will also yield carrier concentrations, I–V curves, and a wealth of other data impossible to obtain by direct measurements.

17.4 Circuit Simulation

This section must necessarily read differently than the previous ones on process and device simulators, because one circuit simulator, Berkeley's SPICE quickly came to dominate the field and most modern proprietary codes are derivatives of it. A brief historical development is presented next, followed by a discussion of several special purpose simulators and parameter extraction programs. For an in-depth history of circuit simulation and extensive practical advice on running simulations see *The Spice Book* by Vladimirescu [542]. The survey article by Pederson [430] and Nagel's dissertation also make interesting reading.

The initial implementations of computer programs for circuit simulation occurred in the late 1960s and early 1970s. With the development of the integrated circuit, the concept of "breadboarding" a new design for testing and development quickly became outmoded. The need for numerical simulation resulted in BIAS-3 [368] and SLIC for the analysis of circuits containing bipolar devices; CIRPAC [491] at Bell Labs and TRAC at Rockwell were written for large-signal time-domain simulation of circuits such as oscillators. The latter evolved into TIME [266] and later MTIME at Motorola.

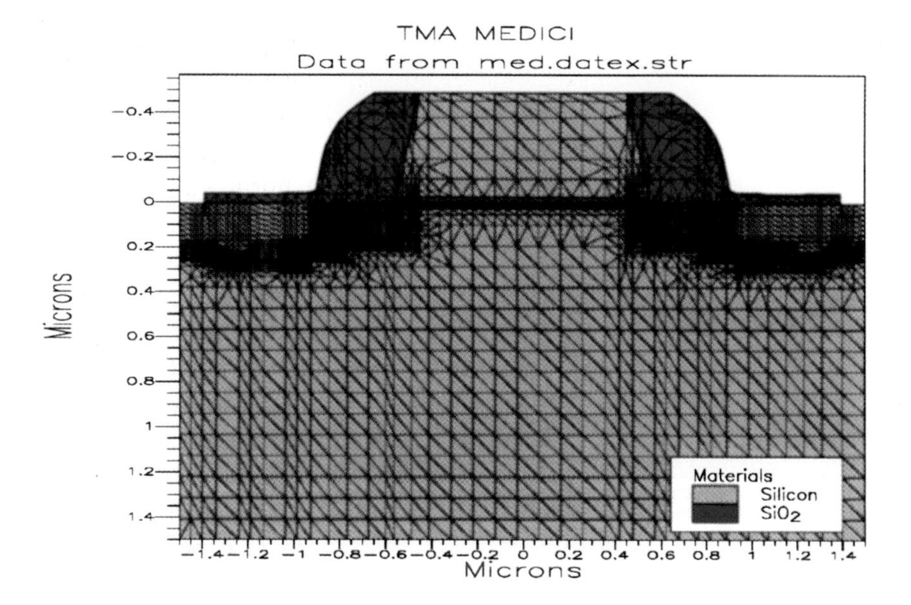

Figure 17.3: The mesh used for device simulation of the LDD MOSFET.

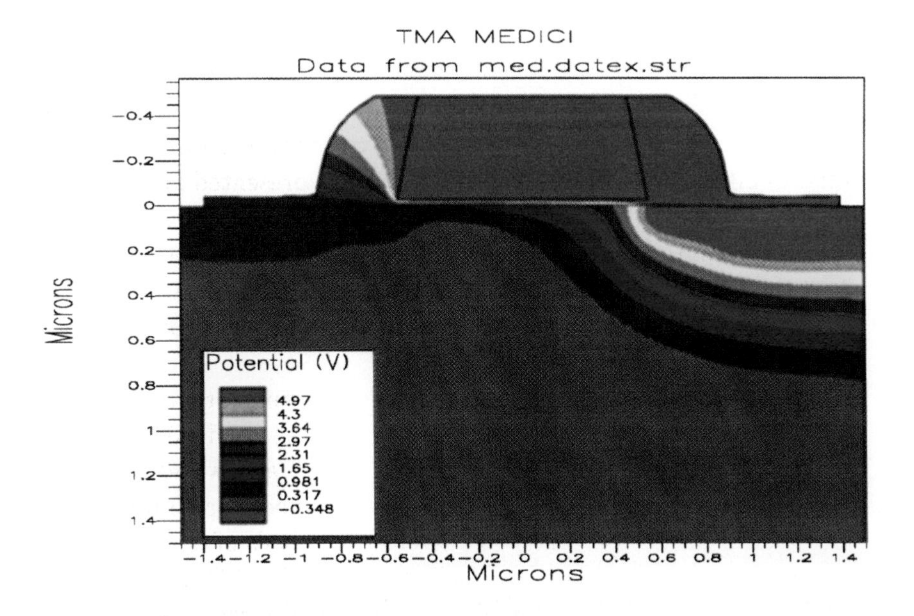

Figure 17.4: Potential contours for the LDD structure obtained from MEDICI.

CANCER [399] written by Nagel and Rohrer at Berkeley in 1971 used the 18 parameter Ebers–Moll [151] model for the bipolar transistor. The user was limited to 100 nodes, 100 transistors/diodes, and a total of 400 components. Numerical methods for this first generation of circuit simulators included implicit integration techniques using fixed timesteps, Newton–Raphson to solve the nonlinear algebraic system, and direct sparse matrix techniques for the linear solves.

The new version of CANCER was named SPICE, Simulation Program with Integrated-Circuit Emphasis, and distributed publicly in May of 1972 as SPICE1 [398]. It implemented the Gummel–Poon model for bipolar transistors, which substantially improved upon Ebers–Moll by including high-level injection and low-level recombination effects, as well as the Shichman–Hodges [492] model for JFETs and MOSFETs. The development of SPICE2 [400] in 1975 was a major milestone in the development of circuit simulation, because it had applicability to a much wider range of circuits and again was available free from the University of California at Berkeley. SPICE2 performed nonlinear DC analysis, small signal AC analysis, and transient analysis; it contained models for all the silicon semiconductor devices in widespread use. It used modified nodal analysis (MNA) to represent the circuit, and variable-step integration with backward Euler/trapezoidal schemes, as well as the variable-order BDF methods of Gear [200]. Dynamic memory allocation removed internal limitations on problem size – now limits were set only by available computer resources. This was very important, for the size and complexity of ICs had grown exponentially during the 1970s. A major upgrade to SPICE2 was released by the Berkeley research group in 1981, SPICE2G6 [543]. Many commercial circuit simulators, such as ISPICE, HSPICE and the PC-based PSPICE, were developed as derivatives of this version of SPICE. At the same time most large corporations with an interest in TCAD had groups working on their own versions of SPICE to be used internally. They put an enormous amount of effort into improving the reliability and robustness of the algorithms in SPICE, and enhancing the device models. In 1989, Berkeley produced SPICE3, an updated version written in the C programming language.

There are other programs in the area of circuit simulation that deserve note. The ASTAP simulator developed at IBM employs the sparse tableau circuit representation which uses all the state variables, in contrast to MNA. Timing simulators can take advantage of MOSFET characteristics to achieve decreased simulation times: many circuits have a high degree of latency in which only a small portion of the devices on the chip are active at any time. This led to waveform relaxation techniques (RELAX2 [554]) and iterated

SPICE NETLIST FOR CMOS RING OSCILLATOR	
M1 2 1 0 0 tn W=5U L=1U	C5 5 0 50f
M2 2 1 8 8 tp W=10U L=1U	VDD 8 0 5.0V
M3 3 2 0 0 tn W=5U L=1U	.IC V(1)=0V
M4 3 2 8 8 tp W=5U L=1U	
M5 4 3 0 0 tn W=5U L=1U	.model tn nmos level=1 vt0=0.64
M6 4 3 8 8 tp W=10U L=1U	+kp=2.57e-5 lambda=0.0105 gamma=0.4
M7 5 4 0 0 tn W=5U L=1U	+phi=0.65 tox=200e-10 ld=0.1u ....
M8 5 4 8 8 tp W=10U L=1U	.model tp pmos level=1 vt0=-0.72
M9 1 5 0 0 tn W=5U L=1U	+kp=1.53e-5 lambda=0.012 gamma=0.4
M10 1 5 8 8 tp W=10U L=1U	+phi=0.65 tox=200e-10 ld=0.12u ....
C1 1 0 50f	
C2 2 0 50f	.PLOT TRAN V(5)
C3 3 0 50f	.TRAN 1ns 40ns
C4 4 0 50f	.END

Table 17.3: SPICE input deck for the CMOS ring oscillator in Figure 17.5. Each of the two `.model` statements contains some thirteen additional parameters which are not listed here.

timing analysis (SPLICE [296]). SPECTRE is a frequency-domain-based simulator, which finds the large-signal steady-state response of a nonautonomous nonlinear circuit. If a circuit is only mildly nonlinear with few harmonics, SPECTRE can be significantly more efficient and accurate than traditional time-domain simulators. This would be the case, for example, if the circuit is high-Q, narrow-band, has slowly responding bias networks, or contains distributed components. The techniques of parameter extraction discussed in Chapter 6 have also been implemented in the programs SUXES and TECAP. Several programs – including MEDUSA and SIFCOD – integrate device and circuit simulation into a single package. Although computationally very expensive, the accuracy of such a mixed-mode simulator is sometimes required as when analyzing an alpha particle soft-error during transient analysis of a six transistor CMOS latch.

We conclude this section with an example that uses process simulation to model a CMOS device, device simulation to extract current-voltage data from the device, parameter extraction to fit the data to a circuit simulation model, and finally circuit simulation to analyze the behavior of a simple CMOS ring oscillator circuit. SSUPREM4 was used to simulate a CMOS process; Figure 17.2 shows the impurity concentration profile at the drain of an n-channel MOSFET. The device simulation of the SSUPREM4 structure was carried out with MEDICI, and a sample device simulation result is shown in Figure 17.4. Tables of drain current versus the terminal voltages

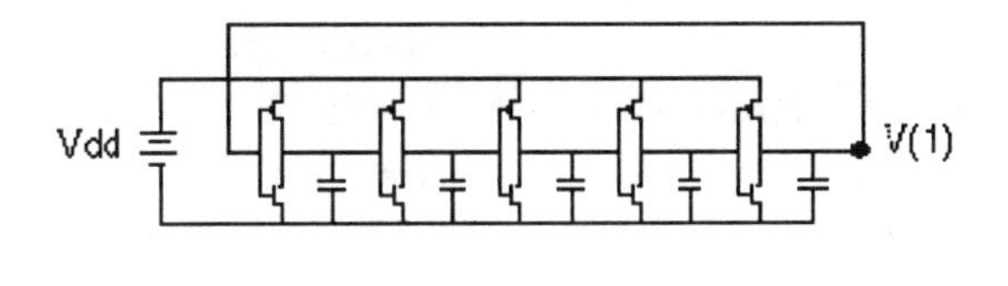

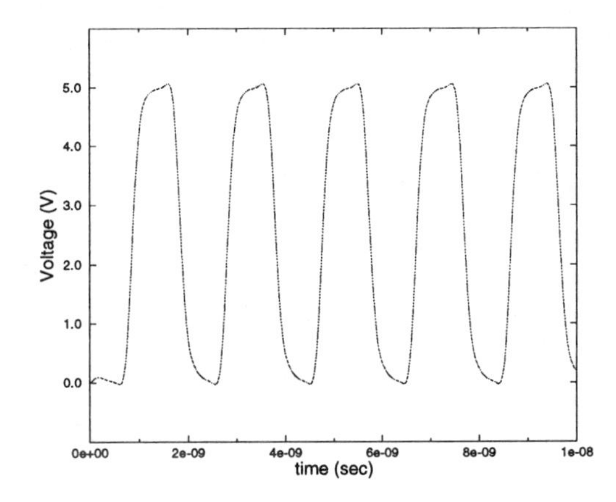

Figure 17.5: Circuit simulation of a ring oscillator under nominal processing conditions.

were generated by MEDICI, and these data were then used to extract SPICE model parameters for the Shichman–Hodges MOSFET model, according to the methodology outlined in Chapter 6. It should be noted that the Shichman–Hodges model is not adequate for careful modeling of submicron devices; it is used here only for illustrative purposes.

The SPICE netlist for the ring oscillator circuit is shown in Table 17.3. M1 through M10 are MOSFETs, C1 through C5 are capacitors, and VDD is a voltage source. The source and drain perimeter and area for each MOSFET is equal to 3 μm+2W and 1.5 $\mu\text{m}\times\text{W}$, respectively. The .IC statement sets an initial condition on the voltage at node 1, which forces the oscillator to start up. The two .MODEL statements contain the Shichman-Hodges parameters of the n-channel and p-channel MOSFETs. The .TRAN statement instructs SPICE to simulate for 40 ns. (see [542] for further details regarding specific syntax for SPICE). Figure 17.5 shows a schematic of the circuit, along with the output of the oscillator.

With this tightly coupled process, device, and circuit simulation, we can study, for example, the effect of a 10 percent variation of the time and temperature of the MOSFET source-drain diffusion on the final frequency of the ring oscillator. Process simulations indicate that such a variation results in approximately $\pm$ 0.02 μm change in the source and drain under lap diffusion. Device simulations with these variations indicate a change in the maximum drive current for the n-channel device of approximately $\pm$ 6 percent. Circuit simulation with the extracted best and worst case model parameters results is a variation of the final ring oscillator frequency from 448 MHz to 512 MHz.

17.5 Practical Considerations

Of great practical importance is the question of when to use a 1D, 2D, or 3D code. The higher dimensional model may be justified on the grounds of characterizing the physics more accurately, but there will be a corresponding increase in cost if the same resolution is required. We face the speed versus accuracy trade-off that was previously encountered when choosing an iterative method. Remember that reasonable grids will require n, n^2, and n^3 points where $1/n$ is at least the minimum feature size and that these result in an operation count for the linear solver of roughly n^k, $(n^2)^k$, and $(n^3)^k$ where $1 < k \leq 3$. The exact value of k will depend upon the linear solver used — banded direct, sparse direct, frontal, or iterative; and of course will be different for a sequential versus a parallel code. Not only do run times increase exponentially as dimension is increased, but the problems of data entry — initial data, boundary conditions, material interfaces, etc. — and the display of results become much more difficult. Visualization of large scientific data sets in 3D is still an area of active research with many open questions.

In older technology, feature sizes were large enough to get accurate results with a 1D simulation. As device dimensions shrink and LOCOS, fringing effects, parasitic capacitances, CMOS Latch-up, and narrow channel effects become more important, there has been a move toward increased use of 2D and 3D simulations. Currently, one might estimate that 1D simulations are used for 60% of the analyses, 2D for 35%, and 3D for 5%, in the design process. If enough symmetry is available, it can be used to significantly reduce the modeling cost. A sequence of 1D simulations can sometimes give as much information as a single 2D run. As a rule of thumb, if the device has a very irregular geometry with tight coupling between different regions or processes, i.e. diffusion and oxidation, and feature size in each direction is roughly the same, then a full 2D or 3D simulation is required. Even the 2D problem may initially consume more than a day in setting up the simulation and displaying output, while the 3D could take a week for a very complicated geometry, depending on the available pre- and post-processing software. More sophisticated preprocessing and graphics are needed to facilitate 2D and 3D simulation.

In recent years, the proliferation of process, device, and parameter extraction programs, as well as photolithography modeling codes such as SAMPLE, has brought about a need for integrating the various programs from many different vendors. One approach has been to design a platform in which many different codes can be called sequentially by the user with the output from one step forming the input to the next, in much the same fashion as individual process steps are performed in SUPREM III, with intermediate data conversions, regridding, etc., being transparent to the user.

A second approach involves the creation of a standard format for transmitting profiles of semiconductor structures between sites and between tools. The proposed process interchange format (PIF) would be a flexible, hierarchical data format to store detailed descriptions of the geometry, attribute profiles, material properties, and transient behavior of a single structure or collection of structures. Here one encounters the same issues raised in database design for higher level CAD: a need for different descriptions of the same data when used by different programs, a need for compression of the voluminous amounts of 3D data, and a need for rapid access to certain parts of the data that are used repeatedly. Furthermore, the PIF needs to be machine, operating system, and language independent, and should meet the requirement to interface easily with commercial graphics packages. For details on the current status of PIF see [150].

Since it may be some time before integrated Technology CAD codes or a standard PIF are available, the question of when to regrid should be addressed. Increasingly better methods are available for grid selection and refinement, as discussed in the chapter on oxidation. It is still true, however, that "gridding" is both an art and a science. Not surprisingly, a grid that works quite well for the process problem may not be suitable for device modeling, e.g. in the case of pinch-off in a channel.

17.6 Summary

The design of today's integrated circuits having millions of devices and submicron feature sizes requires the increased use of simulation at all levels of the design cycle. The previous sixteen chapters discussed in detail the mathematical and numerical methods used in today's simulators, while Chapter 17 discussed representative software in each of the three basic areas of TCAD: process, device, and circuit simulation.

At the macroscopic level linear and nonlinear elements are combined with Kirchoff's laws to yield a mathematical model of an integrated circuit. Numerical techniques used to implement this coupled system of ODEs include Euler and trapezoidal integration, adaptive stepsize, and Newton's method. Microscopically, current flow within a transistor is modeled using the drift-diffusion and hydrodynamic PDE systems. To solve these systems, device simulators employ both finite difference and finite element methods for spatial discretizations, as well as the ideas on grid generation, refinement, upwinding, and multilevel schemes, discussed in earlier chapters. At the fabrication level, physical processes such as diffusion, oxidation, and crystal growth are modeled using the multiple-species reaction-diffusion-convection equations of Chapters 12 to 16.

There are numerous challenges in TCAD simulation for the next decade. As the mathematical models become more complex, the number of equations increases, and there is a need to move to more complicated geometries for smaller feature sizes, several trends will become apparent. There will be an increasing emphasis on all aspects of the grid on which the approximate solution is represented: refinement and coarsening, adaptively generating a grid that evolves with the solution, and choosing data structures which are efficient not only from a numerical standpoint, but also as regards postprocessing of the data. There will also be a greater need for object-oriented codes that permit the user to easily modify the sets of partial differential equations being solved. Finally, the computational overhead for these systems will necessitate increased use of parallel processing throughout TCAD.

17.7 Exercises

1. Obtain a copy of the '64 article, "Self-Consistent Iterative Scheme for One-Dimensional Steady State Transistor Calculations," by H. K. Gummel. Analyze the scheme in terms of a fixed-point iteration.

2. Consider n resistors connected in series. The first and last node are grounded and from each interior node there is a capacitor connected to ground. At time zero each capacitor has a charge Q. Show that this is a circuit analogy of the diffusion equation with appropriate initial and boundary conditions; solve the latter by using SPICE.

3. Carry out the analysis steps for the nonlinear oscillator problem assuming 5 percent variation and 15 percent variation. Compare your findings with those stated in this chapter under the assumption of a 10 percent variation. Comment on the relative sensitivity of the respective process, device, and circuit simulation steps for this application.

Bibliography

[1] Acovic, A., D. K. Sadana, B. Davari, D. Grutzmacher and F. Cardone, "Arsenic source and drain implant induced degradation of short-channel effects in nmosfets," *IEEE Electron Device Letters*, **14**: 345–7, 1993.

[2] Adler, R. B., A. C. Smith and R. L. Longini, *Introduction to Semiconductor Physics SEEC series, Vol. 1*, Wiley, 1964.

[3] Alvarez, A. R., B. L. Abdi, D. L. Young, H. D. Weed, J. Teplik and E. R. Herald, "Application of statistical design and response surface methods to computer-aided VLSI device design," *IEEE Trans Computer-Aided Design*, **7**(2): 272–288, February 1988.

[4] Anderson, D., "Evaluation of diffusion coefficients from nonlinear impurity profiles," *J. Electrochem. Soc.*, **132**(6): 1409–1412, 1985.

[5] Anderson, D. and K. O. Jeppson, "Nonlinear two-step diffusion in semiconductors," *J. Electrochem. Soc.*, **13**(11): 2675–2679, 1984.

[6] Antell, G. R., "The diffusion of silicon in gallium arsenide," *Solid-State Electronics*, **8**: 943–946, 1965.

[7] Antognetti, P. and G. Massobrio, *Semiconductor Device Modeling with SPICE*, McGraw-Hill, 1987.

[8] Antoniadis, D. A., "Diffusion in silicon," in Antognetti, Antoniadis, Dutton and Oldham (eds.), *Process and Device Simulation for MOS VLSI Circuits*, pp. 1–47, Martinus Nijhoff, Boston, 1983.

[9] Antoniadis, D. A. and R. W. Dutton, "Models for computer simulation of complete IC fabrication process," *IEEE Trans. Electron Devices*, **ED-26**(4): 490–500, 1979.

[10] Antoniadis, D. A., A. M. Lin and R. W. Dutton, "Oxidation enhanced diffusion of arsenic and phosphorus in near intrinsic $< 100 >$ silicon," *Appl. Phy. Lett.*, **33**: 1030–1033, 1978.

[11] Antoniadis, D. A. and I. Moskowitz, "Diffusion of substitutional impurities in silicon at short oxidation times: An insight into point defect kinetics," *J. Appl. Phy.*, **53**(10): 6788–6796, 1982.

[12] Apte, D. R. and M. E. Law, "Comparison of iterative methods for AC analysis in PISCES-IIB," *IEEE Trans. Computer-Aided Design*, **11**: 671–3, 1992.

[13] Arienzo, W. A. O., R. Glang, R. F. Lever, R. K. Lewis and F. F. Morehead, "Boron diffusion in silicon at high concentrations," *J. Appl. Phy.*, **63**: 116–120, 1988.

[14] Armigliato, A., D. Nobili, P. Ostoja, M. Servidori and S. Solmi, "Solubility and precipitation of boron in silicon and supersaturation resulting by thermal predeposition," in *Proceedings of Spring Meeting*, pp. 551–3. Electrochem. Soc., 1977.

[15] Arney, D. C. and J. E. Flaherty, "An adaptive mesh-moving and local refinement method for time-dependent PDE's," *ACM Trans. Math Software*, **16**: 48–71, 1990.

[16] Arnold, N., R. Schmidt and K. Heime, "Diffusion in III-IV semiconductors from spin-on film sources," *J. Phys. D: Appl. Phys.*, **17**: 443–474, 1984.

[17] Ashcraft, C., R. Grimes, J. Lewis, B. Peyton and H. Simon, "Recent progress in sparse matrix methods for large linear systems on vector supercomputers," *Inter. J. Supercomputing Appl.*, **1**: 10–30, 1987.

[18] Ashcroft, N. W. and N. D. Mermin, *Solid State Physics*, Holt, Rinehart, and Winston, New York, 1976.

[19] Asom, M. T., J. L. Benton, R. Sauer and L. C. Kimerling, "Interstitial defect reactions in silicon," *Appl. Phys. Lett.*, **51**: 256–258, 1987.

[20] Axelrad, V., "A general purpose device simulator including carrier energy balanced based on PISCES-2B," *VPAD*, pp. 54–55, 1991.

[21] Azoff, E., "Semiclassical high-field transport equations for non-parabolic heterostructure degenerate semiconductors," *J. Appl. Phys.*, **64**(5): 2439, 1988.

[22] Babuska, I., J. Chandra and J. E. Flaherty (eds.), *Adaptive Computational Methods for PDE's*, *SIAM*, 1983.

[23] Babuska, I. and A. Miller, "The post processing approach in the finite element method," *IJNME*, **20**: 1085–1109, 1984.

[24] Babuska, I. and W. Rheinboldt, "*A Posteriori* error estimates for the finite element method," *Int. J. Num. Meth. Eng.*, (12): 1597–1615, 1978.

[25] Babuska, I., B. A. Szabo and I. N. Katz, "The *p*-version of the finite element method," *SIAM J. Numer. Anal*, **18**(3): 515–454, 1981.

[26] Babuska, I., O. C. Zienkiewicz, J. Gago and E. R. de A. Olivera, *Accuracy Estimates and Adaptive Refinements in Finite Element Computations*, Wiley, 1986.

[27] Baccarani, G., R. Guerrieri, P. Ciampolini and M. Rudan, "HFIELDS: a highly flexible 2-D semiconductor device analysis program," *NASECODE IV*, pp. 3–13, 1985.

[28] Baccarani, G. and M. R. Wordeman, "An investigation of steady-state velocity overshoot effects in Si and GaAs devices," *Solid-St. Electron*, **28**: 407–416, 1985.

[29] Baker, K. R., *Semiconductor Process Simulation with Matrix Preconditioning*, Master's thesis, University of Texas at San Antonio, 1993.

[30] Bank, R. E., *PLTMG: A Software Package for Solving Elliptic Partial Differential Equations, User's Guide 6.0*, SIAM, Philadelphia, 1990.

[31] Bank, R. E., W. M. C. Jr., W. Fichter, E. H. Grosse, D. J. Rose and R. K. Smith, "Transient simulation of silicon devices and circuits," *IEEE Trans. Electron Devices*, **ED-32**: 1992, 1985.

[32] Bank, R. E. and D. J. Rose, "Parameter selection for newton-like methods applicable to nonlinear parial differential equations," *SIAM J. Numerical Analysis*, **17**: 806–822, 1980.

[33] Bank, R. E. and D. J. Rose, "Global approximate Newton methods," *Num. Math*, **37**: 279–295, 1981.

[34] Bank, R. E. and A. H. Sherman, "A refinement algorithm and dynamic data structure for finite element meshes," Tech. Rep. TR-166, Center for Numerical Analysis, The University of Texas at Austin, 1980.

[35] Bank, R. E. and A. Weiser, "Some *a posteriori* error estimates for elliptic pde's," *Math. Comp.*, **44**: 283–301, 1985.

[36] Barragy, E. and G. F. Carey, "A parallel element-by-element solution scheme," *IJNME*, **26**: 2367–2382, 1988.

[37] Barry, R., "An optimal ordering of electronic circuit equations for a sparse matrix solution," *IEEE Trans. on Circuit Theory*, **CT-18**: 40–50, January 1971.

[38] Becker, E. B., G. F. Carey and J. T. Oden, *Finite Elements - An Introduction*, volume 1, Prentice Hall, 1981.

[39] Bell, T. E., "The quest for ballistic action," *IEEE Spectrum*, **23**: 36–38, 1986.

[40] Bénard, H., "Les tourbillons cellulaires dans une nappe liquide," *Rev. Gén. Sci. Pures Appl.*, **11**: 1261, 1900.

[41] Bennemann, K. H., "New method for treating lattice point defects in covalent crystals," *Phys. Rev.*, **137**(5A): A1497–A1514, May 1965.

[42] Berger, M. J., "Data structures for adaptive grid generation," *SIAM J. Sci. Stat. Comp.*, **7**(3): 904–906, 1986.

[43] Biersack, J. P. and L. G. Haggemark, "A Monte Carlo computer program for the transport of energetic ions in amorphous targets," *Nuclear Instruments and Methods*, **174**: 257, 1980.

[44] Bird, R. B., W. E. Stewart and E. N. Lightfoot, *Transport Phenomena*, Wiley, NY, 1960.

[45] Biswas, R. and G. A. J. Amaratunga, "Parallel computational techniques for simulating dopant diffusion in silicon," *IEEE Proc. G, Circuits Devices Syst.*, **136**(3): 135–137, June 1989.

[46] Blackmore, J. S., *Semiconductor Statistics*, Dover, 1962.

[47] Bløtekjær, K., "Transport equations for electrons in two-valley semiconductors," *IEEE Trans. Electron Devices*, **ED-17**: 38–47, 1970.

[48] Bokshtein, B. S., S. Z. Bokshtein and A. A. Zhukhovitskii, *Thermodynamics and Kinetics of Diffusion in Solids*, Oxonian Press, Calcutta.

[49] Boltaks, B. I., *Diffusion in Semiconductors*, Infosearch Ltd., London, 1963.

[50] Boltzmann *Wied. Ann.*, **53**: 959, 1894.

[51] Bordelon, T. J., *Evaluation and development of the hydrodynamic transport model for submicron silicon device analysis*, Ph.D. thesis, The Univ. of Texas at Austin, 1992.

[52] Bordelon, T. J., V. M. Agostinelli, X.-L. Wang, C. M. Maziar and A. F. Tasch, "The relaxation time approximation and mixing of hot and cold electron populations," *Electronic Letters*, pp. 1173–1175, 1992.

[53] Bordelon, T. J., X. L. Wang, C. M. Maziar and A. F. Tasch, "An efficient non-parabolic formulation of the hydrodynamic model for Silicon device simulation," *IEDM*, pp. 353–356, 1990.

[54] Borucki, L., H. H. Hansen and K. Varahramyan, "FEDDS: A 2D semiconductor fabrication process simulator," *IBM J. Res. Dev.*, **29**(3): 263–276, 1985.

[55] Botta, E. F. F. and A. E. P. Veldman, "On local relaxtion methods and their applications to convection-diffusion equations," *J. Comput. Phys.*, **48**: 127–149, 1981.

[56] Bourgin, J. C., P. M. Mooney and F. Poulin, "Irradiation-induced defects in germanium," in *Inst. Phys. Conf.*, 59, pp. 33–43, September 1980.

[57] Bova, S. and G. F. Carey, "Mesh generation/refinement using fractal concepts and iterated function systems," *IJNME*, **33**: 287–305, 1992.

[58] Bova, S. and G. F. Carey, "A Taylor-Galerkin finite element method for the hydrodynamic semiconductor equations," *IEEE Trans. on CAD*, 1995, in press.

[59] Bowyer, M. D. J., D. G. Ashworth and R. Oven, "Representation of ion implantation projected range profiles by Pearson distribution curves for silicon technology," *Solid-State Electronics*, **35**(8): 1151–1166, 1992.

[60] Box, G., W. Hunter and J. Hunter, *Statistics for Experimenters*, John Wiley, New York, 1978.

[61] Boyle, G. R., B. M. Cohn, D. O. Peterson and J. E. Solomon, "Macromodeling of integrated circuit operational amplifiers," *IEEE J. Solid State Circuits*, **9**: 353–363, 1974.

[62] Brackbill, J. U. and J. S. Saltzman, "Adaptive zoning for singular problems in two dimensions," *J. of Comp. Phys.*, **46**: 342–368, 1982.

[63] Bramble, J. H., J. E. Pasciak and A. H. Schatz, "The construction of preconditioners for elliptic problems by Substructuring, I," *Math. Comp.*, **47**: 103–134, 1986.

[64] Bramble, K. H., J. E. Pasciak and A. H. Schatz, "An iterative method for elliptic problems on regions partitioned into substructures," *Math Comp*, **46**: 361–370, 1986.

[65] Brandt, A., "Multi-level adaptive techniques (MLAT) for fast numerical solution to boundary value problems in Proc. Third International Conference on Numerical Methods in Fluid Mechanics, Paris 1972," in H. Cabannes and R. R. Teman (eds.), *Lecture Notes In Physics*, 18, pp. 82–89, Springer-Verlag, Berlin, 1972.

[66] Brandt, A., "Multi-level adaptive solutions to boundary-value problems," *Math. Comp.*, **31**: 333–391, 1977.

[67] Brayton, R. K., F. G. Gustavson and G. D. Hachtel, "A new efficient algorithm for solving differential-algebraic systems using implicit backward differentiation formulas," *Proc. of IEEE*, **60**(1), January 1972.

[68] Bridgeman, P. W., "The compressibility of several artificial and natural glasses," *Amer. J. Sci*, **10**: 359–367, 1925.

[69] Bridgeman, P. W. and I. Simon, "Effects of very high pressure on glass," *J. Appl. Phys.*, **24**(4): 405–413, 1953.

[70] Bronner, G. B. and J. D. Plummer, "Impurity diffusion and gettering in silicon," in R. Fair and C. Pearce (eds.), *Materials Research Society Proceedings*, volume 36, New York, 1985, North Holland.

[71] Brooks, A. and T. J. R. Hughes, "Streamline upwind petrov-galerkin formulations for convection dominated flows with particular emphasis on the incompressible navier-stokes equations," *CMAME*, **32**: 199–259, 1982.

[72] Brown, P. N. and A. C. Hindmarsch, "Matrix-free methods for stiff systems of ODEs," *SIAM J. Numer. Anal.*, **24**: 610, 1987.

[73] Brown, P. N. and Y. Saad, "Hybrid Krylov methods for nonlinear systems of equations," *SIAM J. Sci. Stat. Comp.*, **11**: 450, 1990.

[74] Butcher, J. C., *The Numerical Analysis of Ordinary Differential Equations Runga-Kutta and General Linear Methods*, Wiley, 1987.

[75] Buturla, E., P. Cottrell, B. Grossman and K. Salsburg, "Finite element analysis of semiconductor devices: the FIELDAY program," *IBM J. Res. Dev.*, **25**: 218–231, 1981.

[76] Byrne, D. and A. C. Hindmarsh, "Experiments in numerical methods for a problem in combustion modeling," *Appl. Numer. Math.*, **1**: 29, 1985.

[77] Byrne, G. D. and A. C. Hindmarsch, "Stiff ODE solvers: A review of current and coming attractions," *J. Comp. Phys.*, **70**: 1–62, 1987.

[78] Carey, G. F., "A mesh refinement scheme for finite element computations," *CMAME*, **7**: 93–105, 1976.

[79] Carey, G. F., "Adaptive refinement in nonlinear fluid problems," *Comput. Methods Appl. Mech. Eng.*, **18**: 541, 1980.

[80] Carey, G. F., "Exponential upwinding and integrating factors for symmetrization," *Comm. App. Num. Meth*, **1**: 57–60, 1985.

[81] Carey, G. F., "Mesh refinement and redistribution," in R. Lewis and K. Morgan (eds.), *Numerical Methods for Transient and Coupled Problems*, Wiley, 1987.

[82] Carey, G. F., *Grid Generation, Refinement and Redistribution*, Wiley, 1995.

[83] Carey, G. F. and E. Barragy, "Basis function selection and preconditioning high-degree finite element and spectral methods," *BIT*, **29**: 794–804, 1989.

[84] Carey, G. F., E. Barragy, R. McLay and M. Sharma, "Element-by-element vector and parallel computations," *Comm. Appl. Numer. Meth.*, **4**(3): 299–307, 1988.

[85] Carey, G. F. and B. Jiang, "Least-squares finite elements for first-order hyperbolic systems," *Int. J. Num. Meth. Eng.*, **26**: 81–93, 1988.

[86] Carey, G. F. and B. N. Jiang, "Element-by-element linear and nonlinear solution schemes," *Comm. in Applied Num. Methods*, **2**: 145, 1986.

[87] Carey, G. F. and P. Murray, "Compressibility effects in modeling thin silicon dioxide films for semiconductors," *J. of the Electrochemical Society*, **136**(9): 2666–2673, 1989.

[88] Carey, G. F. and P. Murray, "Determination of interfacial stress during thermal oxidation of siliconxo," *J. of Applied Physics*, **65**(9): 3467–3670, 1989.

[89] Carey, G. F. and J. T. Oden, *Finite Elements: Computational Aspects, Vol. 3*, Prentice-Hall, New Jersey, 1984.

[90] Carey, G. F. and J. T. Oden, *Finite Elements: Fluid Mechanics*, Prentice Hall, New York, 1986.

[91] Carey, G. F. and A. Pardhanani, "Multigrid solution and grid redistribution for convection-diffusion," *IJNME*, **27**(3): 655–664, 1989.

[92] Carey, G. F. and P. Patton, "Toward expert systems in finite-element analysis," *Comm. Applied Num. Meth.*, **3**: 527–533, 1987.

[93] Carey, G. F. and M. Sharma, "Semiconductor device modelling using flux upwind finite element," *COMPEL*, **8**(4): 219–224, 1989.

[94] Carey, G. F. and M. Sharma, "Semiconductor device simulation using adaptive refinement and flux upwinding," *IEEE Transactions of CAD of Integrated Circuits and Systems*, **8**(6): 590–598, 1989.

[95] Carey, G. F., M. Sharma and K. C. Wang, "A class of data structures for 2-D and 3-D adaptive mesh refinement," *IJNME*, **26**: 2607–2622, 1988.

[96] Carey, V., "Adaptive tetrahedral refinement," personal communication.

[97] Carrier, G. F., M. Krook and C. E. Pearson, *Functions of a Complex Variable*, McGraw Hill, New York, 1966.

[98] Carrier, G. F. and C. E. Pearson, *Partial Differential Equations*, Academic Press, 1976.

[99] Carslaw, H. S. and J. C. Jaeger, *Conduction of Heat in Solids*, Clarendon Press, Oxford, 1959.

[100] Cauchy, *Theory of Steepest Decent*, 1840.

[101] Cea, S. and M. E. Law, "Two dimensional simulation of silicide growth and flow," in *Proceedings of NUPADS V*, pp. 113–6. IEEE, 1994.

[102] Chang, L. L. and A. Koma, "Interdiffusion between GaAs and AlAs," *Appl. Phys. Lett.*, **29**(3): 138–141, August 1976.

[103] Chantre, A., M. Kechouane and D. Bois, "Vacancy-diffusion model for quenched-in E-centers in cw laser annealed virgin silicon," *Physica*, **116B**: 547–552, 1983.

[104] Chin, D., S. Oh and R. W. Dutton, "A general solution method for two-dimensional non-planar oxidation," *IEEE Trans. Electron Devices*, **ED-30**(9): 993–998, September 1983.

[105] Christel, L. A., J. F. Gibbons and T. W. Sigmon, "Displacement criterion for amorphization of silicon during ion implantation," *J. Appl. Phy.*, **52**: 7143, 1981.

[106] Cline, A. K. and R. Renka, "A storage efficient method for construction of a Thiessen triangulation," *Rocky Mtn. J. Math*, **4**(1): 119–139, 1984.

[107] Coke, R., "Numerical simulation of hot-carrier transport in silicon bipolar transistors," *IEEE Trans. Electron Devices*, **ED-30**(9): 1103, 1983.

[108] Comas, J. and R. G. Wilson, "Channeling and random equivalent depth distributions of 150 keV Li, Be, and B implanted Si," *J. Appl. Phy.*, **51**: 3697, 1980.

[109] Connelly, J. A. and P. Choi, *Macromodeling with SPICE*, Prentice-Hall, Englewood Cliffs, New Jersey, 1992.

[110] Conte, S. D. and C. de Boor, *Elementary Numerical Analysis – An Algorithmic Approach Third Edition*, McGraw-Hill, New York, 1980.

[111] Cook, R. K. and J. Frey, "An efficient technique for two-dimensional simulation of velocity overshoot effects in si and gaas devices," *COMPEL*, **1**(2): 65–87, 1982.

[112] Cook, R. K. and J. Frey, "Two-dimensional numerical simulation of energy transport effects in Si and GaAs MESFET's," *IEEE Trans. Electron Devices*, **ED-29**: 970–977, 1982.

[113] Corbett, J. W., G. D. Watkins and R. S. McDonald, "New oxygen infared bands in annealed irradiated silicon," *Phys. Rev.*, **135**(5A): A1381–A1385, August 1964.

[114] Coughran, W. M., E. H. Grosse and D. J. Rose, "CaZM: A circuit analyzer with macromodeling," *IEEE Trans. Electron Devices*, **30**: 1207–1213, 1983.

[115] Courant, R. and K. O. Friedrichs, *Supersonic Flow and Shock Waves*, Springer-Verlag, New York, 1948.

[116] Crandle, T. L. and B. J. Mulvaney, "An ion-implantation model incorporating damage calculations in crystalline targets," *IEEE Electron Device Letters*, **11**(1): 42–44, January 1990.

[117] Crandle, T. L., W. B. Richardson and B. J. Mulvaney, "A kinetic model for anomalous diffusion during post-implant annealing," *Inter. Electron Devices Meeting Technical Digest*, pp. 636–639, December 1988.

[118] Crank, J., *The Mathematics of Diffusion*, Clarendon Press, Oxford, 1975.

[119] Crochet, M., F. Dupret and Y. Ryckmans, "Numerical simulation of crystal growth in a vertical Bridgeman furnace," *J. Crystal Growth*, **97**: 173–185, 1989.

[120] Crochet, M. J., A. R. Davies and K. Walters, *Numerical Simulaion of Non-Newtonian Flow*, Elsevier, New York, 1984.

[121] Crochet, M. J., F. T. Geyling and J. J. van Schaftingen, "Finite element method for calculating the horizontal Bridgeman growth of semiconductor crystals," in R. H. Gallagher *etal* (ed.), *Finite Elements and Fluids*, volume 17, pp. 321–328, 1985.

[122] Cunnell, F. A. and C. H. Gooch, "Diffusion of zinc in gallium arsenide," *J. Phys. Chem. Solids*, **15**: 127–133, 1960.

[123] Curtice, W., "A mesfet model for use in the design of GaAs integrated circuits," *IEEE Tran. Microwave Theory and Technique*, **28**: 448–456, 1980.

[124] Darwish, M. N., M. C. Dolly and C. A. Goodwin, "Modeling of radiation induced burnout in dmos transistors," in *IEDM Tech. Dig.*, pp. 508–511. IEEE, 1988.

[125] Darwish, M. N., M. A. Shibib, M. R. Pinto and J. L. Titus, "Single event gate rupture of power dmos transistors," in *IEDM Tech. Dig.*, pp. 671–674. IEEE, 1993.

[126] Davis, B., Phd dissertation, University of Texas at Austin, 1996, (in preparation).

[127] Davis, B. and G. F. Carey, "Multilevel solution of the augmented drift-diffusion equation," *COMPELL*, 1995, in press.

[128] Deal, B. E. and A. S. Grove, "General relationship for the thermal oxidation of silicon," *J. Appl. Phy.*, **36**(12): 3770–3778, December 1965.

[129] Deal, B. E. and A. S. Grove, "General relationship for the thermal oxidation of silicon," *J. Appl. Phys.*, **36**(12): 3770–3778, 1965.

[130] Demkowicz, L. and J. T. Oden, "On a mesh optimization method based on a minimization of interpolation error," *IJES*, **24**: 55–68, 1986.

[131] Demkowicz, L., J. T. Oden and P. Devloo, "On an h-type mesh refinement strategy based on minimization of interpolation errors," *Comp. Meth. Appl. Mech. Eng.*, **53**: 67–89, 1985.

[132] Dendy, J., B. Swarz and B. Wendorff, "Computing traveling wave solutions of a nonlinear heat equation," Technical Report LA-UR-76-1842, Los Alamos Scientific Laboratories, 1976.

[133] Dennis, J. R. and E. B. Hale, "Crystalline to amorphous transformation in ion implanted silicon: A composite model," *J. Appl. Phy.*, **49**: 1119, 1978.

[134] Denorme, S., D. Mathiot, P. Dollfus and M. Mouis, "Two-dimensional modeling of the enhanced diffusion in thin base n-p-n bipolar transistors after lateral ion implantations," *IEEE Trans. Electron Devices*, **ED-42**: 523–527, 1995.

[135] Deppe, D. G., N. Holonyak, Jr. and J. E. Baker, "Sensitivity of Si diffusion in GaAs to column IV and VI donor species," *Appl. Phys. Lett.*, **52**(2): 129–131, 1988.

[136] Deppe, D. G., N. Holonyak, Jr., F. A. Kish and J. E. Baker, "Background doping dependence of silicon diffusion in p-type GaAs," *Appl. Phys. Lett.*, **50**(15): 998–1000, April 1987.

[137] DeSalvo, A. and R. Rosa, "A comprehensive computer program for ion penetration in solids," *Rad. Eff.*, **47**: 117, 1980.

[138] Desko, J. C., M. N. Darwish, M. C. Dolly, C. A. Goodwin, J. W. R. Dawes and J. L. Titus, "Radiation hardening of a high voltage IC technology (BCDMOS)," *IEEE*, pp. 2083–2088, 1990.

[139] Deutch, J. and A. Newton, "A multiprocessor implementation of relaxation based electrical circuit simulation," in *Proc. 21st Design Automation Conference*, pp. 350–357, 1984.

[140] Deutch, J. T., *Algorithms and Architecture for Multiprocessor Based Circuit Simulation*, Ph.D. thesis, University of California, Berkeley, 1985.

[141] Devloo, P., J. T. Oden and T. Strouboulis, "Implementation of an adaptive refinement technique for the SUPG algorithm," *Comp. Meth. Appl. Mech. Eng.*, **61**: 339–358, 1987.

[142] Dirks, H. K. and K. M. Eickhoff, "Numerical models and table models for MOS circuit analysis," in *Proc. of the Fourth Int. Conf. on the Numerical Analysis of Semiconductor Devices and Integrated Circuits - NASCODE IV*, pp. 13–23, Dublin, Ireland, 1985.

[143] Dongarra, J. J., J. R. Bunch, C. B. Moler and G. W. Stewart, *LINPACK User's Guide*, SIAM, Philadelphia, 1979.

[144] Doremus, R. H., "Oxidation of silicon: strain and linear kinetics," *Thin Solid Films*, **122**: 191–196, 1984.

[145] Duff, I., A. Erisman and J. K. Reid, *Direct Methods for Sparse Matrices*, Oxford University Press, 1987.

[146] Dunham, S. T. and J. Plummer, "Point defect generation during oxidation of silicon in dry oxygen. II. Comparison to experiment," *J. Appl. Phy.*, **59**(7): 2551–2561, April 1986.

[147] Dunham, S. T. and C. D. Wu, "Atomistic models of vacancy-mediated dopant diffusion in silicon at high doping levels," in *Proceedings of NUPAD V Conference*, pp. 101–4. IEEE, 1994.

[148] Duton, R. W. and Z. Yu, *IC Processes and Devices*, Kluwer, Boston, 1993.

[149] Dutton, R. W., "Modeling of the silicon integrated-circuit design and manufacturing process," *IEEE Trans. Electron Devices*, **ED-30**(9), September 1983.

[150] Duvall, S. G., "An interchange format for process and device simulation," *IEEE Trans. Computer-Aided Design*, **7**(7): 741, July 1988.

[151] Ebers, J. J. and J. L. Moll, "Large signal behavior of bipolar transistors," *Proceedings IRE*, **42**: 1761–1772, 1954.

[152] Ebers, J. J. and J. L. Moll, "Large signal behavior of bipolar transistors," *Proc. IRE*, **42**: 1761–1772, 1954.

[153] Echenique, P. M., R. M. Nieminen, J. C. Ashley and R. H. Ritchie, "Nonlinear stopping power of an electron gas for slow ions," *Phys. Rev. A*, **33**(2): 897–903, 1986.

[154] Edwards, S. P., A. M. Howland and P. J. Mole, "Initial guess strategy and linear algebra techniques for a coupled two-dimensional semiconductor equation solver," *NASECODE IV*, pp. 272–280, 1985.

[155] Eernisse, E. P., "Stress in thermal SiO_2 during growth," *Appl. Phys. Lett.*, **35**(1): 8–10, 1979.

[156] Eickhoff, K. M., "Table models for efficient MOS circuit simulation on vector processors," in *Proc. European Conference on Design Automation*, pp. 212–218, Brussels, Belgium, 1992.

[157] Einspruch, N. G. and W. R. Wisseman (eds.), *VLSI Electronics Microstructure Science: Volume 11 GaAs Microelectronics*, Academic Press, New York, 1985.

[158] Engl, W. L., H. K. Dirks and B. Meinerzhagen, "Device modeling," *Proc. IEEE*, **71**(1): 10–33, 1983.

[159] Engl, W. L., R. Laur and H. K. Dirks, "Medusa - a simulator for modular circuits," *IEEE Trans. Computer Aided Design*, **1**(1): 85–93, April 1982.

[160] et al., E., "Medusa – a simulator for modular circuits," *IEEE Trans. Computer-Aided Design*, p. 85, 1982.

[161] *et al*, T. J. B., "Relaxation time approximation and mixing of hot and cold electron populations," *Electronics Letters*, **28**: 1173–1174, 1992.

[162] Etherington, G., K. H. Jack and J. C. Kennedy, "The viscosity of vitreous silica," *Phys. Chem. Glasses*, **5**(5): 130–136, 1964.

[163] Ewing, R. E., "*A Posteriori* error estimation," *CMAME*, **82**: 59–72, 1990.

[164] Fahey, P. and R. W. Dutton, "Dopant diffusion under conditions of thermal nitridation of Si and SiO_2," in H. R. Huff and T. Abe (eds.), *Semiconductor Silicon — 1986*, pp. 571–582, Pennington, N.J., 1986, The Electrochemical Society.

[165] Fahey, P., R. W. Dutton and S. M. Hu, "Supersaturation of self-interstitials and undersaturation of vacancies during phosphorus diffusion in silicon," *Appl. Phy. Lett.*, **44**: 777, 1984.

[166] Fair, R. B., "Concentration profiles of diffused dopants in silicon," in F. F. Y. Wang (ed.), *Impurity Doping Processes in Silicon*, chapter 7, North-Holland, New York, 1981.

[167] Fair, R. B., "Concentration profiles of diffused dopants in silicon," in F. F. Y. Wang (ed.), *Impurity Doping Processes in Silicon*, chapter 7, North-Holland, New York, 1981.

[168] Fair, R. B., "Oxidation, impurity diffusion, and defect growth in silicon: An overview," *J. Electrochem. Soc.*, **128**: 1360–1368, 1981.

[169] Fair, R. B. and J. C. C. Tsai, "The diffusion of ion-implanted arsenic in silicon," *J. Electrochem. Soc.*, **122**(12): 1689–1696, 1975.

[170] Fair, R. B. and J. C. C. Tsai, "A quantitative model for the diffusion of phosphorus in silicon and the emitter dip effect," *J. Electrochem. Soc.*, **124**: 1107–1118, 1977.

[171] Fane, R. W. and A. J. Goss, "The diffusion of tin and selenium in gallium arsenide," *Solid-State Electronics*, **6**: 383–387, 1963.

[172] Fargeix, A. and G. Ghibaudo, "Role of stress on the parabolic kinetic constant for dry silicon oxidation," *J. Appl. Phys.*, **56**(2): 589–591, 1984.

[173] Farley, C. W. and B. G. Streetman, "Simulation of anomalous acceptor diffusion in compound semiconductors," *J. Electrochem. Soc.*, **134**(2): 453–458, February 1987.

[174] Fatemi, E., C. L. Gardner, J. W. Jerome, S. Osher and D. J. Rose, "Simulation of a steady-state electron shock wave in a submicron semiconductor device using high-order upwind methods," in *Computational Electronics: Semiconductor Transport and Device Simulation*, Kluwer Academic Publishers, Boston, 1991.

[175] Fatemi, E., J. W. Jerome and S. Osher, "Solution of the hydrodynamic device model using high-order non-oscillatory shock capturing algorithms," *IEEE Trans. Computer Aided Design*, in press.

[176] Fatunla, S. O., *Numerical methods for initial value problems in ordinary differential equations*, Academic Press, New York, 1988.

[177] Feller, W., *An Introduction to Probability Theory and its Applications, Vols. I and II, 3rd Edition*, Wiley, New York, 1968.

[178] Feng, Y. K. and A. Hintz, "Simulation of submicrometer GaAs MESFET's using a full dynamic transport model," *IEEE Trans. Electron Devices*, **35**: 1419–1431, 1988.

[179] Feynman, R. P., R. B. Leighton and M. Sands, *The Feynman Lectures on Physics - vol. II*, Addison-Wesley Publishing Company, 1965.

[180] Fichtner, W. D., D. J. Rose and R. E. Bank, "Semiconductor device simulation," *IEEE Trans. Electron Devices*, **ED-30**, 1983.

[181] Fischetti, M. V. and M. Rudan, "Hydrodynamic and monte carlo simulation of an $n^+ - n - n^+$ submicron device," *IEEE Trans. Electron Devices*, 1994.

[182] Fletcher, F., *Practical Methods of Optimization – Unconstrained Optimization*, volume 1, Wiley, 1980.

[183] Fletcher, F., "Conjugate gradient methods for indefinite systems," in *NATO ASI on Process and Device Simulation for MOS-VLSI Circuit*, 1982.

[184] Forghier, A., R. Guerrieri, P. Ciampolini, A. Gnudi, M. Rudan and G. Baccarani, "A new discretization strategy of the semiconductor equations comprising momentum and energy balance," *IEEE Trans. CAD*, **7**(2): 231, 1988.

[185] Fourier, J. B., *Theorie Analytique de la Chaleur*, Œuvres de Fourier, 1822.

[186] Freund, R. W. and N. Nachtigal, "(qmr): A quasi-minimal residual method for non-hermitian linear systems," *Numerische Mathematik*, **60**: 315–339, 1991.

[187] Fried, I., "Conditions of finite element matrices generated from nonuniform meshes," *AIAA Journal*, **10**: 219–221, 1972.

[188] Friedman, A., *Partial Differential Equations of Parabolic Type*, Prentice-Hall Inc., Englewood Cliffs, N.J., 1964.

[189] Friedrichs, K. O., "Symmetric positive linear differential equations," *Comm. Pure Appl. Math.*, **11**: 333–418, 1958.

[190] Fukui, Y., H. Yoshida and S. Higona, "Supercomputing of circuit simulation," in *Proc Supercomputing 1989*, ACM Publication Order Number 415892, pp. 81–85, 1990.

[191] Fukuma, M. and W. Lui, "MOSFET substrate current model including energy transport," *IEEE Elec. Dev. Lett*, **EDL-8**: 214–216, 1987.

[192] Fukuma, M. and R. H. Uebbing, "Two-dimensional MOSFET simulation with energy transport phenomena," *NASECODE*, pp. 428–433, 1989.

[193] Furakawa, S., H. Matsumura and H. Ishiwara, "Theoretical considerations on lateral spread of implanted ions," *Jpn. J. Appl. Phy.*, **11**: 134, 1972.

[194] Gamba, I. M., "Stationary transonic solutions for a 1-D hydrodynamic model for semiconductors," *Comm. P.D.E.*, **17**(3): 553–577, 1992.

[195] Gamba, I. M., "Viscosity approximating solutions to ODE systems that admit shocks, and their limits," *Adv. in App. Math.*, **15**: 129–182, 1994.

[196] Gardner, C. L., "Numerical simulation of a steady-state electron shock wave in a submicron semiconductor device," *IEEE Trans. Electron Devices*, **38**: 392–398, 1991.

[197] Gardner, C. L., "The quantum hydrodynamic model for semiconductor devices," *SIAM J. Applied Math*, **54**: 409–427, 1994.

[198] Gardner, C. L., "Resonant tunneling in the quantum hydrodynamic model," *VLSI Design*, 1994.

[199] Gardner, C. L., J. W. Jerome and D. J. Rose, "Numerical methods for the hydrodynamic device model: subsonic flow," *IEEE Trans. Computer Aided Design*, **8**: 501–507, 1989.

[200] Gear, C. W., "Numerical integration of stiff ordinary differential equations," Technical Report Dept. of Computer Science, Report 221, Univ. of Illinois - Urbana, 1967.

[201] Gear, C. W., "The automatic integration of ordinary differential equations," *Comm. ACM*, **14**: 176, 1971.

[202] Gear, C. W., "The automatic integration of ordinary differential equations," *Comm. ACM*, **14**: 176, 1971.

[203] Gear, C. W., *Numerical Initial Value Problems in Ordinary Differential Equations*, Prentice-Hall, New Jersey, 1971.

[204] Gear, C. W., *Numerical Initial Value Problems for Ordinary Differential Equations*, Prentice-Hall, 1974.

[205] Gear, C. W. and Y. Saad, "Iterative solution of linear equations in ODE codes," *SIAM J. Sci. Stat. Comp.*, **4**: 583, 1983.

[206] Gelinas, R., S. Doss and K. Miller, "The moving finite element method: Applications to general partial differential equations with multiple large gradients," *J. Comp. Phys.*, **40**: 202–249, 1981.

[207] Ghoshtagore, R. N., "Low concentration diffusion in silicon under sealed tube conditions," *Solid-State Electronics*, **15**: 1113–1120, 1972.

[208] Gibbons, J. F., W. S. Johnson and S. W. Mylroie, *Projected Range Statistics*, Halstead Press, Strandsberg, 1975.

[209] Gibbons, J. F., W. S. Johnson and S. W. Mylroie, *Projected Range Statistics*, Halstead Press, Stransdberg, 1975.

[210] Giles, M. D., "Defect-coupled diffusion at high concentrations," *IEEE Trans. Computer-Aided Design*, **8**(5): 460–467, 1989.

[211] Giles, M. D., D. S. Boning, G. R. Chin, J. W. C. Dietrich, M. S. Karasick, M. E. Law, P. K. Mozumder, L. R. Nackman, V. T. Rajan, D. M. H. Walker, R. H. Wang and A. S. Wong, "Semiconductor wafer representation for TCAD," *IEEE Trans. Computer-Aided Design*, **13**: 82–95, 1994.

[212] Glasstone, S. K., K. J. Laidler and H. Eyring, *The Theory of Rate Processes*, McGraw-Hill, 1941.

[213] Goldsman, N. and J. Frey, "Efficient and accurate use of the energy transport method in device simulation," *IEEE Trans. Electron Devices*, **35**(9): 1524, 1988.

[214] Goldsman, N. and J. Frey, "Electron energy distribution for calculation of gate leakage current in MOSFETs," *Solid-St. Electron*, **31**(6): 1089, 1988.

[215] Goldstein, H., *Classical Mechanics*, Addison-Wesley, Reading, Mass., 1950.

[216] Golemshtok, G. M., V. A. Panteleev and N. A. Agodchikov, "Numerical simulation of diffusion doping in the fabrication of local p-n junctions," *Optoelectron. Instrum. Data Process (USA)*, **3**: 37–42, 1988, Trans. of Avtometriga (USSR), 3:35-40, 1988.

[217] Golub, G. H. and C. F. V. Loan, *Matrix Computations*, John Hopkins University Press, Baltimore, 1989.

[218] Gosele, U., "Current understanding of diffusion mechanisms in silicon," in H. R. Huff and T. Abe (eds.), *Semiconductor Silicon 1986: Proc. Fifth Intern. Symp. on Silicon Materials Science and Technology*, pp. 541–555, Pennington, N.J., 1986, The Electrochemical Society.

[219] Gosele, U. and F. Morehead, "Diffusion of zinc in gallium arsenide: A new model," *J. Appl. Phys.*, **52**: 4617, 1981.

[220] Greiner, M. E. and J. F. Gibbons, "Diffusion of silicon in gallium arsenide using rapid thermal processing: Experiment and model," *Appl. Phys. Lett.*, **44**(8): 750–752, 1984.

[221] Greiner, M. E. and J. F. Gibbons, "Diffusion and electrical properties of silicon-doped gallium arsenide," *J. Appl. Phys.*, **57**(12): 5181–5187, June 1985.

[222] Griffin, P. B., S. T. Ahn, W. A. Tiller and J. D. Plummer, "Model for bulk effects on Si interstitial diffusivity in silicon," *Appl. Phys. Lett.*, **51**(2): 115–117, July 1987.

[223] Grindrod, P., *Patterns and Waves: The Theory and Applications of Reaction-Diffusion Equations*, Oxford University Press, Oxford, 1991.

[224] Grove, A. S., *Physics and Technology of Semiconductor Devices*, Wiley, 1967.

[225] Grubin, H. L. and J. P. Kreskovsky, "Quantum moment balance equations and resonant tunneling structures," *Solid State Elecron.*, **32**: 1971–1975, 1989.

[226] Guerrero, E., W. Jungling, H. Potzl, U. Gosele, L. Mader, M. Grasserbauer and G. Stingeder, "Determination of the retarded diffusion of antimony by SIMS measurements and numerical simulations," *J. Electrochem. Soc.*, **33**: 2181–2185, 1986.

[227] Guerrero, E., H. Potzl, R. Tielert, M. Grasserbauer and G. Stingeder, "Generalized model for the clustering of As dopants in Si," *J. Electrochem. Soc.*, **129**: 181, 1982.

[228] Gummel, H. K., "A self-consistent iterative scheme for one-dimensional steady state transistor calculations," *IEEE Trans. Electron Devices*, **ED-11**: 455–465, 1964.

[229] Gummel, H. K. and H. C. Poon, "An integral charge control model of bipolar transistors," *Bell Sys. Tech. Journal*, **49**: 827–852, 1970.

[230] Hageman, L. and D. Young, *Applied Iterative Methods*, Academic Press, New York, 1981.

[231] Hajj, I. N., P. Yang and T. N. Trick, "Avoiding zero pivots in the modified nodal approach," *IEEE J. of Circuits and Systems*, **CAS-28**(4), 1981.

[232] Hall, R. N., "Electron-hole recombination in germanium," *Phys. Rev.*, **87**: 387–388, 1952.

[233] Hansch, W., *The Drift-Diffusion Equation and Its Applications in MOSFET Modeling*, Springer-Verlag, New York, 1991.

[234] Hansch, W. and M. Miura-Mattausch, "The hot-electron problem in small semiconductor devices," *J. Appl. Phys*, **60**(2): 650, 1986.

[235] Hänsch, W. and S. Selberherr, "MINIMOS 3: A MOSFET simulator that includes energy balance," *IEEE Trans. Electron Devices*, **ED-34**: 1074–78, 1987.

[236] Harland, S., A. F. Tasch and C. M. Maziar, "A new structural approach for reducing punchthrough current in deep submicron MOSFETs and extending MOSFET scaling," *Electronics Lett.*, **29**: 1894–1896, 1993.

[237] Harris, R. M. and D. A. Antoniadis, "Silicon self-interstitial supersaturation during phosphorus diffusion," *Appl. Phy. Lett.*, **43**: 937, 1983.

[238] Hearne, M. T., *Diffusion Models for the Doping of Semiconductor Crystals*, University of Nottingham, U.K., 1988.

[239] Heath, M. T., E. Ng and B. W. Peyto, "Parallel algorithms for sparse linear systems," in *Parallel Algorithms for Matrix Computations*, pp. 83–197, SIAM, 1990.

[240] Heilblum, M., M. I. Nathan, D. C. Thomas and C. M. Knoedler, "Direct observation of ballistic transport in GaAs," *Phys. Rev. Lett.*, **55**: 2200–2203, 1985.

[241] Henrici, P. K., *Discrete Variable Methods for Ordinary Differential Equations*, John Wiley, New York, 1962.

[242] Hestenes, M., *Conjugate Direction Methods in Optimization*, Springer-Verlag, New York, 1980.

[243] Hindmarsh, A. C., "ODEPACK, a systematized collection of ODE solvers," in R. S. Stepleman and et al. (eds.), *Scientific Computing*, pp. 55–64, North Holland, Amsterdam, 1983.

[244] Hitschfield, N., P. Conti and W. Fichtner, "Mixed element trees: a generalization of modified octrees for the generation of meshes for the simulation of complex 3-D semiconductor device structures," *IEEE Trans. CAD*, **12**(11): 1714, 1993.

[245] Ho, C. P. and J. Plummer, "Si/SiO_2 interface oxidation kinetics: A physical model for the influence of high substrate doping levels," *J. Electrochem. Soc.*, **126**: 1516–1522, 1979.

[246] Ho, C. P., J. D. Plummer, S. E. Hansen and R. W. Dutton, "VLSI process modeling - SUPREM III," *IEEE Tran. Electron Devices*, **ED-30**(11): 1438–1453, 1983.

[247] Hobler, G. and H. Potzl, "Simulation of two dimensional implantation profiles with a large concentration range in crystalline silicon using an advanced Monte Carlo method," in *Int. Electron Devices Meeting, Technical Digest*, pp. 693–696, IEEE, New York, 1991.

[248] Hobler, G. and S. Selberherr, "Two dimensional modeling of ion implantation induced point defects," *IEEE Trans. Computer-Aided Design*, **CAD-7**: 174–180, 1988.

[249] Hobler, G. and S. Selberherr, "Two dimensional modeling of ion implantation induced point defects," *IEEE Trans. Computer Aided Design*, **7**: 174–180, 1988.

[250] Hofker, W. K., "Concentration profiles of Boron implantations in amorphous and polycrystalline silicon," *Phillips Research Reports*, **8**: 41–57, 1975.

[251] Householder, A., *Theory of Matrices in Numerical Analysis*, Dover, 1975.

[252] Hsieh, H. Y. and M. Ghausi, "A probabilistic approach to optimal pivoting and prediction of fill-in for random sparse matrices," *IEEE Trans. on Circuit Theory*, **CT-19**: 329–336, July 1972.

[253] Hu, S. M., "General theory of impurity diffusion in semiconductors via the vacancy mechanism," *Phys. Rev.*, **180**(3): 773–784, 1969.

[254] Hu, S. M., P. Fahey and R. W. Dutton, "On models of phosphorus diffusion in silicon," *J. Appl. Phys.*, **54**(12): 6912–6922, 1983.

[255] Huang, T. H., H. Kinoshita and D. L. Kwong, "Multi-zone model for the transient enhanced diffusion of ion implanted impurities in silicon during rapid thermal annealing," in *1993 International Symposium on VLSI Technology, Systems, and Applications. - Proceedings*, pp. 315–19. IEEE, 1993.

[256] Huber, A. M., G. Morillot, N. T. Linh, P. N. Favennec, B. Deveaud and B. Toulouse, "Chromium profiles in semi-insulating GaAs after annealing with a Si_3N_4 encapsulant," *Appl. Phys. Lett.*, **34**(12): 858–859, June 1979.

[257] Hughes, T. J. R. and A. Brooks, "A theoretical framework for Petrov-Galerkin methods with discontinuous weighting functions. application to the streamline upwind procedure," in R. H. Gallagher *et al.* (ed.), *Finite Elements in Fluids*, volume 55, pp. 47–65, 1982.

[258] Hwang, C. and R. Dutton, "Substrate current model for submicrometer MOSFET's based on mean free path analysis," *IEEE Trans. Electron Devices*, **36**(7): 1348, 1989.

[259] Irene, E. A., "New results on low-temperature thermal oxidation of silicon," *Phil Mag B*, **55**(2): 131–145, 1987.

[260] Irene, E. A., E. Tierney and J. Angilello, "A viscous flow model to explain the appearance of high density thermal SiO_2 at low oxidation temperatures," *J. Electrochem Soc.*, **129**(11): 2549–2697, 1982.

[261] Ishikawa, Y., Y. Sakina, H. Tanaka, S. Matsumoto and T. Niimi, "The enhanced diffusion of arsenic and phosphorus in silicon by thermal oxidation," *J. Electrochem. Soc.*, **129**: 644–648, 1982.

[262] Ismail, R. and G. Amaratunga, "Application of local Neumann error criteria for remeshing in dopant diffusion problems," in *Sim. of Semi. Dev. and Processes*, volume 3, pp. 549–560, Bologna, Italy, September 1988, Proc. of 3rd Inter. Conf.

[263] Jacoboni, C. and L. Reggiani, "The monte carlo method for the solution of charge transport in semiconductors with applications to covalent materials," *Rev. Mod. Phys.*, **55**: 645–705, 1983.

[264] Jain, R. K. and R. J. V. Overstraeten, "Theoretical calculations of the Fermi level and of other parameters in phosphorus doped silicon at diffusion temperatures," *IEEE Trans. Electron Devices*, **ED-21**(2): 155–165, February 1974.

[265] Jallepalli, S., C. Yeap, S. Krishnarnurthy, C. M. Maziar, A. F. Tasch and X. Wang, "Application of hierarchical transport models for the studyof deep submicron silicon n-MOSFETs," in *Proc. of the 3rd International Workshop on Computational Electronics*, Portland, OR, 1994.

[266] Jenkins, F. S. and S. P. Fan, "TIME - an nonlinear DC and time domain circuit simulation program," *IEEE J. of Solid State Circuits*, **6**: 182–188, 1971.

[267] Jensen, K. F. and D. B. Graves, "Modeling and analysis of low pressure CVD reactors," *J. Electrochem. Soc.*, **130**(9): 1950–1957, September 1983.

[268] Jeppson, K. O., D. Anderson, G. Amaratunga and C. P. Please, "Analytical modeling of nonlinear diffusion of arsenic in silicon," *J. Electrochem. Soc.*, **134**(9): 2316–2319, September 1987.

[269] Joe, B., "Tetrahedral mesh generation in polyhedral regions based on convex polyhedron decomposition," *IJNME*, **37**(4): 693–713, 1994.

[270] Joe, B. and R. B. Simpson, "Triangular meshes for regions on complicated shapes," *IJNME*, **23**: 751–778, 1986.

[271] Johnson, C., "Adaptive finite elements for diffusion and convection problems," *CMAME*, **82**: 301–322, 1990.

[272] Jones, S. K. and C. Hill, "Modeling dopant diffusion in polysilicon," in *Sim. of Semi. Dev. and Processes*, volume 3, pp. 26–28, Bologna, Italy, September 1988.

[273] Jordan, A. S., "An analysis of the anomalous outdiffusion of Mn in GaAs," in S. Makram-Ebeid and B. Tuck (eds.), *Semi-Insulating III-V Materials*, pp. 253–262, Shiva, Cheshire, 1982, Evian.

[274] Jordan, A. S. and G. A. Nikolakopoulou, "A numerical study of manganese redistribution in GaAs employing an interstitial- substitutional model," *J. Appl. Phys.*, **55**(12): 4194–4207, 1984.

[275] Kamgar, A., F. A. Baiocchi and T. T. Sheng, "Kinetics of arsenic activation and clustering in high dose implanted silicon," *Appl. Phy. lett.*, **48**: 1090–1092, 1986.

[276] Kao, D. B., J. P. McVittie, W. D. Nix and K. C. Saraswat, "Two-dimensional silicon oxidation experiments and theory," *IEDM Tech Digest*, pp. 338–391, 1985.

[277] Kasahara, J., M. Arai and N. Watanabe, "Effect of arsenic partial pressure on capless anneal of ion-implanted GaAs," *J. Electrochem. Soc.*, **126**(11): 1997–2001, November 1979.

[278] Kasahara, J., Y. Kato, M. Arai and N. Watanabe, "The effect of stress on the redistribution of implanted impurities in GaAs," *J. Electrochem. Soc.*, **130**(11): 2275, November 1983.

[279] Kaslhio, T. and K. Kato, "A new dopant diffusion model based on point defect kinetics," in *Ext. Abs. of 20th Conf. on Solid State Devices and Materials*, Tokyo, Japan, 1988.

[280] Kavanagh, K. L. and C. W. Magee, "Diffusion of Ge in GaAs at SiO_2 -encapsulated Ge-GaAs interfaces," *Can. J. Phys.*, **65**: 987, 1987.

[281] Kavanagh, K. L., S. W. Mayer, C. W. Magee, J. Sheets, J. Tong and J. M. Woodall, "Silicon diffusion at polycrystalline-Si/GaAs interfaces," *Appl. Phys. Lett.*, **47**(11): 1208–1210, December 1985.

[282] Kazmierski, K. and B. de Cremoux, "Double zinc diffusion fronts in InP: Correlation with models of varying charge transfer during interstitial-substitutional interchange," *Jpn. J. Appl. Phys.*, **24**(2): 239–242, February 1985.

[283] Kazmierski, K., F. Launay and B. de Cremoux, "An extra diffusion of Zn into GaAs in low concentration range: Application of the model of varying charge transfer," *Jpn. J. Appl. Phys.*, **26**(10): 1630–1633, October 1987.

[284] Kelessoglou, T. and D. O. Pederson, "Nectar – a knowledge-based environment to enhance SPICE," *IEEE J. of Solid State Circuits*, **SC-24**: 452–457, 1989.

[285] Kelly, D. W., "The self-equilibration of residuals and complementary *a posteriori* error estimates," *IJNME*, **20**: 1491–1506, 1984.

[286] Khoo, G. S. and C. K. Ong, "The nature of the charged self-interstitial in silicon," *J. Phys. C*, **20**: 5037–5043, 1987.

[287] Kimerling, L. C., H. M. DeAngelis and C. P. Carnes, "Annealing of electron-irradiated n-type silicon. I. Donor concentration dependence," *Phys. Rev. B*, **3**(2): 427–433, January 1971.

[288] King, J. R., "Interacting dopant diffusions in crystalline silicon," *SIAM J. Appl. Math.*, **48**(2): 405–415, 1988.

[289] King, J. R., "The isolation oxidation of silicon," *SIAM J. Appl. Math*, **49**(1): 264–280, 1989.

[290] King, J. R., "The isolation oxidation of silicon: The reaction- controlled case," *SIAM J. Appl. Math*, **49**(4): 1064–1080, 1989.

[291] King, J. R., "On the diffusion of point defects in silicon," *SIAM J. Appl. Math*, **49**(4): 1081–1101, 1989.

[292] King, J. R. and S. D. Houison, "Explicit solution to six free-boundary problems in fluid flow and diffusion," *SIAM J. Appl. Math.*, **43**(2): 155–175, 1989.

[293] Kingery, W. D., "Surface tension of some liquid oxides and their temperature coefficients," *J. Am. Ceram. Soc.*, **42**(1): 6–10, 1959.

[294] Kinoshita, H. and D. L. Kwong, "Physical model for the diffusion of ion implanted boron and BF_2 during rapid thermal annealing," in *International Electron Devices Meeting 1992. Technical Digest*, pp. 165–168. IEEE, 1992.

[295] Kittel, C., *Introduction to Solid State Physics- 5th edition*, Wiley, 1976.

[296] Kleckner, J. E., "Advanced mixed-mode simulation techniques," Technical Report ERL Memo. No. UCB/ERL M84/48, University of California, Berkeley, 1984.

[297] Klein, K. M., C. Park and A. F. Tasch, "Local electron concentration-dependent electronic stopping power model for Monte Carlo simulation of low energy ion implantation in silicon," *Applied Physics Letter*, **57**(25): 2701–2703, December 1990.

[298] Klein, K. M., C. Park and A. F. Tasch, "Monte Carlo simulation of boron implantation into single-crystal silicon," *IEEE Trans. Electron Devices*, **39**(7): 1614–1620, July 1992.

[299] Klein, P. B., P. E. R. Nordquist and P. G. Siebenmann, "Thermal conversion of GaAs," *J. Appl. Phys.*, **51**(9): 4861–4869, September 1980.

[300] Kobayashi, T. and K. Saito, "Two-dimensional analysis of velocity overshoot effects in ultrashort-channel Si MOSFET's," *IEEE Trans. Electron Devices*, **ED-32**(4): 788, 1985.

[301] Kobeda, E. and E. A. Irene, "SiO_2 film stress distribution during thermal oxidation of si," *J. Vac. Sci. Tech. B*, **6**(2): 574–578, 1959.

[302] Kol'dyaev, V. I., V. A. Moroz and S. A. Nayarov, "Two-dimensional simulation of the doping and oxidation," *Optoelectron. Instrum. Data Process (USA)*, **3**: 50–59, 1988, Trans. of Avtometriga (USSR), 3:35-40.

[303] Kreskovsky, J. P., "A hybrid central difference scheme for solid-state device simulation," *IEEE Trans. Electron Devices*, **ED-34**: 1128–1133, 1987.

[304] Kroger, F. A., "Remarks on "A quantitative model for the diffusion of phosphorus in silicon and the emitter dip effect"," *J. Electrochem. Soc.*, **125**: 995, 1978.

[305] Kull, G. M., L. W. Nagel, S.-W. Lee, P. Lloyd, E. J. Prendergast and H. Dirks, "A unified model for bipolar transistors including quasi-saturation effects," *IEEE Trans. Elect. Devices*, **32**(6), June 1985.

[306] Kump, M. and R. W. Dutton, "Two-dimensional process simulation — SUPRA," Technical Report G-201-13, Stanford Elec. Lab., July 1982.

[307] Kump, M. R. and R. W. Dutton, "The efficient simulation of coupled point defect and impurity diffusion," *IEEE Trans. Computer-Aided Design*, **7**(2): 191–204, 1988.

[308] Kunikiyo, T., K. Mitsui, M. Fujinaga, T. Uchida and N. Kotani, "Reverse short-channel effect due to lateral diffusion of point-defect induced by source/drain ion implantation," *IEEE Trans. Computer-Aided Design*, **13(4)**: 507–514, 1994.

[309] Lau, F., "Modeling of polysilicon diffusion sources," in *IEDM Tech. Dig.*, pp. 737–740. IEEE, 1990.

[310] Lau, F. and U. Gosele, "Two-dimensional diffusion properties of phosphorus in MOS source/drain structures," in *Proc. Fifth Intern. Symp. on Silicon Materials Science and Technology*, pp. 606–617. The Electrochemical Society, 1986.

[311] Law, M. E., "Parameters for point-defect diffusion and recombination," *IEEE Trans. Computer-Aided Design*, **10**: 1125–31, 1991.

[312] Law, M. E. and R. W. Dutton, "Verification of analytic point defect models using SUPREM-IV," *IEEE Trans. Computer-Aided Design*, **CAD-7**: 181–190, 1988.

[313] Law, M. E. and J. R. Pfiester, "Low-temperature annealing of arsenic/phosphorus junctions," *IEEE Trans. Electron Devices*, **38**: 278–84, 1991.

[314] Lax, P., "Hyperbolic systems of conservation laws," *Comm. Pure Appl. Math.*, **10**, 1957.

[315] Lax, P., "Hyperbolic systems of conservation laws and the mathematical theory of shock waves," in *SIAM*, Philadelphia, PA, 1972.

[316] Lee, J. W. and W. D. Laidig, "Diffusion and interdiffusion in Zn-disordered AlAs-GaAs superlattices," *J. Electronic Materials*, **13**(1): 147–165, 1984.

[317] Lee, S. C., T. W. Tang and D. Navon, "Transport models for MBTE," in *Proc. NASECODE VI*. Boole Press, 1989.

[318] Lee, S. W. and R. C. Rennick, "A compact IGFET model - ASIM," *IEEE Trans. Computer Aided Design*, **7**(9): 952–975, September 1988.

[319] Levenberg, K., "A method for the solution of certain nonlinear problems in least squares," *Q. Appl. Math.*, **2**: 164, 1944.

[320] Liang, M. and M. E. Law, "An object-oriented approach to device simulation-FLOODS," *IEEE Trans. Computer-Aided Design*, **13**(10): 1235–1240, 1994.

[321] Lie, L. N., R. R. Razouk and B. E. Deal, "High pressure oxidation of silicon in dry oxygen," *J. Electrochem. Soc.*, **129**: 2828–2834, 1982.

[322] Ligenza, J. R., "Effect of crystal orientation on oxidation rates of silicon in high pressure steam," *J. Phys. Chem.*, **65**: 2011–2014, 1961.

[323] Lin, A. M., D. A. Antoniadis and R. W. Dutton, "The oxidation rate dependence of oxidation enhanced diffusion of boron and phosphorus in silicon," *J. Electrochem. Soc.*, **128**: 1131–1137, 1981.

[324] Lin, C. C. and M. E. Law, "Mesh adaption and flux discretizations for dopant diffusion modeling," in *Proceedings of NUPADS V*, pp. 151–4. IEEE, 1994.

[325] Lin, C. C., M. E. Law and R. E. Lowther, "Automatic grid refinement and higher order flux discretization for diffusion modeling," *IEEE Trans. Computer-Aided Design*, **12**: 1209–16, 1993.

[326] Lindhard, L. and M. Scharff, "Energy dissipation by ions in the keV region," *Phys. Rev.*, **124**: 128, 1961.

[327] Linton, T., Phd dissertation, University of Texas at Austin, Austin, Tx, 1989.

[328] Litovski, V. B. and Z. M. Mrcarica, "Macromodeling with SPICE's nonlinear controlled sources," *IEEE Circuits and Devices*, pp. 14–17, November 1993.

[329] Lohner, R., "Some useful data structures for the generation of unstructured grids," *CANM*, **4**: 123–135, 1988.

[330] Lohner, R., K. Morgan and O. C. Zienkiewicz, "An adaptive finite element procedure for high speed flows," *Comp. Meth. Appl. Mech. Eng.*, **51**: 441–465, 1985.

[331] Longini, R. L., "Rapid zinc diffusion in gallium arsenide," *Solid-State Electronics*, **5**: 127–130, 1962.

[332] Lorenz, J., J. Pelka, H. Ryssel, A. Sachs, A. Seidl and M. Svoboda, "COMPOSITE — A complete modeling program of silicon technology," *IEEE Trans. Computer-Aided Design*, **CAD-4**(4), October 1985.

[333] Lorenz, J. and M. Svoboda, "ASWR-method for the simulation of dopant redistribution in silicon," in *Sim. of Semi. Dev. and Processes*, volume 3, p. 243, Bologna, Italy, September 1988.

[334] Loualiche, S., C. Lucas, P. Baruch, J. P. Gaillard and J. C. Pfister, "Theoretical model for radiation enhanced diffusion and redistribution of impurities," *Phys. Stat. Sol. A*, **69**: 663–676, 1982.

[335] Lu, N., L. Gerzberg, C. Lu and J. Meindl, "Modeling and optimization of monolithic polysilicon resistors," *IEEE Trans. Electron Devices*, **ED-28**: 818–830, 1981.

[336] Lu, N., L. Gerzberg and J. D. Meindl, "A quantitative model of the effect of grain size on the resistivity of polycrystalline silicon," *IEEE Electron Device Letters*, **EDL-1**: 38, 1980.

[337] Lucas, R., *Solving Planar Systems of Equations on Distributed-Memory Multiprocessors*, Ph.D. thesis, Stanford University, 1987.

[338] Lundstrom, M., *Transport Fundamentals for Device Applications*, Addison-Wesley, Reading, MA, 1990.

[339] Lyumkis, E. D., B. Polsky, A. I. Shur and P. Visocky, "Transient semiconductor device simulation including energy balance equation," *COMPEL*, **11**: 311–328, 1992.

[340] Makris, J. S. and B. J. Masters, "Phosphorus isoconcentration diffusion studies in silicon," *J. Electrochem. Soc.*, **120**(9): 1252–1255, September 1973.

[341] Maldonado, C. D., "ROMANS-II: A two-dimensional process simulator for modeling and simulation in the design of VLSI devices," *Appl. Phy.*, **A31**: 119–138, 1983.

[342] Malkovich, R. S., "Analytic relationships describing the profile of an impurity in a semiconductor established by concentration-dependent diffusion," *Sov. Phys. Semicond*, **20**(8): 941–943, 1986.

[343] Mandel, J. and S. McCormick, "A multi-level variational method on composite grids," *J. Comp. Phys.*, **80**: 442–450, 1989.

[344] Mandurah, M. M., K. C. Saraswat, C. R. Helms and T. I. Kamins, "Dopant segregation in polycrystalline silicon," *J. Appl. Phys.*, **51**(11): 5755–5763, November 1980.

[345] Mandurah, M. M., K. C. Saraswat and T. I. Kamins, "Arsenic segregation in polycrystalline silicon," *Appl. Phys. Letter.*, **36**: 683–685, 1980.

[346] Mantooth, H. A. and M. Fiegenbaum, *Modeling with an Analog Hardware Description Language*, Kluwer, Boston, 1995.

[347] Marchiando, J. F. and J. Albers, "Effects of ion-implantation damage on two-dimensional boron diffusion in silicon," *J. Appl. Phys.*, **61**(4): 1380–1391, February 1987.

[348] Marcus, R. B. and T. T. Sheng, "The oxidation of shaped silicon surfaces," *J. Electrochem Soc.*, **129**(6): 1278–1282, 1982.

[349] Markowich, P., C. Ringhofer and S. Selberherr, "A singular perturbation approach for the analysis of the fundamental semiconductor equations," *IEEE Trans. Electron Devices*, **ED-30**: 1165–1180, 1983.

[350] Markowitz, H. M., "The elimination form of the inverse and its application to linear programming," *Management Csi.*, **3**: 255–269, April 1957.

[351] Marquardt, D. W., "An algorithm for least squares estimation of nonlinear parameters," *SIAM Journal*, **11**: 431, 1963.

[352] Massoud, H. Z., J. D. Plummer and E. A. Irene, "Thermal oxidation of silicon in dry oxygen: Accurate determination of kinetic rate constants," *J. Electrochem. Soc.*, **132**: 1745–1753, 1985.

[353] Massoud, H. Z., J. D. Plummer and E. A. Irene, "Thermal oxidation of silicon in dry oxygen: Growth-rate enhancement in the thin regime," *J. Electrochem. Soc.*, **132**: 2685–2693, 1985.

[354] Masu, K., M. Konagai and K. Takahashi, "Diffusion of beryllium into GaAs during liquid phase epitaxial growth of p-$Ga_{0.1}Al_{0.8}As$," *J. Appl. Phys.*, **54**(3): 1574–1578, March 1983.

[355] Maszara, W. P. and G. A. Rozgonyi, "Kinetics of damage production in silicon during self-implantation," *J. Appl. Phy.*, **60**: 2310, 1986.

[356] Matano*Jap. J. Phys.*, **8**: 109, 1933.

[357] Mathiot, D., "Diffusion modeling from a fundamental viewpoint," in H. R. Huff and T. Abe (eds.), *Proc. Fifth Intern. Symp. on Silicon Materials Science and Technology*, pp. 556–570, Pennington, N.J., 1986, The Electrochemical Society.

[358] Mathiot, D., "Thermal donor formation in silicon: A new kinetic model based on self-interstitial aggregation," *Appl. Phys. Lett.*, **51**(12): 904–906, September 1987.

[359] Mathiot, D. and J. C. Pfister, "High concentration diffusion of P in Si: A percolation problem?" *J. Physique Letters*, **43**: L453–L459, June 1982.

[360] Mathiot, D. and J. C. Pfister, "Influence of the nonequilibrium vacancies on the diffusion of phosphorus into silicon," *J. Appl. Phys.*, **53**(4): 3053–3058, April 1982.

[361] Mathiot, D. and J. C. Pfister, "Diffusion of arsenic in silicon: Validity of the percolation model," *Appl. Phys. Lett.*, **42**(12): 1043–1044, June 1983.

[362] Mathiot, D. and J. C. Pfister, "Dopant diffusion in silicon: A consistent view involving nonequilibrium defects," *J. Appl. Phys.*, **55**: 3518–3530, 1984.

[363] Mathiot, D. and J. C. Pfister, "Point defect kinetics and dopant diffusion during silicon oxidation," *Appl. Phys. Lett.*, **48**(10): 627–629, 1986.

[364] Matsumoto, S., E. Arai, H. Nakamura and T. Niimi, "Phosphorus diffusion in silicon free from the surface effect under extrinsic conditions," *Jpn. J. Appl. Phys.*, **14**(11): 1665–1672, November 1975.

[365] Mayaran, K., J.-H. Chern and P. Yang, "Algorithms for transient three-dimensional mixed-level circuit and device simulation," *IEEE Trans. Computer Aided Design*, **12**(2): 1726–1733, November 1993.

[366] Mayergoyz, I. D., "Solution of the nonlinear Poisson equation of semiconductor device theory," *J. Appl. Phys.*, **59**: 195–9, 1986.

[367] McAndrew, C. C., B. K. Bhattacharyya and O. Wing, "A single-piece c-infinity continuous MOSFET model including subthreshold conduction," *IEEE Elect. Device Lett.*, **12**: 565–567, 1991.

[368] McCalla, W. J. and W. G. Howard, "Bias-3: A program for nonlinear DC analysis of bipolar transistors," *IEEE Trans. on Circuit Theory*, **6**: 14–19, 1971.

[369] McCamant, A. J., G. D. McCormack and D. H. Smith, "An improved GaAs MESFET model for SPICE," *IEEE Tran. Microwave Theory and Technique*, **38**: 822–824, 1990.

[370] McCormick, S. F., *Multilevel Adaptive Methods for Partial Differential Equations*, SIAM, Philadelphia, 1989.

[371] Mei, L. and R. W. Dutton, "A process simulation model for multilayer structures involving polycrystalline silicon," *IEEE Trans. Elect. Dev.*, **ED-29**: 1726–1734, 1982.

[372] Meinerzhagen, B. and W. Eng, "The influence of the thermal equilibirum approximation on the accuracy of classical two-dimensional numerical modeling of silicon submicrometer MOS transistors," *IEEE Trans. Electron Devices*, **ED-35**(5): 689, 1988.

[373] Michel, A., W. Rausch and P. A. Ronsheim, "Implantation damage and the anomalous transient diffusion of ion implanted boron," *Appl. Phys. Lett.*, **51**: 487–489, 1987.

[374] Michel, A., W. Rausch, P. A. Ronsheim and R. H. Kastl, "Rapid annealing and the anomalous diffusion of ion implanted boron into silicon," *Appl. Phys. Lett.*, **50**: 416–418, 1987.

[375] Michel, A. E., R. H. Kastl, S. R. Mader, B. J. Masters and J. A. Gardner, "Channeling in low energy boron ion implantation," *Appl. Phy. Lett.*, **44**: 404, 1984.

[376] Miller, J. N., D. M. Collins and N. J. Moll, "Control of Be diffusion in molecular beam epitaxy GaAs," *Appl. Phys. Lett.*, **46**(10): 960–962, May 1985.

[377] Miller, K. and R. Miller, "Moving finite elements I," *SIAM J. Numer. Anal.*, **18**(6): 1019–1032, 1981.

[378] Miller, K. and R. Miller, "Moving finite elements II," *SIAM J. Numer. Anal.*, **18**(6): 1033–1057, December 1981.

[379] Miyake, M., "Oxidation-enhanced diffusion of ion-implanted boron in silicon in extrinsic conditions," *J. Appl. Phys.*, **57**(6): 1861–1868, March 1985.

[380] Mock, M. S., *Analysis of Mathematical Models of Semiconductor Devices*, Boole press, Dublin, 1983.

[381] Morehead, F. F. and R. T. Hodgson, "A simple model for the transient, enhanced diffusion of ion-implanted phosphorus in silicon," *Mat. Res. Soc. Symp. Proc.*, **35**: 341–346, 1985.

[382] Morehead, F. F. and R. F. Lever, "Enhanced "tail" diffusion of phosphorus and boron in silicon: self-interstitial phenomena," *Appl. Phy. Lett.*, **48**(2): 151–153, January 1986.

[383] Morehead, F. F. and R. F. Lever, "A new model of tail diffusion of phosphorus and boron in silicon," *Mat. Res. Soc. Symp. Proceedings*, **52**: 49–56, 1986.

[384] Morehead, F. F. and R. F. Lever, "The steady-state model for coupled defect-impurity diffusion in silicon," *J. Appl. Phys.*, **66**: 5349, December 1989.

[385] Morin, F. J. and J. P. Maita, "Conductivity and Hall effect in the intrinsic range of germanium," *Phys. Rev.*, **94**: 1525, 1954.

[386] Morita, T., J. Kobayashi, T. Takamori, A. Takamori, E. Miyauchi and H. Hashimoto, "Thermal diffusion of buried beryllium and silicon layer in GaAs doped by focused ion beam implantation," *Jpn. J. Appl. Phys.*, **26**(8): 1324–1327, August 1987.

[387] Morooka, M. and M. Yoshida, "Calculation of substitutional impurity distribution by dissociative mechanism," *Jpn. J. Appl. Phys.*, **21**: 1409, 1982.

[388] Mueller, A. and G. F. Carey, "Continuously deforming finite elements," *IJNME*, **21**: 2099–2126, 1985.

[389] Mulvaney, B. J. and W. B. Richardson, "Model for defect-impurity pair diffusion in silicon," *Appl. Phys. Lett.*, **51**(18): 1439–1441, 1987.

[390] Mulvaney, B. J. and W. B. Richardson, "PEPPER – a process simulator for VLSI with a non-equilibrium kinetic diffusion model," in *Workshop on Numerical Modeling of Processes and Devices for Integrated Circuits – NUPAD II*. IEEE, 1988.

[391] Mulvaney, B. J. and W. B. Richardson, "The effect of concentration-dependent defect recombination reactions on phosphorus diffusion in silicon," *J. Appl. Phys.*, **67**(6): 3197–3198, 1990.

[392] Mulvaney, B. J. and W. B. Richardson, "Physical models for impurity diffusion in silicon," in *Seventh International Conference on the Numerical Analysis of Semiconductor Devices and Integrated Circuits (NASECODE VII)*, 1991.

[393] Mulvaney, B. J., W. B. Richardson and T. L. Crandle, "PEPPER - A process simulator for VLSI," *IEEE Trans. Computer-Aided Design*, **8**(4): 336–349, 1989.

[394] Mulvaney, B. J., W. B. Richardson, G. Siebers and T. L. Crandle, "PEPPER 1.2 user's manual," Technical Report CAD-239-90 (Q), Microelectronics and Computer Technology Corporation, Austin, Texas, 1989.

[395] Mun, J. (ed.), *GaAs Integrated Circuits*, MacMillan, New York, 1988.

[396] Murray, P. and G. F. Carey, "Determination of interfacial stress during thermal oxidation of silicon," *J. Appl. Phys.*, **65**(9): 3467–3670, 1989.

[397] Murray, W. D. and F. Landis, "Numerical and machine solutions of transient heat conduction problems involving melting or freezing," *J. Heat Transfer*, **81**: 106, May 1959.

[398] Nagel, L. and D. O. Peterson, "SPICE (simulation program with integrated circuit emphasis)," Technical Report Electronics Research Laboratory Report No. ERL M382, University of California, Berkeley, April 1973.

[399] Nagel, L. and R. Rohrer, "Computer analysis of nonlinear circuits,excluding radiation (CANCER)," *IEEE J. of Solid State Circuits*, **SC-6**: 166–182, 1971.

[400] Nagel, L. W., "SPICE2 : A computer program to simulate semiconductor circuits," Technical Report Electronics Research Laboratory Report No. UCB/ERL M75/520, University of California, Berkeley, May 1975.

[401] Nakata, T., N. Tanabe, N. Kajilara, S. Matsushita, H. Onozuka, Y. Asano and N. Koike, "CENJU: A multiprocessor system for modular circuit simulation," in *Computer Systems in Engineering*, volume 1, pp. 101–109, 1990.

[402] Nakhla, M., K. Singhal and J. Vlach, "An optimal pivoting order for the solution of sparse systems of equations," *IEEE J. of Circuits and Systems*, **CAS-21**: 222–225, March 1974.

[403] Neuberger, J. W., "Steepest descent and differential equations," *J. Math. Soc. Japan*, **37**(2): 187–195, 1985.

[404] Nevanlinna, O., "Waveform relaxation always converges for RC-circuits," in *NASECODE VII*, pp. 65–66, 1991.

[405] Newton, A. R. and A. L. Sangiovanni-Vincentelli, "Relaxation-based electrical simulation," *IEEE Trans. Electron Devices*, **ED-30**(9): 1184–1206, September 1983.

[406] Nishi, K., K. Sakamoto, S. Kuroda, J. Ueda, T. Miyoshi and S. Ushio, "A general-purpose two-dimensional process simulator — OPUS — for arbitrary structures," *IEEE Trans. Computer-Aided Design*, **8**(1): 23–32, 1989.

[407] Nobili, D., A. Armigliato, M. Finetti and S. Solmi, "Precipitation as the phenomenon responsible for the electrically inactive phosphorus in silicon," *J. Appl. Phys.*, **53(3)**, 1982.

[408] Nogami, M. and M. Tomozawa., "Effect of stress on water diffusion in silica glass," *J. Amer. Ceram. Soc.*, **67**(2): 151–154, 1984.

[409] O'Brien, R. R., C. M. Hsieh, J. S. Moore, R. F. Lever, P. C. Murley, K. W. Brannon, G. R. Srinivasan and R. W. Knepper, "Two-dimensional process modeling: A description of the safepro program," *IBM J. Res. Develop.*, **29(3)**: 229–41, 1985.

[410] Odanaka, S., H. Umimoto, M. Wakabayashi and H. Esaki, "SMART-P: Rigorous three-dimensional process simulator on a supercomputer," *IEEE Trans. Computer-Aided Design*, **7**: 675, 1988.

[411] Odeh, F., M. Rudan and J. White, "Numerical solution of the hydrodynamic model for a one-dimensional semiconductor device," *COMPEL*, **6**: 11551–170, 1987.

[412] Oden, J. T. and G. F. Carey, *Finite Elements – Mathematical Aspects Vol. IV*, Prentice-Hall, Englewood Cliffs, New Jersey, 1983.

[413] Oliger, J. and A. Sundstrom, "Theoretical and physical aspects of some initial value problems in fluid dynamics," *SIAM J. Appl. Math.*, **35**: 419–446, 1978.

[414] Onuma, T., T. Hirao and T. Sugawa, "Study of encapsulants for annealing Si-implanted GaAs," *J. Electrochem. Soc.*, **129**(4): 837–840, April 1982.

[415] Oppe, T. C., W. D. Joubert and D. R. Kincaid, "NSPCG User's Guide Version 1.0," Technical Report CNA-216, Center for Numerical Analysis, University of Texas, Austin, Texas, 1987.

[416] Orlowski, M., "Advanced diffusion models for sibmicron technologies," in *Technical Conference Proc. IEDM-88*, 1988.

[417] Orlowski, M., "A model for the dynamics between paired and unpaired diffusion for a binary system," August 1988, preprint.

[418] Orlowski, M., "Unified model for impurity diffusion in silicon," *Appl. Phys. Lett.*, **53**(14): 1323–1325, 1988.

[419] Orlowski, M., "Impurity and point defect redistribution in the presence of crystal defects," in *IEDM Tech. Dig.*, pp. 729–32. IEEE, 1990.

[420] Ortega, J. M. and W. C. Rheinboldt, *Iterative Solution of Nonlinear Equations in Several Variables*, Academic Press, New York, 1970.

[421] Osher, S. and S. Charkravarthy, "Upwind schemes and boundary conditions with applications to Euler equations in general geometries," *J. Comp. Phys.*, **50**: 445–481, 1983.

[422] Ou, H. H. and T. W. Tang, "Numerical modeling of hot carriers in submicrometr silicon BJT's," *IEEE Trans. Electron Devices*, **ED-34**(7): 1533, 1987.

[423] P. Sandborn *et al.*, "An assessment of approximate non-stationary change transport models for GaAs device modeling," *IEEE Trans.*, **ED-36**: 1244, 1989.

[424] Paap, K. L., M. Dehlwisch, R. Jendges and B. Klaassen, "Modeling mixed systems with SPICE3," *IEEE Circuits and Devices*, pp. 7–11, September 1993.

[425] Pardhanani, A. and G. F. Carey, "Optimization of computational grids," *Numerical Methods and PDE's*, **4**: 95–117, 1988.

[426] Pardhanani, A. and G. F. Carey, "Efficient numerical simulation of a non-parabolic hydrodynamic model for submicron semiconductor devices," 1995, Submitted to IEEE Trans. CAD.

[427] Park, C., K. M. Klein and A. F. Tasch, "Efficient modeling parameter extraction for dual Pearson approach to simulation of implanted impurity profiles in silicon," *Solid State Electronics*, **33**(6): 645–650, June 1990.

[428] Park, H., K. S. Jones, J. A. Slinkman and M. E. Law, "The effects of strain on dopant diffusion in silicon," in *IEDM Tech. Dig.*, pp. 303–6. IEEE, 1993.

[429] Park, H. J., P. K. Ko and C. Hu, "A charge sheet capacitance model of short channel MOSFETs for SPICE," *IEEE Trans. Computer Aided Design*, **10**(3): 376–389, March 1991.

[430] Pederson, D. O., "Historical review of circuit simulation," *IEEE J. of Circuits and Systems*, **CAS-31**: 103–111, 1984.

[431] Penumalli, B. R., "A comprehensive two-dimensional VLSI process simulation program, BICEPS," *IEEE Trans. Electron Devices*, **ED-30**(9): 986–992, 1983.

[432] Petzold, L., "Differential/algebraic equations are not ODE's," *SIAM J. Sci. Stat. Comp.*, **3**(3), September 1982.

[433] Pichler, P., W. Jungling, S. Selberherr, E. Guerrero and H. Potzl, "Simulation of critical IC-fabrication steps," *IEEE Trans. Electron Devices*, **ED-32**(10): 1940–1953, 1985.

[434] Pilling, M. J., *Reaction Kinetics*, Clarendon Press, Oxford, 1975.

[435] Pinto, M. R., C. C. Rafferty, H. R. Yeager and R. W. Dutton, "PISCES IIB - poisson and continuity equation solver," Technical Report Stanford Integrated Electronics Lab, Stanford Univeristy, Feb 1994.

[436] Please, C. P., "A mathematical model of diff blasting," *SIAM J. Appl. Math.*, **47**(1): 117–127, 1987.

[437] Press, W. H., S. A. Teukolsky, W. T. Vetterling and B. P. Flannery, *Numerical Recipes in C*, Cambridge University Press, Cambridge, 1992.

[438] Probst, V., H. J. Bohm, H. Schaber, H. Oppolzer and I. Weitzel, "Analysis of polysilicon diffusion sources," *J. Electrochem. Soc.*, **135**: 671–676, 1988.

[439] Quarles, T. L., *SPICE3 Version 3C1 User's Guide - ERL Memo No. UCB/ERL M89/42*, Univ. California, Berkeley, 1989.

[440] Rabinowitz, P. H., *Applications of Biffurcation Theory*, Academic Press, New York, 1977.

[441] Rafferty, C., M. Pinto and R. Dutton, "Iterative methods in semiconductor device simulation," *IEEE Trans. Electron Devices*, **ED-32**(10): 2018–2027, Ocotber 1985.

[442] Rafferty, C. S., M. Law, P. Griffin, J. Shott, R. Dutton and J. Plummer, "Modeling LOCOS effects on diffusion," in H. R. Huff and T. Abe (eds.), *Semiconductor Silicon - 1986*, pp. 426–436, Pennington, N.J., 1986, The Electrochemical Society.

[443] Rank, E. and O. C. Zienkiewicz, "A simple error estimator in the finite element method," *CMAME*, pp. 243–249, 1987.

[444] Razouk, R. R., L. N. Lie and B. E. Deal, "Kinetics of high pressure oxidation of silicon in pyrogenic steam," *J. Electrochem. Soc.*, **128**: 2214–2220, 1981.

[445] Reichl, L., *A Modern Course in Statistical Physics*, University of Texas Press, Austin, TX, 1980.

[446] Renyolds, S., D. W. Vook and J. F. Gibbons, "Open-tube Zn diffusion in GaAs using diethylzinc and trimethylarsenic: Experiment and model," *J. Appl. Phys.*, **63**(4): 1052–1058, 1988.

[447] Rheinboldt, W. C. and C. K. Mesztenyi, "On a data structure for adaptive finite element mesh refinements," *ACM Trans. Math. Soft.*, **6**(2): 166–187, June 1980.

[448] Richardson, W. B., "Steepest descent and the least C for Sobolev's inequality," *Bull. London Math. Soc.*, **18**: 478–484, 1986.

[449] Richardson, W. B., "Krylov subspace methods for 3-d process simulation," in *Seventh International Conference on the Numerical Analysis of Semiconductor Devices and Integrated Circuits (NASECODE VII)*, 1991.

[450] Richardson, W. B., G. F. Carey and B. J. Mulvaney, "Modeling phosphorus diffusion in three dimensions," *IEEE Trans. Computer-Aided Design*, **11**(4): 487–96, 1992.

[451] Richardson, W. B. and B. J. Mulvaney, "Plateau and kink in P profiles diffused into Si: A result of strong bimolecular recombination?" *Appl. Phys. Lett.*, **53**(20): 1917–1919, 1988.

[452] Richardson, W. B. and B. J. Mulvaney, "Nonequilibrium behavior of charged point defects during phosphorus diffusion in silicon," *J. Appl. Phys.*, **65**(6): 2243–2247, 1989.

[453] Rivana, M. C., "Selective refinement/de-refinement algorithms for sequences of nested triangulations," *IJNME*, **28**: 2889–2906, 1989.

[454] Rivara, M. C., "Design and data structure of fully adaptive, multigrid, finite element software," *ACM Trans. Math. Soft.*, **10**(3): 242–264, September 1984.

[455] Robinson, M. T. and I. M. Torrens, "Computer simulation of atomic-displacement cascades in solids in the binary-collision approximation," *Phys. Rev.*, **B9**: 5008, 1974.

[456] Roda, G. C., F. Santarelli and G. C. Sarti, "A simplified viscoelastic model for the thermal growth of thin s_io_2 films," *J. Electrochem. Soci.*, **132**(8): 1909–1913, 1985.

[457] Rodriguez, R.(private communication).

[458] Rollins, J. G. and J. J. Choma, "Mixed-mode pisces-spice coupled circuit and device solver," *IEEE Trans. Computer-Aided Design*, **7**: 862–867, 1988.

[459] Ronquist, E. M. and A. T. Patera, "Spectral element multigrid formulation and numerical results," *J. Scient. Comp.*, **2**(4): 389, 1987.

[460] Roosbroeck, W. V., "Theory of the flow of electrons and holes in germanium and other semiconductors," *Bell Sys. Tech. J*, **29**: 560, 1950.

[461] Rudan, M. and F. Odeh, "Multi-dimensional discretization scheme for the hydrodynamic model of semiconductor devices," *COMPEL*, **5**: 149–183, 1986.

[462] Rudan, M., F. Odeh and J. White, "Numerical solution of the hydrodynamic model for a one-dimensional semiconductor device," *COMPEL*, **6**(3): 151, 1987.

[463] Runge, H., "Distribution of implanted ions under arbitrary shaped mask edges," *Phys. Stat. Sol. A*, **39**: 595, 1977.

[464] Runyan, W. R. and K. E. Bean, *Semiconductor Integrated Circuit Processing Technology*, Addison-Wesley, 1990.

[465] Rysell, H. and J. Biersack, "Ion implantation models for process simulators," in W. Engl (ed.), *Process and Device Modeling*, p. 31, North-Holland, Amsterdam, 1986.

[466] Rysell, H., K. Haberger, K. Hoffmann, G. Prinke, R. Dumcke and A. Sachs, "Simulation of doping processes," *IEEE Trans. Electron Dev.*, **ED-27**: 1484–1492, 1980.

[467] Saad, Y., "Krylov subspace methods for solving large unsymmetric linear systems," *Math. of Comp.*, **37**(155): 105–126, 1981.

[468] Saad, Y., "Practical use of some Krylov subspace methods for solving indefinite and nonsymmetric linear systems," *SIAM J. Sci. Stat. Comp.*, **5**: 203, 1984.

[469] Sahul, Z., "Grid and geometry techniques for multilayer process simulation," in *Proc. SISDEP*, Springer Verlag, 1993.

[470] Saito, T., H. Yamakawa, S. Komiya, H. Kang and R. Shimizu, "Dynamic simulation of ion implantation with damaging processes included," *Nucl. Inst. Methods*, **B21**: 456, 1987.

[471] Saleh, R. A., D. L. Rhodes, E. Christen and B. A. A. Antao, "Analog hardware description languages," in *Proc. IEEE 1994 Custom Integrated Circuits Conference*, pp. 349–356, San Diego, California, 1994.

[472] Salsburg, K. A. and H. H. Hansen, "Fedss – finite-element diffusion-simulation system," *IEEE Trans. Electron Devices*, **ED-30**: 1004–1011, 1983.

[473] Sarma, K., R. Dalby, K. Rose, O. Aina, W. Katz and N. Lewis, "Ge diffusion at Ge/GaAs heterojunctions," *J. Appl.Phys.*, **56**(10): 2703–2707, November 1984.

[474] Schaake, H. F., "The diffusion of phosphorus in silicon from high surface concentrations," in *Mat. Res. Soc. Symp. Proc.*, volume 36, pp. 131–136, 1985.

[475] Scharfetter, D. L. and H. K. Gummel, "Large-signal analysis of a silicon Read diode oscillator," *IEEE Trans. Electron Devices*, **ED-16**: 64–77, 1969.

[476] Schichman, H. and D. A. Hodges, "Modeling and simulation of insulated gate field effect transistor switching circuits," *IEEE J. Solid-State Circuits*, **3**: 285–289, September 1968.

[477] Seeger, A. and K. P. Chik, "Diffusion mechanisms and point defects in silicon and germanium," *Phys. Stat. Sol.*, **29**: 455–542, 1968.

[478] Seidel, T. E., C. S. Pai, D. J. Lischner, D. M. Maher, R. V. Knoell, J. S. Williams, B. R. Penumalli and D. C. Jacobson, "Rapid thermal annealing in Si," *Mat. Res. Soc. Sym. Proc.*, **35**: 329–340, 1984.

[479] Selberherr, S., *Analysis and Simulation of Semiconductor Devices*, Springer-Verlag, New York, 1984.

[480] Selberherr, S., "Process modeling," *Microelectron. Eng.*, **9**: 605–610, May 1989.

[481] Selberherr, S., A. Schütz and H. W. Pötzl, "MINIMOS – a two-dimensional MOS transistor analyzer," *IEEE Trans. Electron Devices*, **ED-27**: 1540–50, 1980.

[482] Selberherr, S., A. Schutz and H. W. Potzl, "MINIMOS- a two-dimensional MOS transistor analyzer," *IEEE Trans. Electron Devices*, **27**(8): 1540–1550, 1980.

[483] Sharma, M., *Finite Element Modelling of Semiconductor Devices*, Ph.D. thesis, The University of Texas at Austin, 1988.

[484] Sharma, M. and G. F. Carey, "Semiconductor device simulation using adaptive refinement and flux upwinding," *IEEE Trans. Computer Aided Design*, **8**: 590–598, 1989.

[485] Shaw, D. (ed.), *Atomic Diffusion in Semiconductors*, Plenum Press, 1973.

[486] Shaw, D., "Self- and impurity diffusion in Ge and Si," *Phys. Stat. Sol. B*, **72**: 11–39, 1975.

[487] Shaw, D., "Alternative mechanisms for the diffusion of Sn and Zn in GaAs," *Phys. Stat. Sol. A*, **86**: 629–635, 1984.

[488] Shaw, D. and S. R. Showan, "The diffusion of Zn in the III-IV semiconductor compounds," *Phys. Stat. Sol.*, **32**: 109–118, 1969.

[489] Shephard, H. and R. H. Gallagher, "Finite element grid optimization," ASME Monograph PVP-38, 1980.

[490] Sheu, B. J., D. L. Scharfetter, P. K. Ko and M.-C. Jung, "BSIM: Berkeley short channel IGFET model for MOS transistors," *IEEE J. Solid-State Circuits*, **22**: 558–566, August 1987.

[491] Shichman, H., "Computation of DC solutions for bipolar transistor networks," *IEEE Trans. on Circuit Theory*, **16**: 460–466, 1969.

[492] Shichman, H. and D. A. Hodges, "Modeling and simulation of insulated gate field effect transistor circuits," *IEEE J. of Solid State Circuits*, **SC-3**: 285–289, September 1968.

[493] Shieh, T. and R. L. Carter, "RAPS-a rapid thermal processor simulation program," *IEEE Trans. Electron Devices*, **ED-36**: 19–24, 1989.

[494] Shockley, W., *Electrons and Holes in Semiconductors*, Van Nostrand, Princeton, NJ, 1950.

[495] Shockley, W. and W. T. Read, Jr., "Statistics of the recombinations of holes and electrons," *Phys. Rev.*, **87**(5): 835–842, September 1952.

[496] Shur, M., "Influence on non-uniform field distribution on frequency limits of GaAs field-effect transistors," *Electronics Letters*, **12**: 615, 1976.

[497] Shur, M. and E. F. Eastman, "Ballistic transport in semiconductor at low temperature for low-power high-speed logic," *IEEE Trans. Electron Device*, **ED-26**: 1677–1683, 1979.

[498] Sitkowski, M., "Macromodeling of phase locked loops for the SPICE simulator," *IEEE Circuits and Devices*, pp. 11–15, March 1991.

[499] Skoryatina, E. A., "Diffusion of manganese in extrinsic p-type GaAs," *Sov. Phys. Semicond.*, **20**(10): 1177–1178, October 1986.

[500] Sloan, S. W., "A FORTRAN program for profile and wavefront reduction," *IJNME*, **28**: 2651–2679, 1989.

[501] Sloan, S. W. and G. T. Houlsby, "An implementation of Watson's algorithms for computing 2d Delaunay triangulation," *Adv. in Eng. Software*, **6**(4): 192–197, 1984.

[502] Slotboom, J. W., "Computer aided two-dimensional analysis of bipolar transistors," *IEEE Trans. Electron Devices*, **ED–20**: 669–679, 1973.

[503] Smeid, I., N. Grillemot and G. Kamarino, "OSIRIS11 — A two-dimensional process simulator for SIMOX structures using both analytical and numerical resolution technique," in *Sim. of Semi. Dev. and Processes*, pp. 267–275, Bologna, Italy, September 1988, Proc. of 3rd Inter. Conf.

[504] Smith, R. K. and J. W. M. Coughran, "Computational challenges in simulations of ulsi semiconductor devices," in *Proceedings of NUPAD V Conference*, pp. 7–15. IEEE, 1994.

[505] Smoller, J., *Shock-Waves and Reaction-Diffusion Equations*, Springer-Verlag, New York, 1983.

[506] Sosman, R. B., *The Properties of Silica*, Chemical Catalog Company, N. Y., 1927.

[507] Statz, H., P. Newman, I. W. Smith, R. A. Pucel and H. A. Haus, "Gaas FET device and circuit simulation in SPICE," *IEEE Trans. Electron Devices*, **34**: 160–169, 1987.

[508] Stefan, J., "Über die theorie der der eisbildung, insbesondere über die eisbildung im polarmeere," *Annalen der Physik und Chemie*, **42**, 1891.

[509] Stone, H. L., "Iterative solutions of implicit appproximations of multidimensional partial differential equations," *SIAM J. Numer. Anal.*, **5**: 530–558, 1968.

[510] Strang, G., *Linear Algebra and its Applications*, Harcourt Brace Jovanovich, 1988.

[511] Strang, G., *Calculus*, Wellesley-Cambridge, 1991.

[512] Stratton, P., "Diffusion of hot and cold electrons in semiconductor barriers," *Phys. Rev.*, **126**(6): 2002, 1962.

[513] Swalin, R. A., "Theoretical calculations of the enthalpies and entropies of diffusion and vacancy formation in semiconductors," *J. Phys. Chem. Solids*, **18**(4): 290–296, 1961.

[514] Sze, S. M., *Physics of Semiconductor Devices*, John Wiley, 1981.

[515] Tang, T. W., "Extension of the Scharfetter-Gummel algorithm to the energy balance equation," *IEEE Trans. Electron Devices*, **ED-31**: 1912–1914, 1984.

[516] Taniguchi, K., Y. Shibata and C. Hamaguihi, "Universal model for impurity diffusion and oxidation of silicon," in *Ext. Abs. of 21st Conf. on Solid State Dev. and Materials*, Tokyo, Japan, August 1989.

[517] Tasch, A. F., "Scaling the MOS transistor to ultrasmall dimensions," in *Proc. of the International Conf. on Advanced Microelectronics, Devices and Processing*, p. 277, Sendai, Japan, 1994.

[518] Tasch, A. F., "Non-parabolic energy band models," personal communication.

[519] Tasch, A. F., H. Shin, C. Park, J. Alvis and S. Novak, "An improved approach to accurately model shallow B and BF_2 implants in silicon," *J. Electrochemical Soc.*, **136**(3): 810–814, March 1989.

[520] Tatti, S. R., "Amphoteric impurity diffusion in GaAs; Simulation of anomalous outdiffusion of Mn in GaAs; Zn diffusion in GaAs," Technical report, MCC, 1988.

[521] Tatti, S. R., S. Mitra and J. P. Stark, "A proposed model to explain impurity induced layer disordering in AlAs-GaAs heterostructures," .

[522] Thompson, J., *Numerical Grid Generation*, Elsevier, 1982.

[523] Ting, C. H. and G. L. Pearson, "Time-dependence of zinc diffusion in gallium arsenide under a concentration gradient," *J. Electrochem. Soc.*, **118**(9): 1454–1458, September 1971.

[524] Tomizawa, M., K. Yokoyama and A. Yoshi, "Nonstationary carrier dynamics in quarter-micron Si MOSFET's," *IEEE Trans. CAD*, **7**(2): 254, 1988.

[525] Toyabe, T., H. Masuda, Y. Aoki, H. Shukuri and T. Hagiwara, "Three-dimensional device simulator CADDETH with highly convergent matrix solution algorithms," *IEEE Trans. Electron Devices*, **4**(4): 482–488, 1985.

[526] Trumbore, F. A., "Solid solubilities of impurity elements in germanium and silicon," *Bell System Tech. J.*, **39**: 205, 1960.

[527] Tsai, J. C. C., D. G. Schimmel, R. E. Ahrens and R. B. Fair, "Point defect generation during phosphorus diffusion in silicon. I. concentrations below solid solubility, ion implanted phosphorus," *J. Electrochem. Soc.*, **134**(9): 2348–2356, September 1987.

[528] Tsai, M. Y., F. F. Morehead, J. E. Baglin and A. E. Michel, "Shallow junctions by high-dose *As* implants in *Si*: Experiments and modeling," *J. Appl. Phys.*, **51**(6): 3230–3235, June 1980.

[529] Tsivides, Y. P., *Operation and Modeling of the MOS Transistor*, McGraw-Hill, 1989.

[530] Tsividis, Y. P. and K. Suyama, "MOSFET modeling for analog circuit CAD: Problems and prospects," *IEEE J. Solid State Circuits*, **29**(3): 210–216, March 1994.

[531] Tsoukalas, D. and P. Chenevier, "Boron diffusion in silicon in inert and oxidizing ambient and extrinsic conditions," *Phys. Stat. Sol. A*, **100**: 461–465, 1987.

[532] Tuck, B., *Introduction to diffusion in semiconductors*, Peter Peregrinus, Ltd., 1974.

[533] Tuck, B. and M. A. H. Kadhim, "Anomalous diffusion profiles of zinc in GaAs," *J. Materials Science*, **7**: 585–591, 1972.

[534] Tuck, B. and R. G. Powell, "High-temperature diffusion of GaAs," *J. Phys. D: Appl. Phys.*, **14**: 1317, 1981.

[535] van Ommen, A. H., "Examination of models for Zn diffusion in GaAs," *J. Appl. Phys.*, **54**(9): 5055–5058, September 1983.

[536] Vechten, J. A. V., "Enthalpy of vacancy migration in Si and Ge," *Phys. Rev. B*, **10**(4): 1482–1506, August 1974.

[537] Vechten, J. A. V., "Divacancy binding enthalpy in semiconductors," *Phys. Rev. B*, **11**(10): 3910–3917, May 1975.

[538] Vechten, J. A. V., "Simple ballistic model for vacancy migration," *Phys. Rev. B*, **12**(4): 1247–1251, August 1975.

[539] Vechten, J. A. V., "Point defects and deep traps in III-IV compounds," *Czech. J. Phys. B*, **30**: 388–394, 1980.

[540] Vechten, J. A. V. and C. D. Thurmond, "Entropy of ionization and temperature variation of ionization levels of defects in semiconductors," *Phys. Rev. B*, **14**(8): 3539–3550, October 1976.

[541] Vlach, J. and K. Singhal, *Computer Methods for Circuit Analysis and Design Second Edition*, Van Nostrand Reinhold, New York, 1994.

[542] Vladimirescu, A., *The SPICE Book*, Wiley, 1994.

[543] Vladimirescu, A., K. Zhang, A. R. Newton, D. O. Peterson and A. L. Sangiovanni-Vincentelli, "SPICE version 2g users guide," Technical report, University of California, Berkeley, August 1981.

[544] W. M. Coughran, J., J. Cole, P. Lloyd and J. K. White (eds.), *Semiconductors, Parts I and II*, Springer-Verlag, New York, 1994.

[545] W. M. Coughran Jr., E. H. Grosse and D. J. Rose, "CAZM: A circuit analyzer with macro-modeling," *IEEE Trans. Electron Devices*, **ED-30**: 1207–1213, 1983.

[546] Wada, Y. and S. Nishimatsu, "Grain growth mechanism of heavily phosphorus implanted polycrystalline silicon," *J. Electrochem. Soc.*, **125**: 1499–1504, 1978.

[547] Wang, S., *Fundamentals of Semiconductor Theory and Device Physics*, Prentice Hall, 1989.

[548] Warren, M. S. and J. K. Salmon, "Astrophysical N-body simulations using hierarchical tree data structures," in *Proceedings of Supercomputing '92*, pp. 570–576. IEEE Computer Society Press, 1992.

[549] Wathen, A., "Spectral bounds and preconditioning methods using element-by-element analysis for galerkin finite element equations," Technical Report Report No. AM-87-04, Dept. of Mathematics, University of Bristol, Bristol England, 1987.

[550] Watkins, G. D., "EPR studies of the lattice vacancy and low-temperature damage processes in silicon," in *Inst. Phys. Conf.*, 23, pp. 1–22, 1975.

[551] Watkins, G. D., A. P. Chatterjee and R. D. Harris, "Negative-U for point defects in silicon," in *Inst. Phys. Conf.*, 59, pp. 199–204, 1981.

[552] Weisberg, L. R. and J. Blanc, "Diffusion with interstitial-substitutional equilibrium. Zinc in GaAs," *Phys. Rev.*, **131**(4): 1548–1552, August 1963.

[553] Westermann, M., N. Strecker, P. Regli and W. Fichtner, "Reliable solid modeling for three-dimensional semiconductor process and device simulation," in *Proceedings of NUPAD V Conference*, pp. 49–52. IEEE, 1994.

[554] White, J. K. and A. Sangiovanni-Vincentelli, *Relaxation Techniques for the Simulation of VLSI Circuits*, Kluwer Academic Publishers, 1987.

[555] Willoughby, A., A. Evans, P. Champ, K. Yallup, D. Godfrey and M. Dowsett, "Diffusion of boron in heavily doped $n-$ and $p-$type silicon," *J. Appl. Phys.*, **59**: 2392, 1986.

[556] Wilson, R. G., "The Pearson IV distribution and its application to ion implanted depth profiles," *Radiation Effects*, **46**: 141–148, 1980.

[557] Wilson, R. G., "Channeling of 20-800 keV arsenic ions in the $< 110 >$ and the $\langle 100 \rangle$ directions of silicon, and the roles of electronic and nuclear stopping," *J. Appl. Phy.*, **52**: 3985, 1981.

[558] Wilson, R. G., "Boron, fluroine, and carrier profiles for B and BF_2 implants into crystalline and amorphous Si," *J. Appl. Phys.*, **54**(12): 6879–6889, 1983.

[559] Wolf, S., *Silicon Processing for the VLSI Era*, volume 2: *Process Integration*, Lattice Press, Sunset Beach, California, 1990.

[560] Yang, P. and P. K. Chatterjee, "Spice modeling for small geometry MOSFET circuits," *IEEE Trans. Computer Aided Design*, **1**: 169–182, 1982.

[561] Yang, P., B. D. Epler and P. K. Chatterjee, "An investigation of the charge conservation problem for MOSFET circuit simulation," *IEEE J. of Solid State Circuits*, **SC-18**(1): 128–138, 1983.

[562] Yang, P., I. Hajj and T. Trick, "Slate: A circuit simulation program with latency exploitation and node tearing," in *Proc. Int. Conf. on Circuits and Computers*, October 1980.

[563] Yeager, H. R. and R. W. Dutton, "An approach to solving multiparticle diffusion exhibiting nonlinear stiff coupling," *IEEE Trans. Electron Devices*, **ED-32**(10): 1964–1976, 1985.

[564] Yokoyama, K., M. Tomizawa, A. Yoshii and T. Sudo, "Semiconductor device simulation at NTT," *IEEE Trans. Electron Devices*, **ED-32**(10): 2008–2017, 1985.

[565] Yoshida, M., "Criterion for the assumptions of thermal equilibrium of interstitial atoms and of vacancies in dissociative diffusion," *Jpn. J. Appl. Phys.*, **8**(10): 1211–1216, October 1969.

[566] Yoshida, M., "Diffusion of Group V impurity in silicon," *Jpn. J. Appl. Phys.*, **10**(6): 702–713, June 1971.

[567] Yoshida, M., "Numerical solution of phosphorus diffusion equation in silicon," *Jpn. J. Appl. Phys.*, **18**(3): 479–489, 1979.

[568] Yoshida, M., "General theory of phosphorus and arsenic diffusions in silicon," *Jpn. J. Appl. Phys.*, **19**(12): 2427–2440, December 1980.

[569] Yoshida, M., "Diffusion mechanism of phosphorus and arsenic in silicon," in *Inst. Phys. Conf.*, 59, pp. 557–562, 1981.

[570] Yoshida, M., E. Arai, H. Nakamura and Y. Terunuma, "Excess vacancy generation mechanism at phosphorus diffusion into silicon," *J. Applied Phys.*, **45**: 1498–1506, 1974.

[571] Yoshida, M., S. Matsumoto and Y. Ishikawa, "Solutions of simultaneous equations for oxidation enhanced and retarded diffusions and oxidation stacking fault in silicon," *Jpn. J. Appl. Phys.*, **25**(7): 1031–1035, July 1986.

[572] Yoshida, M. and K. Saito, "Dissociative diffusion of nickel in silicon and self-diffusion of silicon," *Jpn. J. Appl. Phys.*, **6**(5): 573–581, May 1967.

[573] Young, A. B. Y. and G. L. Pearson, "Diffusion of sulfur in gallium phosphide and gallium arsenide," *J. Phys. Chem. Solids*, **31**: 517–527, 1970.

[574] Young, D. M., *Iterative Solution of Large Linear Systems*, Academic Press, 1971.

[575] Young, D. M. and K. C. Jea, "Generalized conjugate gradient acceleration of nonsymmetrizable iterative methods," *J. of Linear Algebra and Its Applications*, **34**: 159–194, 1980.

[576] Yu, P. W., "Iron in heat-treated gallium arsenide," *J. Appl. Phys.*, **52**(9): 5786–5791, September 1981.

[577] Yue, C., V. Agostinelli, G. Yeric and A. F. Tasch, "Improved universal MOSFET electron mobility degradation models simulation," *IEEE Trans. on CAD/ICAS*, **12**: 1542–1545, 1993.

[578] Zahari, M. D. and B. Tuck, "Effect of vacancy reduction on diffusion in semiconductors," *J. Phys. D: Appl. Phys.*, **15**: 1741–1750, 1982.

[579] Zahari, M. D. and B. Tuck, "Substitutional-interstitial diffusion with bulk vacancy generation in semiconductors," *J. Phys. D: Appl. Phys.*, **16**: 635–644, 1983.

[580] Ziegler, J. F., J. Biersack and U. Littmark, *The Stopping and Ranges of Ions in Solids - Vol. 1*, Pergamon Press, New York, 1985.

[581] Zienkiewicz, O. C. and J. Z. Zhu, "A simple error estimator and adaptive procedure for practical engineering analysis," *INJME*, **24**: 337–357, 1987.

Index